# Control System Engineering

# Control System Engineering

**M. E. El-Hawary**

*Professor of Engineering*
*Technical University of Nova Scotia*
*Halifax, Nova Scotia*
*Canada B3J 2X4*

**Reston Publishing Company, Inc.**
***A Prentice-Hall Company***
**Reston, Virginia**

**Library of Congress Cataloging in Publication Data**

El-Hawary, M. E.
Control system engineering.

1. Automatic control. 2. Systems engineering.
I. Title.
TJ213.E48 1984 629.8 83-24705
ISBN 0-8359-1015-6


*A Prentice-Hall Company*
Reston, Virginia 22090



10 9 8 7 6 5 4 3 2 1

Printed in the United States of America

## Dedication

To the City of Alexandria,
home of the first known
feedback control device.

"I know thy works: behold, I have set before thee an open door, and no man can shut it: for thou hast a little strength, and hast kept my word, and hast not denied my Name."

*The Revelation of St. John the Divine, 2:8.*

# TABLE OF CONTENTS

# PREFACE

This book is intended to provide an up-to-date introduction to a number of important topics in control systems engineering. The text is designed to be of primary interest for a number of purposes: (1) as an upper division (junior or senior level) course textbook in control systems for electrical, chemical, industrial, mechanical and agricultural engineering programs; (2) as an upper division course text for a baccalaureate engineering technology program and (3) as a book for control engineers and technologists who deal with combinations of design, development and operation of control devices and systems. The reasonably large number of example problems and solved problems should enhance the suitability of this book for independent study.

There are ten chapters in the book. The first chapter is intended to provide an account of major historical landmarks in the development of the fascinating area of control systems engineering. This historical review is followed by an outline of the text. The format of the rest of the book is such that at the conclusion of the chapters there are a number of solved problems that the reader should consult to focus on the concepts and techniques developed in the unit. An additional set of unsolved (but certainly solvable) problems is provided to conclude each chapter.

In Chapter 2 some fundamental concepts and techniques that are useful for control systems analysis and design are introduced. Under the heading "Control Systems Modeling and Representation," Laplace transform and its inverse are reviewed. This is followed by a discussion of elementary physical system models. The concept of a transfer function for linear time-invariant dynamic systems is treated next. Block diagram and signal flow representations are considered to conclude the chapter. The title of Chapter 3 is "Features of Feedback Control Systems." The chapter begins by providing definitions and terms used in feedback systems. The structure of a typical configuration is used to illustrate the terms. Control actions are defined and the advantages of feedback are discussed. A section is devoted to a study of typical examples to show how feedback arises either as a natural phenomenon or a design choice. The chapter is concluded by a comprehensive treatment of the topic of system sensitivity to changes in parameters.

Concepts and issues in state-space representation are dealt with in Chapter 4. A brief review of linear algebraic prerequisites is given prior to a well-balanced treatment of techniques for writing the state equations. The solution of the state equations and the concept of a state transition matrix are discussed. System diagonalization, eigen-values, and eigen-vectors are also introduced. A number of practical examples are incorporated in the text material.

Chapter 5 is concerned with the time-domain analysis of control systems. Test signals, error coefficients, and constants are discussed. The time responses of first-order, second-order, and higher-order systems are treated. Performance specifications for second-order systems are introduced in this chapter. In Chapter 6 we use the knowledge gained in Chapters 4 and 5 to consider the important issue of system stability. The Hurwitz, Lienard-Chipart, and Routh criteria are introduced and typical examples are considered. Chapter 7 is a natural extension of Chapter 6 and is devoted to Evans' root-locus method.

In Chapter 8 we deal with frequency-response characteristics and stability criteria. Nyquist, Bode, and Nichols diagrams are introduced and concepts of gain and phase margins are considered. Chapter 9 is devoted to design and compensation considerations and is based on tools developed earlier in the text.

Digital control systems are treated in Chapter 10. Following the necessary definitions, we introduce the z-transform and its inverse. Stability analysis is also discussed. The chapter is intended to provide a firm background for further study in this fascinating area.

In writing this book, I attempted to assume the role of an inquisitive reader who desired a thorough and detailed analysis of the subject discussed. As a result, it is hoped that the book contains a minimum amount of the sometimes irksome "skipped details." In the process of learning and subsequent practice of control systems engineering, many of us discover the joy of putting abstract mathematical notions to the fruitful development of actual functioning devices and systems. *It is hoped that this book will aid the reader in an early discovery of the joy of control systems engineering and that the joy will still be felt for many years to come.*

The book is an outgrowth of courses taught by the author at Memorial University of Newfoundland and more recently at the Technical University of Nova Scotia. I am indebted to my many colleagues and students who have contributed significantly to the development of this text. I would like to acknowledge my appreciation of the continuing support of President J. Clair Callaghan, Dean Donald A. Roy, of the Technical University of

Nova Scotia. Without these two gentlemen's support, this book would not have been possible. The drafting of the manuscript involved the patient and able typing skills of Ms. Verilea Ellis and Ms. Wanda Roy of Nova Scotia Tech, whose work is greatly appreciated. My sincere thanks to Ben Wentzell, Greg Michael, and Dan McCauley of Reston Publishing Company for their excellent work on this book. Finally, the patience and understanding of my wife, Ferial, and children are appreciated.

**M. E. El-Hawary**

*Halifax, Nova Scotia*

*October, 1983*

# Chapter 1

# INTRODUCTION

The purpose of this chapter is twofold. We first provide a brief historical perspective on the development of control system engineering. This is not intended to be a complete and detailed historical account, but rather to provide highlights of conceptual developments in this important area. The historical review is covered in Sections 1.1 through 1.8 and is followed by an outline of the text. A reference list pertaining to main references on the history of control systems engineering is provided at the end of the chapter.

## 1.1 Definitions and Early History

A *control system* is a combination of components that are interconnected in a configuration designed to provide a specified system response. The concept of feedback is central to control systems discussed in this text. It is appropriate at this stage to seek some definitions of the concept. The reader should keep in mind that accepted definitions and terminology of control systems engineering are covered in Chapter 3.

Norbert Wiener provides a concise definition of feedback in his 1954 book entitled *The Human Use of Human Beings: Cybernetics and Society.* Wiener describes feedback as a method of controlling a system by reinserting into it the results of its past performance. In 1951, the American Institute of Electrical Engineers (AIEE) published a committee report entitled "Proposed Symbols and Terms for Feedback Control Systems," which defines a *feedback control system* as a control system that tends to maintain a prescribed relationship of one system variable to another by comparing functions of these variables and using the difference as a means of control. The British Standards Institution in its 1967 "Glossary of Terms Used in Automatic Controlling and Regulating Systems" defined *feedback* as "the transmission of signal from a later to an earlier stage." The word "feedback" seems to have been used first in 1920 by personnel at Bell Telephone Laboratories and has since overtaken the British term "reset" and the American term "closed cycle" used earlier to describe the same action.

The concept of feedback control can be argued to have originated with the beginning of life forms, for all living entities are marvelous complexes of feedback control systems designed to react to their individual environmental conditions. One can also see evidence of feedback in the words of Genesis:

> And God said, let there be light and there was light. And God saw the light, that it was good: and God divided the light from the darkness.
>
> *(Genesis: 1 : 3–4)*

Indeed, life itself is a continuing process of feedback. However, we traditionally perceive feedback control as beginning with its first conscious use by human beings.

About 2100 years before the birth of Christ, King Hammurabi of Babylonia ordered a set of 282 laws codified and inscribed on stone. In this earliest known law code of the civilized world, one finds references to rules governing Babylonian irrigation practices. A penalty was prescribed for flooding one's neighbor's field in the process of irrigating his own field. A Babylonian practiced feedback control in irrigating his land as he consciously regulated the moisture content of the soil by opening and closing irrigation canals originating in the rivers Tigris and Euphrates, and in doing so he had to be careful not to flood his neighbor's field.

In this book, however, we are interested in automatic feedback control systems. The word *automatic* signifies the absence of the intervention of a human operator. The earliest feedback devices known to our civilization were employed in clepsydrae (water clocks) believed to have been invented in the Hellenistic period. Ktesibios, a native of Alexandria, and a contemporary of such men as Aristarchos, Euclid, and Archimedes, under King Ptolemy II Philadelphus (285–247 B.C.), is credited (among other things) with the invention of the water clock.

The water clock is thought to have employed the *first float valve regulator*, as shown in Figure 1.1. The water level in the small regulating

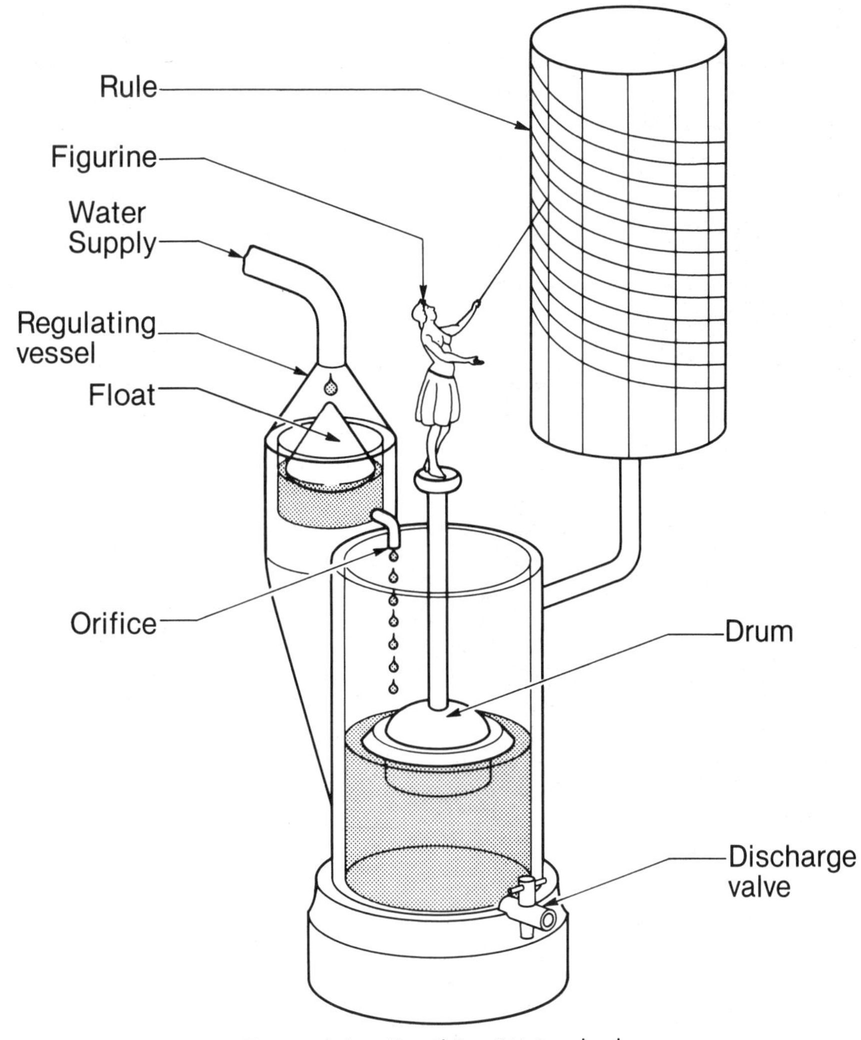


**Figure 1.1.** Ktesibian Waterclock

vessel is the controlled variable in the process. The float senses the water level and opens the valve (its tip) when the level falls and closes when the level rises. The flow of the water from the regulating vessel to the large receiving vessel is thus maintained at a constant rate. The figurine rises at a constant rate with the rise of the drum, thus making it possible to read the elapsed time on a rule.

Another Hellenistic early device that employed the concept of feedback in controlling liquid levels is Philon's oil lamp, shown in Figure 1.2. A wick is placed into the lower reservoir. When the lamp is lit, oil in the reservoir is consumed by burning. When the oil level has dropped low enough to expose the lower end of the vertical riser in the middle, air will rise into the oil reservoir, permitting oil to flow out through the capillary tubes. As a result, the oil level in the lower reservoir will rise until it reaches the lower end of the risers, stopping the upward flow of air and thus the downward flow of oil.

The works of Heron of Alexandria incorporate a number of gadgets that employ the concept of feedback. In Heron's Inexhaustible Goblets, float valves are employed to maintain constant liquid levels in two connected vessels. The ancient inventions of the float valve remained an integral part of water clocks constructed by the Arabs. Among the latest known references to water clocks is the book by Ibn Al-Saati (Son of the Clock-Maker, 1203) entitled "Book on the Construction of the Clock and on Its Use." We also have Al-Jazari's "On the Knowledge of Ingenious Engineering Mechanisms," which devotes a considerable portion of this work

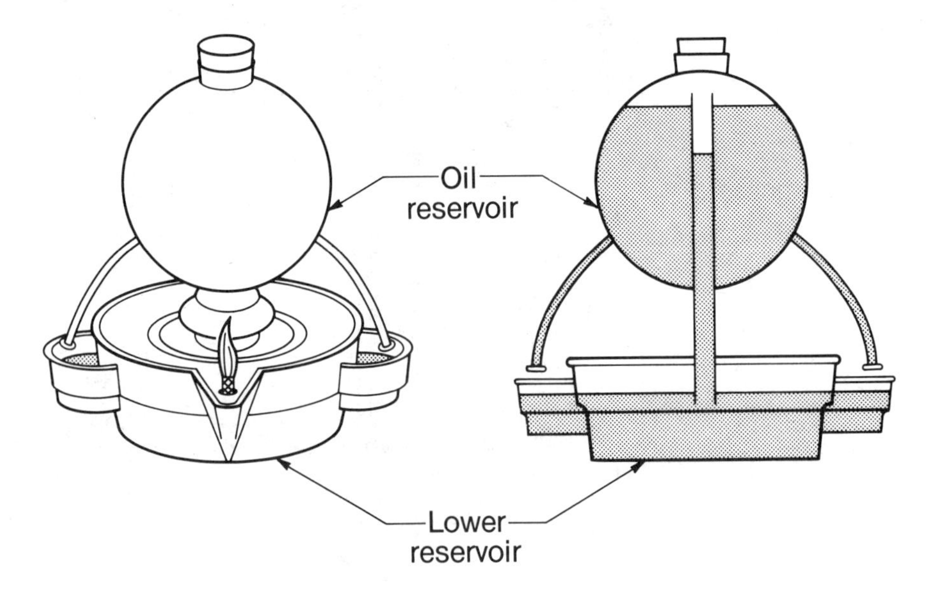


**Figure 1.2.** Philon's Oil Lamp

to describing the construction of water clocks. The golden era of water clocks ended in 1258 with the taking of Baghdad during the Mongolian invasion.

## 1.2 Reemergence of Feedback

The gadgets of antiquity that employed feedback in their functioning disappeared with the Mongolian invasion of Baghdad, as mentioned in the preceding section. It was not until the sixteenth century that interest in performing functions requiring feedback reemerged. This time the scene was modern Europe.

The temperature regulator of Cornelis Drebbel (1572–1633), a Dutch mechanic and chemist who spent most of his life in the service of English royalty, is the first feedback system invented in modern Europe. Drebbel's temperature regulator was evidently used in a chicken incubator and in the athanor, a general furnace for the chemist. The artificial hatching of chickens using incubators was an activity that occupied the noted physicist René A. F. de Réaumur (1683–1757). On seeing the incubators, Prince de Conti suggested an improvement that amounts to a temperature regulator. The initial suggestion was to insert some device that would open an air vent as soon as a certain temperature was reached to let warm air escape and bring the temperature down.

Let us now shift our attention to colonial America of the eighteenth century. William Henry (1729–1786) of Lancaster, Pennsylvania, reported on his invention of a temperature regulator in the first *Transactions* of the American Philosophical Society of Philadelphia in 1771. Henry's Sentinel Register is shown in Figure 1.3. The temperature feeler D controls the flue damper A. The air in vessel C expands with the rising temperature and by exerting pressure on the water level, it forces the water to rise in the vertical pipe. The float D transfers this motion through linkages to the damper A.

British millwrights of the eighteenth century invented a number of feedback mechanisms. The oldest of these devices is the fan-tail (Figure 1.4), which is an accessory to windmills. It consists of a small wind wheel mounted at right angles to the main wheel. The purpose of this mechanism is to point the mill continuously into the wind. The invention of the fan-tail is due to Edmund Lee of Lancaster in 1745. Andrew Meikle is often falsely credited with its invention in 1750.

Self-regulating windmill sails were patented by E. Lee and became popular in the nineteenth century. Figure 1.5 shows a drawing of a windmill employing a fan-tail and self-regulating sails. The four sails are pivoted to tilt around their lengthwise axes. As a result, the center of wind pressure

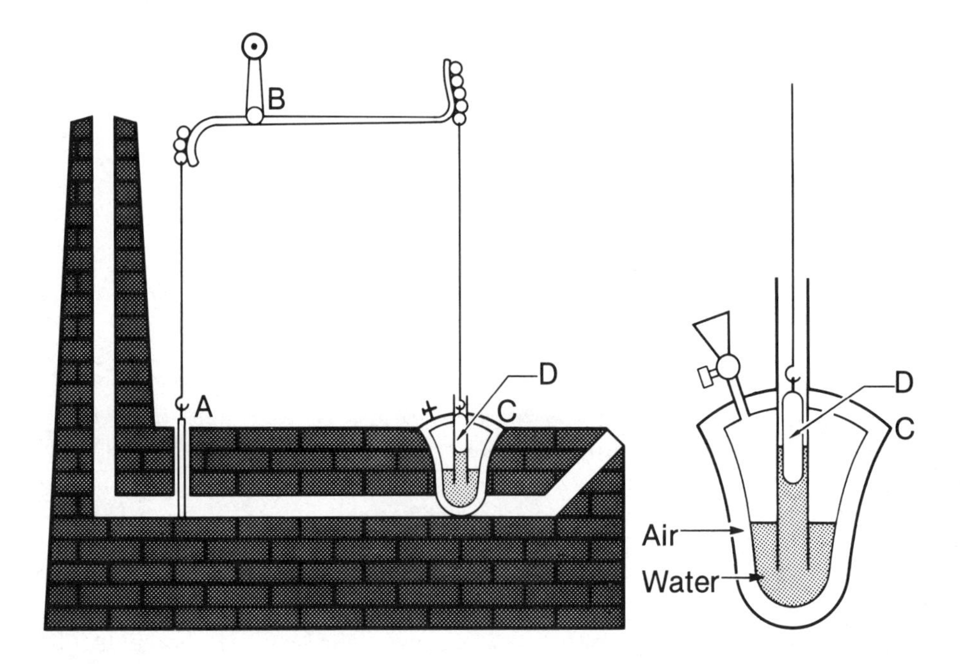


**Figure 1.3.** William Henry's Sentinel Register and Schematic of the Temperature Feeler

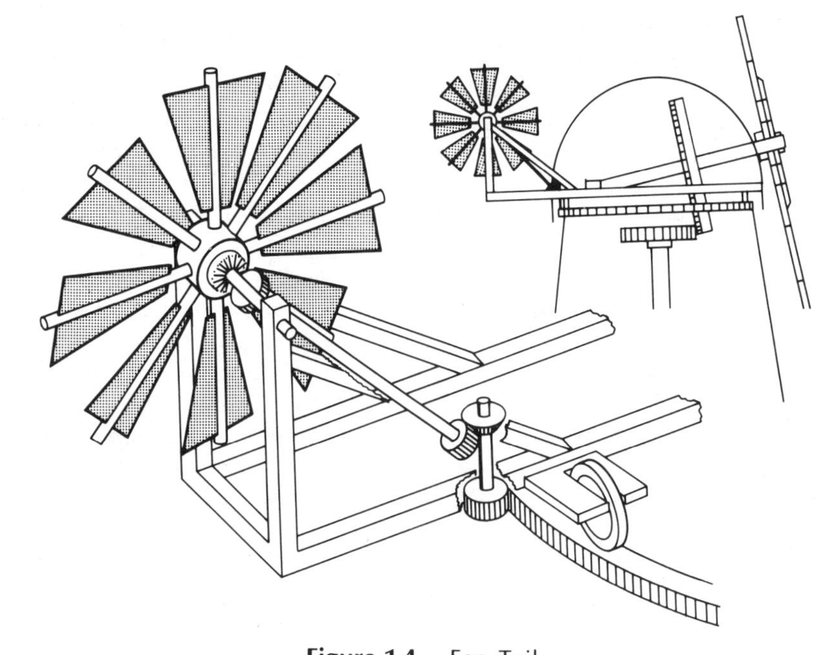

**Figure 1.4.** Fan-Tail

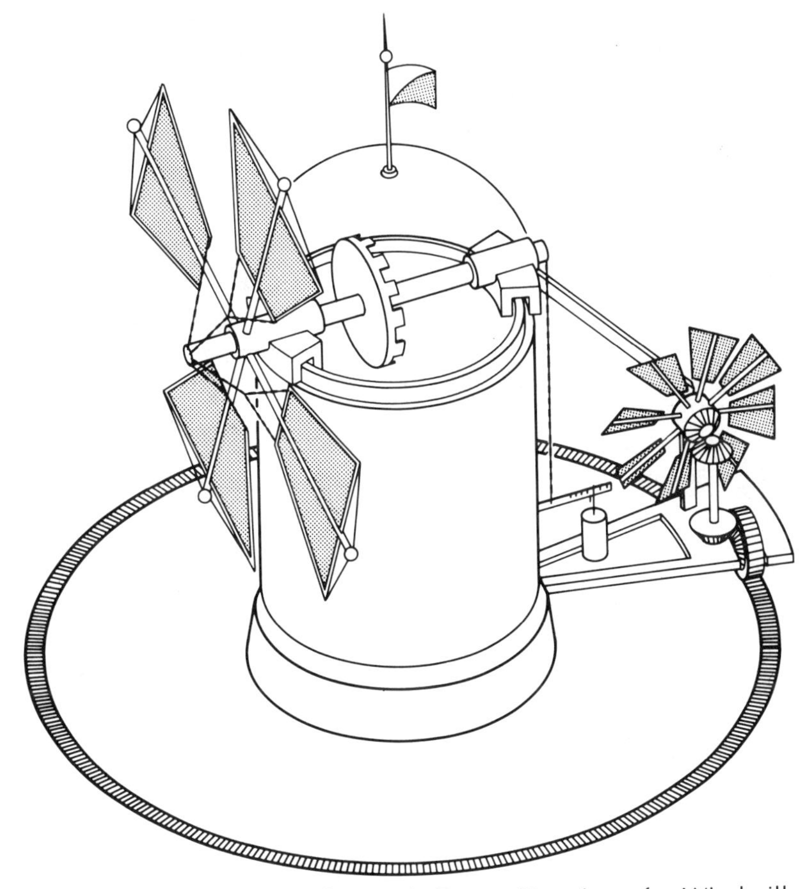

**Figure 1.5.** An adaptation of E. Lee's Patent Drawing of a Windmill with Fan-Tail and Self-Regulating Sails

lies behind the axis. A set of chains running through the hollow main shaft of the mill connects the trailing edges of the sails with a counterweight in the rear. The weight draws the trailing edges of the sails forward toward the wind. As the wind increases, the sails move backward, projecting a smaller area to the wind so that the torque driving the mill will barely increase in spite of the growing wind. Andrew Miekle modified the idea by replacing the counterweight with springs.

Float valve regulators reappeared in steam boiler design as patented in 1758 by James Brindley and by Sutton Thomas Wood of Oxford in 1784. The first feedback contribution claimed by the Russians is the boiler level regulator of I. I. Polzunov (1763). Pressure regulators for steam engines were invented in the same era as were boiler-level regulators. Denis Papin (1647–1712) patented a pressure cooker in which the pressure was regulated by a weight-loaded valve. A more elaborate design of a pressure regulator was patented by Robert Delap in 1799 and an improvement on it by Matthew Murray was patented three months later.

# 1.3 The Centrifugal Governor

A basic problem in operating a windmill is to maintain the gap between the millstones as constant as possible to ensure uniform-quality flour. The flow rate of grain feed increases as the mill speed increases, with the consequence that the stones tend to separate from each other. The solution is a device that controls the gap and is commonly known as a lift-tenter. In 1787, Thomas Mead took out a patent on a lift-tenter, shown in Figure 1.6. The speed of rotation was measured in terms of the centrifugal motion of a

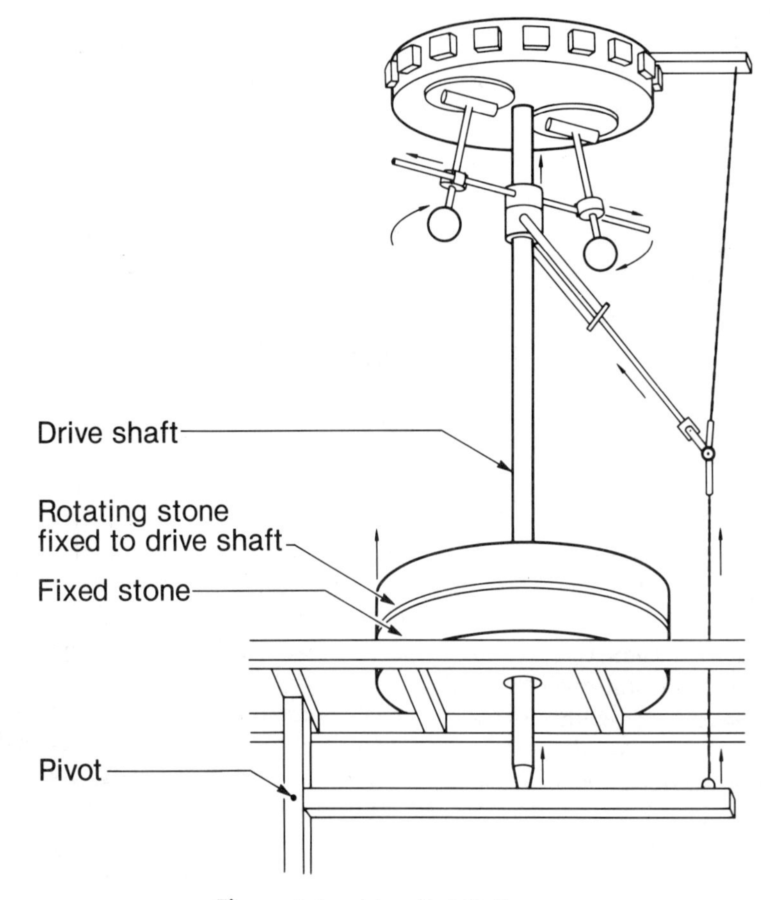


**Figure 1.6.** Mead's Lift-Tenter

revolving pendulum. An arrangement of sleeves and levers transmitted this motion to lifting and lowering of the rotating millstone. The centrifugal pendulum is also used in Mead's speed regulator for windmills, shown in Figure 1.7. The idea here is that with rising speed the centrifugal weights swing outward and, through the pull of the rope, force the sleeve to slide toward the rear of the mill. The canvas is thus rolled up to reduce the sail area.

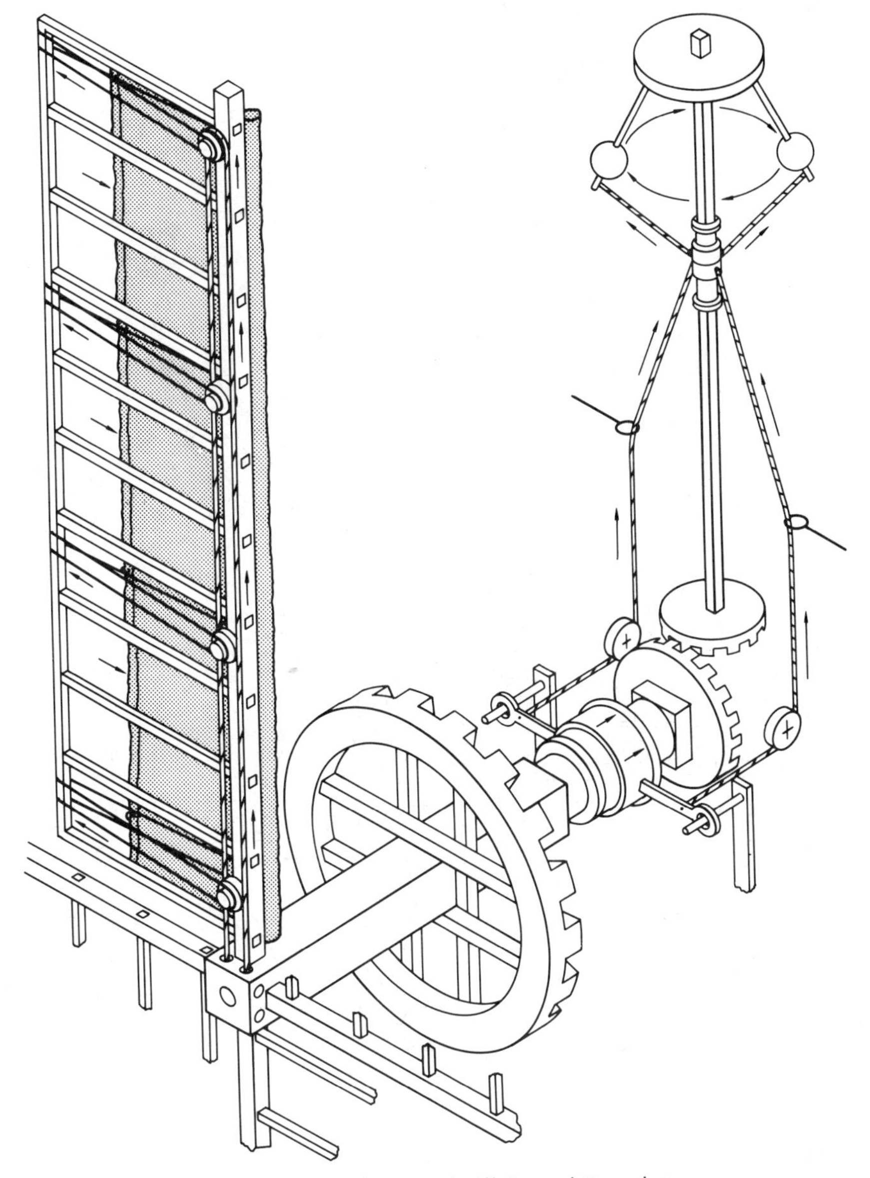

**Figure 1.7.** Mead's Windmill Speed Regulator

Matthew Boulton and James Watt were active in the early 1780s in promoting the rotary steam engine for milling uses. The centrifugal governor was introduced in 1788 in the engine's design as a result of Boulton's suggestion to Watt following the former's inspection of milling machinery. Figure 1.8 shows a Watt steam engine with a centrifugal governor, and Figure 1.9 provides details of the centrifugal (or flyball) governor action. The two balls swing outward with increasing speed due to centrifugal force action. The swing is translated into linear motion through the lever system

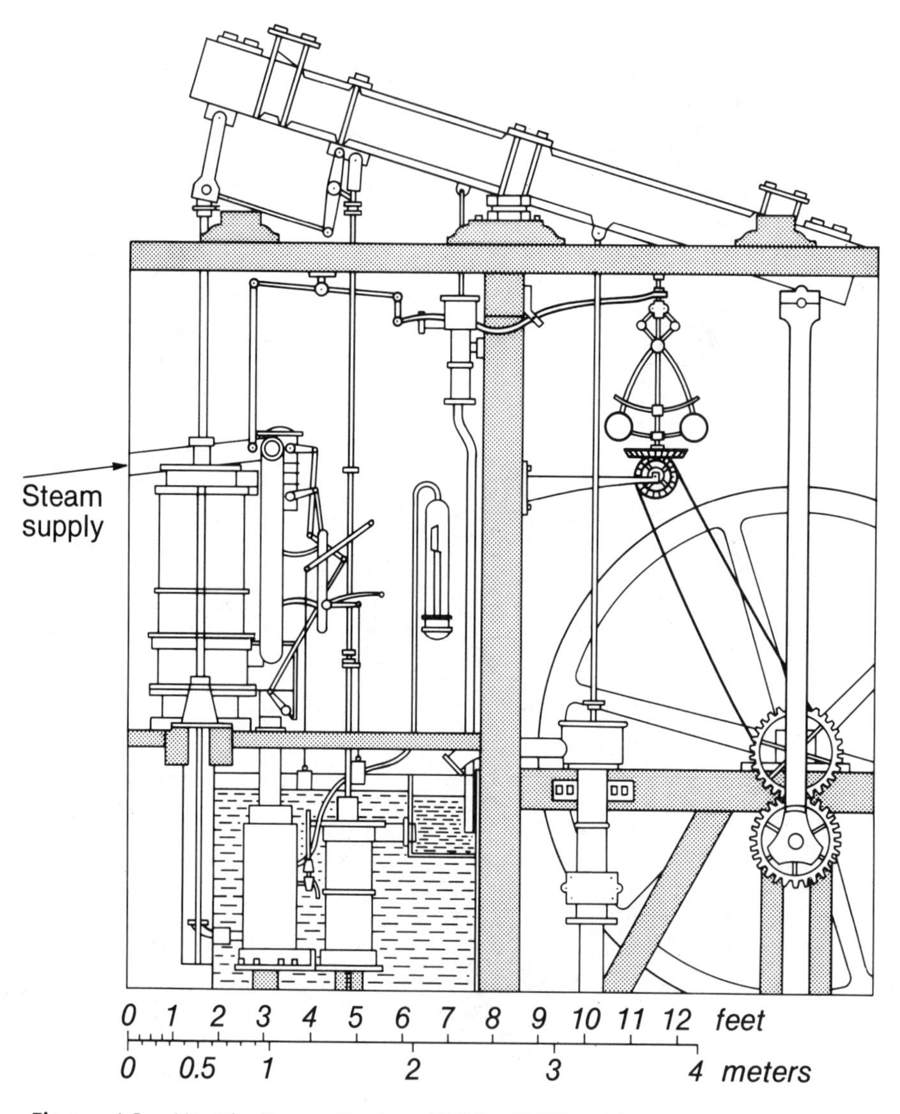


**Figure 1.8.** Watt's Steam Engine (1789 – 1800) with Centrifugal or Flyball Governor

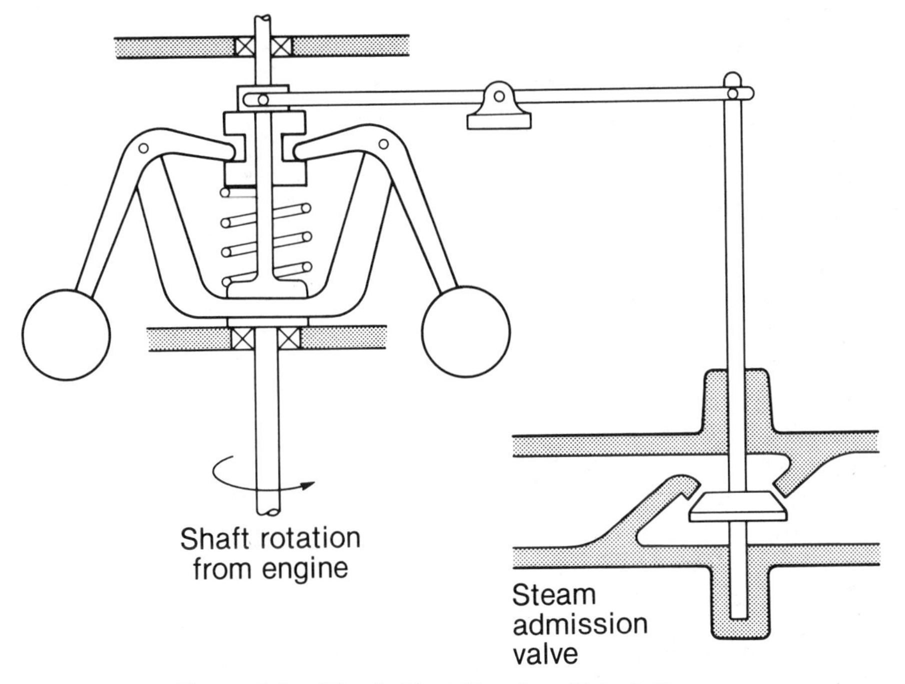


**Figure 1.9.** Watt's Centrifugal or Flyball Governor

to the steam valve. If the load on the engine is increased, the speed will decrease so that the balls drop, causing a wider opening in the steam admission valve. This will cause an increase in torque input and an acceleration of the engine. The balls fly outward again, reducing the valve opening. Ultimately, the engine will reach an equilibrium at a new speed.

The invention of the centrifugal governor led to a surge of interest at centers of academic excellence at the theoretical level. This was a turning point in the history of control systems, for analysis was now at hand.

## 1.4 The Dawn of Analysis

The development and improvements in the centrifugal governor were followed by an interest in the analysis of the dynamic behavior of this important device. Thomas Young gave a formula for the height of the balls in Volume 1 of his book "Lectures on Natural Philosophy and the Mechanic Arts" in 1807. Equilibrium equations were derived by Jean-Victor Poncelet in 1826 and published in his "Cours de Mécanique." Sir George Biddell Airy

(1801–1892), who published over 500 papers and 11 books, took interest in modeling governors and drew attention to the problem of instability of dynamic systems. William Thomson and J. Clerk Maxwell became interested in engine governors due to their involvement, together with William Siemens, in a committee established by the British Association for the Advancement of Science in 1861. The committee designed an experiment for the determination of the ohm, which required the rotation of a coil at constant speed. A governor for the purpose was designed and attracted Maxwell's interest from a theoretical point of view as an example of a dynamic system. It should be noted that earlier, in 1856, Maxwell had won the Adams Prize for an essay on the stability of Saturn's rings. It should be no surprise then that Maxwell's paper of 1868, entitled "On Governors," utilized a linear differential equation approach and derived stability conditions for a third-order system. Maxwell was unable, however, to extend the conditions to higher-order systems.

The Maxwell stability condition for a third-order system was incorporated in the second edition of Routh's "Rigid Dynamics" in 1868. Edward John Routh, born January 20, 1831, in Quebec, Canada, was affiliated with Peterhouse College of Cambridge from 1855 to 1888. In 1874, Routh was able to determine a complete set of stability criteria for a fifth-order system. Further generalization can be found in the Adams Prize–winning essay by Routh entitled "A Treatise on the Stability of a Given State of Motion" in 1876.

The Russian engineer Wischnegradski presented an analysis of the governor to the Académie des Sciences in Paris in 1876. His model was a third-order one in which graphical stability criteria were developed. The Wischnegradski criteria were used by Aurel B. Stodola of the Federal Polytechnic of Zurich in 1893 in studying dynamics involved in the regulation of a high-pressure water turbine. In extending the work to include other effects, the order of the system increased from third to seventh. The problem was then presented to the mathematician Adolf Hurwitz. A solution was obtained by Hurwitz and reported in 1895. It should be noted that Routh and Hurwitz were unaware of the other's work. It was Bompiani in 1911 who established the equivalence of the Hurwitz and Routh criteria. This topic is covered in Chapter 6.

# 1.5 The Electrical Age

Electricity was shown to produce light when Sir Humphry Davy demonstrated an arc lamp at a public lecture at the Royal Institution in 1809. However, commercial use of the arc lamp did not materialize until the

1870s. With this new technology new control problems emerged. Constant current control was one such requirement, and so was the control distance between electrodes. Arc-length regulation fascinated nineteenth-century inventors, and hundreds of patents were granted, with very few being successful. The early designs employed an electromagnet in series with a lamp. The current in the circuit decreased as the electrodes burned away, and thus the magnetic pull was reduced, allowing the electrodes to come together. These designs were suited for one arc lamp in the circuit driven by a voltaic cell.

In the mid-1870s the self-excited direct-current (dc) generator provided a cheap source of electrical energy and hence made economical the use of arc lamps in street lighting. The Thomson–Houston arc-length control system was patented in 1879 and enabled the control of a string of lamps on the same circuit. Arc lamps eventually gave way to the incandescent lamp.

With the emergence of the new electric technology, many innovations in voltage and current regulation appeared. Later, as electric generators were interconnected, the importance of frequency control was realized. Clearly, the concept of feedback played an important role in developing systems designed to meet the requirements of this new technology. Electric motor-speed and position-control problems received considerable attention. This started when H. Ward Leonard patented the now well-known system in 1891. Most of these developments were concerned with steady-state behavior; controllers were developed by empirical methods and there was little analysis.

## 1.6 The Link To Communication Engineering

On December 12, 1901, Guglielmo Marconi successfully demonstrated his wireless telegraphy by sending a signal from Poldhu in Cornwall to St. John's in Newfoundland, a distance of 2700 km. Marconi's signal detector was a glass tube filled with metal fillings which he called a "coherer tube." John Ambrose Fleming, a consultant to Marconi's company, motivated by a hearing impairment, sought a means for visually detecting radio signals. Fleming was able to detect radio signals using a two-electrode lamp which he patented under the name "thermoionic diode" in 1904. Two years later Lee De Forest added a third filament, the grid, producing what he called the Audion, patented in 1908. The Audion, or triode valve, offered the much-needed improved sensitivity of radio receivers. It was in the summer of 1912 that Edwin Armstrong, an undergraduate at Columbia University (age 22), discovered that the gain of the Audion could be increased if part of the output signal was coupled back to the input circuit. Thus was born positive feedback.

Long-distance telephony became a practical reality with the availability of the vacuum-tube amplifier. Repeater amplifiers were used along the communication path to boost the signal. The systems that existed in the early 1920s suffered from distortion and noise effects. In 1927, Harold S. Black, a recent graduate of Worcester Polytechnic, discovered the negative-feedback amplifier as a solution to repeater amplifier distortion. Black's findings as a member of the staff of Bell Telephone Laboratories were published in 1934 in the *Bell System Technical Journal.*

In investigating the performance of the negative-feedback amplifier, it was discovered that the amplifier had a tendency to generate self-oscillations at a certain frequency. It was also obvious that an investigation into the stability of the amplifier using a differential equation approach involved an inordinate amount of work since an amplifier had close to 50 independent energy-storing elements. Harry Nyquist (1889–1976), a native of Sweden who came to the United States at the age of 18 and worked at Bell Labs for most of his career, pioneered a frequency-response approach to the problem of the stability of the negative-feedback amplifier, which is documented in his famous paper "Regeneration Theory," published in the *Bell System Technical Journal* in 1932.

The negative-feedback amplifier investigations at Bell Labs provided the control systems engineering area with another major contribution in the form of the work of Hendrik W. Bode. In the *Bell System Technical Journal* in 1940, a paper by Bode entitled "Relation between Attenuation and Phase in Feedback Amplifier Design" introduced logarithmic units of amplitude gain and logarithmic frequency in the design process named after him. Transfer of the communications engineer approach to other areas of control systems analysis and design was made possible through the 1942 contribution of H. Harris of the Massachusetts Institute of Technology. In his report, Harris advocates the use of transfer functions and block diagrams. Needless to say, this contribution can be argued to be the first major step toward a unified treatment of control systems.

## 1.7 World War II Development

The war effort created an urgent requirement for high-performance feedback control systems. As a result, an immense pooling of scientific and engineering talent was called for. This lead to dissemination of results available in one area to the other, with the resulting emergence of unified theories applicable to a wide range of activities in the chemical, mechanical, aeronautical, naval, and electrical fields. The foundation for the unified area of control systems engineering was laid in the textbooks and reports that appeared following World War II.

Digital control systems have their roots in fire control systems developed during World War II. The rotating antenna of a radar system receives a return signal from a target only periodically. As a result, the concept of an ideal sampler and the study of sampled data control systems were developed to meet this need. The later development of digital computers and their incorporation in control systems drew on theories developed for sampled data systems. This involved the $z$-transform theory developed to match the Laplace transform methodology adopted for continuous control systems, and the extension of frequency-response methods to discrete-time control systems.

# 1.8 The Recent Past and the Future

Frequency-response methods have dominated the control systems engineering area since the early 1950s as analysis and design tools. The Nyquist and Bode diagrams became indispensable for assessing closed-loop stability of feedback control systems. Evans' root-locus method brought the complex-valued-based approach to linear feedback systems to a fully developed state by 1948. A mature classical control system engineering area thus evolved.

Further advances in control systems engineering resulted from the space program and associated challenges. Launching, maneuvering, guidance, and tracking of space vehicles were problems that required pooling of talent by engineers and scientists devoted to the analysis and design of the multitudes of subsystems involved in the overall program. The emergence of a modern control theory based on state-space ideas is generally traced back to the late 1950s. The state-space concept itself is much older than this. In 1844, F. L. N. M. Moigno in "Leçons de Calcul d'après Cauchy" had shown that any $n$th-order ordinary differential equation may be reduced to an equivalent set of first-order differential equations through simple substitutions. Moigno's concept was incorporated by H. Poincaré in his book "Méthodes Nouvelles de la Mécanique Céleste" published in 1892. It is through Poincaré's work on celestial mechanics that control engineers involved in aerospace research and development work adopted the state-space representation of dynamical systems.

Advances in control systems engineering on the theoretical as well as the development side continue, and the saying "the sky is the limit" seems appropriate. It is obvious that advances in computing and electronics provide a continuing impetus for new products and processes in which control systems methodology plays a central role. The majority of the innovations are directed toward improving productivity and enhancing the human working and leisure environment. A notable area is industrial robotics, which is now in the realm of fact rather than fiction, exemplified by R2-D2 and C3P0 of George Lucas' *Star Wars*, *The Empire Strikes Back*, and *The Return of the Jedi*. We now return to our main mission, which begins with an outline of this text.

## 1.9 Outline of the Text

There are nine chapters in this book in addition to this introductory chapter. The format of each chapter is such that at the conclusion of the chapter there are a number of solved problems that the reader should consult to focus on the concepts and techniques developed in the unit. An additional set of unsolved (but certainly solvable) problems is provided to conclude each chapter.

In Chapter 2 we introduce some fundamental concepts and techniques that are useful for control systems analysis and design. Under the heading "Control Systems Modeling and Representation," we provide a review of the Laplace transform and its inverse. This is followed by a discussion of elementary physical system models. The concept of a transfer function for linear time-invariant dynamic systems is treated next. Block diagram and signal flow representations are considered to conclude the chapter. The title of Chapter 3 is "Features of Feedback Control Systems." The chapter begins by providing definitions and terms used in feedback systems. The structure of a typical configuration is used to illustrate the terms. Control actions are defined and the advantages of feedback are discussed. A section is devoted to a study of typical examples to show how feedback arises either as a natural phenomenon or a design choice. The chapter is concluded by a comprehensive treatment of the topic of system sensitivity to changes in parameters.

Concepts and issues in state-space representation are dealt with in Chapter 4. A brief review of linear algebraic prerequisites is given prior to a well-balanced treatment of techniques for writing the state equations. The solution of the state equations and the concept of a state transition matrix are discussed. System diagonalization, eigenvalues, and eigenvectors are also introduced. A number of practical examples are incorporated in the text material.

Chapter 5 is concerned with the time-domain analysis of control systems. Test signals, error coefficients, and constants are discussed. The time responses of first-order, second-order, and higher-order systems are treated. Performance specifications for second-order systems are introduced in this chapter. In Chapter 6 we use the knowledge gained in Chapters 4 and 5 to consider the important issue of system stability. The Hurwitz, Lienard–Chipart, and Routh criteria are introduced and typical examples are considered. Chapter 7 is a natural extension of Chapter 6 and is devoted to Evans' root-locus method.

In Chapter 8 we deal with frequency-response characteristics and stability criteria. Nyquist, Bode, and Nichols diagrams are introduced and concepts of gain and phase margins are considered. Chapter 9 is devoted to design and compensation considerations and is based on tools developed earlier in the text.

Digital control systems are treated in Chapter 10. Following the necessary definitions, we introduce the $z$-transform and its inverse. Stability analysis is also discussed. The chapter is intended to provide a firm background for further study in this fascinating area.

# References

Bennett, S. *A History of Control Engineering, 1800–1930.* London: Peter Peregrinus, 1979.

MacFarlane, A. G. J. *Frequency-Response Methods in Control Systems.* New York: IEEE Press, 1979.

Mayr, O. *The Origins of Feedback.* Cambridge, Mass.: MIT Press, 1970.

Rörentrop, K. *Entwicklung der modernen Regelungstechnik.* Munich: Oldenbourg, 1971.

# Chapter 2

# CONTROL SYSTEMS MODELING AND REPRESENTATION

## 2.1 Introduction

This chapter lays the groundwork for the study of control systems engineering, starting with some prerequisites from the theory of the Laplace transform and its inverse. This invaluable tool provides a methodology of analysis that is appealing to the control systems engineer. This is followed by a discussion of aspects of modeling control system components and an introduction to the concept of transfer functions. Representation of control systems utilizing block diagrams and signal flow graphs is covered to conclude the chapter.

## 2.2 The Laplace Transform

The *Laplace transform* is one of the most indispensable tools at the disposal of the control systems engineer. The fundamental definition of the Laplace transform $X(s)$ of a time-varying function $x(t)$ is given by the basic relationship

$$X(s) = \int_0^\infty x(t)e^{-st}\,dt \tag{2-1}$$

Note that the operation on the right-hand side of Eq. (2-1) involves the multiplication of the available function $x(t)$, which is assumed to be defined from $t = 0$ to infinity by the factor $e^{-st}$. The new variable $(s)$ is called the *Laplace operator*. The resulting expression is clearly a function of $s$ only, as the integration is between zero and infinity. A commonly used notation to express the Laplace transform operation is

$$X(s) = \mathscr{L}\{x(t)\} \tag{2-2}$$

The Laplace transforms of some common functions are obtained in the following examples. Table 2.1 lists the Laplace transform of functions encountered in control systems engineering.

EXAMPLE 2.1

Find the Laplace transform of the function

$$x(t) = e^{\lambda t} \qquad t \geq 0$$

SOLUTION

Using Eq. (2-1), we can write

$$X(s) = \int_0^\infty e^{\lambda t}e^{-st}\,dt$$

Thus

$$X(s) = \int_0^\infty e^{-(s-\lambda)t}\,dt$$

$$= \left.\frac{e^{-(s-\lambda)t}}{s-\lambda}\right|_\infty^0$$

$$= \frac{1}{s-\lambda}$$

**TABLE 2.1**
**SOME LAPLACE TRANSFORM PAIRS**

| $x(t)$ | $X(s)$ |
|---|---|
| Unit impulse $\delta(t)$ | $1$ |
| Unit step $u(t)$ | $\frac{1}{s}$ |
| $t$ | $\frac{1}{s^2}$ |
| $t^n$ | $\frac{n!}{s^{n+1}}$ |
| $e^{-\lambda t}$ | $\frac{1}{s+\lambda}$ |
| $t^n e^{-\lambda t}$ | $\frac{n!}{(s+\lambda)^{n+1}}$ |
| $\sin \omega t$ | $\frac{\omega}{s^2+\omega^2}$ |
| $\cos \omega t$ | $\frac{s}{s^2+\omega^2}$ |
| $e^{-\lambda t}\sin \omega t$ | $\frac{\omega}{(s+\lambda)^2+\omega^2}$ |
| $e^{-\lambda t}\cos \omega t$ | $\frac{s+\lambda}{(s+\lambda)^2+\omega^2}$ |

We can thus conclude that

$$\mathscr{L}\{e^{\lambda t}\} = \frac{1}{s-\lambda} \tag{2-3}$$

The result of Example 2-1 is useful in many ways. First, the Laplace transform of the exponential function is of interest in its own right. Second, since the exponential is a generating function of many other interesting functions, we can use Eq. (2-3) to provide the Laplace transform of these other functions. To start, note that a step function $u(t)$ is defined as

$$\begin{aligned} u(t) &= 1 \qquad t \geq 0 \\ &= 0 \qquad t < 0 \end{aligned}$$

Note that

$$u(t) = e^{0t} \tag{2-4}$$

Thus substituting $\lambda = 0$ into Eq. (2-3) gives us

$$\mathscr{L}\{u(t)\} = \frac{1}{s} \tag{2-5}$$

Before proceeding with further uses of Eq. (2-3), we show a linear property of the Laplace transform. Consider two functions $x_1(t)$ and $x_2(t)$

and two scalars $a_1$ and $a_2$. Define the sum $x_3(t)$ by

$$x_3(t) = a_1 x_1(t) + a_2 x_2(t)$$

The Laplace transform of $x_3(t)$ is given by

$$X_3(s) = \int_0^\infty [a_1 x_1(t) + a_2 x_2(t)] e^{-st}\, dt$$
$$= a_1 \int_0^\infty x_1(t) e^{-st}\, dt + a_2 \int_0^\infty x_2(t) e^{-st}\, dt$$

Thus we conclude that

$$X_3(s) = a_1 X_1(s) + a_2 X_2(s)$$

This proves the linear property of the Laplace transform stated as

$$\mathscr{L}\{a_1 x_1(t) + a_2 x_2(t)\} = a_1 \mathscr{L}\{x_1(t)\} + a_2 \mathscr{L}\{x_2(t)\} \tag{2-6}$$

We can combine (2-6) with the results of Example 2.1 to derive the Laplace transform of other functions, as shown in the following examples.

EXAMPLE 2.2

Find the Laplace transform of

$$x(t) = A \sin(\omega t + \alpha)$$

SOLUTION

First recall *Euler's identity,*

$$e^{j\theta} = \cos\theta + j\sin\theta$$

Thus

$$\cos\theta = \frac{e^{j\theta} + e^{-j\theta}}{2}$$
$$\sin\theta = \frac{e^{j\theta} - e^{-j\theta}}{2j}$$

where $j = \sqrt{-1}$. We can thus write

$$x(t) = \frac{A}{2j}[e^{j(\omega t+\alpha)} - e^{-j(\omega t+\alpha)}]$$
$$= a_1 e^{\lambda_1 t} + a_2 e^{\lambda_2 t}$$

Clearly, we have

$$a_1 = \frac{A}{2j} e^{j\alpha}$$
$$a_2 = \frac{-A}{2j} e^{-j\alpha}$$
$$\lambda_1 = j\omega$$
$$\lambda_2 = -j\omega$$

On the basis of Eqs. (2-6) and (2-3), we thus write

$$X(s) = \frac{a_1}{s - \lambda_1} + \frac{a_2}{s - \lambda_2}$$

The above reduces to

$$\begin{aligned} X(s) &= \frac{A}{2j}\left(\frac{e^{j\alpha}}{s - j\omega} - \frac{e^{-j\alpha}}{s + j\omega}\right) \\ &= \frac{A}{2j}\frac{s(e^{j\alpha} - e^{-j\alpha}) + j\omega(e^{j\alpha} + e^{-j\alpha})}{s^2 + \omega^2} \\ &= A\frac{s \sin \alpha + \omega \cos \alpha}{s^2 + \omega^2} \end{aligned}$$

Thus

$$\mathscr{L}\{A \sin(\omega t + \alpha)\} = A\frac{s \sin \alpha + \omega \cos \alpha}{s^2 + \omega^2} \tag{2-7}$$

If $\alpha = 0$, we obtain

$$\mathscr{L}\{A \sin \omega t\} = A\frac{\omega}{s^2 + \omega^2} \tag{2-8}$$

For $\alpha = \pi/2$, we obtain

$$\mathscr{L}\{A \cos \omega t\} = A\frac{s}{s^2 + \omega^2} \tag{2-9}$$

## Some Useful Theorems

A number of results concerning the Laplace transform of derivatives, integrals, and shifts of functions are briefly reviewed. To begin with, let us perform the integration indicated in Eq. (2-1) by parts to conclude that

$$\begin{aligned} X(s) &= -\left.\frac{x(t)e^{-st}}{s}\right|_0^\infty + \int_0^\infty \frac{dx(t)}{dt}\frac{e^{-st}}{s}dt \\ &= \frac{x(0)}{s} + \frac{1}{s}\mathscr{L}\frac{dx(t)}{dt} \end{aligned}$$

As a result, we have the *differentiation theorem* given by Eq. (2-10):

$$\mathscr{L}\{\dot{x}(t)\} = sX(s) - x(0) \tag{2-10}$$

In general, we can state for the $n$th derivative of $x(t)$,

$$\begin{aligned} \mathscr{L}\left\{\frac{d^n}{dt^n}x(t)\right\} &= s^n X(s) - s^{n-1}x(0) - s^{n-2}\dot{x}(0) \\ &\quad - \cdots - sx^{(n-2)}(0) - x^{n-1}(0) \end{aligned} \tag{2-11}$$

where $x(0)$, $\dot{x}(0)$, $\ddot{x}(0), \ldots, x^{n-2}$, and $x^{n-1}$ denote the values of

$x(t)$, $\dot{x}(t)$, $\ddot{x}(t), \ldots, (d^{n-2}/dt^{n-2})x(t), (d^{n-1}/dt^{n-1})x(t)$ evaluated at $t = 0$. Note that if all initial values of $x(t)$ and its derivatives are zero, the Laplace transform of the $n$th derivative of $x(t)$ is $s^n X(s)$ and the differentiation operation is equivalent in the Laplace domain to the $s$ operator.

We can arrive at the *integration theorem* through use of Eq. (2-10) as follows. Let $y(t) = \dot{x}(t)$; thus

$$x(t) = \int y(t)\, dt$$

As a result of Eq. (2-10), we have

$$\mathscr{L}\{y(t)\} = s\mathscr{L}\left\{\int y(t)\, dt\right\} - \int y(t)\, dt\Big|_{t=0}$$

Thus

$$\mathscr{L}\left\{\int y(t)\, dt\right\} = \frac{Y(s)}{s} + \frac{y^{-1}(0)}{s} \tag{2-12}$$

where $y^{-1}(0) = \int y(t)\, dt$ evaluated at $t = 0$.

The *shift theorem* deals with the Laplace transform of $x(t - \tau)$, where $\tau$ is a delay given that $x(t)$ is defined and $x(t) = 0$, $t < 0$. Applying the fundamental expression (2-1), we have

$$\begin{aligned} X(s) &= \int_0^\infty x(\sigma)e^{-\sigma s}\, d\sigma \\ &= \int_0^\infty x(t - \tau)e^{-s(t-\tau)}\, dt \\ &= e^{\tau s}\int_0^\infty x(t - \tau)e^{-st}\, dt \end{aligned}$$

Thus

$$\mathscr{L}\{x(t - \tau)\} = e^{-\tau s}X(s) \tag{2-13}$$

The *final-value theorem* can be obtained from the differentiation theorem (2-10) by taking the limits as $s \to 0$, with the result

$$\lim_{t\to\infty} x(t) = \lim_{s\to 0} sX(s) \tag{2-14}$$

The *initial-value theorem* is stated as

$$\lim_{t\to 0} x(t) = \lim_{s\to\infty} sX(s) \tag{2-15}$$

We will consider now an important case involving the *convolution integral* defined for the two functions $h(t)$ and $x(t)$ by the relation

$$y(t) = h(t) * x(t) \tag{2-16}$$

or

$$y(t) = \int_0^t h(t - \tau)x(\tau)\, d\tau \tag{2-17}$$

The asterisk denotes the convolution operation. The Laplace transform of $y(t)$ is obtained using Eq. (2-1) as

$$Y(s) = \int_0^\infty y(t)e^{-st}\,dt \tag{2-18}$$

Let us note that both $h(t)$ and $x(t)$ are assumed to be zero for $t < 0$. We can thus conclude that $h(t-\tau)$ is zero for $\tau > t$. As a result, we can write

$$y(t) = \int_0^\infty h(t-\tau)x(\tau)\,d\tau \tag{2-19}$$

As a result, Eq. (2-18) can be written as

$$Y(s) = \int_0^\infty \left[\int_0^\infty h(t-\tau)x(\tau)\,d\tau\right] e^{-st}\,dt$$

Interchanging the order of integration, we get

$$Y(s) = \int_0^\infty \int_0^\infty e^{-st}h(t-\tau)x(\tau)\,dt\,d\tau$$

$$= \int_0^\infty x(\tau)\left[\int_0^\infty e^{-st}h(t-\tau)\,dt\right]d\tau$$

Consider the integral with respect to $t$ and let $\sigma = t - \tau$; thus

$$\int_0^\infty e^{-st}h(t-\tau)\,dt = \int_{\sigma=-\tau}^\infty e^{-s(\sigma+\tau)}h(\sigma)\,d\sigma$$

$$= e^{-\tau s}\int_0^\infty e^{-s\sigma}h(\sigma)\,d\sigma$$

$$= H(s)e^{-\tau s}$$

Thus we have

$$Y(s) = \int_0^\infty e^{-\tau s}x(\tau)H(s)\,d\tau$$

Since $H(s)$ is independent of $\tau$ we get

$$Y(s) = H(s)\int_0^\infty e^{-s\tau}x(\tau)\,d\tau$$

As a result,

$$Y(s) = H(s)X(s) \tag{2-20}$$

Our conclusion is that the Laplace transform of the convolution of $h(t)$ and $x(t)$ is the product of the Laplace transforms $H(s)$ and $X(s)$.

## 2.3 The Inverse Laplace Transform

Consider the simple differential equation

$$a_2\frac{d^2x}{dt^2} + a_1\frac{dx}{dt} + a_0x = u(t)$$

Application of the Laplace transform to both sides, assuming zero initial conditions, gives us

$$(a_2 s^2 + a_1 s + a_0)X(s) = U(s)$$

This is an algebraic equation which can be written as

$$X(s) = \frac{U(s)}{a_2 s^2 + a_1 s + a_0}$$

Suppose now that the input function $u(t)$ is a unit step; thus

$$U(s) = \frac{1}{s}$$

As a result, the Laplace transform of $x(t)$ is given by

$$X(s) = \frac{1}{s(a_2 s^2 + a_1 s + a_0)}$$

Finding the function $x(t)$ whose transform is as given above is symbolized by the inverse transform operator $\mathscr{L}^{-1}$; thus

$$x(t) = \mathscr{L}^{-1}\{X(s)\} = \mathscr{L}^{-1}\left\{\frac{1}{s(a_2 s^2 + a_1 s + a_0)}\right\}$$

A formal definition of the *inverse Laplace transform* is given by

$$x(t) = \mathscr{L}^{-1}\{X(s)\} = \frac{1}{2\pi j}\int_{a-j\infty}^{a+j\infty} X(s)e^{st}\,ds \tag{2-21}$$

where $a$ is a real constant. It is quite possible (although somewhat involved) to obtain the inverse Laplace transform by performing the integration indicated in Eq. (2-21). A much more effective way is commonly employed in control systems engineering, which relies on performing a partial fraction expansion that results in an expression of $X(s)$ as the sum of the functions $X_1(s), X_2(s), \ldots, X_n(s)$.

$$X(s) = X_1(s) + X_2(s) + \cdots + X_n(s) \tag{2-22}$$

The functions $X_1(s), X_2(s), \ldots$ can then be looked up in a table of Laplace transform pairs and hence we can obtain the corresponding inverses $x_1(t), x_2(t), \ldots, x_n(t)$. The final result is thus

$$x(t) = x_1(t) + x_2(t) + \cdots + x_n(t) \tag{2-23}$$

## Partial Fraction Expansion

In most control systems engineering applications it is desired to find the inverse Laplace transform of a function $F(s)$ expressed as the ratio of two functions $N(s)$ and $D(s)$ according to

$$F(s) = \frac{N(s)}{D(s)} \tag{2-24}$$

The numerator and denominator functions are commonly obtained as polynomials in $s$ of the form

$$N(s) = \sum_{j=0}^{m} b_j s^j \qquad (2\text{-}25)$$

$$D(s) = \sum_{i=0}^{n} a_i s^i \qquad (2\text{-}26)$$

The denominator is an $n$th-degree polynomial, while the numerator is of $m$th degree, with $n > m$. For example, a typical function $F(s)$ is

$$F(s) = \frac{4 + s}{2 + 3s + s^2}$$

Thus

$$N(s) = 4 + s$$
$$D(s) = 2 + 3s + s^2$$

In order to carry out the partial fraction expansion procedure it is necessary to obtain $N(s)$ and $D(s)$ in the following factored form:

$$N(s) = (s - z_1)(s - z_2) \cdots (s - z_m) \qquad (2\text{-}27)$$
$$D(s) = (s - p_1)(s - p_2) \cdots (s - p_n) \qquad (2\text{-}28)$$

In Eq. (2-27), the $z_1, z_2, \ldots, z_m$ are called the *zeros* of the function $F(s)$, while in Eq. (2-28), the $p_1, p_2, \ldots, p_n$ are called the *poles* of the function $F(s)$. Obtaining the poles and zeros from definitions of Eqs. (2-25) and (2-26) may or may not be straightforward. In the preceding example it is easy to see that

$$z_1 = -4$$
$$p_1 = -1$$
$$p_2 = -2$$

This follows since $D(s)$ is a second-order polynomial which can be easily factored.

For higher-order polynomials it is often necessary to resort to an iterative procedure to obtain the necessary factors. Consider, for example, the expression

$$D(s) = s^4 + 9s^3 + 26s^2 + 24s$$

It is easy to see that $s$ is a common factor

$$D(s) = sD_1(s)$$

with

$$D_1(s) = s^3 + 9s^2 + 26s + 24$$

Now $D_1(s)$ is a third-order polynomial and we have to find its roots.

Although there is a well-defined procedure for doing just that, we illustrate the use of an iterative method such as *Newton's method.* We start with an estimate $s^{(0)}$ of the solution, and obtain $s^{(1)}$ using

$$s^{(1)} = s^{(0)} - \frac{D_1(s^{(0)})}{D_1'(s^{(0)})}$$

where $D_1'(s)$ is the derivative of $D_1(s)$. In our example we have

$$D_1'(s) = 3s^2 + 18s + 26$$

Thus take $s^{(0)} = -1$ to obtain

$$s^{(1)} = -1 - \frac{-1 + 9 - 26 + 24}{3 - 18 + 26}$$
$$= -1.55$$

An improvement on $s^{(1)}$ is obtained as $s^{(2)}$:

$$s^{(2)} = s^{(1)} - \frac{D_1(s^{(1)})}{D_1'(s^{(1)})}$$
$$= -1.55 - \frac{1.62}{5.35}$$
$$= -1.85$$

We next obtain $s^{(3)}$ as

$$s^{(3)} = s^{(2)} - \frac{D_1(s^{(2)})}{D_1'(s^{(2)})}$$
$$= -1.97$$

The iterations are continued according to

$$s^{(N+1)} = s^N - \frac{D_1(s^{(N)})}{D_1'(s^{(N)})}$$

We stop when no further improvement is possible. For the present example we get the solution at

$$s = -2$$

Thus $s + 2$ is a factor. Thus we get

$$D_1(s) = (s + 2)D_2(s)$$

For our case

$$D_2(s) = \frac{D_1(s)}{s + 2} = s^2 + 7s + 12$$

This is a second-order polynomial,

$$D_2(s) = (s + 3)(s + 4)$$

We can thus conclude that

$$D(s) = s(s+2)(s+3)(s+4)$$

The factorization process is now complete.

Let us assume that $F(s)$ of Eq. (2-24) is given in the factored forms (2-27) and (2-28) written as

$$F(s) = \frac{(s-z_1)(s-z_2)\cdots(s-z_m)}{(s-p_1)(s-p_2)\cdots(s-p_n)} \tag{2-29}$$

The partial fraction expansion requires writing Eq. (2-29) as

$$F(s) = \frac{A_1}{s-p_1} + \frac{A_2}{s-p_2} + \cdots + \frac{A_n}{s-p_n} \tag{2-30}$$

The task of course is to determine $A_1, A_2, \ldots, A_n$. This can be accomplished as follows. Rewrite Eq. (2-29) as

$$D(s)F(s) = N(s) \tag{2-31}$$

We also write Eq. (2-30) as

$$D(s)F(s) = A_1\frac{D(s)}{s-p_1} + A_2\frac{D(s)}{s-p_2} + \cdots + A_n\frac{D(s)}{s-p_n} \tag{2-32}$$

Thus we have

$$N(s) = A_1\frac{D(s)}{s-p_1} + A_2\frac{D(s)}{s-p_2} + \cdots + A_n\frac{D(s)}{s-p_n} \tag{2-33}$$

Note that

$$\frac{D(s)}{s-p_1} = (s-p_2)(s-p_3)\cdots(s-p_n)$$

Thus let $s = p_1$ in Eq. (2-33) to obtain

$$N(p_1) = A_1(p_1-p_2)(p_1-p_3)\cdots(p_1-p_n)$$

Thus

$$A_1 = \frac{N(p_1)}{(p_1-p_2)(p_1-p_3)\cdots(p_1-p_n)} \tag{2-34}$$

Similarly, let $s = p_2$ to get

$$A_2 = \frac{N(p_2)}{(p_2-p_1)(p_2-p_3)(p_2-p_4)\cdots(p_2-p_n)} \tag{2-35}$$

This process is repeated to find $A_3, \ldots, A_n$.

With $F(s)$ obtained in the form of Eq. (2-30), we can write the inverse Laplace transform of $F(s)$ as

$$f(t) = A_1 e^{p_1 t} + A_2 e^{p_2 t} + \cdots + A_n e^{p_n t} \tag{2-36}$$

Let us remark here that the discussion above assumes that the poles $p_1, p_2, \ldots$ are distinct; that is, $p_1 \neq p_2 \neq p_3 \neq \cdots$. When there are multiple poles our procedure will differ slightly. Before looking at this case, let us take an example.

EXAMPLE 2.3

Consider the function $f(t)$ defined by its Laplace transform

$$F(s) = \frac{1}{(s+2)(s+5)(s+11)}$$

Find $f(t)$ using partial fraction expansion.

SOLUTION

We write

$$F(s) = \frac{A_1}{s+2} + \frac{A_2}{s+5} + \frac{A_3}{s+11}$$

Thus

$$A_1(s+5)(s+11) + A_2(s+2)(s+11) + A_3(s+2)(s+5) = 1$$

Put

$$s = -2; \qquad \text{thus } A_1 = \frac{1}{27}$$

Put

$$s = -5; \qquad \text{thus } A_2 = \frac{-1}{18}$$

Put

$$s = -11; \qquad \text{thus } A_3 = \frac{1}{54}$$

As a result,

$$F(s) = \frac{1}{54}\left(\frac{2}{s+2} - \frac{3}{s+5} + \frac{1}{s+11}\right)$$

The inverse Laplace transform is thus

$$f(t) = \frac{1}{54}(2e^{-2t} - 3e^{-5t} + e^{-11t})$$

Consider now the case when a pole is repeated in the $F(s)$ of Eq. (2-29). Assume that $p_1 = p_2 = \cdots = p_{n_1}$ in Eq. (2-29), so that we have

$$F(s) = \frac{N(s)}{(s-p_1)^{n_1}(s-p_{n_1+1})\cdots(s-p_n)} \tag{2-37}$$

Clearly, we need an alternative expression to Eq. (2-30). The required expression is

$$F(s) = \frac{A_1}{s - p_1} + \frac{A_2}{(s - p_1)^2} + \cdots + \frac{A_{n_1}}{(s - p_1)^{n_1}} + \frac{A_{n_1+1}}{s - p_{n_1+1}} + \cdots + \frac{A_n}{s - p_n} \tag{2-38}$$

The inverse Laplace transform is given by

$$f(t) = A_1 e^{p_1 t} + A_2 t e^{p_1 t} + \cdots + A_{n_1} \frac{t^{n_1 - 1}}{(n_1 - 1)!} e^{p_1 t} + A_{n_1+1} e^{p_{n_1+1} t} + \cdots + A_n e^{p_n t} \tag{2-39}$$

To obtain the coefficients $A_1$, $A_2, \ldots, A_n$ we adopt the procedure outlined previously with a slight modification, as shown in the next two examples.

EXAMPLE 2.4

Consider the function $f(t)$ defined by its Laplace transform

$$F(s) = \frac{1}{s^2(s + 3)}$$

Find $f(t)$ using partial fraction expansion.

SOLUTION

We write

$$F(s) = \frac{A_1}{s} + \frac{A_2}{s^2} + \frac{A_3}{s + 3}$$

Thus

$$A_1 s(s + 3) + A_2(s + 3) + A_3 s^2 = 1$$

Put

$$s = 0; \qquad \text{thus } A_2 = \tfrac{1}{3}$$

Put

$$s = -3; \qquad \text{thus } A_3 = \tfrac{1}{9}$$

To obtain $A_1$ we have to equate coefficient of $s^2$ on both sides to obtain

$$A_1 + A_3 = 0$$

Thus

$$A_1 = -\tfrac{1}{9}$$

As a result,

$$F(s) = \frac{1}{9}\left(-\frac{1}{s} + \frac{3}{s^2} + \frac{1}{s+3}\right)$$

The inverse Laplace transform is thus given by

$$f(t) = \tfrac{1}{9}(-1 + 3t + e^{-3t})$$

EXAMPLE 2.5

Consider the function $f(t)$ defined by its Laplace transform

$$F(s) = \frac{1}{s(s+3)^2}$$

Find $f(t)$ using partial fraction expansion.

SOLUTION

We can write

$$F(s) = \frac{A_1}{s} + \frac{A_2}{s+3} + \frac{A_3}{(s+3)^2}$$

Thus

$$A_1(s+3)^2 + A_2 s(s+3) + A_3 s = 1$$

Put

$$s = 0; \qquad \text{thus } A_1 = \frac{1}{9}$$

Put

$$s = -3; \qquad \text{thus } A_3 = \frac{-1}{3}$$

The coefficient of $s^2$ on both sides yields

$$A_1 + A_2 = 0$$

Thus

$$A_2 = \frac{-1}{9}$$

As a result,

$$F(s) = \frac{1}{9}\left[\frac{1}{s} - \frac{1}{s+3} - \frac{3}{(s+3)^2}\right]$$

The inverse Laplace transform is thus given by

$$f(t) = \frac{1}{9}(1 - e^{-3t} - 3te^{-3t})$$

There are situations where the presence of complex-conjugate poles makes it impossible to find a real root using Newton's method. The following example involves such a situation and proposes a method for treatment.

EXAMPLE 2.6

Find the inverse Laplace transform of

$$F(s) = \frac{1}{s^4 + 6s^3 + 17s^2 + 24s + 18}$$

SOLUTION
The denominator function is given by

$$D(s) = s^4 + 6s^3 + 17s^2 + 24s + 18$$

We attempt to factor $D(s)$ into the product of two second-order polynomials

$$D(s) = D_1(s)D_2(s)$$

where

$$D_1(s) = s^2 + a_1 s + b_1$$
$$D_2(s) = s^2 + a_2 s + b_2$$

Performing the multiplications and equating coefficients of equal powers, we conclude that

$$a_1 + a_2 = 6$$
$$b_1 + b_2 + a_1 a_2 = 17$$
$$a_1 b_2 + a_2 b_1 = 24$$
$$b_1 b_2 = 18$$

The equations above can be combined to yield one equation in $b_1$ given by

$$b_1^6 - 17b_1^5 + 126b_1^4 - 612b_1^3 + 2268b_1^2 - 5508b_1 + 5832 = 0$$

An iterative solution yields

$$b_1 = 3$$

Thus

$$b_2 = \frac{18}{b_1} = 6$$

Hence

$$6a_1 + 3a_2 = 24$$

Also,

$$a_1 + a_2 = 6$$

As a result, we find that

$$a_1 = 2$$
$$a_2 = 4$$

We can thus conclude that

$$D(s) = (s^2 + 2s + 3)(s^2 + 4s + 6)$$

Instead of performing the partial fraction expansion in terms of simple poles, we do it in terms of the second-order terms as

$$F(s) = \frac{A_1 + B_1 s}{s^2 + 2s + 3} + \frac{A_2 + B_2 s}{s^2 + 4s + 6}$$

Since

$$F(s) = \frac{1}{(s^2 + 2s + 3)(s^2 + 4s + 6)}$$

we thus have

$$(A_1 + B_1 s)(s^2 + 4s + 6) + (A_2 + B_2 s)(s^2 + 2s + 3) = 1$$

Expanding, we obtain

$$(B_1 + B_2)s^3 + (A_1 + A_2 + 4B_1 + 2B_2)s^2$$
$$+ (4A_1 + 6B_1 + 2A_2 + 3B_2)s + (6A_1 + 3A_2) = 1$$

Equating the coefficients of equal power in $s$ on both sides, we get

$$B_1 + B_2 = 0$$
$$A_1 + A_2 + 4B_1 + 2B_2 = 0$$
$$4A_1 + 2A_2 + 6B_1 + 3B_2 = 0$$
$$6A_1 + 3A_2 = 1$$

Solving the equations above, we obtain

$$A_1 = -\tfrac{1}{9}$$
$$A_2 = \tfrac{5}{9}$$
$$B_1 = -\tfrac{2}{9}$$
$$B_2 = \tfrac{2}{9}$$

Thus the partial fraction of $F(s)$ is

$$F(s) = \frac{1}{9}\left[\frac{5 + 2s}{s^2 + 4s + 6} - \frac{1 + 2s}{s^2 + 2s + 3}\right]$$

To get the inverse Laplace transform, we write $F(s)$ as

$$F(s) = \frac{1}{9}\left[\frac{1}{(s + 2)^2 + 2} + \frac{2(s + 2)}{(s + 2)^2 + 2} + \frac{1}{(s + 1)^2 + 2} - \frac{2(s + 1)}{(s + 1)^2 + 2}\right]$$

Thus the inverse Laplace transform is obtained as

$$f(t) = \frac{1}{9}\left(\frac{1}{\sqrt{2}}e^{-2t}\sin\sqrt{2}\,t + 2e^{-2t}\cos\sqrt{2}\,t + \frac{1}{\sqrt{2}}e^{-t}\sin\sqrt{2}\,t - 2e^{-t}\cos\sqrt{2}\,t\right)$$

# 2.4 Elementary Physical System Models

Control systems engineers rely on a number of physical laws that describe the interaction between variables of interest in the system under consideration. In many instances the task at hand can be handled through the use of element laws. In this section we review a number of laws in diverse areas of engineering science and indicate how these are integrated for a number of control systems devices. We start by some basics in electric circuit analysis.

Passive elements in an electric circuit are the resistance $R$, the inductance $L$, and the capacitance $C$. The relation between the voltage across $v(t)$ and current through $i(t)$ depends on the element. For a resistive element we have

$$v(t) = Ri(t)$$

For an inductance we have

$$v(t) = L\frac{di(t)}{dt}$$

For a capacitance we have

$$i(t) = C\frac{dv(t)}{dt}$$

Kirchhoff's current and voltage laws are also useful in treating electric circuits. The current law states that the algebraic sum of all currents into a node is zero. The voltage law states that the sum of voltages around a closed loop is zero.

## Direct-Current Generator

Figure 2.1 shows a schematic diagram of a separately exited direct-current (dc) generator. The input voltage $v_i(t)$ produces a current in the field circuit $i_f(t)$. Assuming that the field circuit's resistance is $R_f$ and that its inductance is $L_f$, we can write

$$v_i(t) = R_f i_f(t) + L_f\frac{di_f(t)}{dt}$$

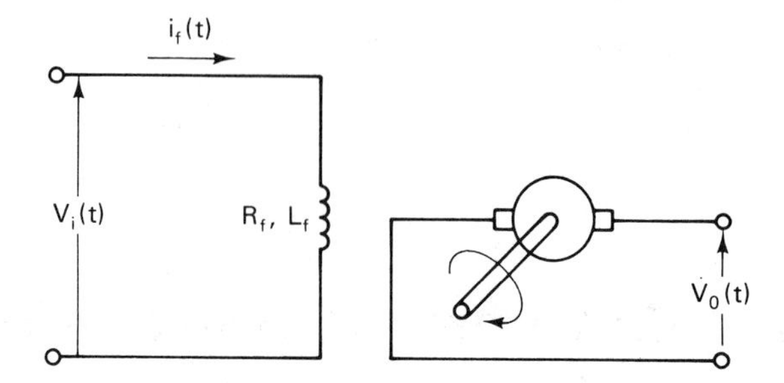


**Figure 2.1.** Schematic of a Direct-Current Generator

Assuming that the magnetization characteristic of the machine is linear in the region of interest and that the generator is driven at a constant speed, we can write the output voltage as

$$v_o(t) = K_g i_f(t)$$

$K_g$ is a proportionality constant.

## Amplidyne

The amplidyne is a two-stage dc generator, as shown in the schematic diagram of Figure 2.2. Assume that the control winding's resistance and inductance are denoted by $R_c$ and $L_c$, that the control voltage is $v_c(t)$, and that the current in the control winding is $i_c(t)$. We have

$$v_c(t) = R_c i_c(t) + L_c \frac{di_c}{dt}$$

The electromotive force (emf) developed in the quadrature axis winding is proportional to the control field current. Thus

$$e_q(t) = K_{cq} i_c(t)$$

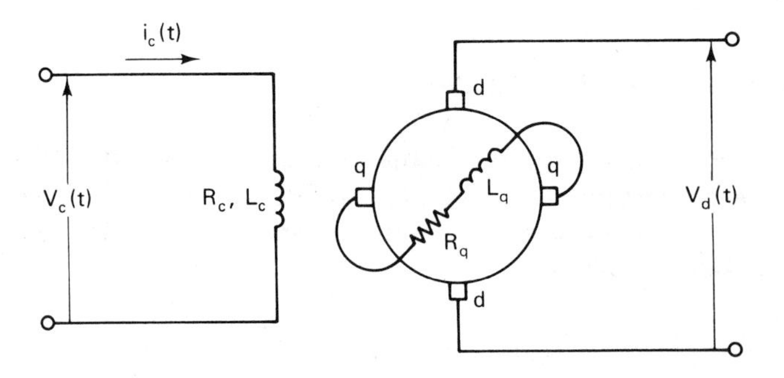


**Figure 2.2.** Schematic Diagram of the Amplidyne

where $K_{cq}$ is a constant for constant generator speed. The quadrature-axis voltage $e_q(t)$ produces a current $i_q(t)$ assuming that $R_q$ and $L_q$ are the resistance and inductance values for the quadrature-axis winding. Thus we write

$$e_q(t) = R_q i_q(t) + L_q \frac{di_q}{dt}$$

The quadrature-axis current sets up a flux, which in turn produces the output voltage $v_d(t)$ according to

$$v_d(t) = K_{dq} i_q(t)$$

## Field-Controlled DC Motor

In a field-controlled dc motor (Figure 2.3), the voltage input $v_i(t)$ is fed to the field, which can be modeled as a resistance $R_f$ in series with the inductance $L_f$. A field current $i_f(t)$ is established in accordance with

$$v_i(t) = R_f i_f(t) + L_f \frac{di_f}{dt}$$

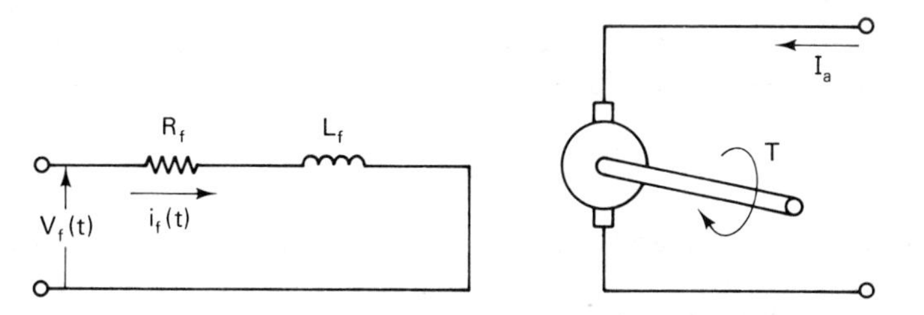


**Figure 2.3.** Schematic of a Field-Controlled DC Motor

The field current sets up a flux, which together with the rotational speed of the motor develops a back electromotive force $e_m(t)$. Assuming a linear (straight-line) relation between $e_m$ and $i_f$ for a given rotational speed $\omega_m$, we can write

$$e_m(t) = K_m i_f(t) \omega_m(t)$$

The parameter $K_m$ is a constant that depends on the motor's design particulars.

The motor's developed electrical power is given in terms of the armature current $I_a$ and the back emf as

$$P_e(t) = I_a e_m(t)$$

Note that the armature current is assumed constant. We can thus write

$$P_e(t) = K_m I_a i_f(t) \omega_m(t)$$

Assuming that the electrical power $P_e(t)$ undergoes no losses in being transmitted as mechanical power $P_m(t)$, we can write

$$P_e(t) = P_m(t)$$

We know that the mechanical power is the product of torque and angular velocity; thus

$$P_m(t) = T(t)\omega_m(t)$$

It is thus clear that under the foregoing assumptions, we can write the developed torque as

$$T(t) = K_1 i_f(t)$$

where

$$K_1 = K_m I_a$$

As a result, we conclude that the motor's developed torque is proportional to the field current.

## Armature-Controlled DC Motor

If the input voltage to the dc motor is fed to the armature circuit while field current is maintained constant, we say that the motor is armature controlled. With reference to Figure 2.4, we can write the loop equation

$$v_i(t) = e_m(t) + R_a i_a(t) + L_a \frac{di_a}{dt}$$

In this equation, $R_a$ and $L_a$ are the resistance and inductance of the armature circuit, respectively. The armature current is denoted by $i_a(t)$, while the motor's back emf is denoted by $e_m(t)$. Of course, the input voltage is denoted by $v_i(t)$. Since the field current is fixed at $I_f$, we write

$$e_m(t) = K_m I_f \omega_m(t)$$

The torque developed by the motor under assumptions similar to those

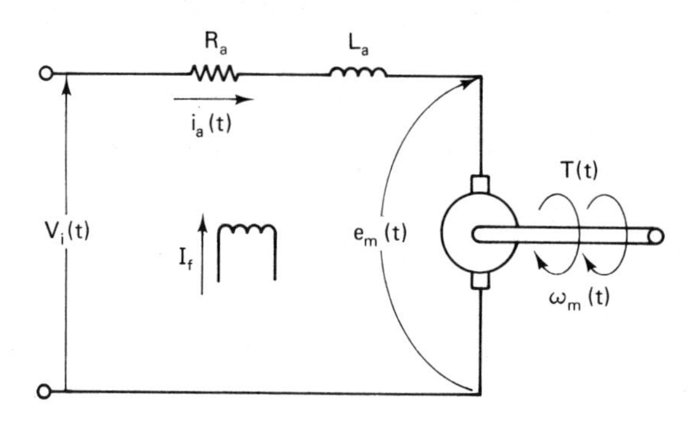


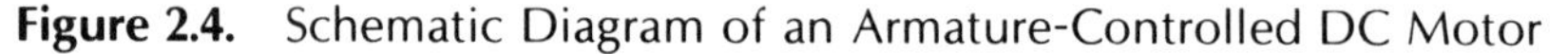
**Figure 2.4.** Schematic Diagram of an Armature-Controlled DC Motor

stated for the field-controlled motor can be obtained from

$$P(t) = e_m(t)i_a(t) = T(t)\omega_m(t)$$

Thus

$$T(t) = K_m I_f i_a(t)$$

Let us define an armature-controlled motor constant $K_a$ by

$$K_a = K_m I_f$$

As a result, we write

$$e_m(t) = K_a \omega_m(t)$$
$$T(t) = K_a i_a(t)$$

Thus the back emf and torque are proportional to the motor's velocity and armature current, respectively.

## Two-Phase Servomotor

In a two-phase servomotor (Figure 2.5a) the torque decreases linearly with the speed for a constant control winding voltage. As the control winding voltage is increased, the torque–speed characteristic is raised as shown in Figure 2.5b. We can thus model the motor's developed torque by

$$T(t) = -K_n \omega(t) + K_c e_c(t)$$

where $K_n$ and $K_c$ are constants.

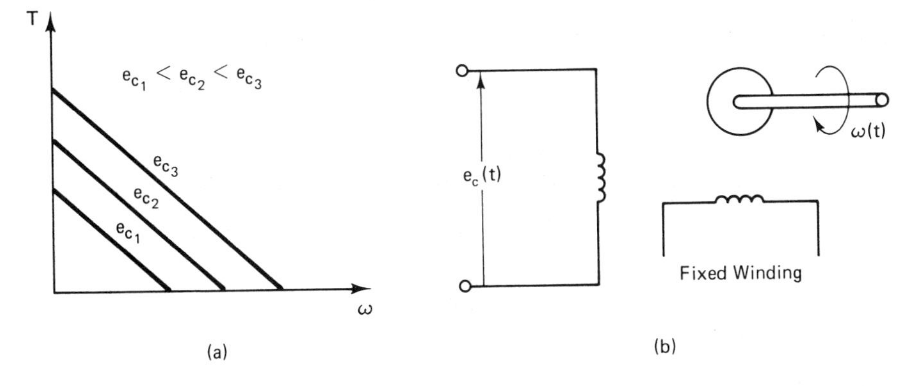


**Figure 2.5.** Two-Phase Servomotor Characteristics and Schematic Diagram

## Mechanical Translation Elements

Newton's law as applied to mechanical systems states that the sum of the forces applied to an element is equal to the sum of the reaction forces. There are three basic elements in a mechanical translation system model. These are the mass $M$, the stiffness $K$, and the damping (viscous friction) $B$.

A force $F$ applied to a mass $M$ produces an acceleration $a$ of the mass. The reaction force to $F$ is given by

$$F = Ma$$

If the translation is $x$, the acceleration is given by

$$a = \frac{d^2x}{dt^2}$$

Thus for the mass we have

$$F_M = M\frac{d^2x}{dt^2}$$

A diagram showing a mass is given in Figure 2.6a.

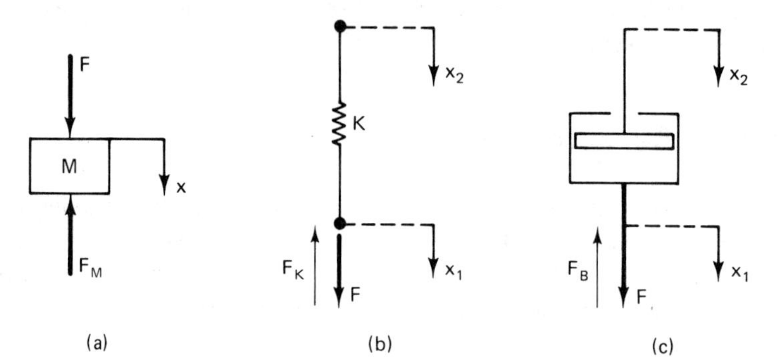


**Figure 2.6.** Elements of Mechanical Translation Systems

The elastance, or stiffness, $K$ of a spring provides a reaction force that is proportional to the deformation $\Delta x$ of the spring. Thus

$$F_K = K\,\Delta x$$

Note that in Figure 2.6b,

$$\Delta x = x_1 - x_2$$

Viscous friction or damping $B$ involves energy absorption. The damping force is proportional to the difference of the velocities of the two bodies (Figure 2.6c)

$$F_B = B\left(\frac{dx_1}{dt} - \frac{dx_2}{dt}\right)$$

## Mechanical Rotational Elements

Rotational systems involve elements and variables that correspond to those in a translation system (Figure 2.7). The body's moment of inertia $J$ corresponds to the mass in writing the dynamical equation. The angular displacement corresponds to the translational displacement $x$, the angular velocity corresponds to the translation velocity $v$, and the angular acceleration corresponds to the translational acceleration. Torque equations for rotational systems are parallel to force equations in translational systems.

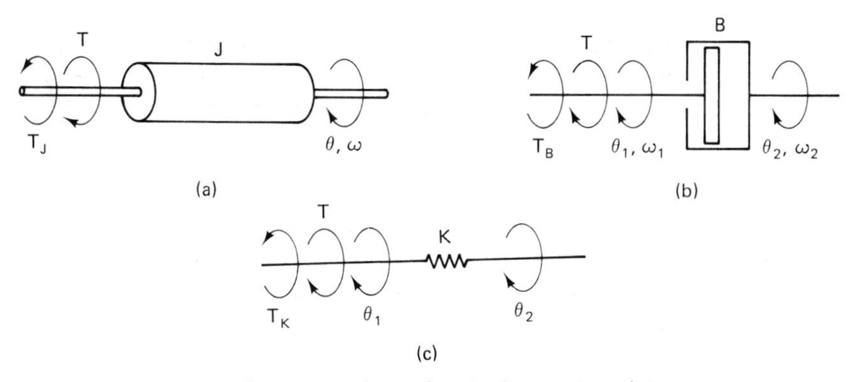


**Figure 2.7.** Elements of Mechanical Rotational Systems

For a body with a moment of inertia $J$, an applied torque $T$ produces an angular acceleration $d^2\theta/dt^2$. The reaction torque $T_J$ is in opposition to $T$, with a torque equation given by

$$T_J = J\frac{d^2\theta}{dt^2}$$

Damping is encountered whenever a body moves through a fluid. The damping torque $T_B$ is equal to the product of damping $B$ and the relative angular velocity of the damper and is in opposition to the applied torque:

$$T_B = B\left(\frac{d\theta_1}{dt} - \frac{d\theta_2}{dt}\right)$$

The stiffness torque $T_K$ is produced by a spring of stiffness $K$ if it is twisted by angle $\Delta\theta$ by an applied torque $T$. We have

$$T_K = K(\theta_1 - \theta_2)$$

## Dealing with Gear Trains

A normal practice in coupling motors to loads is to employ a gear train to transmit the driving torque to the load. An analysis of such a case is important to evaluate the moment of inertia and damping relative to the motor. Figure 2.8 shows a driving motor that supplies an input torque $T_m$ at an angular velocity $\omega_m$ to a load through a gear train having a gear ratio of

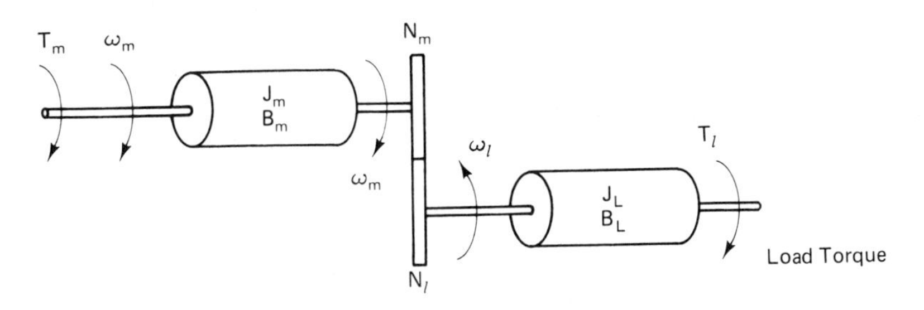


**Figure 2.8.** Schematic of a Gear Train

$N_m/N_l$. The load torque is denoted by $T_l$ and its angular velocity is denoted by $\omega_l$. Assume that the moment of inertia and viscous friction of the motor side are denoted by $J_m$ and $B_m$, respectively. In a similar way $J_l$ and $B_l$ are quantities related to the load.

The torque balance equation on the motor side is given by

$$T_m = J_m\ddot{\theta}_m + B_m\dot{\theta}_m + T_{ml}$$

Thus the torque provided by the motor is equal to the sum of motor inertial and friction torques and the torque $T_{ml}$ transmitted through the gears. Assuming that there is no power loss in the gears, we can write the power equation

$$P_{ml} = T_{ml}\omega_m = \tilde{T}_{ml}\omega_l$$

This is simply a restatement of the fact that power is the product of torque and angular velocity. We also know that

$$\frac{\omega_m}{\omega_l} = \frac{N_m}{N_l}$$

Thus we can write the driving torque on the load side as

$$\tilde{T}_{ml} = T_{ml}\frac{N_m}{N_l}$$

Now on the load side a torque balance equation can be written as

$$\tilde{T}_{ml} = J_L\ddot{\theta}_L + B_L\dot{\theta}_L + T_l$$

In terms of the motor's angular velocity $\dot{\theta}_m$ and angular acceleration we thus have

$$\tilde{T}_{ml} = \left(J_L\ddot{\theta}_m + B_L\dot{\theta}_m\right)\frac{N_l}{N_m} + T_l$$

or

$$T_{ml} = \left(J_L\ddot{\theta}_m + B_L\dot{\theta}_m\right)\left(\frac{N_l}{N_m}\right)^2 + T_l\frac{N_l}{N_m}$$

As a result, we can assert that the motor's driving torque is given by

$$T_m = (J_m + \left(\frac{N_l}{N_m}\right)^2 J_L)\ddot{\theta}_m + (B_m + \left(\frac{N_l}{N_m}\right)^2 B_L)\dot{\theta}_m + T_l\frac{N_l}{N_m}$$

This result shows that an equivalent moment of inertia $J_{\text{eq}}$ and an equivalent viscous friction $B_{\text{eq}}$ are experienced by the motor:

$$J_{\text{eq}} = J_m + \left(\frac{N_l}{N_m}\right)^2 J_L$$

$$B_{\text{eq}} = B_m + \left(\frac{N_l}{N_m}\right)^2 B_L$$

The load torque is seen by the motor as

$$\tilde{T}_l = T_l \frac{N_l}{N_m}$$

Thus the motor's developed torque is given by

$$T_m = J_{eq}\ddot{\theta}_m + B_{eq}\dot{\theta}_m + \tilde{T}_l$$

# 2.5 Transfer Functions

A transfer function is defined as the ratio of the Laplace transform of the output variable in a linear time-invarient dynamic system to the Laplace transform of the input variable with zero initial conditions. A number of examples is appropriate to illustrate the concept.

EXAMPLE 2.7

Consider the $RC$ integrating configured network shown in Figure 2.9. The current at the output terminals is zero, and we can write the input voltage as

$$V_i(s) = \left(R + \frac{1}{Cs}\right)I(s)$$

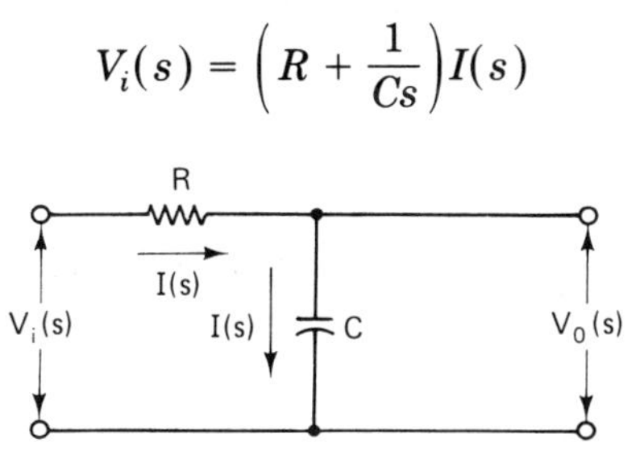


**Figure 2.9.** *RC* Integrating Configuration Network

The output voltage is given by

$$V_o(s) = \frac{1}{Cs}I(s)$$

The transfer function is thus obtained as

$$\frac{V_o(s)}{V_i(s)} = \frac{1}{1 + RCs}$$

EXAMPLE 2.8

For the differentiating configured $RC$ network shown in Figure 2.10, we can write the transfer function as

$$\frac{V_o(s)}{V_i(s)} = \frac{R}{R + 1/Cs} = \frac{RCs}{1 + RCs}$$

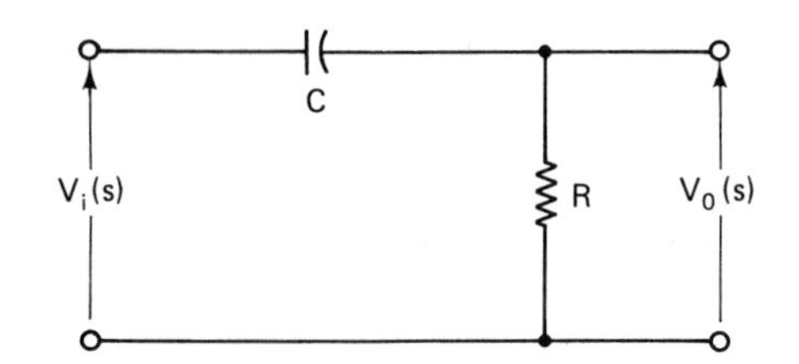


**Figure 2.10.** *RC* Differentiating Configuration Network

EXAMPLE 2.9

For the spring–dashpot system shown in Figure 2.11, we can write a force balance equation as

$$B\dot{x}_o + Kx_o = Kx_i$$

Employing the Laplace transform, we thus have

$$(Bs + K)X_o(s) = KX_i(s)$$

As a result, the transfer function is given by

$$\frac{X_o(s)}{X_i(s)} = \frac{K}{Bs + K}$$

or

$$\frac{X_o(s)}{X_i(s)} = \frac{1}{1 + (B/K)s}$$

Note the similarity of this transfer function and that of Example 2.7.

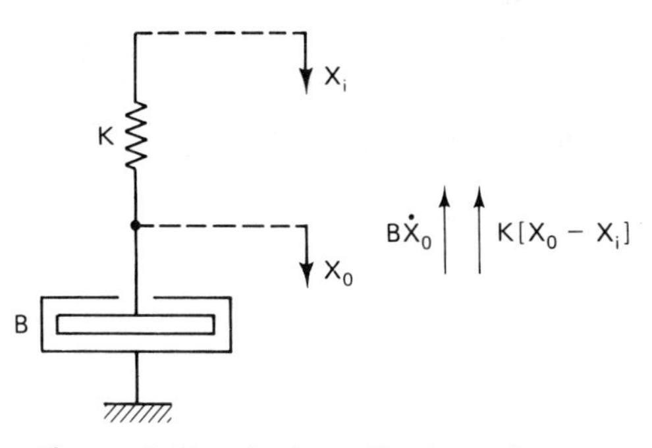


**Figure 2.11.** Spring – Dashpot System

EXAMPLE 2.10

For the system of Example 2.9, if the *K* and *B* are interchanged, we obtain the configuration shown in Figure 2.12. The force balance equation is given by

$$B(\dot{x}_o - \dot{x}_i) + Kx_o = 0$$

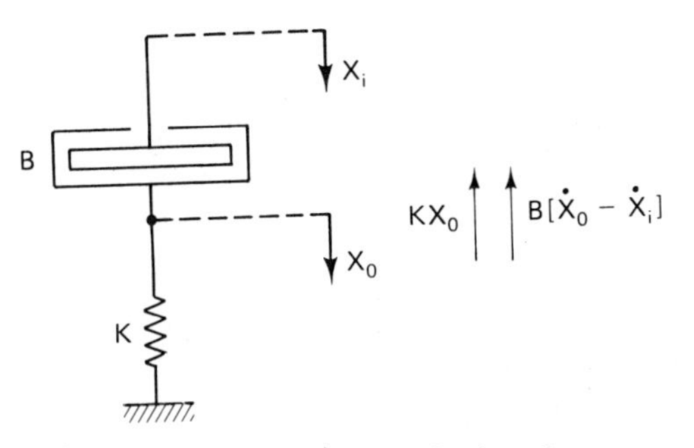


**Figure 2.12.** Dashpot – Spring System

Thus the transfer function can be written as

$$\frac{X_o(s)}{X_i(s)} = \frac{Bs}{Bs + K}$$

This transfer function is similar to that of Example 2.8.

EXAMPLE 2.11

Consider the mechanical rotational system shown in Figure 2.13. We can write

$$J\frac{d\omega}{dt} + B\omega = T$$

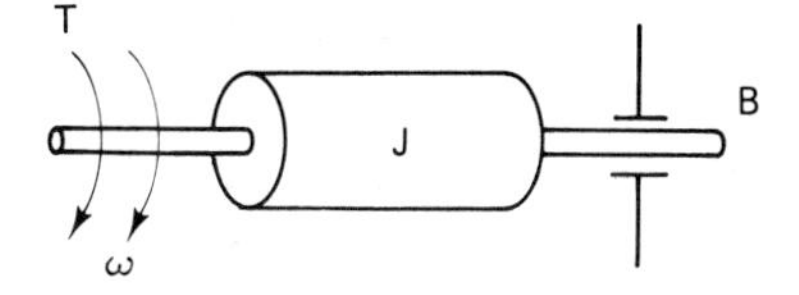


**Figure 2.13.** Rotating Body

Thus

$$\frac{\Omega(s)}{T(s)} = \frac{1}{Js + B}$$

The transfer function obtained relates the transferred output velocity $\Omega(s)$ to the transform of the input torque $T(s)$ in the presence of moment of inertia $J$ and viscous friction $B$.

EXAMPLE 2.12

Consider the mechanical translational system shown in Figure 2.14 consisting of a spring–mass–dashpot combination. We can write the force balance equation as

$$M\ddot{x} + B\dot{x} + Kx = F$$

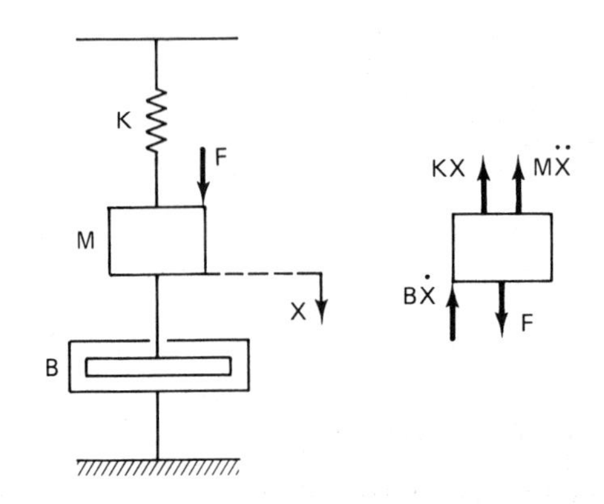


**Figure 2.14.** Spring – Mass – Dashpot System for Example 2.12

As a result, the transfer function is obtained as

$$\frac{X(s)}{F(s)} = \frac{1}{Ms^2 + Bs + K}$$

EXAMPLE 2.13

For the *RLC* circuit shown in Figure 2.15, we can write in the Laplace domain

$$V_i(s) = \left(R + Ls + \frac{1}{Cs}\right)I(s)$$

$$V_o(s) = \frac{1}{Cs}I(s)$$

Thus the transfer function is given by

$$\frac{V_o(s)}{V_i(s)} = \frac{1}{LCs^2 + RCs + 1}$$

This is clearly similar to the transfer function of Example 2.12.

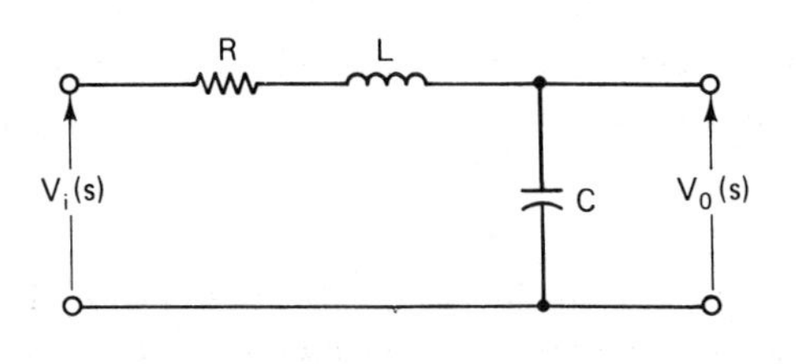


**Figure 2.15.** *RLC* Network

EXAMPLE 2.14

For the lead–lag *RC* network shown in Figure 2.16, we can write the following impedance functions:

$$Z_1 = \frac{R_1/C_1 s}{R_1 + 1/C_1 s}$$

$$= \frac{R_1}{1 + R_1 C_1 s}$$

Let

$$\tau_a = R_1 C_1$$

Thus

$$Z_1 = \frac{R_1}{1 + \tau_a s}$$

Also,

$$Z_2 = R_2 + \frac{1}{C_2 s}$$

$$= \frac{1 + R_2 C_2 s}{C_2 s}$$

Let

$$\tau_b = R_2 C_2$$

Thus

$$Z_2 = \frac{1 + \tau_b s}{C_2 s}$$

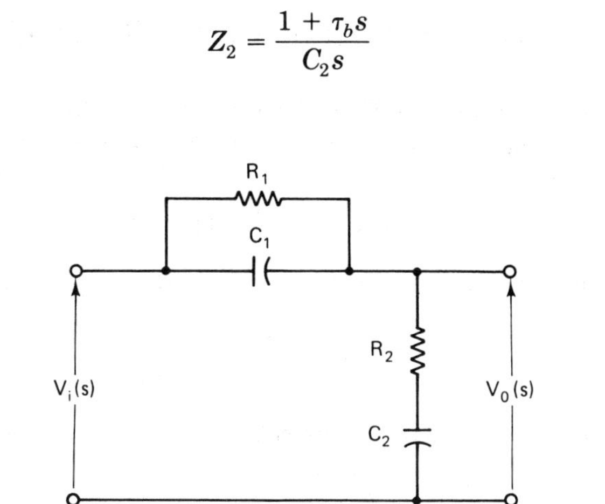


**Figure 2.16.** Lead – Lag *RC* Network

The transfer function is thus given by

$$\frac{V_o(s)}{V_i(s)} = \frac{Z_2}{Z_1 + Z_2}$$

This is written in terms of the network elements as

$$\frac{V_o(s)}{V_i(s)} = \frac{(1 + \tau_a s)(1 + \tau_b s)}{(1 + \tau_a s)(1 + \tau_b s) + R_1 C_2 s}$$

Let $\tau_{ab} = R_1 C_2$; thus

$$\frac{V_o(s)}{V_i(s)} = \frac{(1 + \tau_a s)(1 + \tau_b s)}{\tau_a \tau_b s^2 + (\tau_a + \tau_b + \tau_{ab})s + 1}$$

We can rewrite this as

$$\frac{V_o(s)}{V_i(s)} = \frac{1 + \tau_a s}{1 + \tau_1 s} \frac{1 + \tau_b s}{1 + \tau_2 s}$$

where

$$\tau_1 \tau_2 = \tau_a \tau_b$$

$$\tau_a + \tau_b + \tau_{ab} = \tau_1 + \tau_2$$

The examples above help to illustrate the concept of a transfer function. In the preceding section we dealt with a number of electric motors. Presently, we extend the treatment to arrive at some transfer functions.

## Transfer Function of a Field-Controlled DC Motor

We have seen in Section 2.4 that the developed torque of the field-controlled motor is proportional to its field current. Expressed in the Laplace domain, we have

$$T(s) = K_1 I_f(s)$$

The relation between the field current and input voltage is

$$I_f(s) = \frac{V_i(s)}{R_f + L_f s}$$

As a result, we conclude that a transfer function between the input voltage and output torque can be written as

$$\frac{T(s)}{V_i(s)} = \frac{K_1}{R_f + L_f s}$$

## Transfer Function of an Armature-Controlled DC Motor

From the discussion of the armature-controlled dc motor in Section 2.4, we can conclude that in the Laplace domain, we have

$$V_i(s) - E_m(s) = (R_a + L_a s) I_a(s)$$
$$E_m(s) = K_a \omega_m(s)$$
$$T(s) = K_a I_a(s)$$

Thus the Laplace transform of the developed torque is given by

$$T(s) = \frac{K_a V_i(s)}{R_a + L_a s} - \frac{K_a^2 \omega_m(s)}{R_a + L_a s}$$

The relation above is not a single input–single output relationship and thus a straightforward transfer function cannot be obtained for $T(s)$ in terms of the input voltage. It is possible, however, to obtain a transfer function between $\omega_m(s)$ and $V_i(s)$ as indicated below.

Assume that the motor is driving a load such that no opposing torque $T_l$ exists. We can thus write

$$T(s) = (J_{eq} s + B_{eq}) \omega_m(s)$$

The equation above assumes that the load is connected to the motor through a gear train in accordance with the assumptions delineated in Section 2.4. We can thus eliminate $T(s)$ to obtain

$$\frac{K_a V_i(s)}{R_a + L_a s} = \left[ (J_{eq} s + B_{eq}) + \frac{K_a^2}{R_a + L_a s} \right] \omega_m(s)$$

As a result, a transfer function between $\omega_m(s)$ as output and $V_i(s)$ as input can be written as

$$\frac{\omega_m(s)}{V_i(s)} = \frac{K_a}{K_a^2 + (J_{eq} s + B_{eq})(R_a + L_a s)}$$

In the next section we will have an opportunity to deal with a nonzero load torque.

## Transfer Function of a DC Generator

From the discussion of Section 2.4 it is clear that in the Laplace domain we can write

$$V_i(s) = (R_f + L_f s) I_f(s)$$
$$V_o(s) = K_g I_f(s)$$

As a result, the transfer function between input and output voltages is given by

$$\frac{V_o(s)}{V_i(s)} = \frac{K_g}{R_f + L_f s}$$

## Transfer Function of the Amplidyne

The Laplace domain expressions for the performance of an amplidyne are given by

$$I_c(s) = \frac{V_c(s)}{R_c + L_c s}$$

$$V_q(s) = K_{cq} I_c(s)$$

$$I_q(s) = \frac{V_q(s)}{R_q + L_q s}$$

$$V_d(s) = K_{dq} I_q(s)$$

As a result, we have

$$\frac{V_d(s)}{V_c(s)} = \frac{K_{dq} K_{cq}}{(R_q + L_q s)(R_c + L_c s)}$$

## Transfer Function of the Two-Phase Servomotor

From Section 2.4 we can write

$$T(s) = -K_n \omega(s) + K_c E_c(s)$$

The torque is also expressed as

$$T(s) = (Js + B)\omega(s)$$

Thus we have

$$(Js + B + K_n)\omega(s) = K_c E_c(s)$$

The resulting transfer function is given by

$$\frac{\omega(s)}{E_c(s)} = \frac{K_c}{Js + (B + K_n)}$$

It is thus clear that in general the transfer function of a linear system can be written in the form

$$G(s) = \frac{X(s)}{U(s)} = \frac{\sum_{j=0}^{m} b_j s^j}{\sum_{i=0}^{n} a_i s^i}$$

where $X(s)$ is the Laplace transform of the output variable and $U(s)$ corresponds to the Laplace transform of the input.

## 2.6 BLock Diagrams

A pictorial representation of the relationships between system variables is offered by the block diagram. In a block diagram three ingredients are commonly present.

1. *Functional block.* This is a symbol representing the transfer between the input $U(s)$ to an element and the output $X(s)$ of the element. The block contains the transfer function $G(s)$, as shown in Figure 2.17. The arrow directed into the block

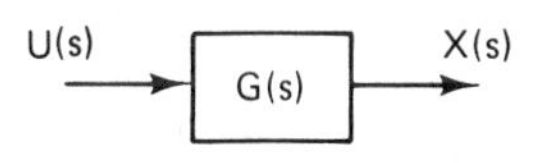


**Figure 2.17.** Functional Block

represents the input $U(s)$, while that directed out of the block represents the output $X(s)$. The block shown represents the algebraic relationship

$$X(s) = G(s)U(s)$$

2. *Summing point.* This is a symbol denoted by a circle, the output of which is the algebraic sum of the signals entering into it. A minus sign close to an input signal arrow denotes that this signal is reversed in sign in the output expression. Figure 2.18 shows the relationship

$$E(s) = R(s) - C(s)$$

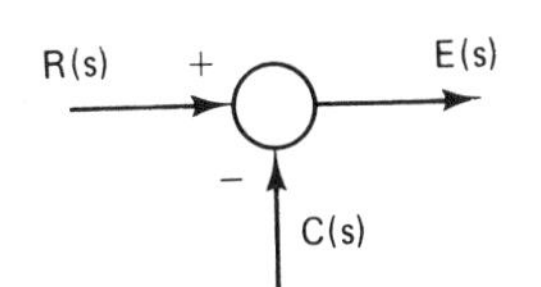


**Figure 2.18.** Summing Point

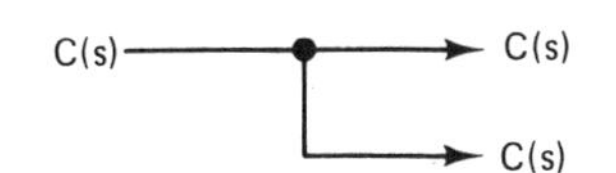


**Figure 2.19.** Takeoff Point

3. *Takeoff point.* A takeoff point on a branch in a block diagram signifies that the same variable is being utilized elsewhere, as shown in Figure 2.19.

A fundamental block diagram configuration is the single-loop feedback system shown in Figure 2.20a. The output variable $C(s)$ is modified by the feedback element with transfer function $H(s)$ to produce the signal $B(s)$:

$$B(s) = C(s)H(s)$$

The signal $B(s)$ is compared to a reference signal $R(s)$ to produce the error

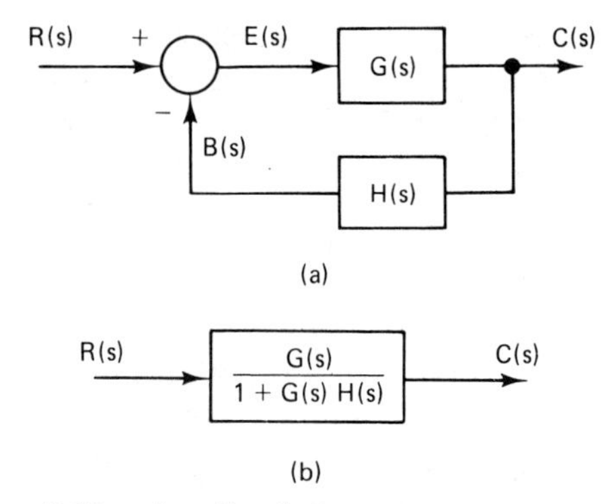


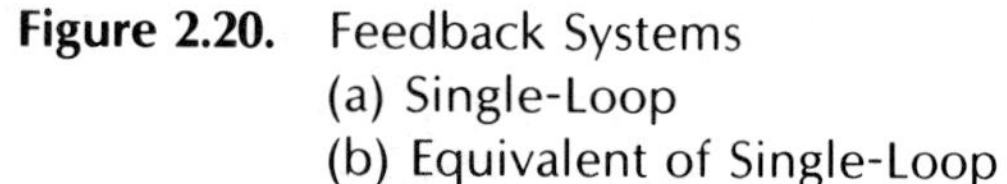


(b)

**Figure 2.20.** Feedback Systems
(a) Single-Loop
(b) Equivalent of Single-Loop

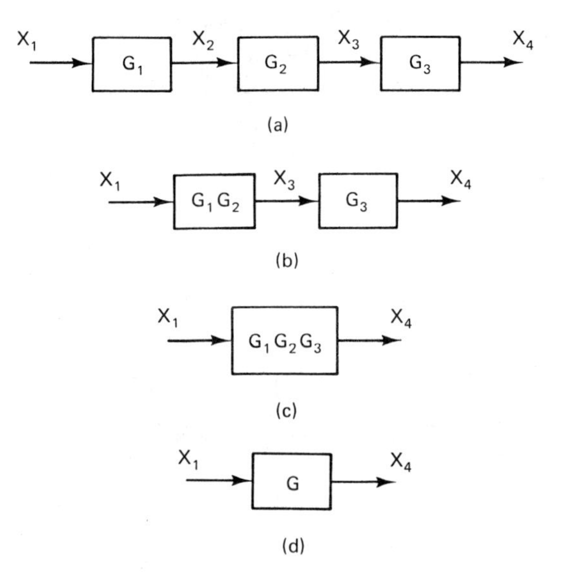


**Figure 2.21.** Cascaded System and the Steps in Its Reduction

**TABLE 2.2**
**BLOCK DIAGRAM REDUCTION AIDS**

| *ORIGINAL CONFIGURATION* | *ALTERNATE CONFIGURATION* |
|---|---|

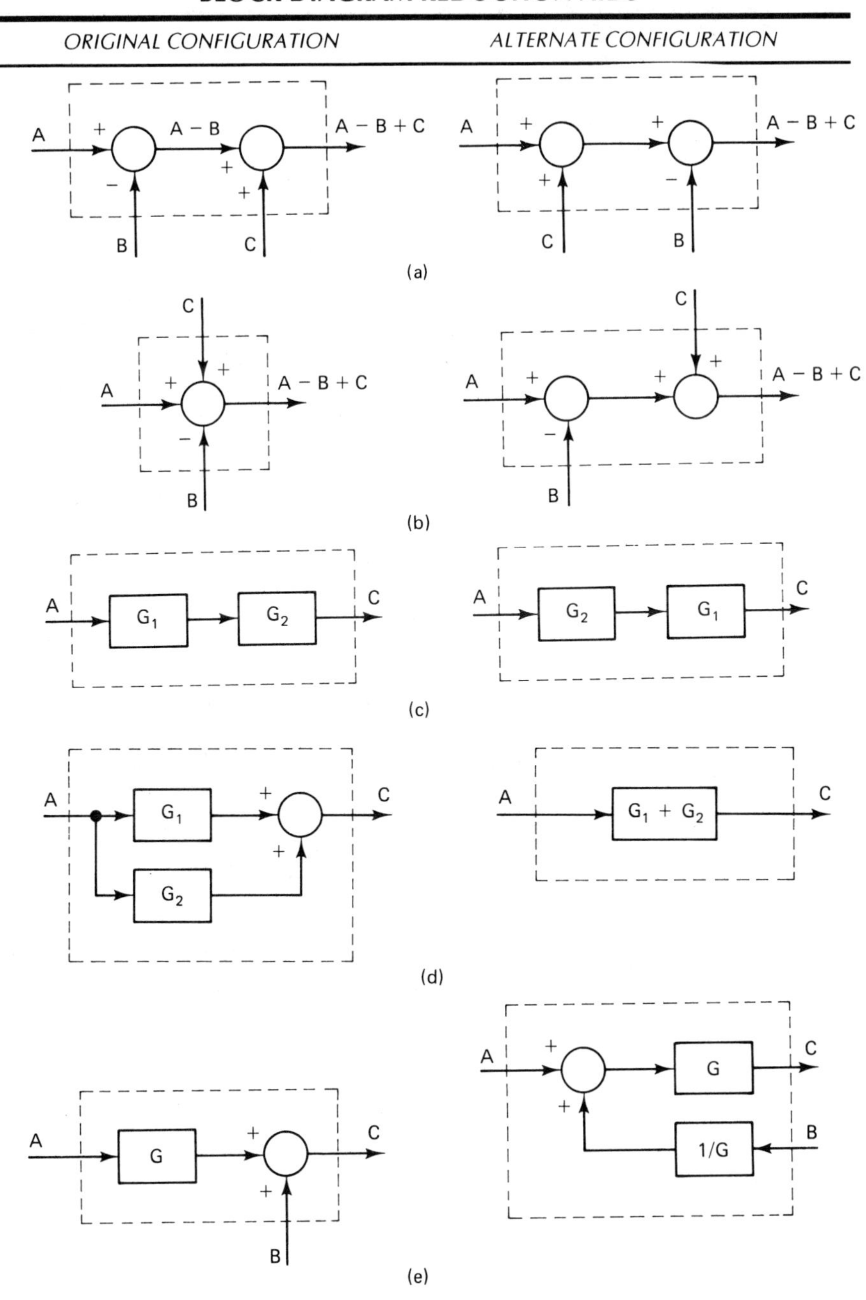


*continued*

**TABLE 2.2 (continued)**
**BLOCK DIAGRAM REDUCTION AIDS**

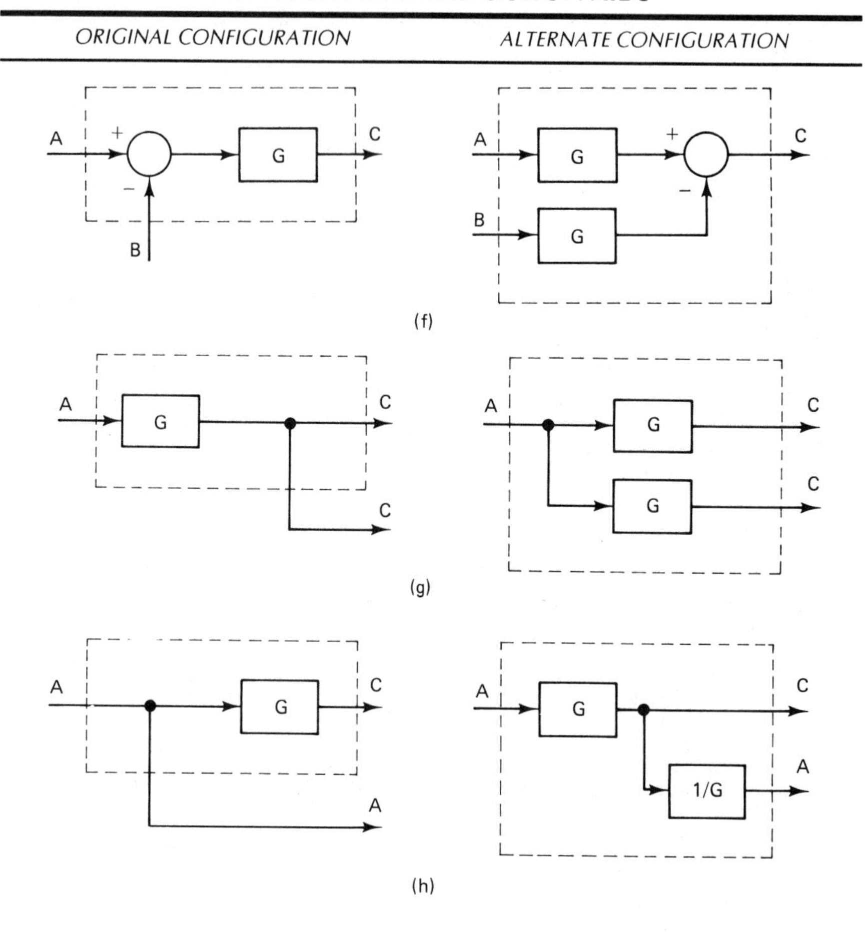


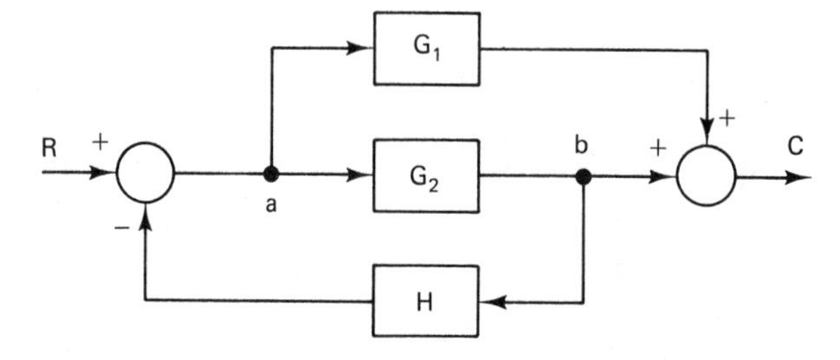


**Figure 2.22.** Block Diagram for Example 2.15

signal $E(s)$:

$$E(s) = R(s) - B(s)$$

The error signal actuates the plant with transfer function $G(s)$ to produce the output $C(s)$:

$$C(s) = G(s)E(s)$$

Combining the equations above, the overall transfer function is obtained as

$$\frac{C(s)}{R(s)} = \frac{G(s)}{1 + G(s)H(s)}$$

As a result, we conclude that Figure 2.20b represents an equivalent of Figure 2.20a.

To obtain the overall relationship between the outputs and inputs of complex systems, we often have to eliminate intermediate variables in the system representation. We consider here the transfer functions of cascaded elements as shown in Figure 2.21a.

We write

$$X_2 = G_1 X_1$$
$$X_3 = G_2 X_2 = G_1 G_2 X_1$$

Thus we can obtain the reduction shown in Figure 2.21b.

We can further write

$$X_4 = G_3 X_3 = G_1 G_2 G_3 X_1$$

Thus a single equivalent as shown in Figure 2.21c is obtained with

$$G = G_1 G_2 G_3$$

Table 2.2 shows some important equivalents in block diagram manipulations.

EXAMPLE 2.15

Simplify the block diagram shown in Figure 2.22.

SOLUTION

First, move the takeoff $b$ to $a$ as shown in Figure 2.23a. Now we can see that $G_1$ and $G_2$ are in parallel and the block diagram reduces to that shown in Figure 2.23b. The feedback loop with a forward gain of 1 and feedback element $H$ can be reduced as shown in Figure 2.23c. Finally, Figure 2.23d shows that the overall transfer between $R$ and $C$.

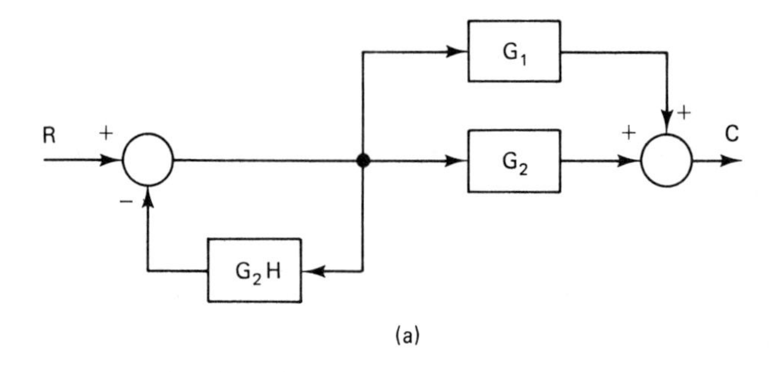


(a)

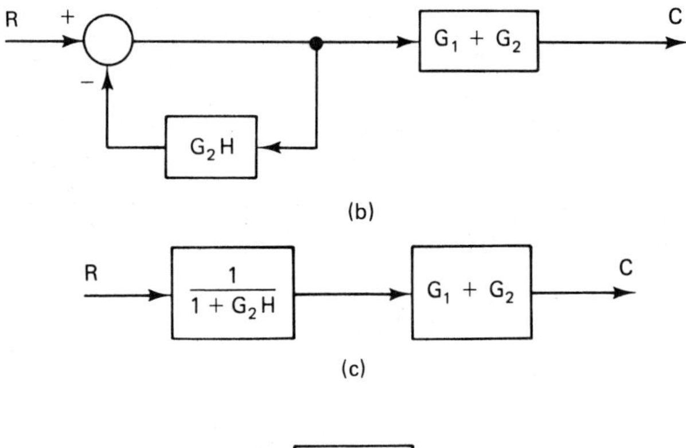


(c)

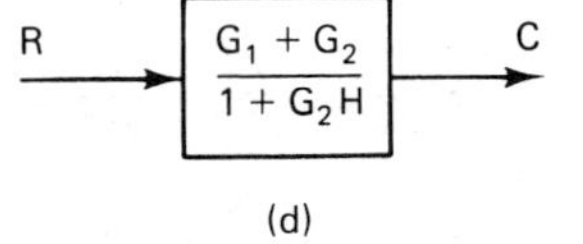


(d)

**Figure 2.23.** Simplified Block Diagrams for Example 2.15

EXAMPLE 2.16

Use block diagram reduction techniques to obtain the ratio $C/R$ for the system shown in the block diagram of Figure 2.24.

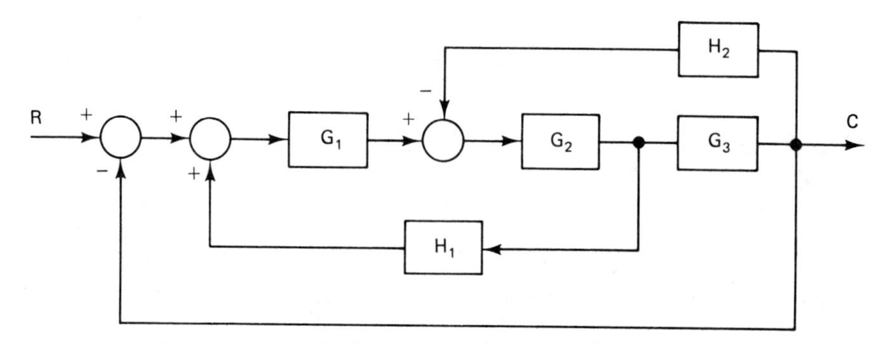


**Figure 2.24.** Block Diagram for Example 2.16

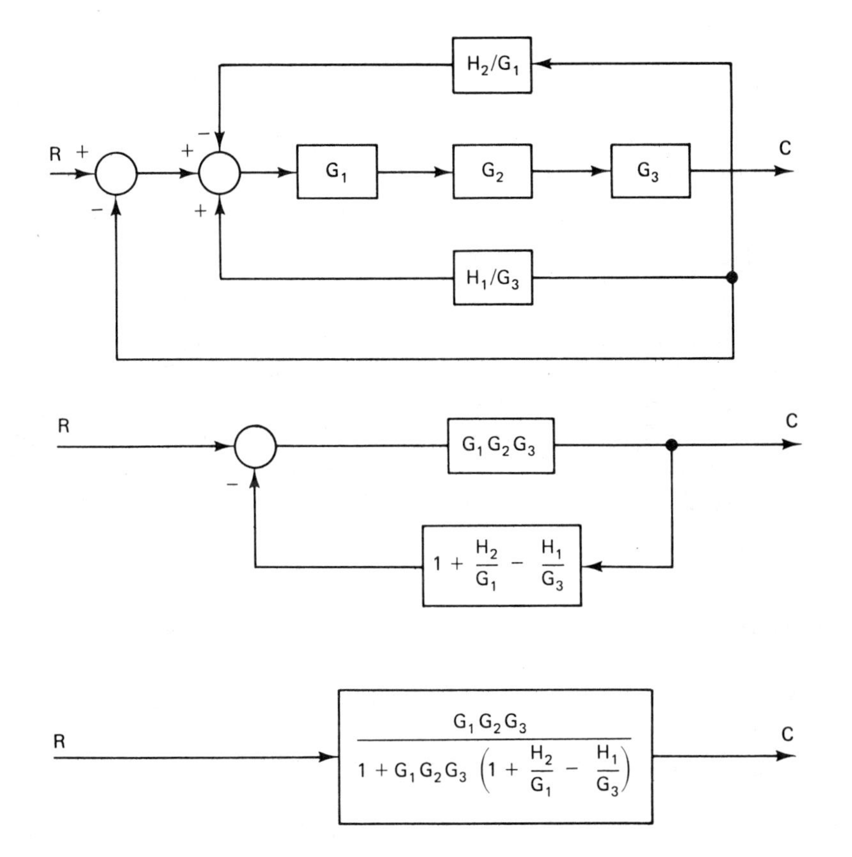


**Figure 2.25.** Steps in the Block Diagram Reduction for Example 2.16

SOLUTION

Figure 2.25 shows the steps of reduction. As a result of this, we conclude that

$$\frac{C}{R} = \frac{G_1G_2G_3}{1 + G_1G_2G_3 + G_2G_3H_2 - G_1G_2H_1}$$

## Block Diagram of a Field-Controlled DC Motor

We have seen that the transfer function between the field voltage and output torque of a field-controlled DC motor is given by

$$\frac{T(s)}{V_i(s)} = \frac{K_1}{R_f + L_f s}$$

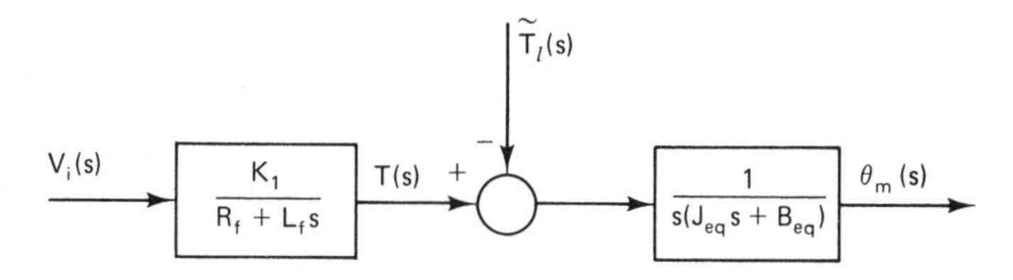


**Figure 2.26.** Block Diagram of a Field-Controlled DC Motor

This is represented in block diagram form in the first block of Figure 2.26.

Assume now that the motor is driving a load torque $T_l$ through a gear train of speed ratio $N_m/N_l$. The motor's inertia and viscous friction are denoted $J_m$ and $B_m$, and the load's inertia and viscous friction are denoted $J_l$ and $B_l$. We have dealt with the relation between the motor's developed torque, load torque, and motor speed in Section 2.4. In the Laplace domain we can write

$$T(s) = \left(J_{eq}s^2 + B_{eq}s\right)\theta_m(s) + \tilde{T}_l(s)$$

The equation above is realized in block diagram form on the right-hand side of Figure 2.26.

## Block Diagram of Armature-Controlled DC Motor

In Section 2.4 we concluded that the developed torque is given by

$$T(s) = \frac{K_a V_i(s)}{R_a + L_a s} - \frac{K_a^2 \omega_m(s)}{R_a + L_a s}$$

With a load torque, we write

$$T(s) = \left(J_{eq}s + B_{eq}\right)\omega_m(s) + \tilde{T}_l(s)$$

A block diagram can be constructed as shown in Figure 2.27a. Note the presence of a feedback path to account for the effect of the motor's velocity on the armature current.

By moving the load torque's summing junction to the left-hand side as shown in Figure 2.27b, we can see that the motor is in actual fact subject to two inputs, $V_i(s)$ and $\tilde{T}_l(s)$.

## Block Diagram of DC Generator

It is a simple matter on the basis of the discussion of Section 2.5 to demonstrate that the block diagram shown in Figure 2.28 represents the transfer function of the DC generator.

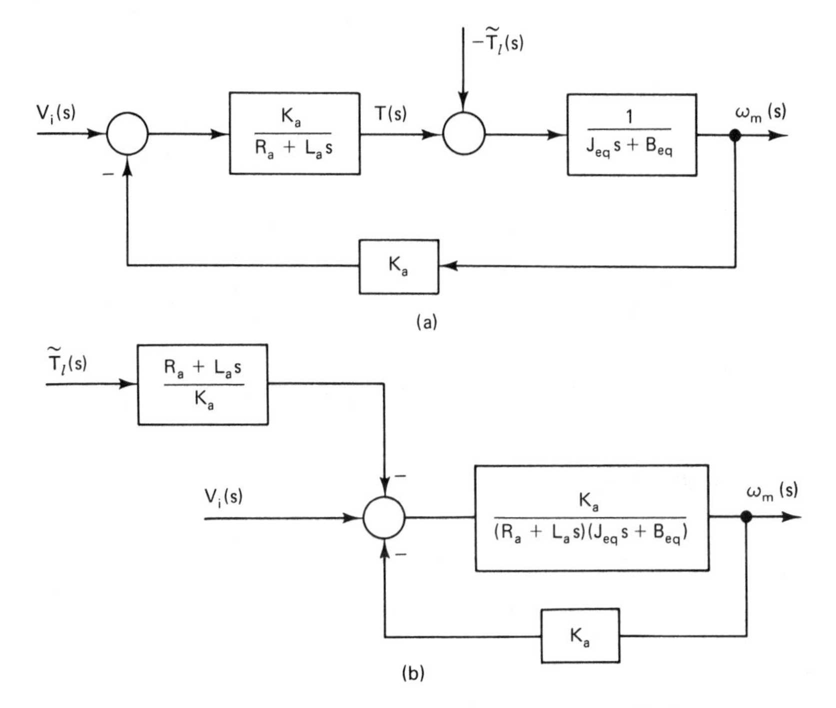


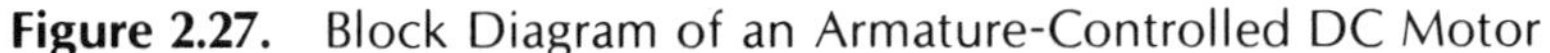
**Figure 2.27.** Block Diagram of an Armature-Controlled DC Motor

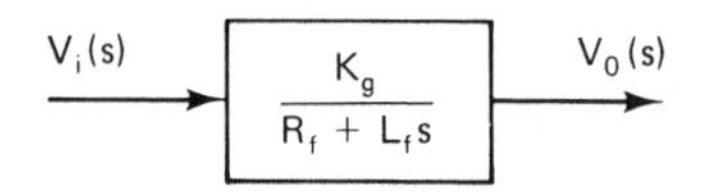


**Figure 2.28.** Block Diagram Representation of a DC Generator

## Block Diagram of Amplidyne

Figure 2.29 shows in block diagram form the amplidyne model as discussed in Section 2.5.

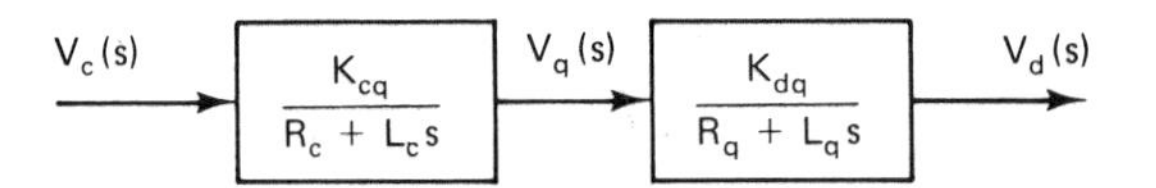


**Figure 2.29.** Block Diagram Representation of the Amplidyne

## Liquid-Level System

Consider the two-reservoir system shown in Figure 2.30. The rate of liquid inflow is denoted by $q_i(t)$ ($m^3$/sec). The water level (head) in the first reservoir is $h_1$ (meters) and the surface area of the vertical-sided reservoir is

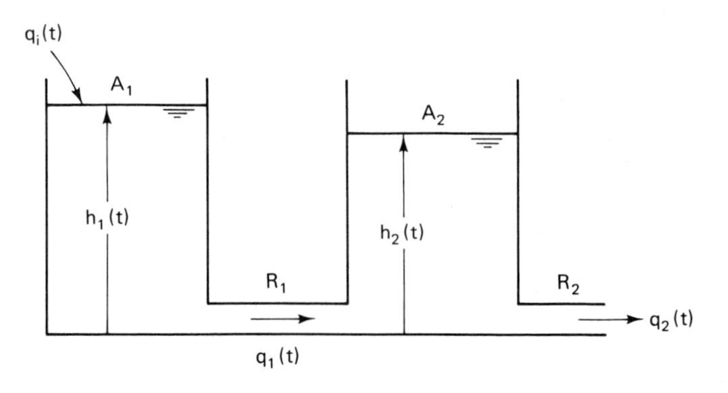


**Figure 2.30.** Liquid-Level System

$A_1$. An outflow $q_1(t)$ is directed to the second reservoir. We can write a balance equation for the first reservoir as

$$q_i(t) - q_1(t) = A_1 \frac{dh_1}{dt}$$

The rate of flow $q_1$ is related to the head difference through the flow resistance $R_1$ by

$$q_1 = \frac{h_1 - h_2}{R_1}$$

The head in reservoir 2 is denoted by $h_2$. The balance equation for reservoir 2 is given by

$$q_1(t) - q_2(t) = A_2 \frac{dh_2}{dt}$$

We also have by the definition of the flow resistance

$$q_2 = \frac{h_2}{R_2}$$

Figure 2.31 shows a block diagram of the process and steps in its reduction. The final transfer function is obtained as

$$\frac{Q_2(s)}{Q_i(s)} = \frac{1}{A_1 R_2 s + (1 + R_1 A_1 s)(1 + R_2 A_2 s)}$$

## Thermal Systems

Thermal systems involve the transfer of heat from one medium to another. For small changes in the variables involved, some linear relations can be used. The rate of heat transferred into a body is denoted by $q$ and is

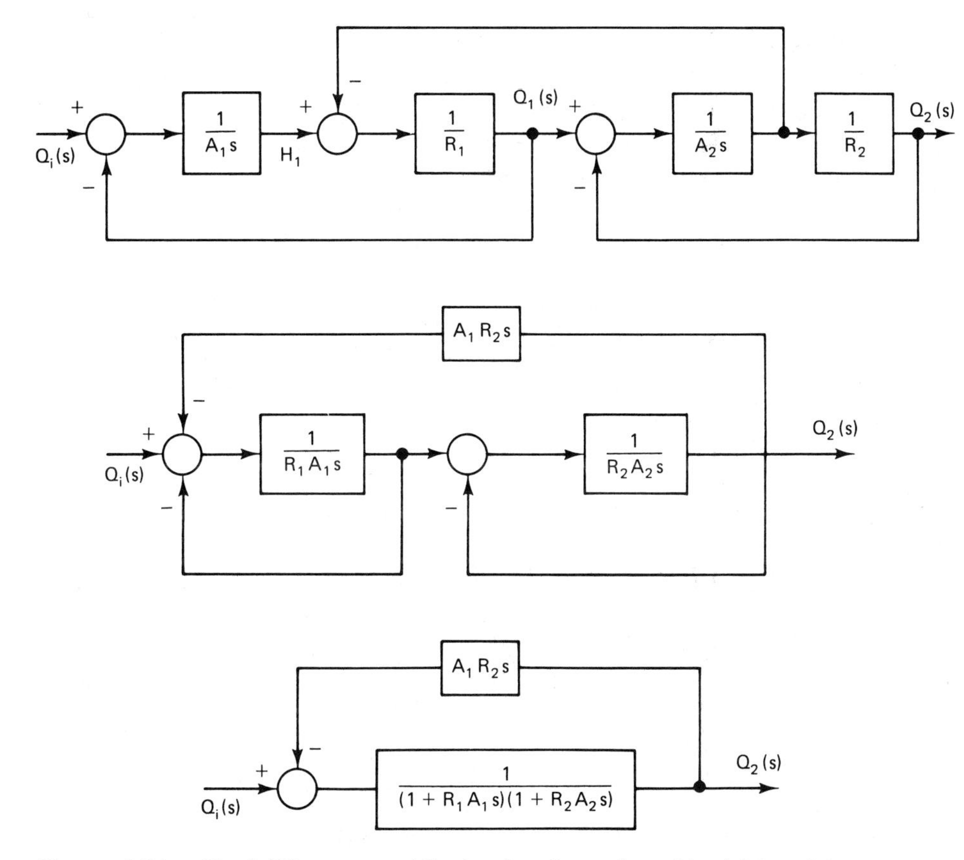

**Figure 2.31.** Block Diagram and Reduction Steps for a Liquid-Level System

proportional to the temperature difference $\Delta\Theta$ across the body

$$q = \frac{\Delta\Theta}{R_T}$$

The thermal resistance is denoted by $R_T$ and is expressed in terms of conductor thickness $\Delta x$, its thermal conductivity $k$, and the area normal to heat flow $A$ as

$$R_T = \frac{\Delta x}{kA}$$

The rate of change of temperature of a body is related to the rate of heat transfer by

$$q = C_T \frac{d\Theta}{dt}$$

The equivalent thermal capacitance $C_T$ is the product of the body's mass $M$ and the average specific heat of the body $C$:

$$C_T = MC$$

## Stirred Tank

Consider the tank shown in Figure 2.32a, which is continually stirred; the liquid temperature is $\Theta$ and the surrounding temperature is $\Theta_1$. Thus the rate of heat transfer is expressed as

$$q = \frac{\Theta_1 - \Theta}{R_T}$$

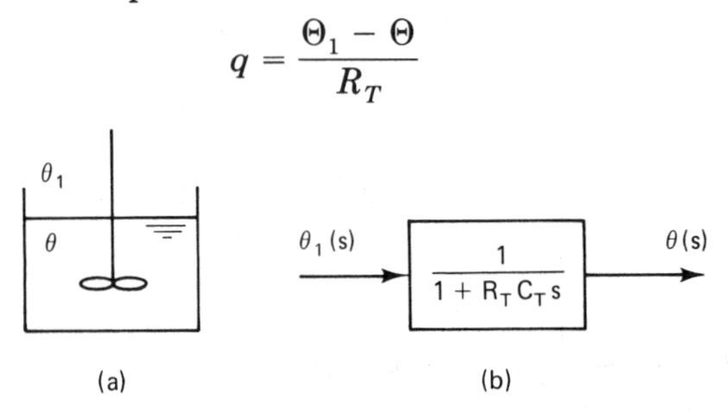


**Figure 2.32.** Stirred Tank and Its Block Diagram Representation

In terms of heat retention we also have

$$q = C_T \frac{d\Theta}{dt}$$

Thus we can state that

$$\frac{\Theta_1(s) - \Theta(s)}{R_T} = C_T s \Theta(s)$$

$$\Theta(s) = \frac{\Theta_1(s)}{1 + R_T C_T s}$$

Figure 2.32b shows a block diagram of the system.

## Water Heater

Figure 2.33 shows an electric water heater in which the electric heating element produces a rate of heat flow $q$. Cooler water inflow into the tank results in heat flow $q_i$ for the assumed incoming temperature $\Theta_i$:

$$q_i = \frac{\Theta_i}{R_T}$$

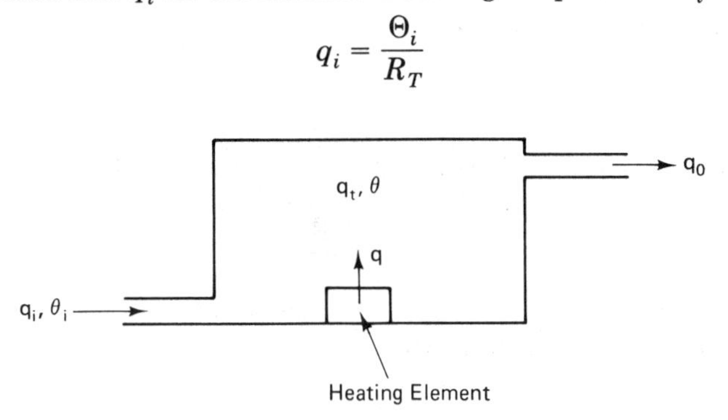


**Figure 2.33.** Water Heater

The tank's water temperature is $\Theta$, and as a result the rate of heat transfer to the tank $q_t$ is given by

$$q_t = C_T \frac{d\Theta}{dt}$$

The outflowing water results in a rate of heat transfer $q_o$ given by

$$q_o = \frac{\Theta}{R_T}$$

A heat-rate balance equation can be written as

$$q_t = q + q_i - q_o$$

Thus the tank's heat is equal to the sum of incoming heat $q$ from the element and the incoming water $q_i$, with the outgoing heat $q_o$ subtracted.

It is on the basis of the analysis above that the block diagram of Figure 2.34a is constructed. Note that there are two inputs to the system. Alternative block diagram representations are given in Figure 2.34b and c.

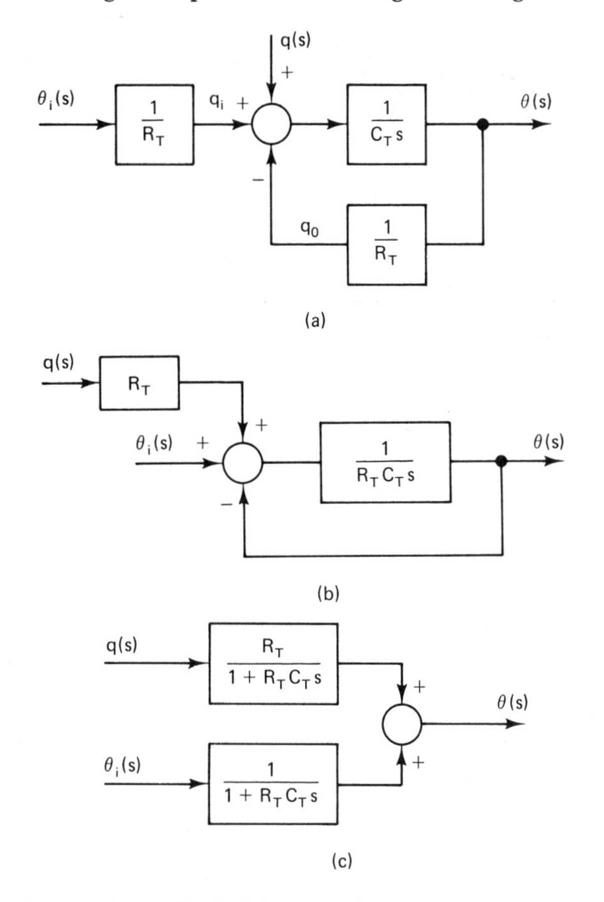


**Figure 2.34.** Block Diagrams for a Water Heater

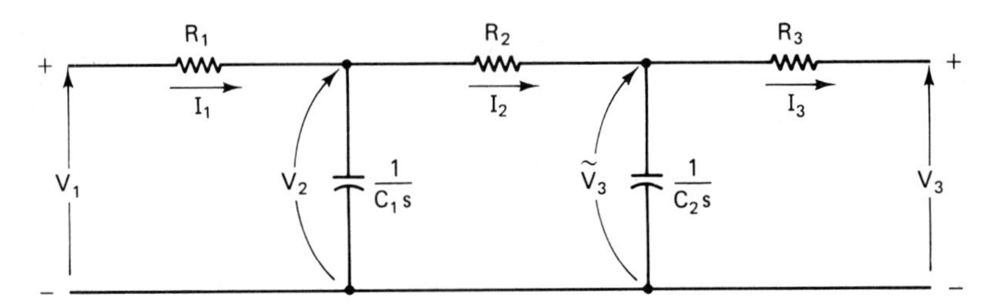


**Figure 2.35.** Block Diagram for Example 2.17

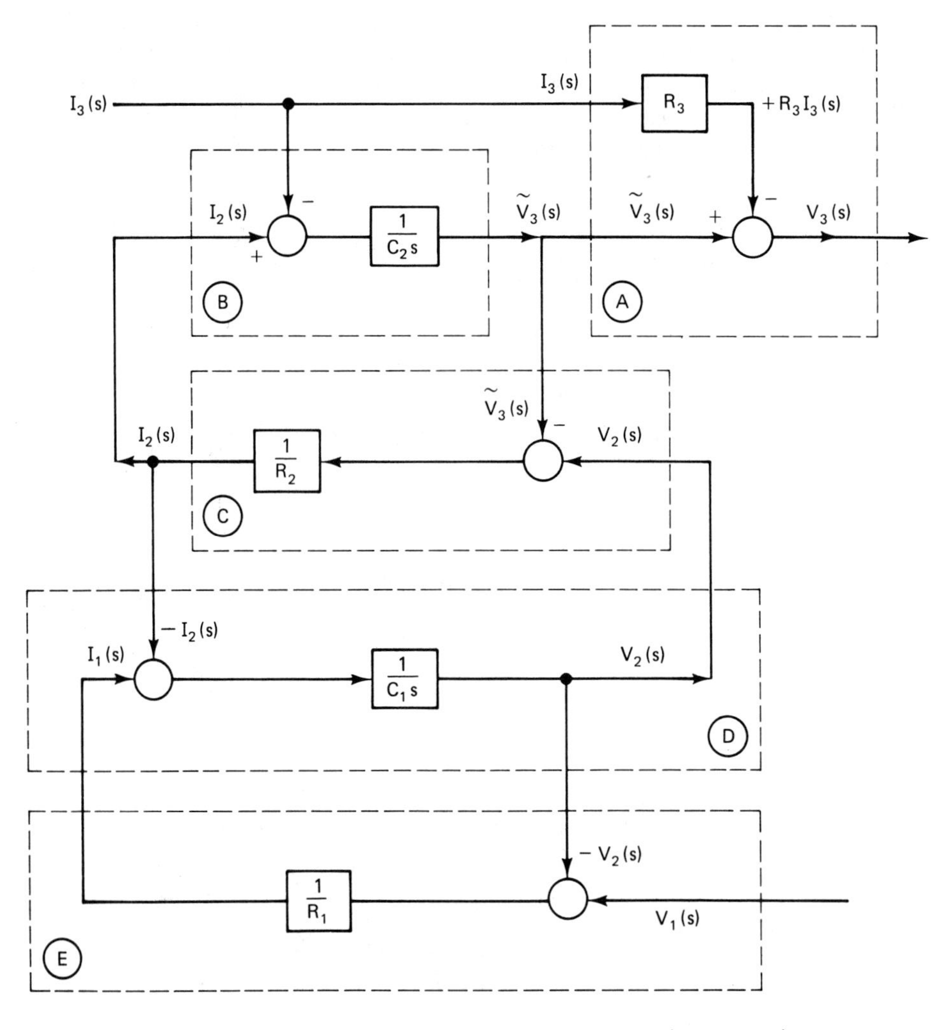


**Figure 2.36.** Construction of the Block Diagram for Example 2.17

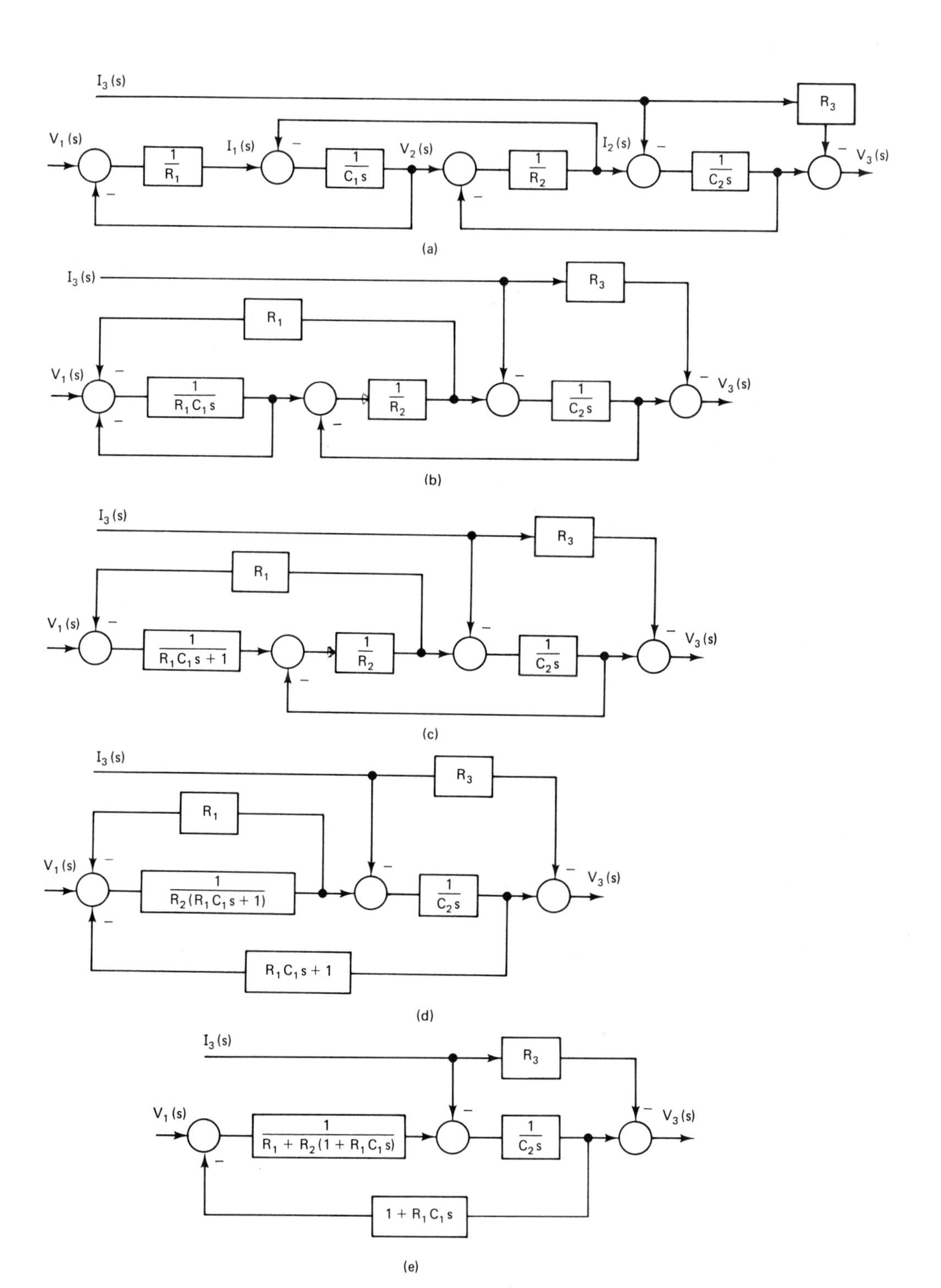


**Figure 2.37.** Steps in the Block Diagram Reduction for Example 2.17

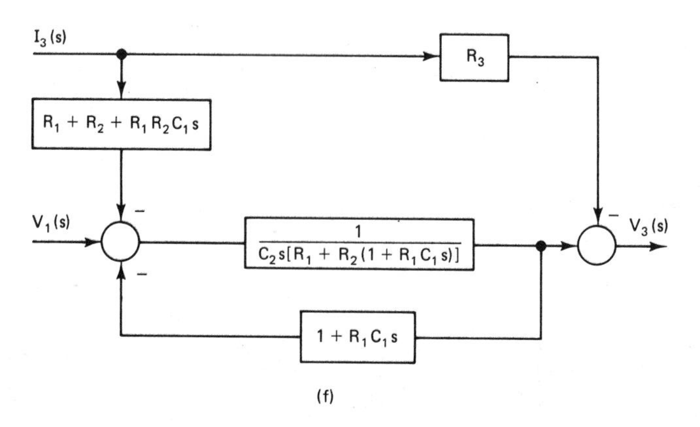


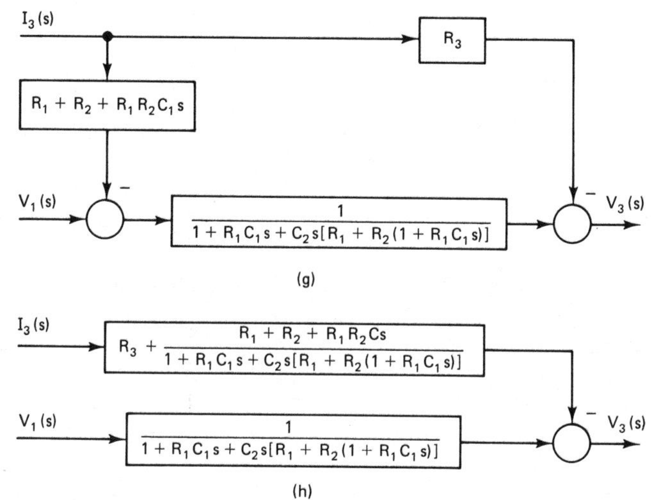


**Figure 2.37 (continued).** Steps in the Block Diagram Reduction for Example 2.17

EXAMPLE 2.17

Construct a block diagram for the network shown in Figure 2.35, then reduce the block diagram to show the effect of $V_1(s)$ and $I_3(s)$ on the output $V_3(s)$.

SOLUTION

We can write the following equations:

$$V_3(s) = \tilde{V}_3(s) - R_3 I_3(s)$$
$$I_2(s) = C_2 s \tilde{V}_3(s) + I_3(s)$$

$$V_2(s) = \tilde{V}_3(s) + R_2 I_2(s)$$
$$I_1(s) = C_1 s V_2(s) + I_2(s)$$
$$V_1(s) = R_1 I_1(s) + V_2(s)$$

Figure 2.36 shows the resulting block diagram. In Figure 2.37, steps in reducing the block diagram are shown.

# 2.7 Signal Flow Diagrams and Mason's Theorem

This method is believed to provide a faster means for determining the response of multiloop systems than do the block diagram reduction techniques discussed in the previous section.

Consider a set of linear equations having the form

$$y_i = \sum_{j=1}^{n} a_{ij} y_j \qquad i = 1, 2, \ldots, n$$

A node is assigned to each variable of interest as shown in Figure 2.38a. A branch between two nodes relates the variables at both ends. In a fashion similar to block diagrams the gain between the variables is indicated on the

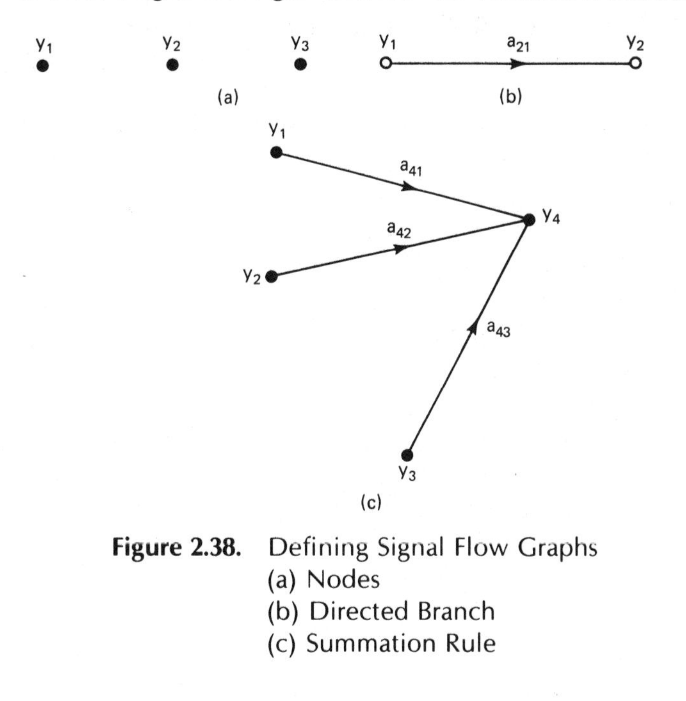


**Figure 2.38.** Defining Signal Flow Graphs
(a) Nodes
(b) Directed Branch
(c) Summation Rule

branch of an associated arrow. Thus in Figure 2.38b we have

$$y_2 = a_{21} y_1$$

The value of a variable at a node is equal to the sum of all incoming signals. Thus in Figure 2.38c,

$$y_4 = a_{41} y_1 + a_{42} y_2 + a_{43} y_3$$

A number of definitions is appropriate at this time. Figure 2.39 is used to illustrate the concepts discussed.

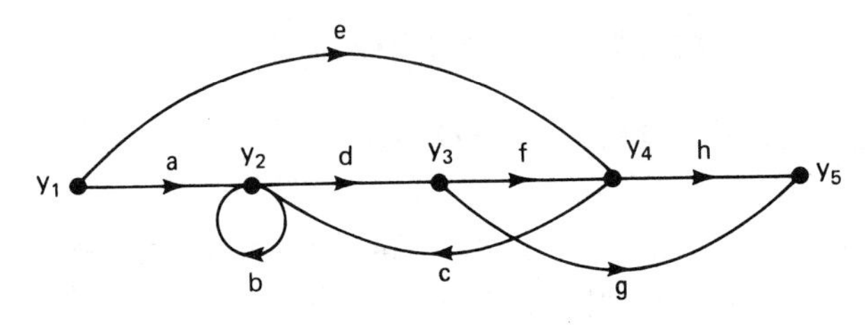


**Figure 2.39.** Signal Flow Graph

- A *source* is a node with only outgoing branches such as $y_1$.
- A *sink* is a node with only incoming branches such as $y_5$.
- A *path* is a group of connected branches having the same sense of direction. For example, in Figure 2.39, *eh*, *adfh*, and *b* are paths.
- A *forward path* is a path originating from a source and terminating in a sink and along which no node is encountered more than once. For example, in Figure 2.39, *eh*, *ecdg*, *adg*, and *adfh* are forward paths.
- A *path gain* is the product of the coefficients associated with the branches along the path.
- A *feedback loop* is a path originating from a node and terminating at the same node. In addition, a node cannot be encountered more than once. Thus in Figure 2.39, *b* and *dfc* are feedback loops.
- A *loop gain* is the product of the coefficients associated with the branches forming a feedback loop.

A signal flow graph can easily be constructed using the available information about the interaction of variables. This will be illustrated using the system of Example 2.16, which is redrawn in block diagram form in Figure 2.40a, identifying various variables needed. From the block diagram we have

$$E = R - C$$
$$E_1 = E + X_4$$

$$X_4 = H_1X_1$$
$$X_3 = G_1E_1$$
$$X_2 = X_3 - X_5$$
$$X_1 = G_2X_2$$
$$C = G_3X_1$$
$$X_5 = H_2C$$

We start by drawing our nodes for the variables $R$, $E$, $E_1$, $X_3$, $X_2$, $X_1$, $X_5$, $X_4$, and $C$. Each of the equations above is used to complete the graph showing the interaction between variables. The graph is shown in Figure 2.40b. Figure 2.41 shows some basic block diagrams and the associated signal flow graphs.

The main advantage of signal flow graphs is that there is a general formula for expressing the gain of the graph based on its topography. The expression is referred to as Mason's gain formula and is treated next.

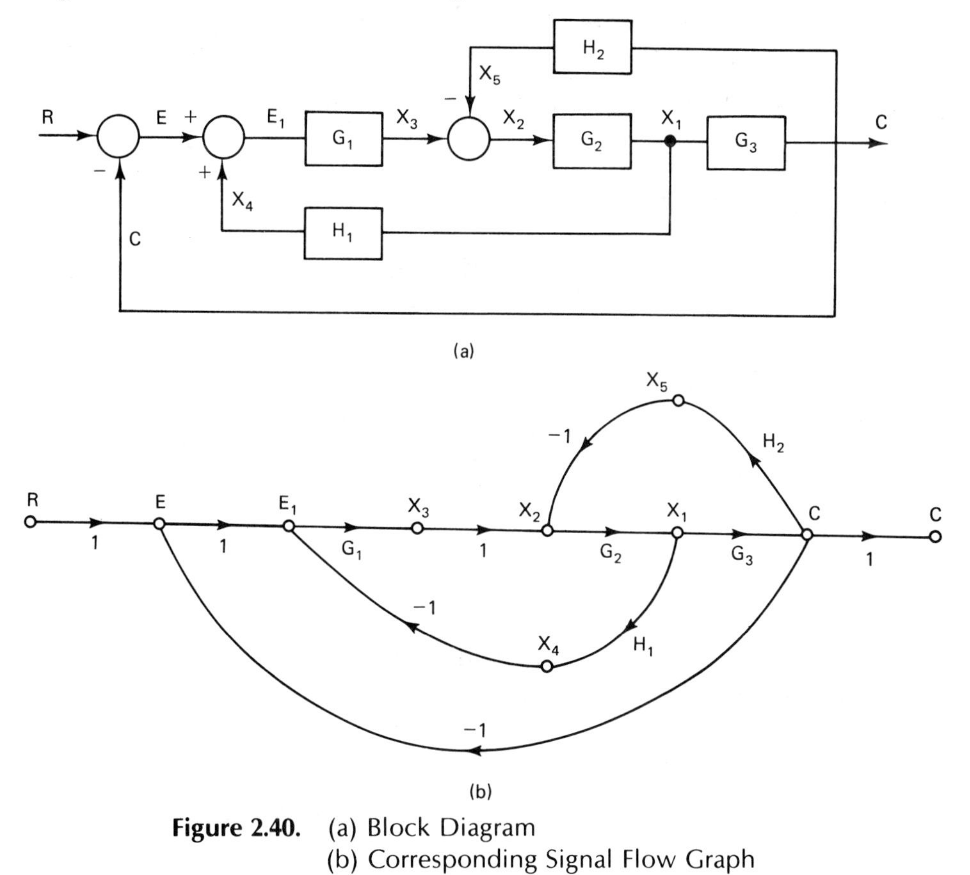


**Figure 2.40.** (a) Block Diagram
(b) Corresponding Signal Flow Graph

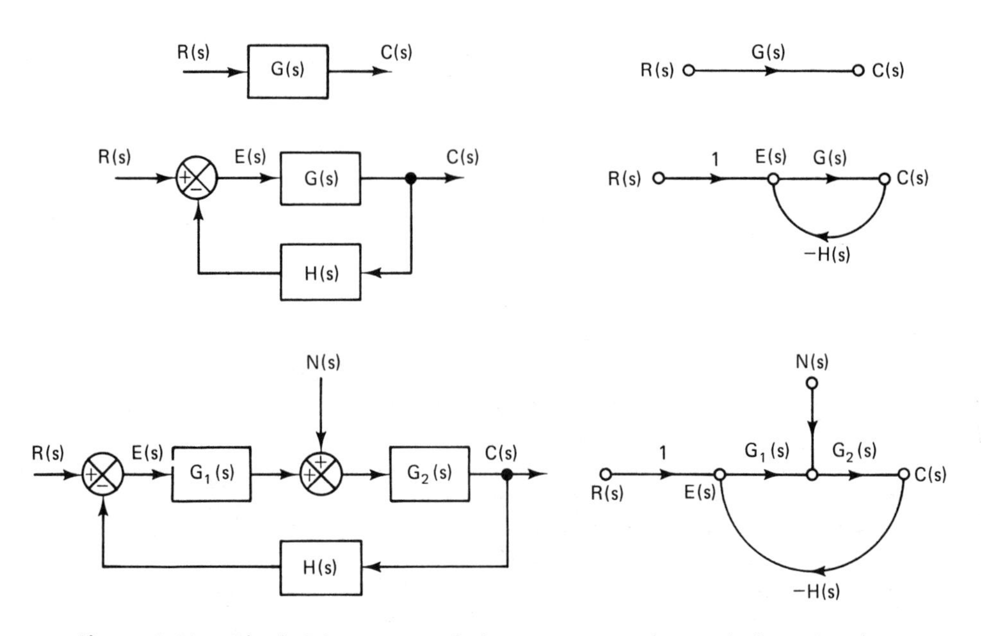


**Figure 2.41.** Block Diagrams and Their Associated Signal Flow Graphs

## Mason's Gain Formula

The general expression for the gain of a graph is given by the expression

$$G = \frac{\sum_i G_i \Delta_i}{\Delta}$$

The summation is carried over all forward paths in the graph, with

$$G_i = \text{gain of the } i\text{th forward path}$$

The diagram determinant is denoted by $\Delta$ and is given by

$$\Delta = 1 - \sum L_1 + \sum L_2 - \sum L_3 + \cdots + (-1)^m \sum L_m$$

In the definition above we have

$L_1$ = gain of each closed loop in the diagram

$L_2$ = product of loop gains of any two nontouching closed loops [loops are considered nontouching if they have no node in common]

⋮

$L_m$ = product of the loop gains of any $m$ nontouching loops

Finally, we have

$\Delta_i$ = value of $\Delta$ for that part of the graph not touching the $i$th forward path (value of $\Delta$ remaining when the path producing $G_i$ is removed)

The utility of Mason's gain formula is illustrated in the following examples.

EXAMPLE 2.18

Consider the block diagram of Example 2.15; a signal flow diagram of the system is shown in Figure 2.42. There are two forward paths with gains $G_1$ and $G_2$ and one loop with gain

$$L_1 = -G_2H$$

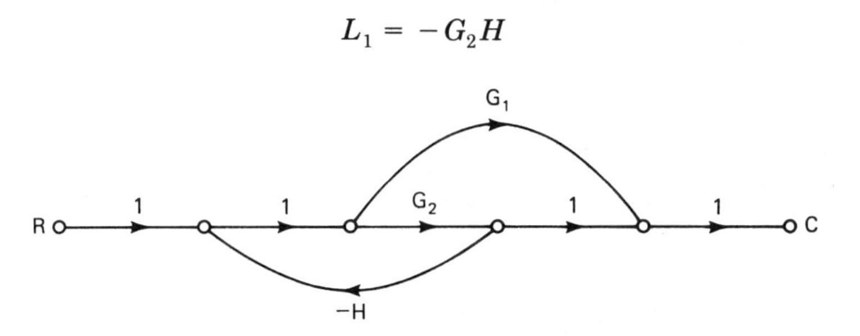


**Figure 2.42.** Signal Flow Graph for Example 2.18

The individual determinants are

$$\Delta_1 = \Delta_2 = 1$$

The diagram's determinant is

$$\Delta = 1 + G_2H$$

In accordance with Mason's gain formula, we get

$$G = \frac{G_1 + G_2}{1 + G_2H}$$

This agrees with our conclusion in Section 2.6.

## Voltage-Regulating System

A schematic diagram of a voltage-regulating arrangement for a dc generator is shown in Figure 2.43. The exciter field circuit is subject to the difference between a reference voltage $V_r$ and the generator terminal voltage $V_t$. As a result, the exciter field current $I_e$ is expressed as

$$I_e(s) = \frac{V_r(s) - V_t(s)}{R_1 + L_1s}$$

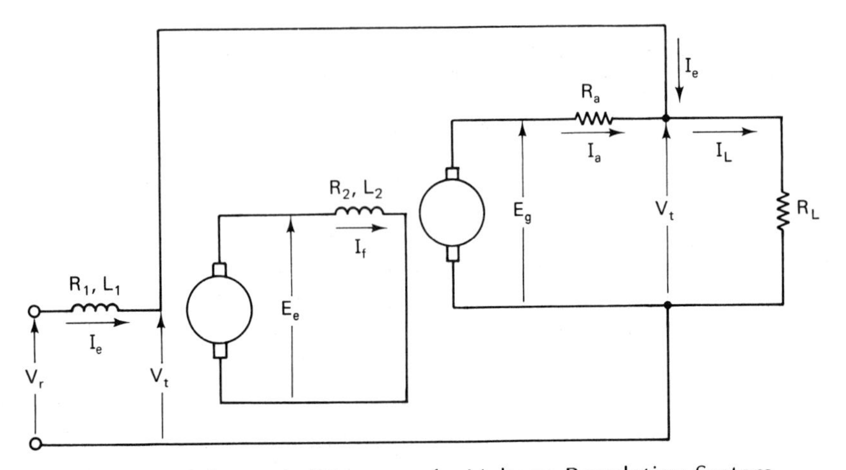


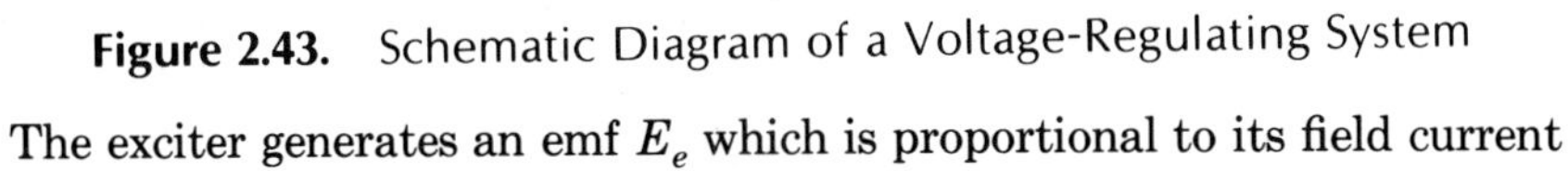

**Figure 2.43.** Schematic Diagram of a Voltage-Regulating System

The exciter generates an emf $E_e$ which is proportional to its field current

$$E_e(s) = K_e I_e(s)$$

The generator's field current $I_f(s)$ is given by

$$I_f(s) = \frac{E_e(s)}{R_2 + sL_2}$$

The generator's emf is given by

$$E_g(s) = K_f I_f(s)$$

The armature current is

$$I_a(s) = \frac{E_g(s) - V_t(s)}{R_a}$$

The load current is

$$I_L(s) = I_e(s) + I_a(s)$$

The terminal voltage is

$$V_t(s) = I_L(s) R_L$$

Figure 2.44 shows a signal flow graph of the system, and Figure 2.45 shows the block diagram for the system.

There are two forward paths in the system. Instead of finding the overall gain, we will be interested in finding the open-loop part denoted by $G$ in Figure 2.44b. We have the gain of the first forward path

$$G_1(s) = \frac{K_e K_f R_L}{R_a} \frac{1}{(R_1 + L_1 s)(R_2 + L_2 s)}$$

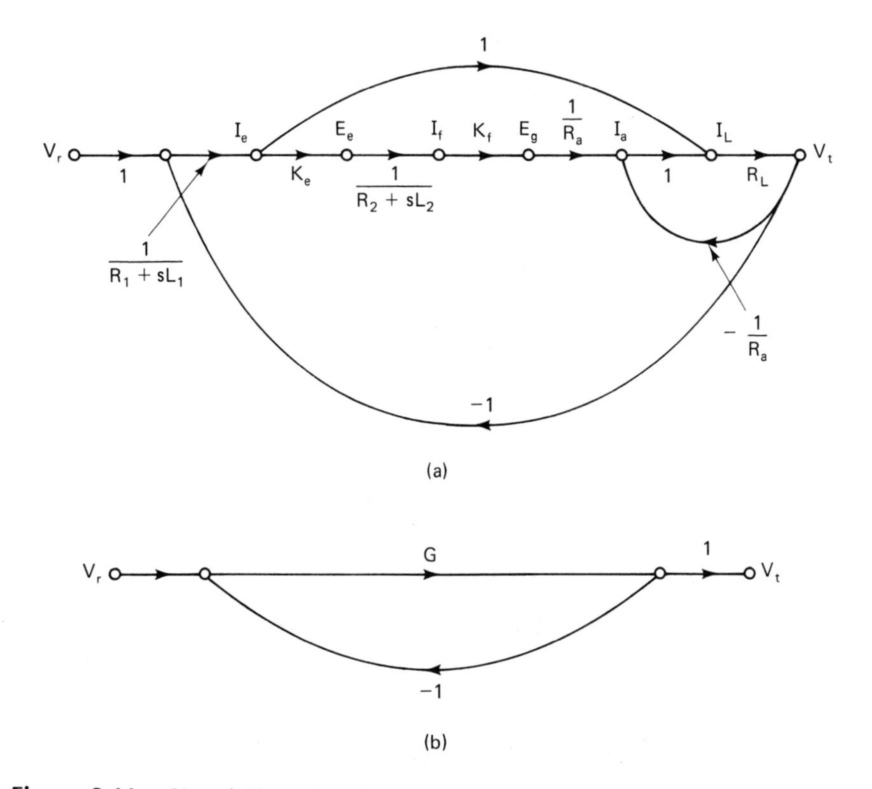


**Figure 2.44.** Signal Flow Graph of a Voltage-Regulating System

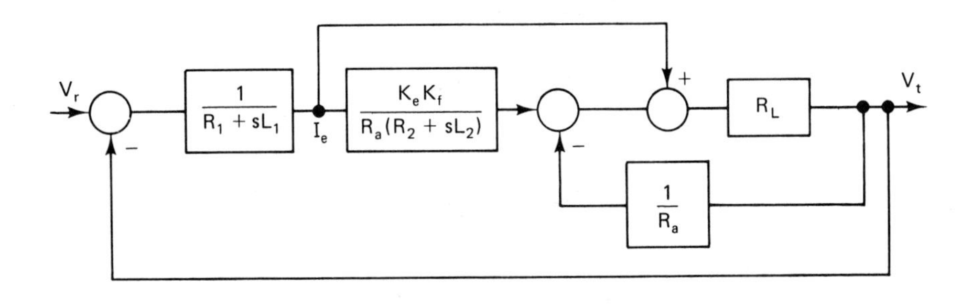




**Figure 2.45.** Block Diagram of a Voltage-Regulating System

The second forward path's gain is

$$G_2(s) = \frac{R_L}{R_1 + L_1 s}$$

There is only one feedback loop,

$$L_1 = -\frac{R_L}{R_a}$$

Thus

$$\Delta = 1 + \frac{R_L}{R_a}$$

We also have

$$\Delta_1 = 1$$

$$\Delta_2 = 1 + \frac{R_L}{R_a}$$

We can thus conclude that

$$G = \frac{K_e K_f R_L}{R_a} \frac{1}{(R_1 + L_1 s)(R_2 + L_2 s)} \frac{1}{1 + R_L/R_a} + \frac{R_L}{R_1 + L_1 s}$$

# Some Solved Problems

PROBLEM 2A-1

Consider the bridged-T network shown in Figure 2.46. Find the transfer function $V_2(s)/V_1(s)$, using block diagram reduction.

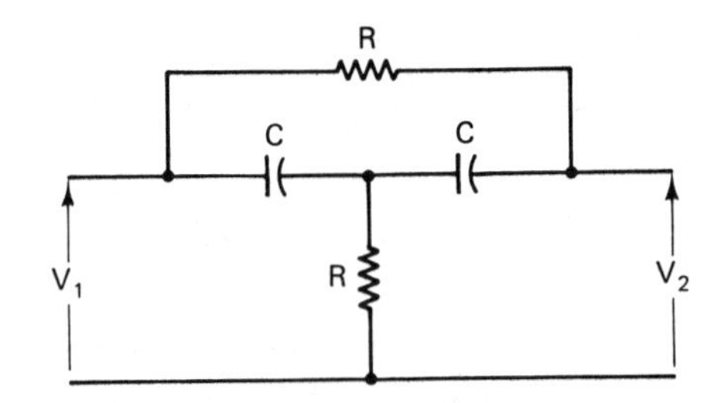


**Figure 2.46.** Bridged-T Network for Problem 2A-1

SOLUTION

We can write a loop equation around *abn* shown in Figure 2.47 given by

$$V_1 = I_1\left(R + \frac{1}{Cs}\right) - \frac{I_2}{Cs} \tag{1}$$

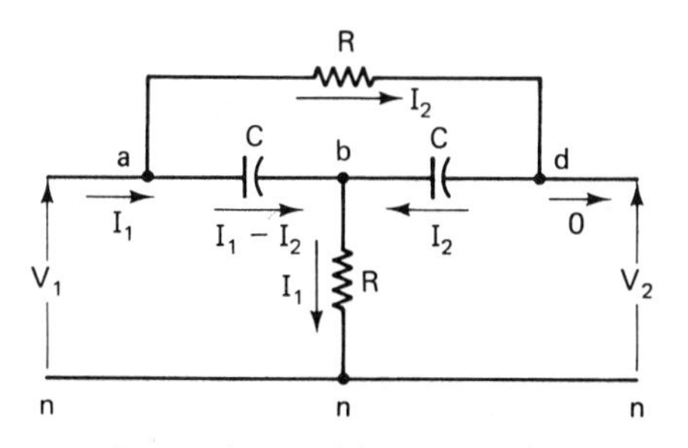


**Figure 2.47.** Nodes and Variables Identified for Problem 2A-1

A second loop equation is written around loop *dbn* given by

$$V_2 = I_1 R + I_2 \frac{1}{Cs} \tag{2}$$

The potential difference between *a* and *d* results in

$$V_1 - V_2 = I_2 R \tag{3}$$

Figure 2.47 identifies the nodes selected as well as the variables selected.

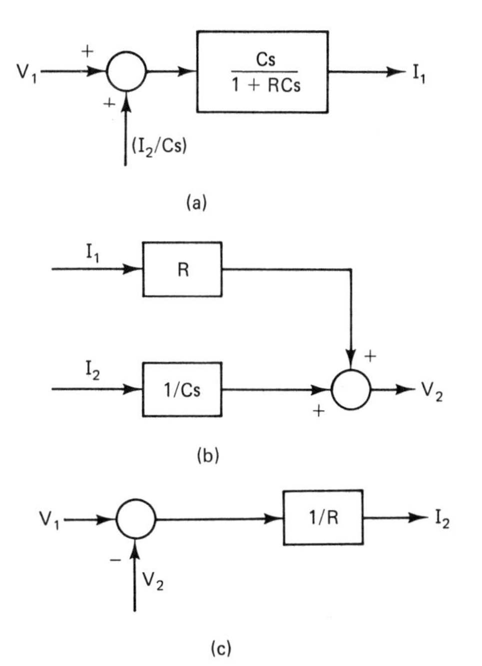


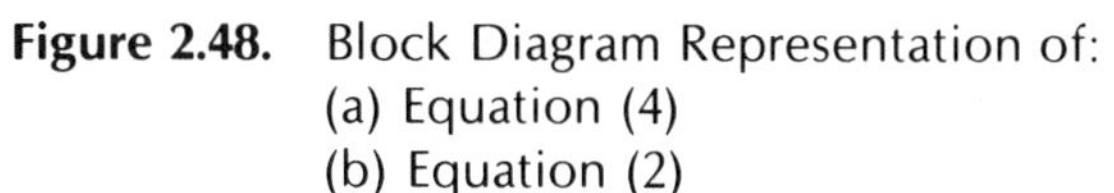


(c)

**Figure 2.48.** Block Diagram Representation of:
(a) Equation (4)
(b) Equation (2)
(c) Equation (5)

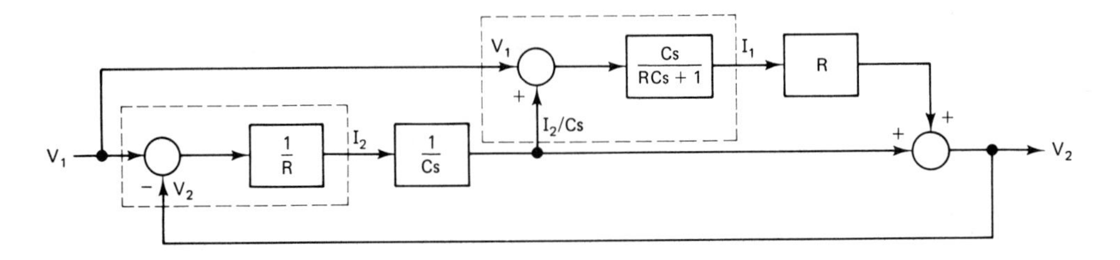


(a)

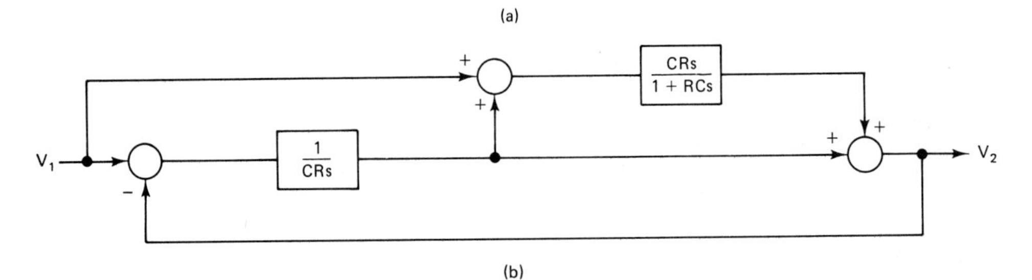


(b)

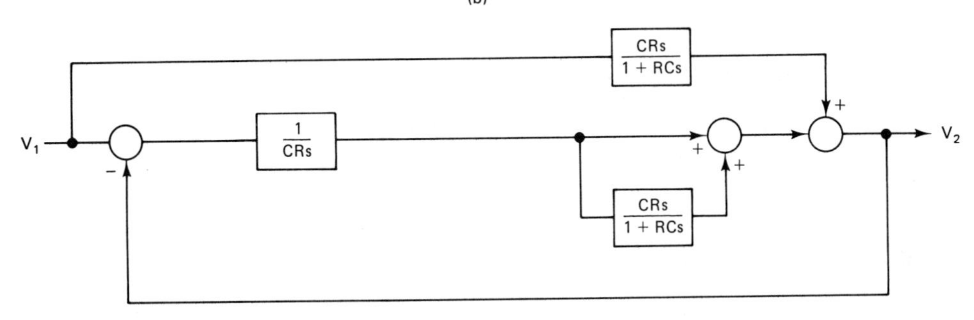


(c)

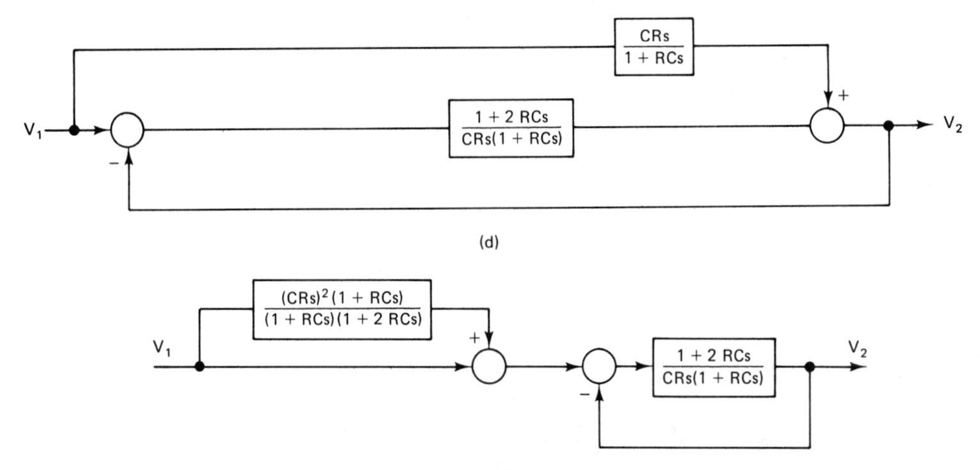


(e)

**Figure 2.49.** Steps in the Reduction of the Block Diagram for Problem 2A-1

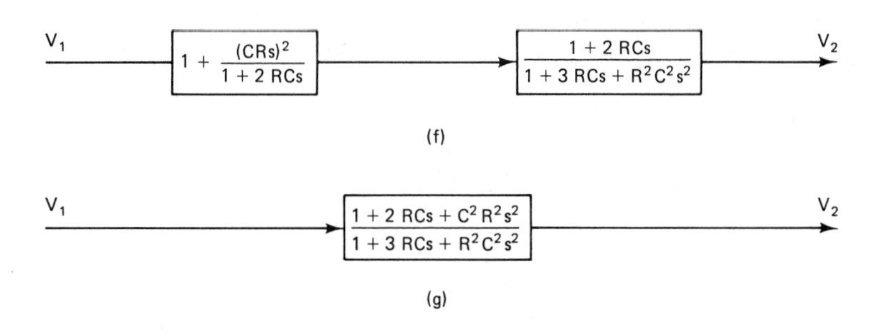


**Figure 2.49 (continued).** Steps in the Reduction of the Block Diagram for Problem 2A-1

Equation (1) can be rearranged so that $I_1$ is the output

$$I_1 = \left(V_1 + \frac{I_2}{Cs}\right)\frac{Cs}{RCs + 1} \tag{4}$$

A block diagram showing this relation is given in Figure 2.48a.

Equation (2) can provide us with $V_2$ as an output, and Figure 2.48b shows a block diagram representation of this relationship. Equation (3) is rearranged to provide $I_2$ as

$$I_2 = \frac{V_1 - V_2}{R} \tag{5}$$

Figure 2.48c shows the block diagram representation of this relationship.

In Figure 2.49 the assembled block diagram is shown together with steps in its reduction. We can conclude thus that the transfer function is given by

$$\frac{V_2(s)}{V_1(s)} = \frac{1 + 2RCs + C^2R^2s^2}{1 + 3RCs + R^2C^2s^2}$$

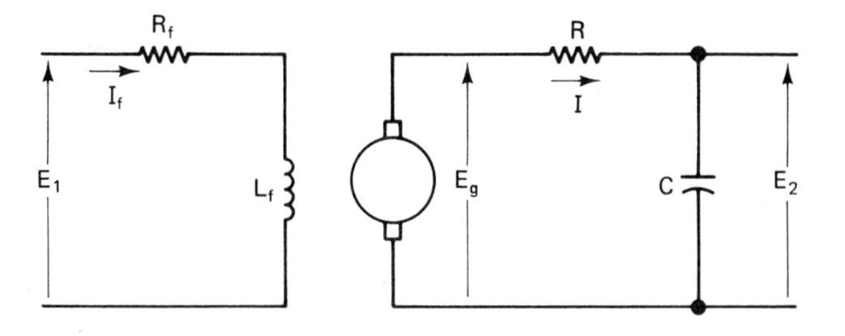


**Figure 2.50.** Schematic for Problem 2A-2

PROBLEM 2A-2

Find the transfer function $E_2(s)/E_1(s)$ for the generator shown in Figure 2.50 commonly used as a voltage amplifier.

SOLUTION

It is a simple matter to write the following equations:

$$I_f = \frac{E_1}{R_f + sL_f}$$

$$E_g = K_g I_f$$

$$I = \frac{E_g}{R + 1/Cs}$$

$$E_2 = \frac{I}{Cs}$$

Figure 2.51 shows the block diagram and the required transfer function.

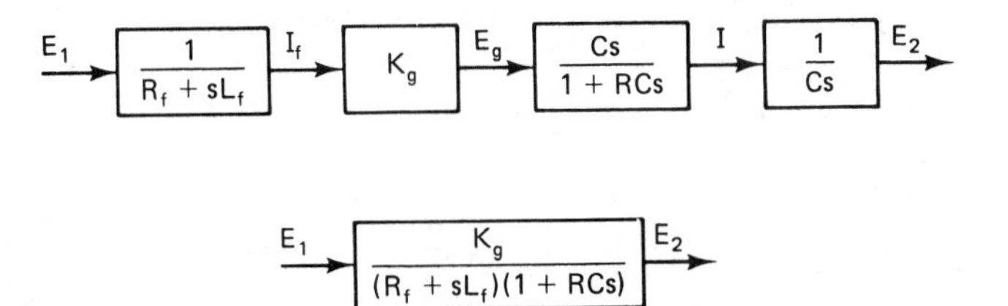


**Figure 2.51.** Block Diagram for Problem 2A-2

PROBLEM 2A-3

Find the transfer function $\omega_m(s)/E_f(s)$ for the system shown in Figure 2.52. The generator, driven at a constant speed, provides the field voltage for the motor. The motor has an inertia $J$.

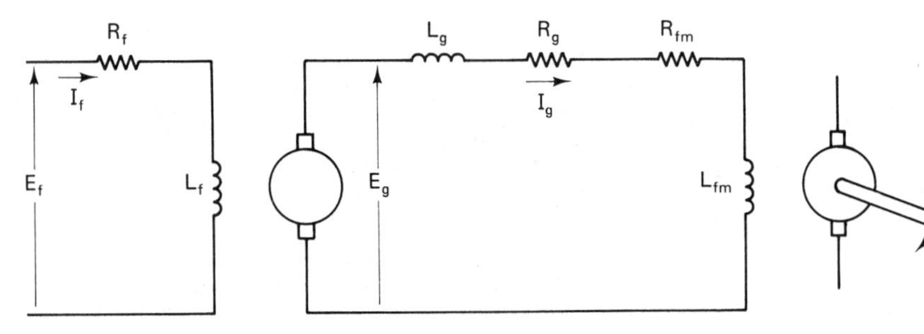


**Figure 2.52.** Schematic for Problem 2A-3

SOLUTION

The following equations apply:

$$I_f = \frac{E_f}{R_f + sL_f}$$

$$E_g = K_g I_f$$

$$I_g = \frac{E_g}{(R_g + R_{fm}) + s(L_g + L_{fm})}$$

$$T_m = K_m I_g$$

$$T_m = Js\omega_m$$

Figure 2.53 shows the required results.

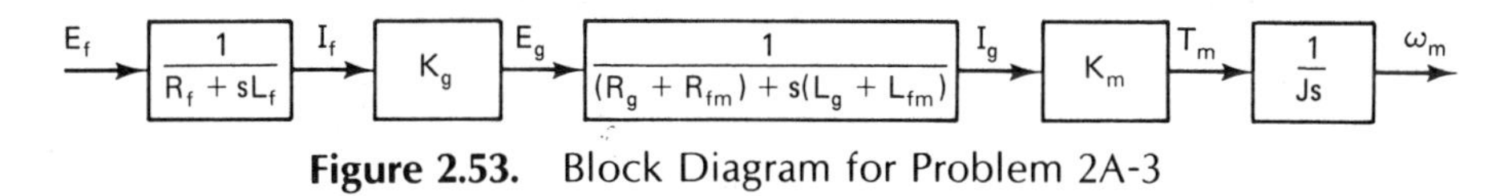


**Figure 2.53.** Block Diagram for Problem 2A-3

PROBLEM 2A-4

For the system shown in Figure 2.54, find the transfer function $\omega_m(s)/V_f(s)$.

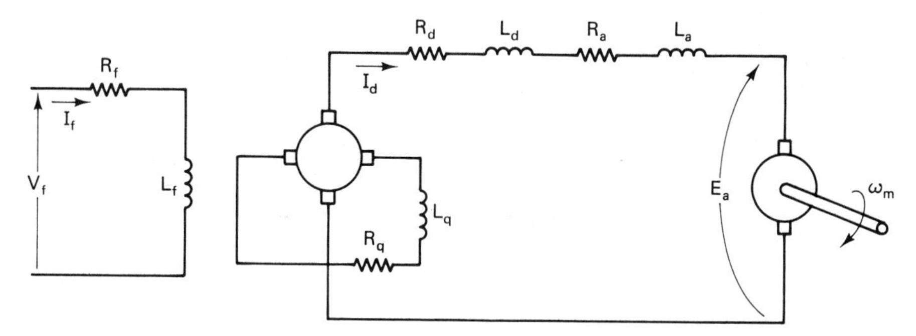


**Figure 2.54.** Schematic for Problem 2A-4

SOLUTION

The following equations can be written

$$I_f = \frac{V_f}{R_f + sL_f}$$

$$E_q = K_{fq} I_f$$

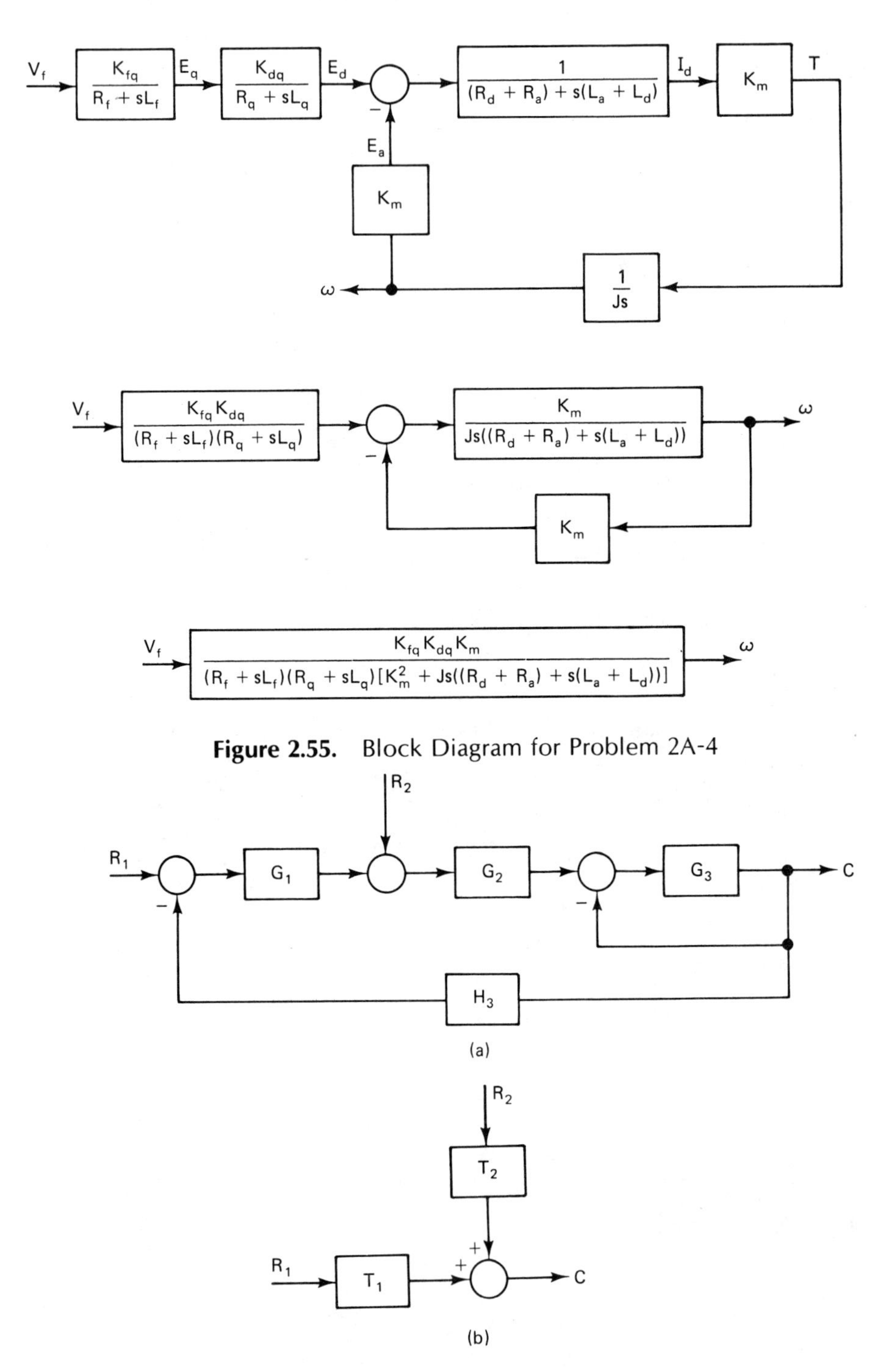


**Figure 2.55.** Block Diagram for Problem 2A-4



**Figure 2.56.** Block Diagrams for Problem 2A-5

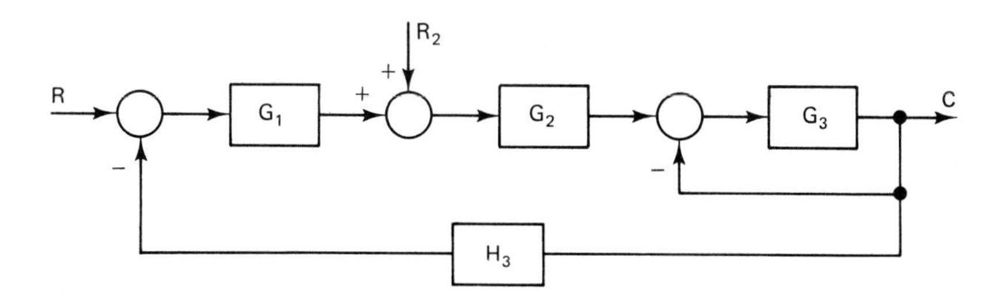


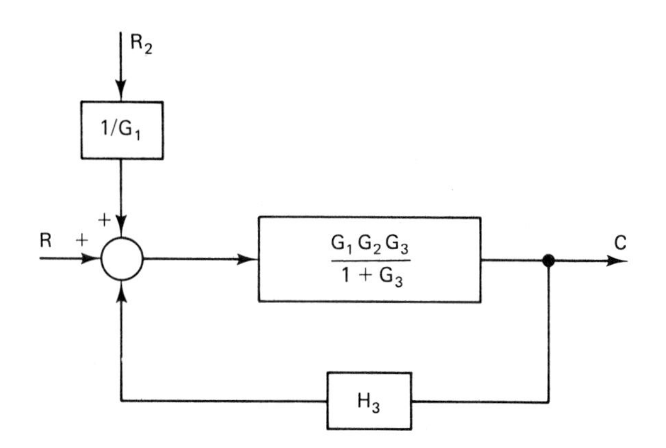


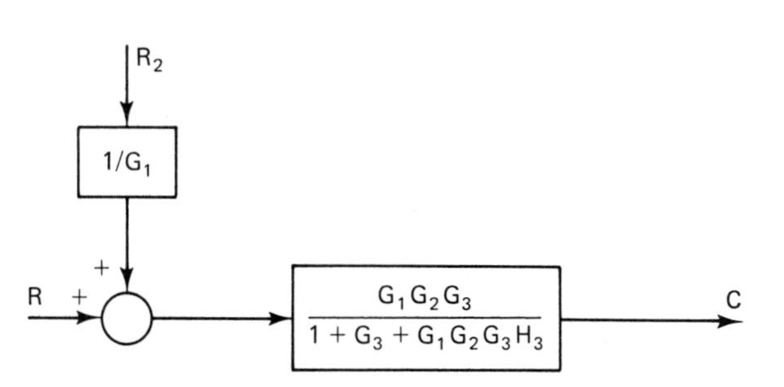


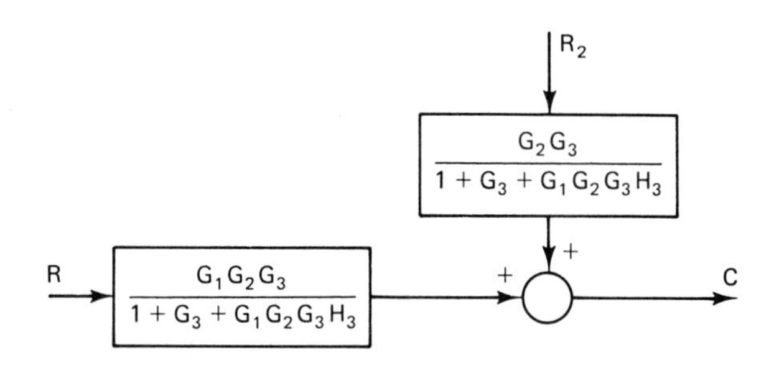


**Figure 2.57.** Solving Problem 2A-5

$$I_q = \frac{E_q}{R_q + sL_q}$$
$$E_d = K_{dq} I_q$$
$$I_d = \frac{E_d - E_a}{(R_d + R_a) + s(L_a + L_d)}$$
$$E_a = K_m \omega$$
$$T = K_m I_d$$
$$\omega = \frac{T}{Js}$$

Figure 2.55 shows the required block diagram and steps in its reduction.

PROBLEM 2A-5

Reduce the system shown in Figure 2.56a to the form shown in Figure 2.56b.

SOLUTION

Figure 2.57 shows the required steps.

PROBLEM 2A-6

Obtain the overall transfer function of the system shown in Figure 2.58 using block diagram reduction.

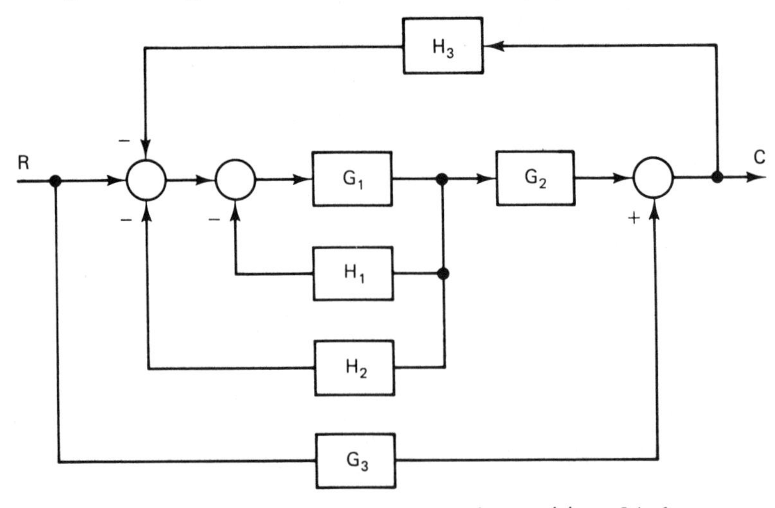


**Figure 2.58.** Block Diagram for Problem 2A-6

SOLUTION

The required steps are shown in Figure 2.59.

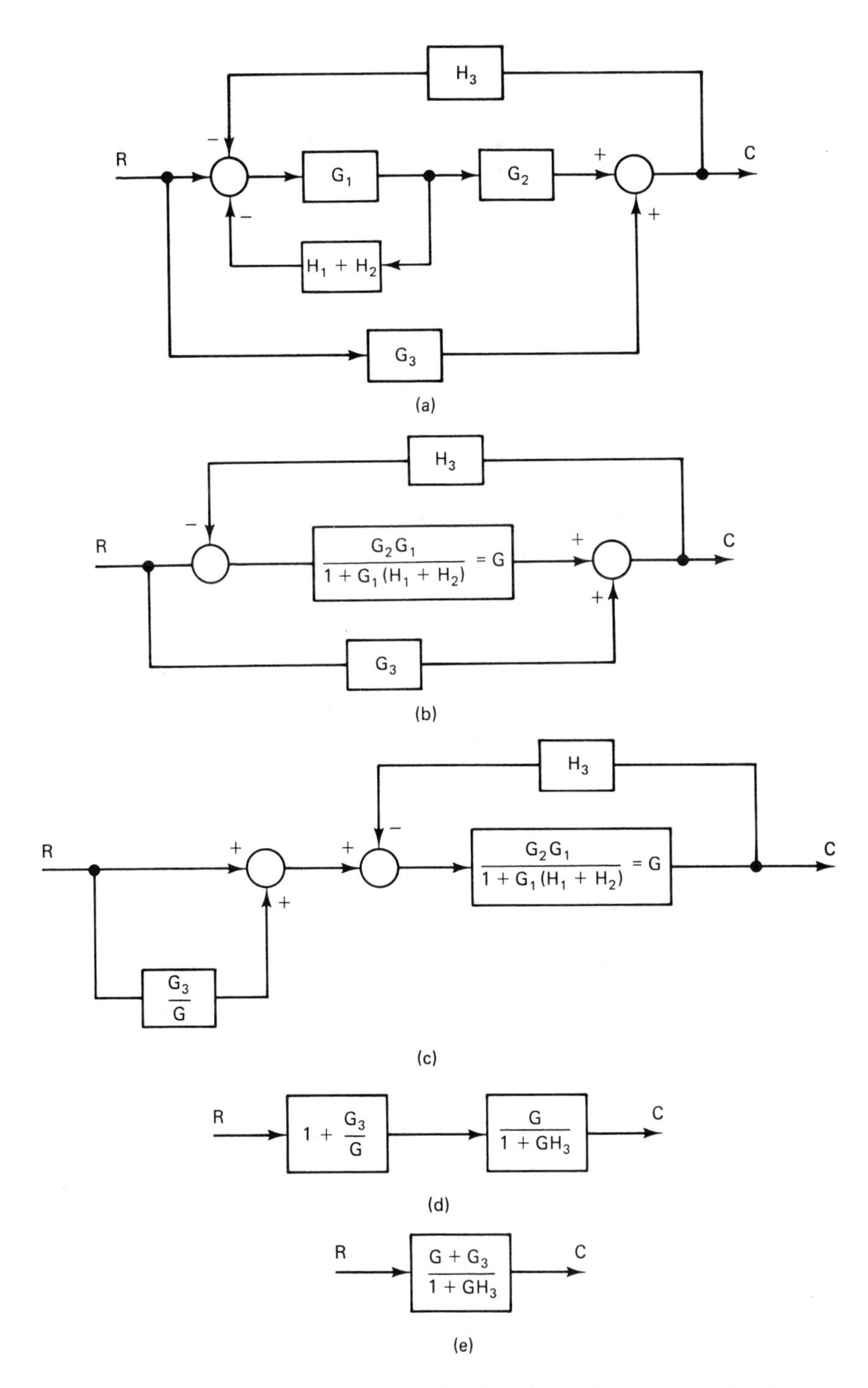


**Figure 2.59.** Block Diagram Reduction Steps for Problem 2A-6

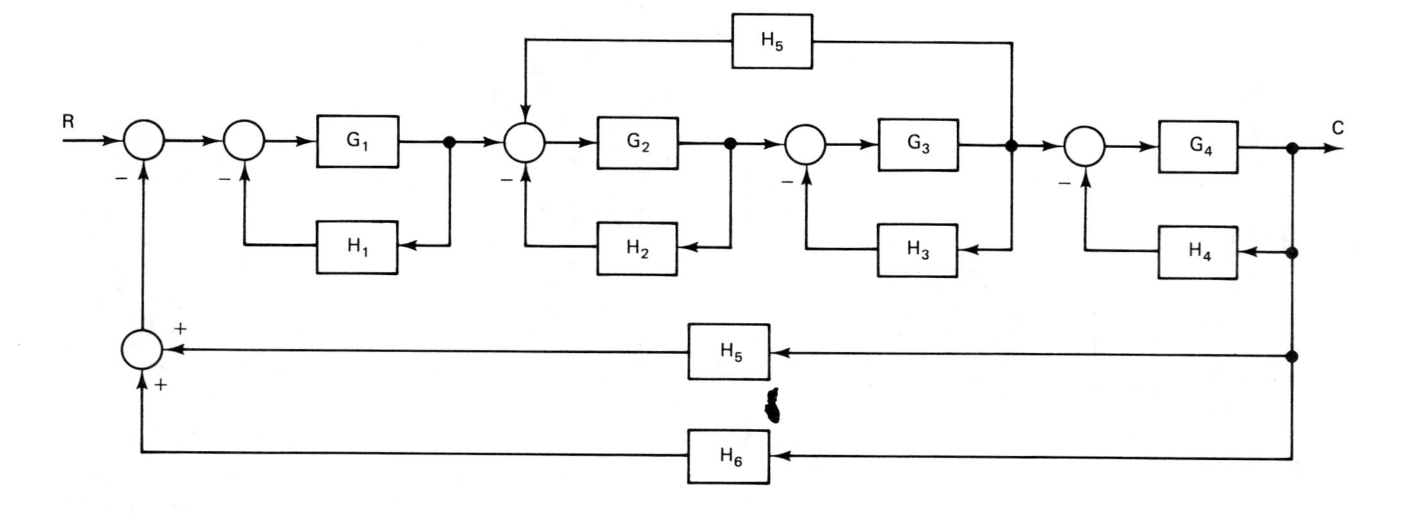


**Figure 2.60.** Block Diagram for Problem 2A-7

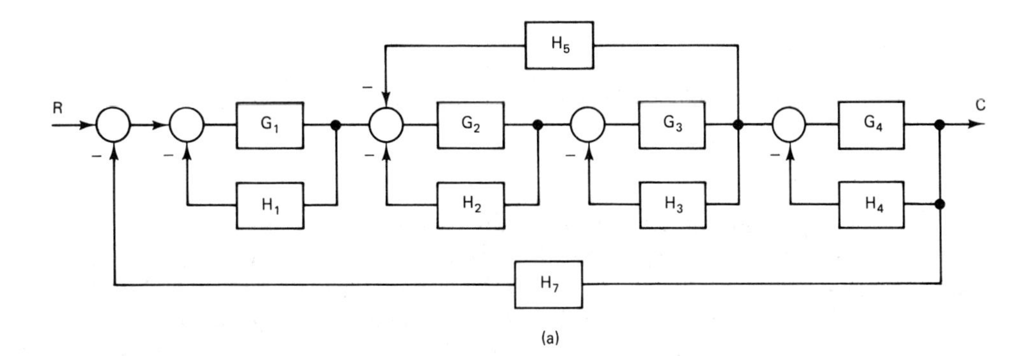


(a)

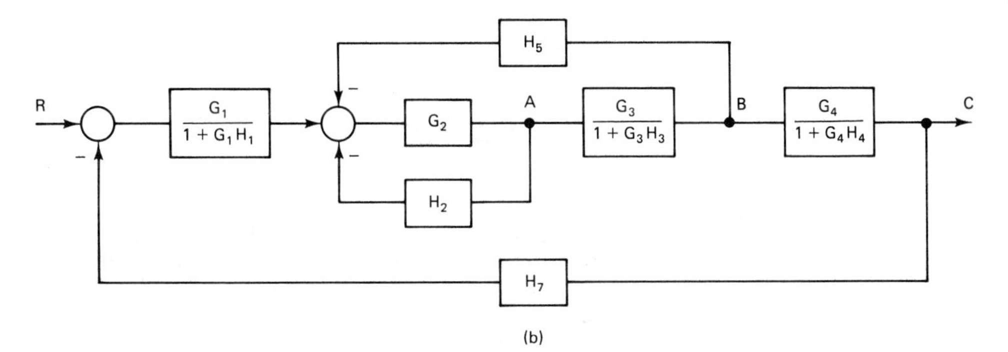


(b)

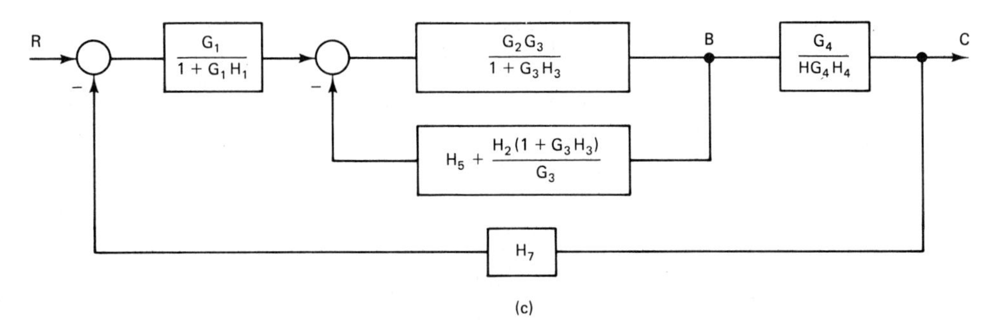


(c)

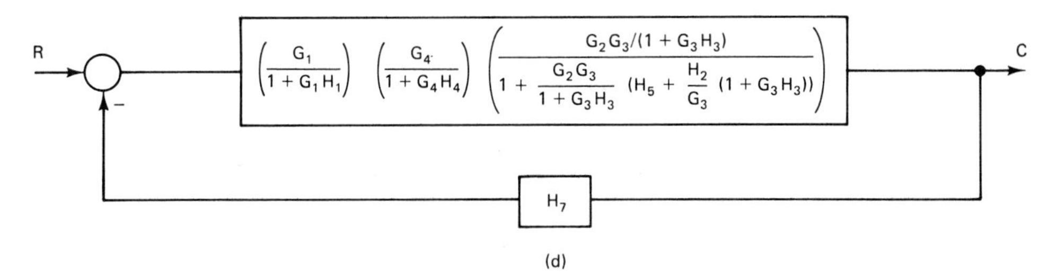


(d)

**Figure 2.61.** Block Diagram Reduction Steps for Problem 2A-7

PROBLEM 2A-7

Reduce the block diagram of Figure 2.60.

SOLUTION

We start by noting that we can replace $H_5$ and $H_6$ by one block as they add to give $H_7 = H_5 + H_6$. This is shown in Figure 2.61a. In part (b), the feedback loops $G_1H_1$, $G_3H_3$, and $G_4H_4$ are reduced using the standard form. The takeoff point $A$ is moved to $B$ with the result shown in part (c). Note also that $H_5$ is added to the result. The final result is shown in part (d).

# *Problems*

PROBLEM 2B-1

Find the Laplace transform of the pulse function shown in Figure 2.62.

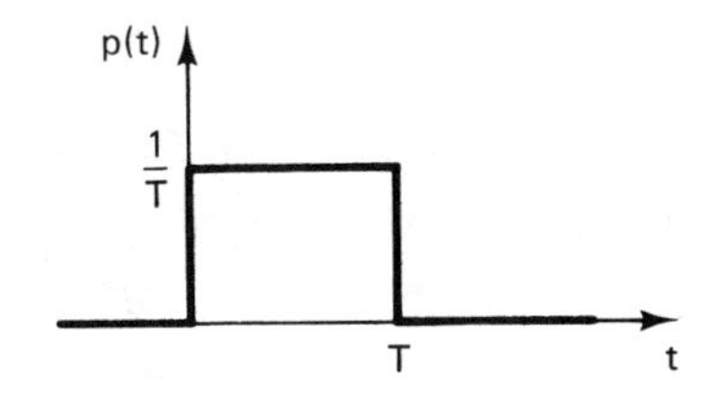


**Figure 2.62.** Pulse Function for Problem 2B-1

PROBLEM 2B-2

Find the Laplace transform of the square wave shown in Figure 2.63.

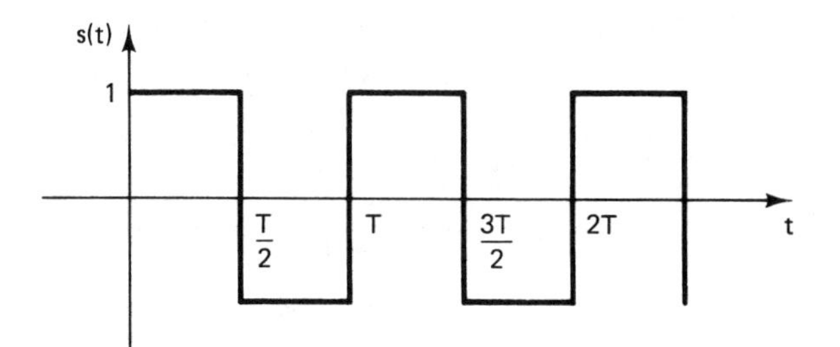


**Figure 2.63.** Square Wave for Problem 2B-2

PROBLEM 2B-3

Find the Laplace transform of the radio-frequency pulse shown in Figure 2.64. The pulse is defined by

$$r(t) = A \sin \omega_c t \qquad 0 \le t \le T$$
$$= 0 \qquad t > T$$

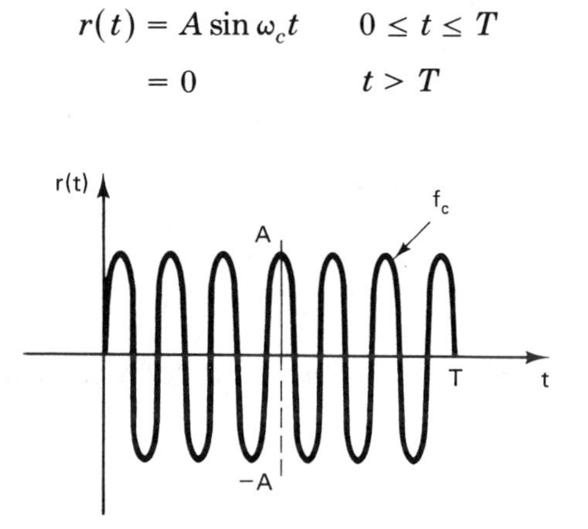


**Figure 2.64.** Radio-Frequency Pulse for Problem 2B-3

PROBLEM 2B-4

Find the Laplace transform of the following functions.

(a) $g(t) = te^{-3t}$

(b) $g(t) = t \sin \omega t$

(c) $g(t) = t \sinh \omega t$

(d) $g(t) = A \sin \omega_1 t(1 + \cos \omega_2 t)$

PROBLEM 2B-5

Find the inverse Laplace transform of the following Laplace transforms.

(a) $C(s) = \dfrac{10}{s(s+7)}$

(b) $C(s) = \dfrac{5}{s(s^2+5s+4)}$

(c) $C(s) = \dfrac{8}{s^2(s+1)(s+3)(s+4)}$

PROBLEM 2B-6

Find the transfer function $G(s) = E_o(s)/E_i(s)$ for the ladder network shown in Figure 2.65.

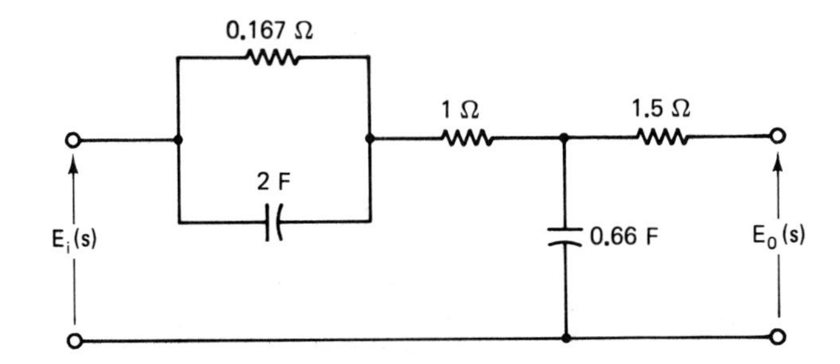


**Figure 2.65.** Ladder Network for Problem 2B-6

PROBLEM 2B-7

Find the output $e_o(t)$ of the $RC$ network shown in Figure 2.66 for an input voltage given by $e_i(t) = p(t)$ defined in Problem 2B-1.

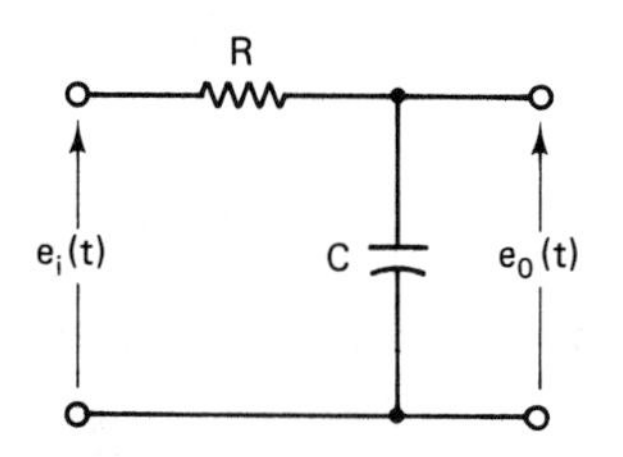


**Figure 2.66.** Network for Problem 2B-7

PROBLEM 2B-8

Find the transfer function of the network shown in Figure 2.67.

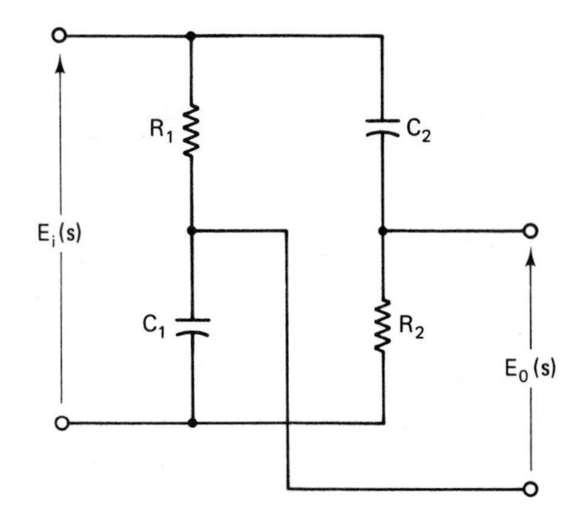


**Figure 2.67.** Network for Problem 2B-8

PROBLEM 2B-9

Use Kirchhoff's current and voltage laws to construct a block diagram of the network shown in Figure 2.68. The diagram should include the currents $I_1$ and $I_2$, the voltage $V_3$, and the input and output voltages $V_1(s)$ and $V_2(s)$. Reduce the block diagram to obtain the transfer function $V_2(s)/V_1(s)$.

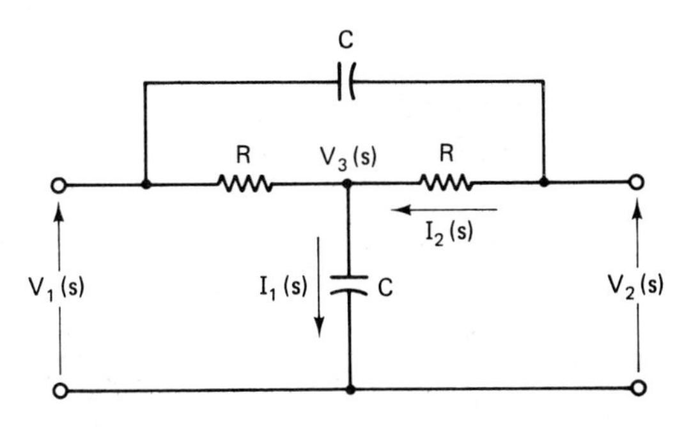


**Figure 2.68.** Network for Problem 2B-9

PROBLEM 2B-10

Obtain the transfer function $E_o(s)/E_i(s)$ for the system shown in Figure 2.69.

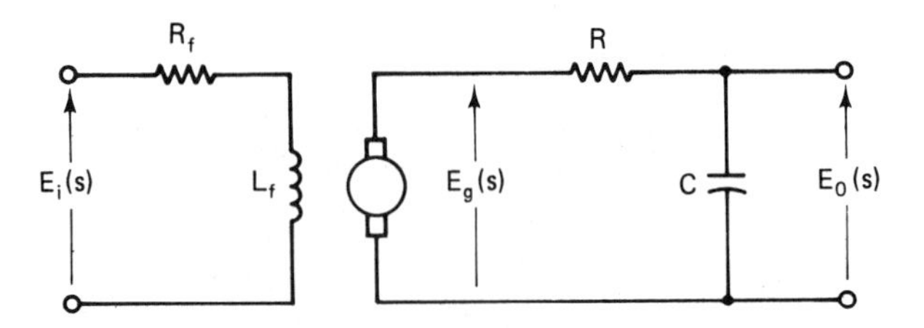


**Figure 2.69.** System for Problem 2B-10

PROBLEM 2B-11

Construct a block diagram for the twin-T network shown in Figure 2.70. Use block diagram reduction to find the transfer function $E_o(s)/E_i(s)$.

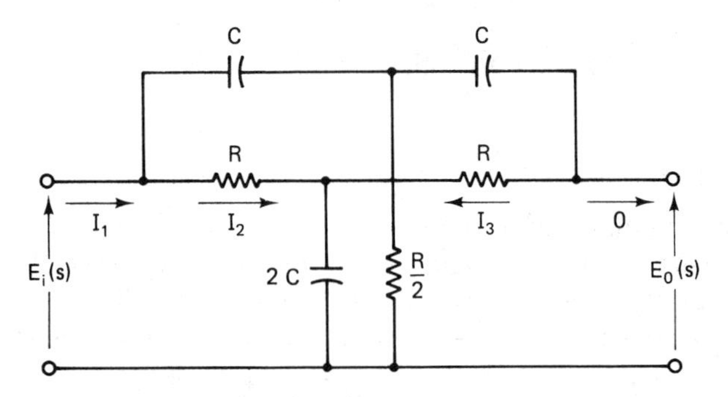


**Figure 2.70.** Twin-T Network for Problem 2B-11

PROBLEM 2B-12

Find the transfer function $C(s)/R(s)$ of the system represented by the block diagram of Figure 2.71.

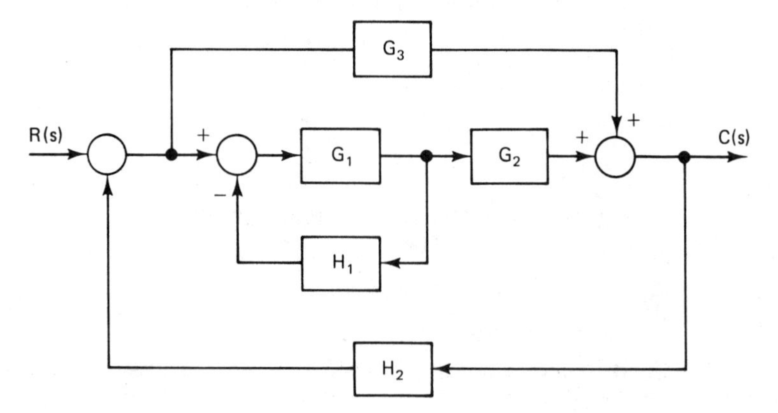


**Figure 2.71.** Block Diagram for Problem 2B-12

PROBLEM 2B-13

Find the transfer function $C(s)/R(s)$ of the system represented by the block diagram of Figure 2.72.

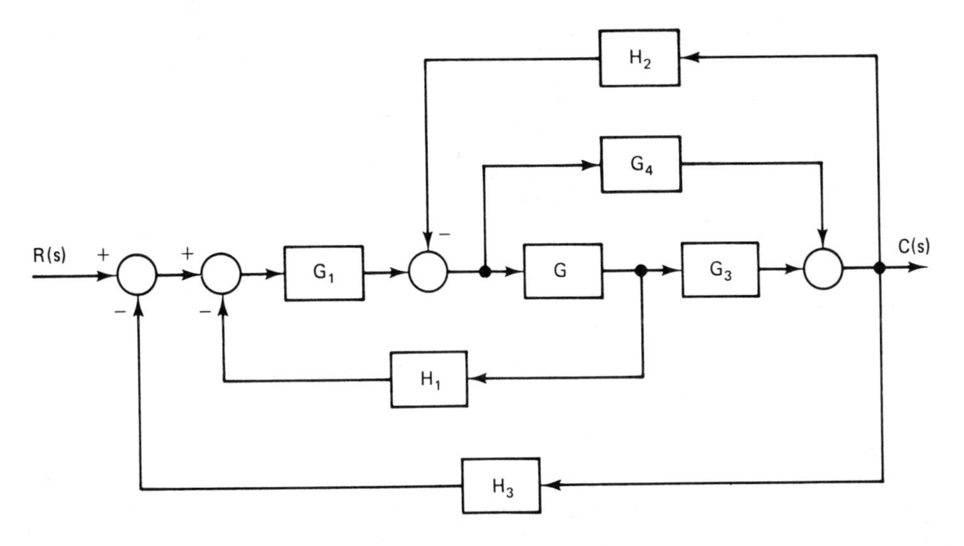


**Figure 2.72.** Block Diagram for Problem 2B-13

PROBLEM 2B-14

Construct a block diagram to represent the ladder network of Problem 2B-6.

PROBLEM 2B-15

Repeat Problem 2B-11 using signal flow graphs.

PROBLEM 2B-16

Draw a signal flow graph to represent the block diagram of Problem 2B-12. Use Mason's gain formula to find the transfer function.

PROBLEM 2B-17

Draw a signal flow graph to represent the block diagram of Problem 2B-13. Use Mason's gain formula to find the transfer function.

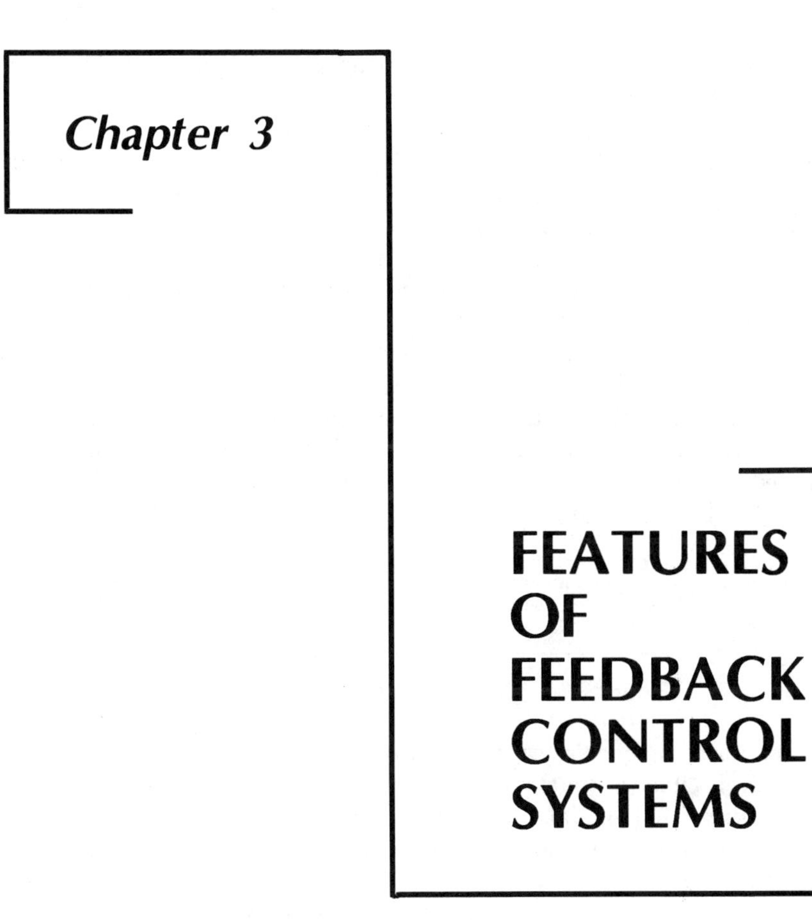


## 3.1 Introduction

The subject of this book is control systems. It is appropriate to start this chapter by defining the term and its key words. The term *system* refers to a collection of interacting and specialized parts or structures and junctions that is an integrated whole intended to perform a function that is not possible with any of the individual parts. The term *control* in the context of this book refers to the methods and means that are employed to govern or influence the performance of a system, apparatus, or machine. It is, therefore, clear that a *control system* is an assembly of control apparatus coordinated to carry out a planned set of controls. Alternatively, a control system is one in which a desired effect is attained by operating on the various inputs to the system until the output (a measure of the desired effects) is brought within an acceptable range of values.

An *automatic control system* is a control system that operates without human intervention. Control systems can be classified either as open loop or closed loop. The IEEE Standard Dictionary (an approved American National Standard, 1978) defines an *open-loop control system* as one in which the controlled quantity $c(t)$ is permitted to vary in accordance with the inherent characteristics of the control system and the controlled power apparatus for any given adjustment of the controller. No function of the controlled variable is used for automatic control of the system. Figure 3.1 shows in block diagram form the concept of an open-loop system.

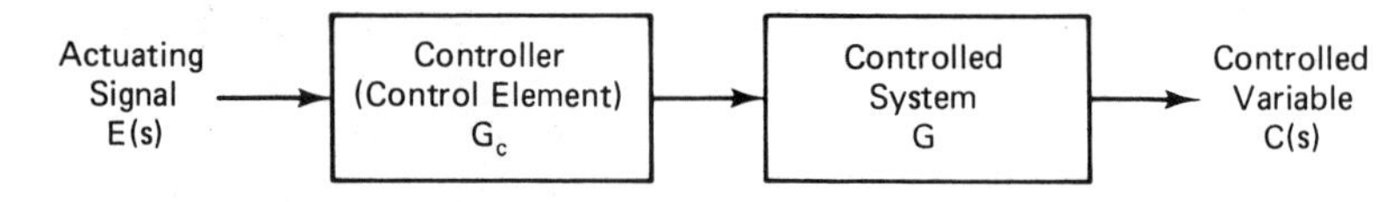


**Figure 3.1.** Open-Loop Control System

The term *feedback* refers to the process of returning a fraction of the controlled variable (output) to the input. A *feedback control system* is a control system that operates to achieve prescribed relationships between selected system variables by comparing functions of these variables and utilizing the comparison to effect control. When a feedback control system operates without human intervention it is referred to as an *automatic feedback control system*. A *closed-loop control system* shown in Figure 3.2 is a control system in which the controlled variable is measured and compared with a standard representing the desired performance. Any deviation from the standard is fed back into the control system in such a sense that it will reduce the deviation of the controlled variable from the standard. A *feedback loop* is a part of a closed-loop system that provides controlled response information allowing comparison with reference command.

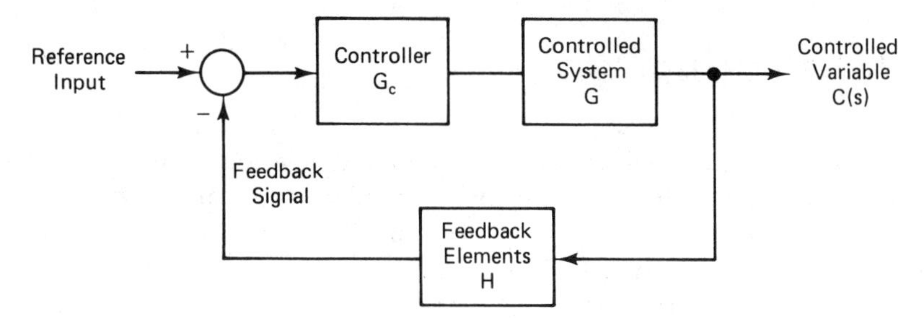


**Figure 3.2.** Closed-Loop Control System

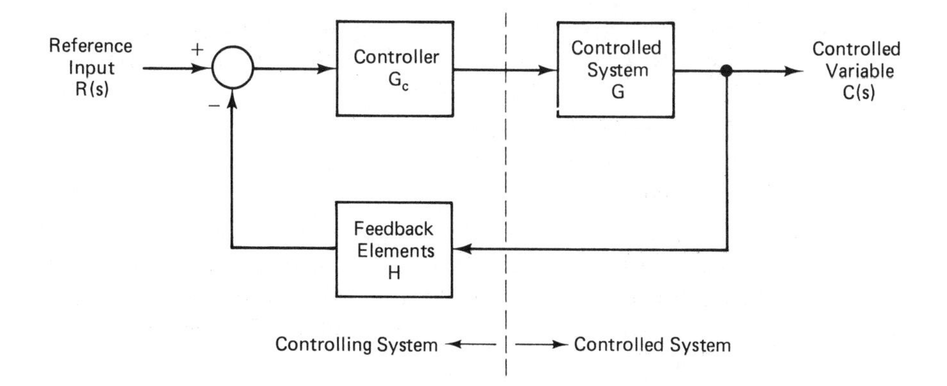


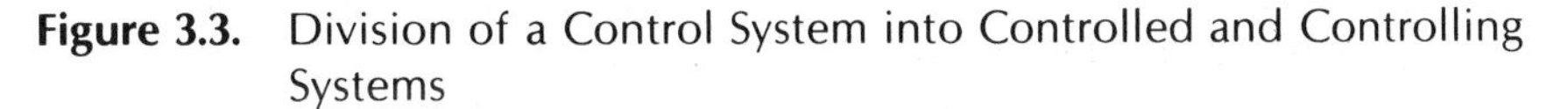


**Figure 3.3.** Division of a Control System into Controlled and Controlling Systems

A control system can be divided as shown in Figure 3.3 into two main systems. The *controlled system* is the apparatus or machine that is used to affect changes in the values of the ultimately controlled variable. In an open-loop system, the *controlling system* is the portion of the control system that manipulates the controlled system. In a feedback system, the controlling system is the portion that compares the functions of a directly controlled variable and command and adjusts a manipulated variable as a function of the difference.

# 3.2 Definitions for Feedback Control Systems

Figure 3.4 shows a standard block diagram of an automatic control system to illustrate definitions of variables and elements in common usage.

The *command* $v(t)$ is an input function that is developed by external means and is independent of the feedback control system.

The *reference input elements* $G_i(s)$ are system components and produce a signal $r(t)$ that is proportional to the command.

The *reference input* $r(t)$ is the signal derived from the command and is the actual signal input to the system.

The *controlled variable* $c(t)$ is the variable that is directly measured and controlled as the output of the controlled system.

The *primary feedback signal* $b(t)$ is derived from the controlled variable and is compared with the reference input to produce the actuating signal $e(t)$.

The *feedback elements* $H(s)$ are system components that act on the controlled variable $c(t)$ to produce the primary feedback signal $b(t)$.

The *actuating signal* $e(t)$ is the output of a comparison measuring device and is the difference between the reference input and the primary feedback signal. The power level of $e(t)$ is relatively low and is the input to the control elements $G_c$.

The *control elements* $G_c(s)$ produce the manipulated variable $m(t)$ from the actuating signal $e(t)$.

The *manipulated variable* $m(t)$ is the output of the control elements and is applied to the controlled system. The power level of $m(t)$ is higher than that of $e(t)$ and may be modified in form.

The *disturbance* $n(t)$ represents undesired signals that tend to affect the controlled system. The disturbance may be introduced into the system at more than one location.

The *controlled system* $G(s)$ is the device or process that is to be controlled.

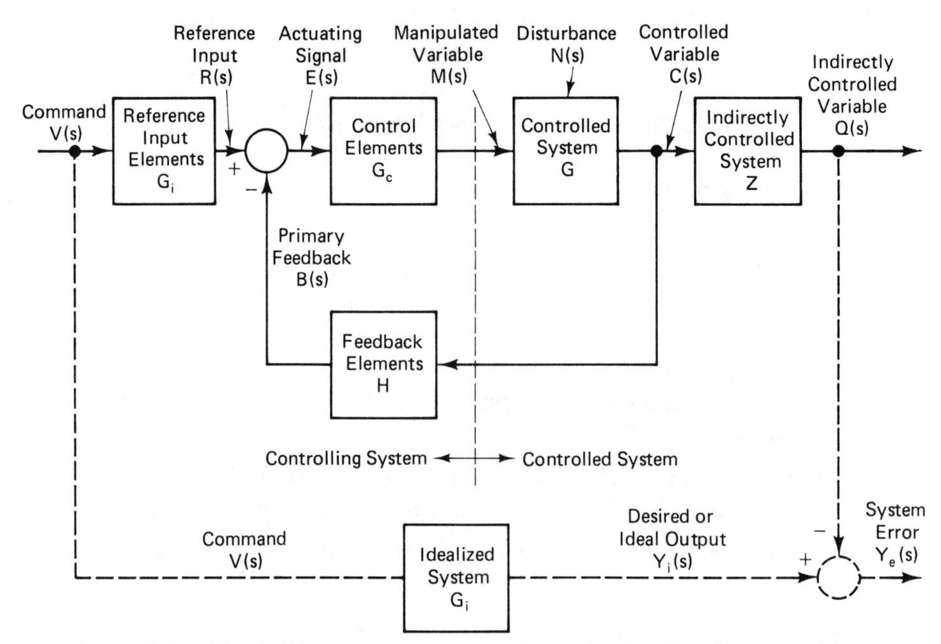


**Figure 3.4.** Block Diagram Representation of a Feedback Control System

The *indirectly controlled system* $Z(s)$ is outside the feedback loop and relates the indirectly controlled variable $q(t)$ to the controlled variable $c(t)$.

The *indirectly controlled variable* $q(t)$ is a variable related to the controlled variable and is not directly measured or controlled.

An idealized or desired equivalent of the system is shown in the dashed-line portion of the block diagram. This represents the relation between the command $v(t)$ and the ideal system output $y_i(t)$. The difference between the value of the indirectly controlled variable and the ideal system output is denoted by $y_e(t)$ and is a measure of the deviation of the control system output from the ideal or desired output. The dashed-line portion does not exist in practice as it is simply a mathematical model of a desirable but physically unattainable system $G_i(s)$.

# 3.3 Control Actions

For a control element (or a controlling system), the term *control action* refers to the nature of the change of the element's (or controlling system's) output affected by the input. The output may be a signal to another element or the value of a manipulated variable $m(t)$. The input may be the closed-loop feedback signal when the command is constant, the actuating signal, or the output of another control element. Control action can be used to effect *compensation*, which is a modifying action intended to improve the performance with respect to some specified characteristic such as system deviation. There are three basic types of control actions.

1. *Proportional control action.* In this type of control action, there is a continuous linear relationship between the output and the input of the control element

$$m(t) = K_p e(t)$$

or, in terms of transformed variables,

$$\frac{M(s)}{E(s)} = K_p$$

A block diagram of a proportional controller is shown in Figure 3.5a.

2. *Derivative control action.* The output of the control element in this type of control system is proportional to the time rate

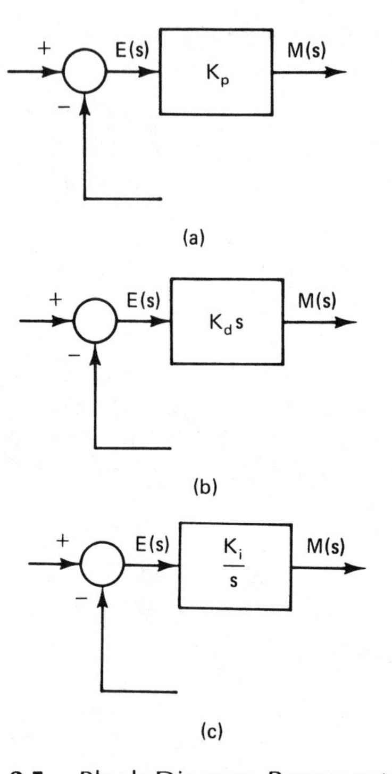


**Figure 3.5.** Block Diagram Representation of:
(a) Proportional Control Action
(b) Derivative Control Action
(c) Integral Control Action

of change of the input:

$$m(t) = K_d \frac{de(t)}{dt}$$

or, in terms of transformed variables,

$$\frac{M(s)}{E(s)} = K_d s$$

A block diagram representation of derivative control action is shown in Figure 3.5b.

3. *Integral control action.* This is a control action in which the output is proportional to the time integral of the input:

$$m(t) = K_i \int_0^t e(t)\, dt$$

or

$$\frac{M(s)}{E(s)} = \frac{K_i}{s}$$

A block diagram representation of integral control action is shown in Figure 3.5c.

Combinations of the actions listed above can be used. Common forms include:

1. *Proportional plus derivative (PD) control action.* The output of the controlling system is proportional to a linear combination of the input and the time rate of change of the input. Figure 3.6a shows a block diagram of a PD controller.

$$\frac{M(s)}{E(s)} = K_p(1 + T_d s)$$

2. *Proportional plus integral (PI) control action.* The output of the controlling system is proportional to a linear combination of the input and the time integral of the input. Figure 3.6b

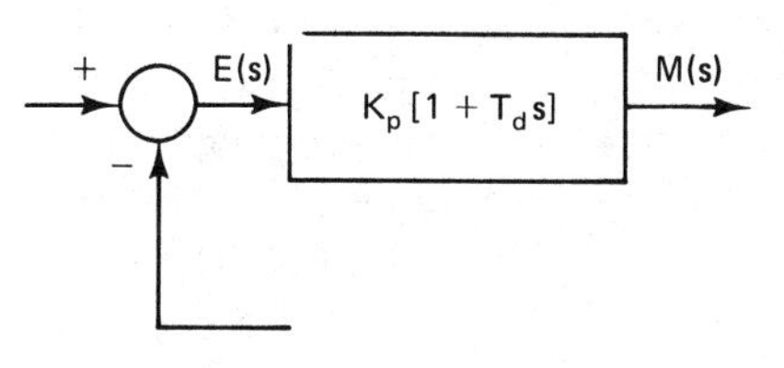


(a)

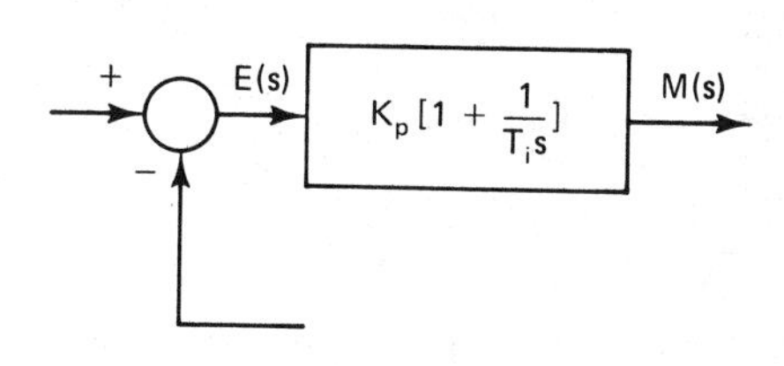


(b)

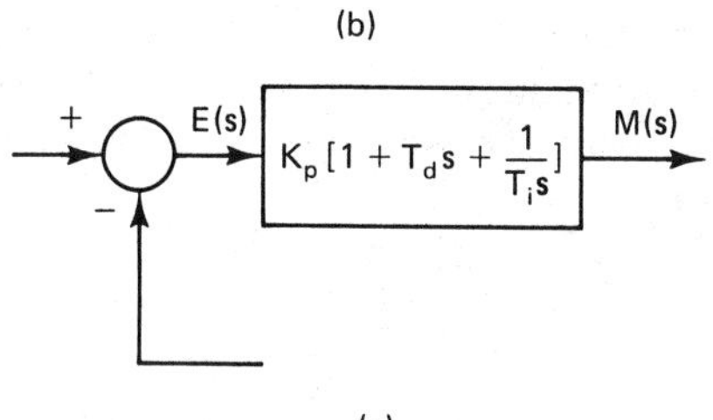


(c)

**Figure 3.6.** Block Diagram Representation of:
(a) PD Controller
(b) PI Controller
(c) PID Controller

shows a block diagram of a PI controller.

$$\frac{M(s)}{E(s)} = K_p\left(1 + \frac{1}{T_i s}\right)$$

3. *Proportional plus integral plus derivative (PID) control action.* The output in this case is proportional to a linear combination of the input, the time integral of the input, and the time rate of change of the input. This is shown in the block diagram in Figure 3.6c.

$$\frac{M(s)}{E(s)} = K_p\left(1 + T_d s + \frac{1}{T_i s}\right)$$

# 3.4 Feedback Control Actions and Effects

The feedback elements can be used to provide any desired action in a fashion similar to that of the forward control actions discussed in the preceding section. We discuss presently some effects of feedback on control system performance. We will gain more knowledge about the desirable properties of feedback as we progress through the text.

## Effect of Feedback on Gain and Time Constant

Consider the open-loop transfer function of a first-order system given by

$$G(s) = \frac{K}{1 + \tau s}$$

The parameter $\tau$ is called the time constant and, as will be seen in Chapter 5, is a measure of how fast the system will track an input. A small time constant is desirable. Assume now that feedback is employed as shown in Figure 3.7b. The closed-loop transfer function is thus given by

$$T(s) = \frac{K/(1 + \tau s)}{1 + [hK/(1 + \tau s)]}$$

$$= \frac{K/(1 + hK)}{1 + [\tau/(1 + hK)]s}$$

We can thus see that $T(s)$ is still a first-order function written as

$$T(s) = \frac{K_C}{1 + \tau_C s}$$

The equivalent gain $K_C$ is given by

$$K_C = \frac{K}{1 + hK}$$

The equivalent time constant $\tau_C$ is given by

$$\tau_C = \frac{\tau}{1 + hK}$$

It is clear that the effect of feedback is to reduce both gain and time constant.

Let us observe here that we can reduce the system's time constant to any desired value by derivative positive feedback, as shown in Figure 3.8. In this case we have

$$T(s) = \frac{K}{(\tau - bK)s + 1} = \frac{K}{\tau_{eq}s + 1}$$

where the equivalent time constant $\tau_{eq}$ is given by

$$\tau_{eq} = \tau - bK$$

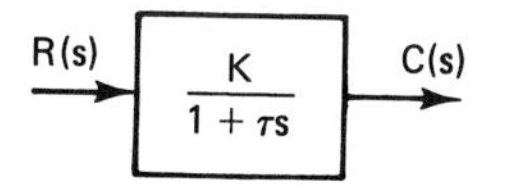


(a)

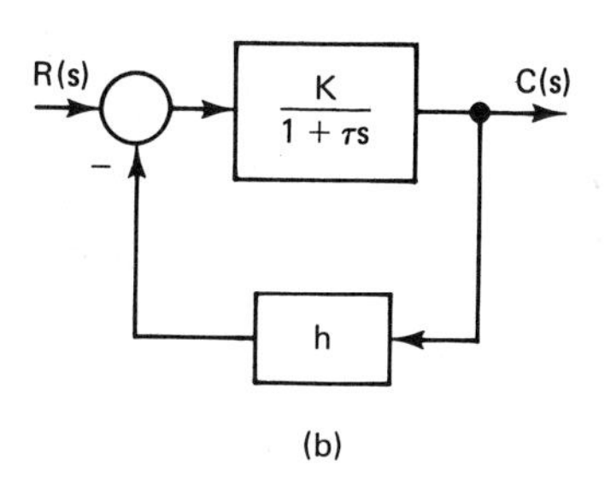


(b)

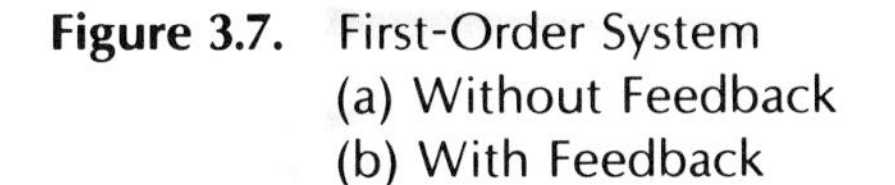

**Figure 3.7.** First-Order System
(a) Without Feedback
(b) With Feedback

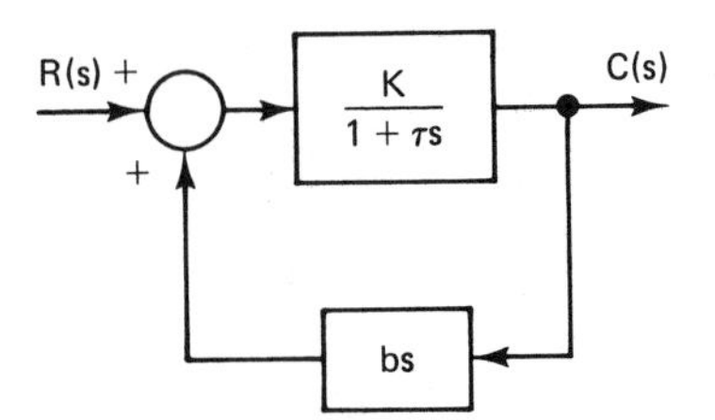


**Figure 3.8.** Employing Derivative Positive Feedback to Reduce the Time Constant

If $b$ is chosen such that

$$b = \frac{\tau}{K}$$

we get

$$\tau_{eq} = 0$$

## Effect of Feedback on External Disturbance

During its operation, a physical system is subject to external inputs (noise), which may cause undesirable effects on the system. The control systems engineer can use feedback substantially to reduce these effects. A simple approach to modeling noise is to assume that it is additive and can be incorporated in the system representation as shown in Figure 3.9a. If we recall our block diagram models of motors treated in Chapter 2, we realize that the load torque $T_l(s)$ can be treated as an additive noise.

Consider the process shown in the block diagram of Figure 3.9a. The output $C_A(s)$ is obtained as the sum of two components:

$$C_A(s) = C_{RA}(s) + C_{NA}(s)$$

Here the output component due to the input $R(s)$ is given by

$$C_{RA}(s) = G_1(s)G_2(s)R(s)$$

The output component due to the noise $N(s)$ is given by

$$C_{NA}(s) = G_2(s)N(s)$$

Consider now the effect of introducing a feedback loop with function $H(s)$, as shown in Figure 3.9b. The output $C_B(s)$ is obtained as the sum of two components:

$$C_B(s) = C_{RB}(s) + C_{NB}(s)$$

The output component due to input $R(s)$ is given by

$$C_{RB}(s) = \frac{G_1(s)G_2(s)}{1 + G_1(s)G_2(s)H(s)}R(s)$$

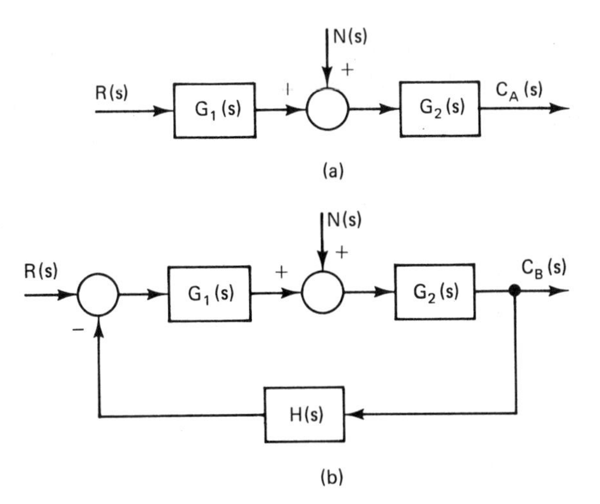


**Figure 3.9.** Control Process Subject to External Disturbance or Noise
(a) Without Feedback
(b) With Feedback

The output component due to the noise $N(s)$ is given by

$$C_{NB}(s) = \frac{G_2(s)}{1 + G_1(s)G_2(s)H(s)} N(s)$$

Let us now compare the noise component in the output without feedback $C_{NA}(s)$ to that with feedback $C_{NB}(s)$. The ratio is

$$\frac{C_{NB}(s)}{C_{NA}(s)} = \frac{1}{1 + G_1(s)G_2(s)H(s)}$$

We can thus see that the noise component in the output is reduced by the factor $[1 + G_1(s)G_2(s)H(s)]$ by the introduction of feedback.

In communication systems applications, the input $R(s)$ is referred to as the signal. An important measure of the quality of the process is the signal-to-noise ratio (SNR), defined by

$$\text{SNR} = \frac{\text{output due to signal}}{\text{output due to noise}}$$

For the case without feedback, we obtain

$$\text{SNR}_A = \frac{C_{RA}(s)}{C_{NA}(s)} = G_1(s)\frac{R(s)}{N(s)}$$

For the feedback case we get

$$\text{SNR}_B = \frac{C_{RB}(s)}{C_{NB}(s)} = G_1(s)\frac{R(s)}{N(s)}$$

Thus the introduction of feedback does not change the signal-to-noise ratio in the absence of other adjusting factors.

Let us assume that together with feedback we adjust the input $R(s)$ to a value $\tilde{R}(s)$ and the plant function $G_1(s)$ to $\tilde{G}_1(s)$ such that

$$C_{\tilde{R}B}(s) = \frac{\tilde{G}_1(s)G_2(s)\tilde{R}(s)}{1 + \tilde{G}_1(s)G_2(s)H(s)} = G_1(s)G_2(s)R(s) = C_{RA}(s)$$

Now with the assumption above we have

$$C_{NB}(s) = \frac{G_2(s)}{1 + \tilde{G}_1(s)G_2(s)H(s)} N(s)$$

The signal-to-noise ratio for this case is

$$\begin{aligned} \text{SNR}_{\tilde{B}} &= \frac{C_{\tilde{R}B}(s)}{C_{NB}(s)} \\ &= \frac{G_1(s)G_2(s)R(s)}{\dfrac{G_2(s)N(s)}{1 + \tilde{G}_1(s)G_2(s)H(s)}} \\ &= G_1(s)\frac{R(s)}{N(s)}\left[1 + \tilde{G}_1(s)G_2(s)H(s)\right] \end{aligned}$$

Recalling that without feedback we have

$$\text{SNR}_A = G_1(s)\frac{R(s)}{N(s)}$$

we conclude that

$$\frac{\text{SNR}_{\tilde{B}}}{\text{SNR}_A} = 1 + \tilde{G}_1(s)G_2(s)H(s)$$

Thus feedback, together with the adjustments to $R$ and $G_1$, increases the SNR for the system.

omit

## 3.5 Some Examples

Having studied some effects of feedback, it is appropriate at this time to consider a number of examples where feedback can arise either naturally or introduced purposely by the designer in the system.

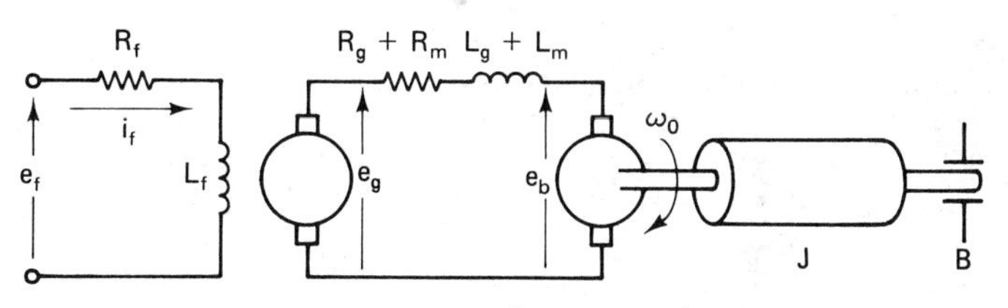


**Figure 3.10.** Ward – Leonard System

## Ward–Leonard System

The conventional Ward–Leonard system is a configuration with a dc generator driving an armature-controlled dc motor coupled to the load. A schematic diagram of the system is shown in Figure 3.10. The following equations describe the performance of the system.

$$E_f(s) = (R_f + L_f s)I_f(s)$$
$$E_g(s) = KI_f(s)$$
$$E_g(s) - E_b(s) = \left[(R_g + R_m) + (L_g + L_m)s\right] I_a(s)$$
$$E_b(s) = K_m\omega_o(s)$$
$$T_m(s) = K_m I_a(s)$$
$$T_m(s) = (Js + B)\omega_o(s)$$

A block diagram of the system is shown in Figure 3.11. Note that feedback arises naturally in this configuration due to the presence of an armature-controlled motor.

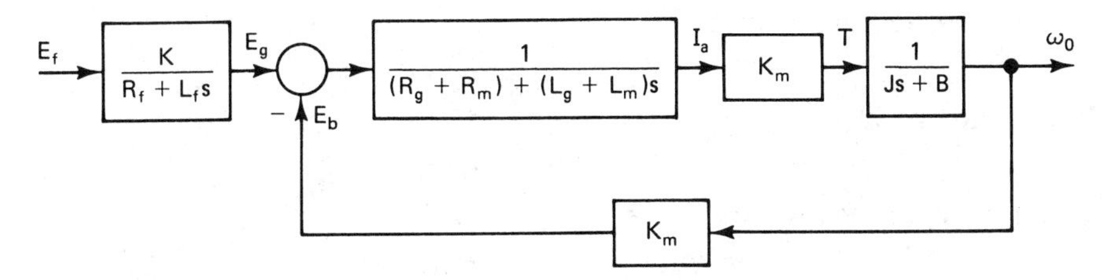


**Figure 3.11.** Block Diagram of the Ward – Leonard System

### Introducing Feedback

The time constant for devices with high inductance such as the armature circuit of the Ward–Leonard system is usually too long for design requirements. Feedback is used to decrease the time constant by inserting a resistor $R_a$ in series with the armature circuit as shown in Figure 3.12.

The following performance equations are used to construct the block diagram of Figure 3.13a. The generator's field circuit can be described by

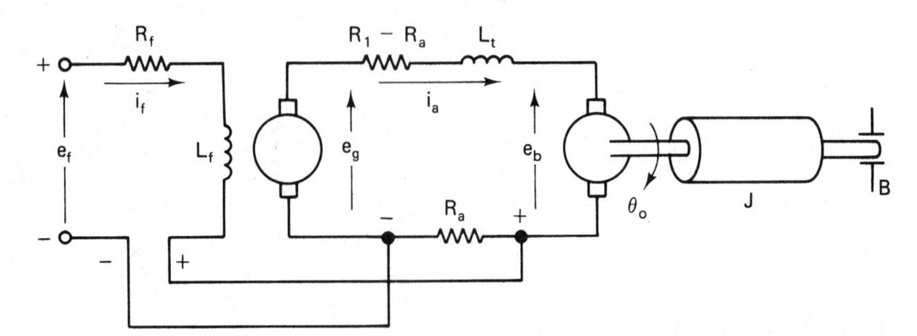


**Figure 3.12.** Block Diagram of the Ward–Leonard System with Feedback to Improve Time Constant

the loop equation

$$E_f(s) - R_a I_a(s) = (R_f + L_f s) I_f(s)$$

The generator's gain is described by

$$E_g(s) = K I_f(s)$$

The armature circuit is described by the loop equation rearranged in the form

$$I_a(s) = \frac{E_g(s) - E_b(s)}{R_t + L_t s}$$

Here

$$R_t = R_g + R_m + R_a$$
$$L_t = L_g + L_m$$

The back emf of the motor is

$$E_b(s) = K_m \omega_o(s)$$

The motor torque is

$$T_m(s) = K_m I_a(s)$$

In terms of speed we also have

$$T_m(s) = (Js + B)\omega_o(s)$$

The output angle $\theta_o(s)$ is related to the speed by

$$\theta_o(s) = \frac{\omega_o(s)}{s}$$

Steps in the block diagram reduction are shown in Figure 3.13b–d. The block identified as $G_A(s)$ is now worth commenting on. We have

$$G_A(s) = \frac{1}{R_t + L_t s + \dfrac{R_a K}{R_f(1 + \tau_f s)}} = \frac{1}{D_A(s)}$$

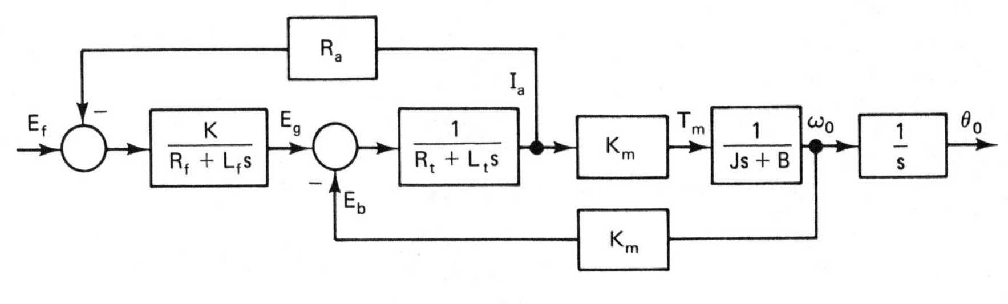


(a)

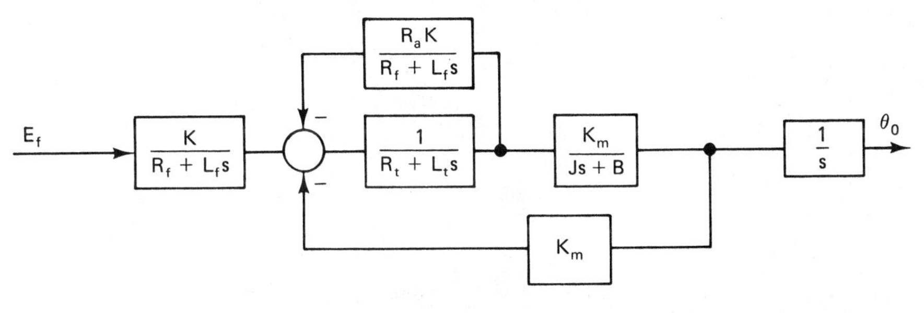


(b)

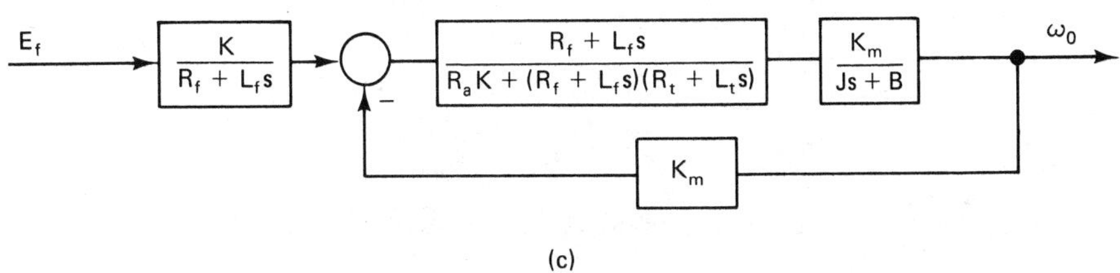


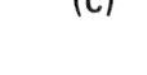

(c)

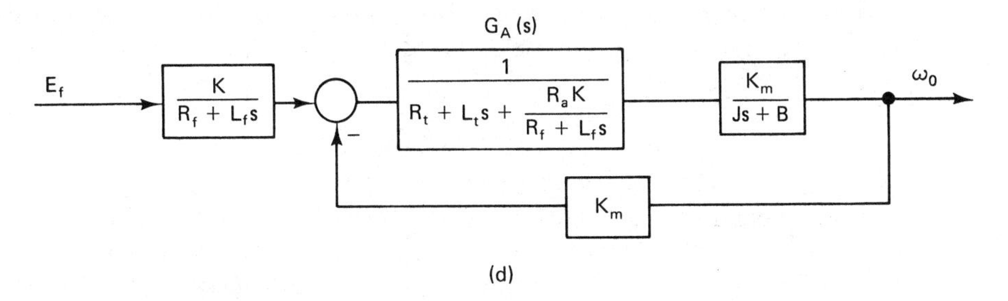


(d)

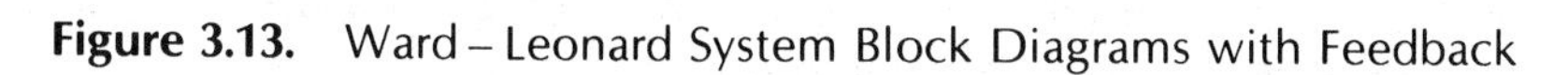

**Figure 3.13.** Ward – Leonard System Block Diagrams with Feedback

where

$$\tau_f = \frac{L_f}{R_f}$$

Consider the denominator $D_A(s)$ written approximately as

$$D_A(s) \simeq R_t + L_t s + \frac{R_a K}{R_f}(1 - \tau_f s)$$

$$= \left(R_t + \frac{R_a K}{R_f}\right) + \left(L_t - \frac{K L_f R_a}{R_f^2}\right) s$$

Let

$$R_A = R_t + \frac{R_a K}{R_f}$$

$$L_A = L_t - \frac{K L_f R_a}{R_f^2}$$

The denominator thus has a time constant $\tau_A$ given by

$$\tau_A = \frac{L_A}{R_A}$$

Clearly,

$$\tau_A < \frac{L_g + L_m}{R_g + R_m}$$

Thus a reduction in the time constant has resulted from feedback.

**Tachometer Feedback**

A dc generator with a permanent magnet supplying a constant air-gap flux is called a dc tachometer, as the induced voltage is proportional to its rotating speed:

$$e = K_t \dot{\theta}$$

The constant $K_t$ is referred to as the tachometer gain.

An ac tachometer is essentially a two-phase induction motor with one of the windings supplied with rated voltage. The second winding is separated by 90° in space from the first. When the rotor shaft is stationary, the voltage on the second winding is zero. With the rotor shaft rotating at $\dot{\theta}$, the output voltage is proportional to the speed.

In Figure 3.14 we show the Ward–Leonard system modified by the introduction of position (proportional) feedback $K_p$, tachometer feedback $K_t$, and current feedback $K_A$. This particular form of feedback is referred to as state-variable feedback.

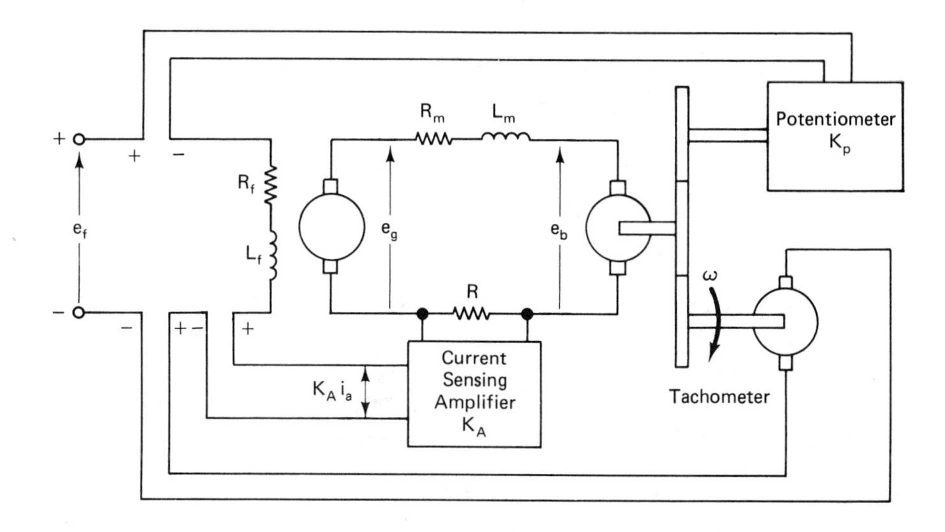


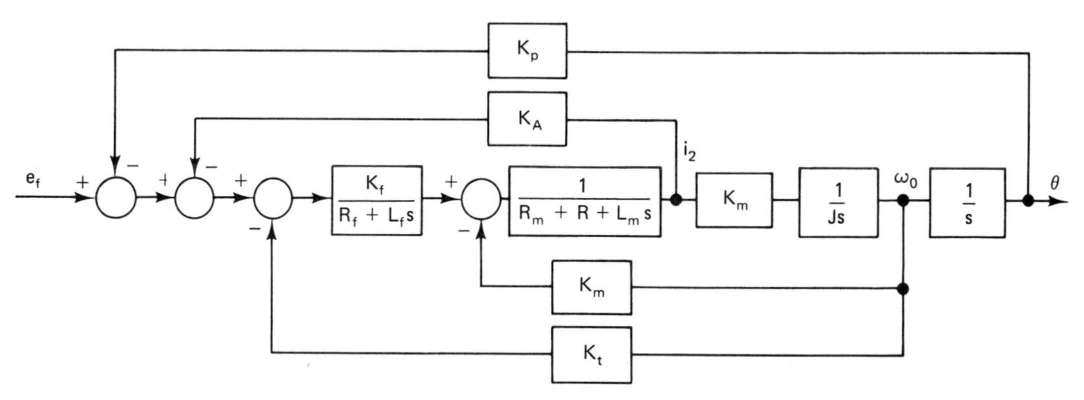


**Figure 3.14.** Ward – Leonard System with Further Feedback

## Phase-Locked Loop

Phase-locked loops (PLLs) are used widely in modern communications, radar, telemetry, command, time and frequency control, ranging, and instrumentation systems. A typical PLL, shown in Figure 3.15, consists of a multiplier (phase discriminator), an amplifier, and a filter (referred to as a loop filter) in the forward path and a voltage-controlled oscillator (VCO) in the feedback path. The input signal includes three components. The first represents the carrier $e_s(t)$, the second is the voltage in the modulation sidebands $e_d(t)$, and the third is the noise voltage $N(t)$. The input signal is multiplied by the output of the VCO and the resulting signal is fed to the amplifier and the loop filter. For our purposes we will neglect $e_d(t)$ and $N(t)$ and center our attention on the carrier signal $e_s(t)$.

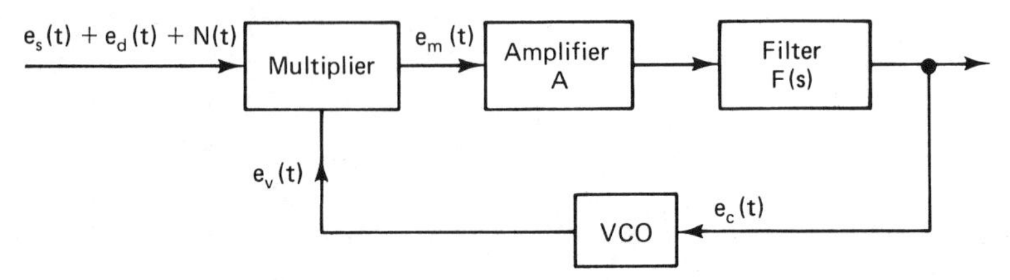


**Figure 3.15.** Typical Phase-Locked Loop

The carrier signal $e_s(t)$ is assumed to be of the form

$$e_s(t) = \sqrt{2}\,E_s \sin(\omega_s t + \theta_1)$$

The angle $\theta_1$ is the carrier phase and $\omega_s$ is the carrier radian frequency. The output of the VCO is denoted by $e_v(t)$ is modeled by

$$e_v(t) = E_o \cos(\omega_s + \theta_2)$$

The multiplier output $e_m(t)$ is given by

$$e_m(t) = K_1 e_s(t) e_v(t)$$

Thus the multiplier output is

$$e_m(t) = K_1 \sqrt{2}\,E_s E_o \sin(\omega_s t + \theta_1)\cos(\omega_s t + \theta_2)$$

Using the trignometric identity

$$\sin\alpha\cos\beta = \tfrac{1}{2}[\sin(\alpha + \beta) + \sin(\alpha - \beta)]$$

yields

$$e_m(t) = \frac{K_1 E_s E_o}{\sqrt{2}}[\sin(2\omega_s t + \theta_1 + \theta_2) + \sin(\theta_1 - \theta_2)]$$

An inherent assumption in designing a PLL is that the filter is of low-pass type. We thus have to consider only the low-frequency component of $e_m(t)$. As a result we have

$$e_m(t) = \frac{K_1 E_s E_o}{\sqrt{2}} \sin(\theta_1 - \theta_2)$$

Define the gain $K$ by

$$K = \frac{K_1 E_s E_o}{\sqrt{2}}$$

Thus

$$e_m(t) = K \sin(\theta_1 - \theta_2)$$

The phase difference $(\theta_1 - \theta_2)$ is assumed small; thus

$$e_m(t) = K(\theta_1 - \theta_2)$$

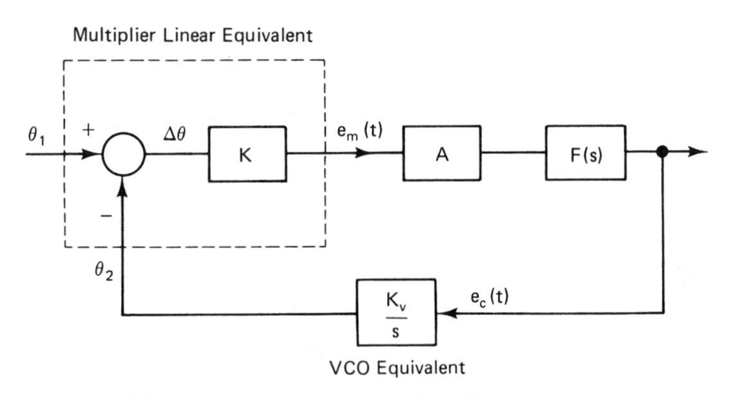


**Figure 3.16.** Linearized PLL Block Diagram

An equivalent block diagram representation of the multiplier is shown in the dashed portion of Figure 3.16. The carrier phase $\theta_1$ is compared with the VCO's phase $\theta_2$. The error $\Delta\theta$ is then amplified by the gain $K$.

The VCO's frequency is given by

$$\omega_{\text{osc}} = \omega_s + \dot{\theta}_2$$

This frequency is controlled by a control voltage $e_c(t)$. It is assumed that

$$\dot{\theta}_2 = K_v e_c$$

Thus the VCO's equivalent transfer function is given by

$$\frac{\theta_2(s)}{E_c(s)} = \frac{K_v}{s}$$

This completes the block diagram representation of the PLL.

## Model of Eye Movement

Techniques of system modeling and control can be used effectively in bioengineering studies. An example application is the model of horizontal eye movement, which involves only two muscles of each eye. The agonist muscle contracts when stimulated and acts in opposite direction to the antagonist muscle which is being lengthened. It should be noted that our presentation of the eye movement control system has to be brief, and we therefore omit many important details of the model development of this complicated system. The interested reader is referred to authoritative references such as Bahill's.*

*Bahill, A. T., "Bioengineering: Biomedical, Medical, and Clinical Engineering," Prentice-Hall, Englewood Cliffs, New Jersey, 1981.

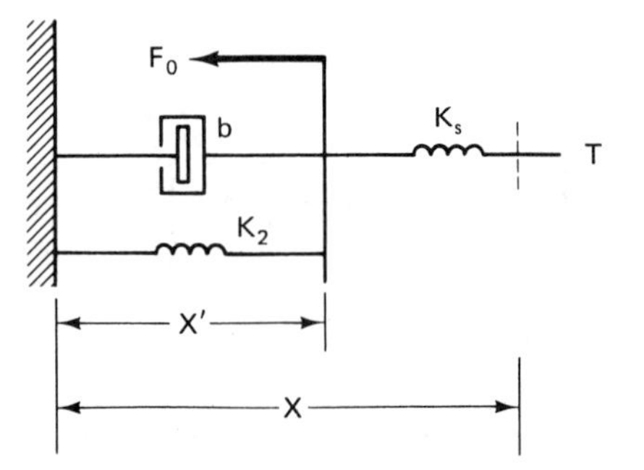


**Figure 3.17.** Muscle Model

A good model for the behavior of a muscle is central to the overall model of the eye movement. It is known that the most important property of a muscle is that it produces a force when it is stimulated. An ideal force generator is employed to model this fact. It is common practice to refer to this force generator $F_o$ using A. V. Hill's 1938 term "active-state tension generator." The force generator is shown in the model schematic of Figure 3.17.

The spring stiffness $K_2$ in the model represents the combined effect of two springs, $K_{\mathrm{PE}}$ and $K'$:

$$K_2 = K_{\mathrm{PE}} + K'$$

The stiffness $K_{\mathrm{PE}}$ accounts for the passive elasticity of a muscle that is known to stretch only by applying a force. The greater the force, the greater the extension. The second stiffness $K'$ accounts for the correction needed when dealing with extraocular muscles. The series elasticity $K_{\mathrm{SE}}$ accounts for the muscle's ability to change its length instantaneously with the abrupt change in load.

The dashpot $b$ in the model is a linear approximation to the force velocity characteristic of the muscle. The force exerted by the tendon on the mass is referred to as the muscle force or muscle tension $T$.

From the muscle model shown in Figure 3.17, we can write for the spring $K_s$,

$$T = K_s(X - X') \tag{3-1}$$

We can also write

$$T = F_o + K_2 X' + b\dot{X}' \tag{3-2}$$

The variable $X'$ can be eliminated, with the result

$$T = \frac{K_s}{K_s + K_2} F_o + \frac{K_s K_2}{K_s + K_2} X + \frac{bK_s}{K_s + K_2} \dot{X}' \tag{3-3}$$

Let us define

$$\alpha = \frac{K_s}{K_s + K_2}$$

$$B = \alpha b$$

As a result, the muscle model is expressed by the two equations

$$T = K_s(X - X') \tag{3-4}$$

$$T = \alpha F_o + \alpha K_2 X + B\dot{X}' \tag{3-5}$$

For the agonist muscle, the tension is denoted by $T_t$, while the active-state tension generator is denoted by $F_t$. The angles $\theta_1$ and $\theta_2$ are related to $X$ and $X'$ by

$$X = -\theta_1$$

$$X' = -\theta_2$$

As a result, we have the performance equations

$$T_t = K_s(\theta_2 - \theta_1) \tag{3-6}$$

$$T_t = \alpha F_t - \alpha K_2 \theta_1 - B_t \dot{\theta}_2 \tag{3-7}$$

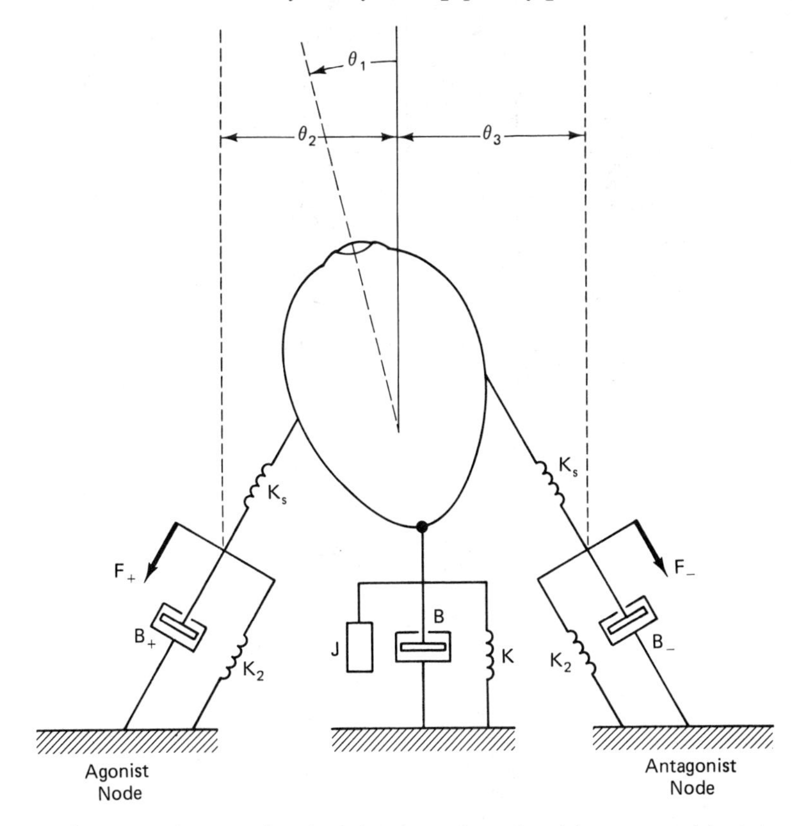


**Figure 3.18.** Mechanical Analog of an Eye Movement Model

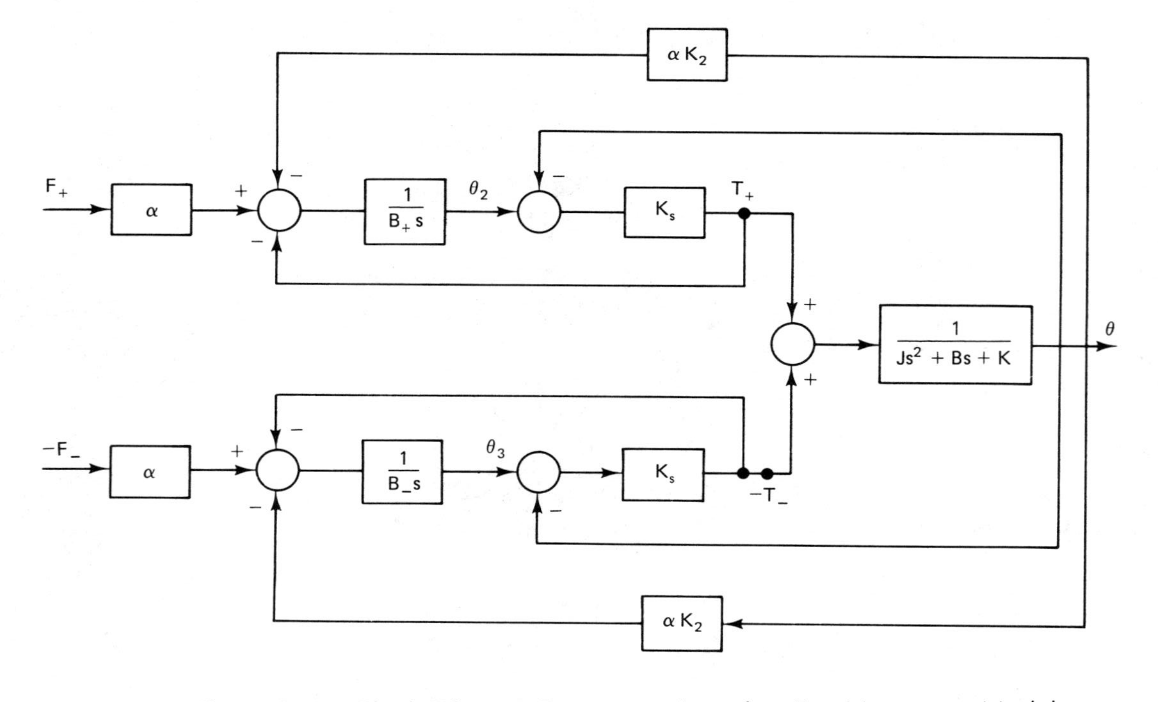

**Figure 3.19.** Block Diagram Representation of an Eye Movement Model

For the antagonist muscle, the tension is denoted by $T_-$ and its active-state generator is denoted by $F_-$. The angles $\theta_1$ and $\theta_3$ are related to $X$ and $X'$ by

$$X = \theta_1$$
$$X' = \theta_3$$

Thus

$$T_- = K_s(\theta_1 - \theta_3) \tag{3-8}$$
$$T_- = \alpha F_- + \alpha K_2 \theta_1 + B_- \dot{\theta}_3 \tag{3-9}$$

The globe and surrounding area are modeled by a system consisting of an inertia $J$, a viscous element $B$, and a spring stiffness (passive elasticity) $K$. The system is subject to an input tension which is the algebraic sum of the agonist tension $T_t$ and the antagonist tension $T_-$. Noting that $T_t$ and $T_-$ act in opposite directions, we write the globe's dynamic equation as

$$T_t - T_- = J\ddot{\theta}_1 + B\dot{\theta}_1 + K\theta_1 \tag{3-10}$$

Figure 3.18 shows a mechanical analog representation of Eqs. (3-6) through (3-10), which are used to construct the block diagram of Figure 3.19.

It is important to note that there are two parallel paths in the block diagram, each corresponding to one muscle action. In each path we note that three feedback loops arise naturally to produce the tensions $T_t$ and $T_-$ on the eye's globe.

## 3.6 System Sensitivity

Control systems are subject to effects that can cause changes in the performance characteristics of most components and hence in the overall performance of the system. These effects include a changing environment, aging, lack of knowledge about the exact values of component parameters, and other factors. The changes are accounted for as parameter changes in the system model represented by its transfer function or any other representation.

The system designer should be able to evaluate and compare the effects of parameter changes on the available design options. For a control system with an overall transfer function $T(s)$ subject to a change in a parameter $p$, a system sensitivity function $S_p^T$ can be defined in the following manner. In accordance with Taylor's expansion, retaining only the first two terms we have

$$T(p^{(0)} + \Delta p, s) = T(p^{(0)}, s) + (\Delta p) \left.\frac{\partial T}{\partial p}\right|_{p^{(0)}}$$

The expression above states that the transfer function $T(s)$ with the parameter $p$ perturbed from $p^{(0)}$ by an incremental change $\Delta p$ to $(p^{(0)} + \Delta p)$ is equal to the sum of its nominal value for $p^{(0)}$, and an incremental change given by

$$\Delta T = (\Delta p) \left.\frac{\partial T}{\partial p}\right|_{p^{(0)}}$$

As we are not interested in absolute values of changes, we consider the relative change

$$\frac{\Delta T}{T(p^{(0)})} = \frac{\Delta p}{p^{(0)}} \left[ \frac{\partial T / \partial p|_{p^{(0)}}}{T(p^{(0)})/p^{(0)}} \right]$$

The quantity in brackets is defined as the system sensitivity,

$$S_p^T(p^{(0)}, s) = \frac{\partial T / \partial p|_{p^{(0)}}}{T(p^{(0)})/p^{(0)}}$$

Thus the system sensitivity is a function of the Laplace operator $s$ and the nominal parameter value $p^{(0)}$. It is obtained as the ratio of the derivative of $T$ with respect to $p$ evaluated at the nominal value $p^{(0)}$ to the ratio of transfer function to the nominal value $p^{(0)}$ evaluated at $p^{(0)}$. In terms of the sensitivity function, we have

$$T(p^{(0)} + \Delta p, s) = T(p^{(0)}, s)\left(1 + \frac{\Delta p}{p^{(0)}} S\right)$$

It is clear that a small value for $S$ is desirable, as this means that the effect of parameter change is not appreciable on the transfer function. The relative change in $T$ is given by

$$\frac{\Delta T}{T_o} = \frac{\Delta p}{p^0} S$$

Assume that the relative change in $p$ is 0.05 and that $S$ is 2; this means that the corresponding relative change in the transfer function is 0.1, that is, double that of the parameter. If $S$ is 0.1, the relative change in the transfer function is 0.005 and the system can be deemed insensitive to changes in $p$.

A plausible interpretation of the sensitivity function is shown in Figure 3.20. Here we treat the relative change in the parameter $\Delta p / p$ as an input (cause) to a process modeled by the sensitivity function with the output (effect) being the relative change in the transfer function $\Delta T / T$.

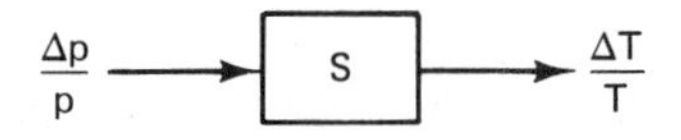


**Figure 3.20.** Interpreting the Sensitivity Function as a Gain Function

## Closed-Loop and Open-Loop Sensitivity: Variations in $G(s)$

Consider the closed-loop system shown in Figure 3.21b. The transfer function $T(s)$ is given by

$$T(s) = \frac{G(s)}{1 + G(s)H(s)} \tag{3-11}$$

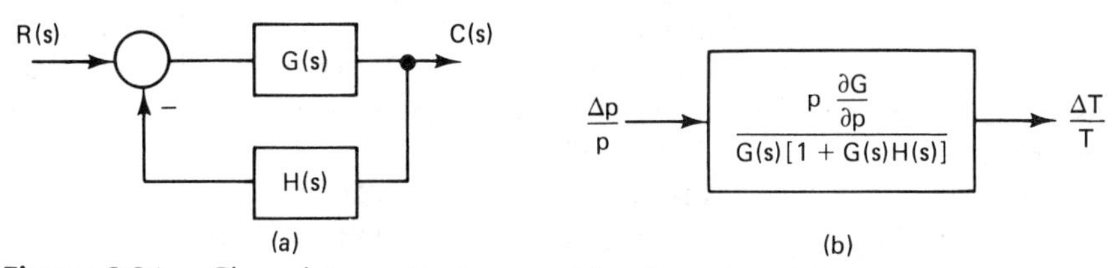


**Figure 3.21.** Closed-Loop System and Its Associated Sensitivity Function for $p$ in $G(s)$

Assume that the parameter $p$ is included in $G(s)$ and that $H(s)$ is not subject to parameter variation. We can then invoke the chain rule to obtain

$$\frac{\partial T}{\partial p} = \frac{\partial T}{\partial G}\frac{\partial G}{\partial p}$$

In the present case

$$\frac{\partial T}{\partial G} = \frac{1}{[1 + G(s)H(s)]^2} \tag{3-12}$$

As a result the sensitivity $S_p^T$ is obtained as

$$S_p^T = \frac{p}{G(s)[1 + G(s)H(s)]}\frac{\partial G}{\partial p} \tag{3-13}$$

Figure 3.21b shows the sensitivity function in block diagram form.

Assume now that we are dealing with an open-loop system such as the one depicted in Figure 3.22a. It is clear that for this system we can apply the result above with $H(s) = 0$. Thus the sensitivity of an open-loop system is

$$S_p^T = \frac{p}{G(s)}\frac{\partial G}{\partial p} \tag{3-14}$$

It can readily be seen that feedback reduces the system sensitivity to changes in the process open-loop function.

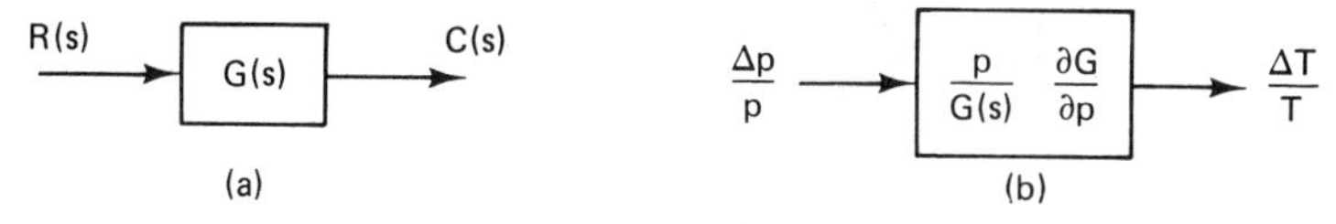


**Figure 3.22.** Open-Loop System and Its Associated Sensitivity Function

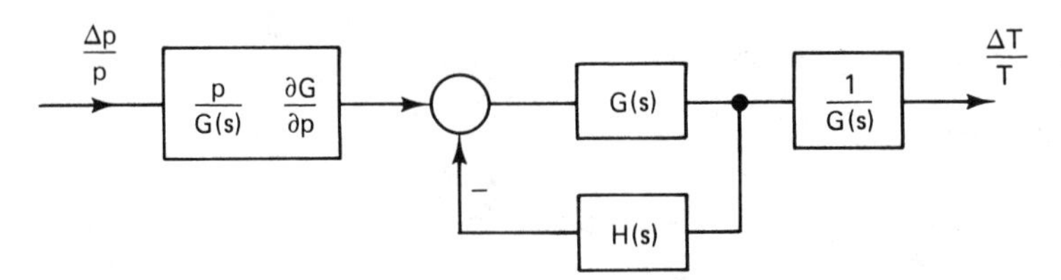


**Figure 3.23.** Sensitivity Function for a Closed-Loop System for the $p$ in $G(s)$ Reinterpreted

Yet another interpretation of the sensitivity of a closed-loop system in terms of its open-loop sensitivity when $p$ is in $G(s)$ is shown in Figure 3.23. As can be seen, the first block is the open-loop sensitivity, which is then operated on by the effect of the feedback.

Observe that the sensitivity function depends on the Laplace operator $s$. Two special cases arise. For $s = 0$, the resulting function is referred to as the steady-state or static sensitivity. For sinusoidal operation with $s = j\omega_f$, where $\omega_f$ is the forcing frequency, the function is referred to as *dynamic sensitivity.*

## Closed-Loop Sensitivity: Variations in *H(s)*

Let us assume that the parameter $p$ is part of the feedback function $H(s)$. In this case we invoke

$$\frac{\partial T}{\partial p} = \frac{\partial T}{\partial H}\frac{\partial H}{\partial p}$$

We now have

$$\frac{\partial T}{\partial H} = \frac{-G^2(s)}{[1 + G(s)H(s)]^2}$$

The sensitivity of the system is thus obtained as

$$S_p^T = -\frac{G(s)}{1 + G(s)H(s)}p\frac{\partial H}{\partial p} \tag{3-15}$$

Figure 3.24 shows a block diagram interpretation of the expression above.

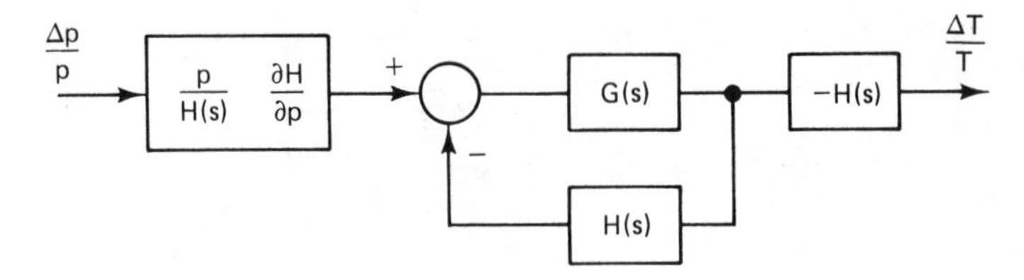


**Figure 3.24.** Sensitivity Function for a Closed-Loop System for $p$ in $H(s)$

EXAMPLE 3.1

Find the open-loop sensitivity of $G(s)$ with respect to changes in the delay parameter $\tau$ for

$$G(s) = e^{-\tau s}\tilde{G}(s)$$

SOLUTION

We have, by definition,

$$S_{\tau}^{G} = \frac{\tau}{G(s)} \frac{\partial G}{\partial \tau}$$

We first obtain

$$\begin{aligned}\frac{\partial G}{\partial \tau} &= -se^{-\tau s}\tilde{G}(s)\\ &= -sG(s)\end{aligned}$$

As a result, the required sensitivity is obtained as

$$S_{\tau}^{G} = -\tau s$$

## Some Open-Loop Sensitivity Expressions

The foregoing discussion pointed out that the sensitivity function for an open-loop system with transfer function $G(s)$ for changes in a parameter $p$ is given by

$$S_{p}^{G} = \frac{p}{G(s)} \frac{\partial G}{\partial p} \tag{3-16}$$

A number of cases are studied here, depending on the role of the parameter $p$ in $G(s)$.

We can assume in general that $G(s)$ is obtained in the following form:

$$G(s) = \frac{KN(s)}{D(s)} \tag{3-17}$$

The parameter $K$ is a gain factor. $N(s)$ and $D(s)$ are numerator and denominator polynomials which are assumed to be available in factored forms. The first case to be studied is where the parameter $p$ is included in $K$. Thus

$$K = \tilde{K}p$$

In this case we obtain

$$\begin{aligned}S_{p}^{G} &= \frac{p}{\tilde{K}p[N(s)/D(s)]}\left[\tilde{K}\frac{N(s)}{D(s)}\right]\\ &= 1\end{aligned} \tag{3-18}$$

The second case is for $p$ in the numerator expression such that

$$N(s) = (s + p)\tilde{N}(s) \tag{3-19}$$

Here we obtain

$$S_p^G = \frac{p}{N(s)} \frac{\partial N}{\partial p}$$

This reduces to

$$S_p^G = \frac{p}{s + p} \tag{3-20}$$

The third case is for $p$ in the denominator expression

$$D(s) = (s + p)\tilde{D}(s)$$

Here we obtain

$$S_p^G = \frac{-p}{D(s)} \frac{\partial D}{\partial p} \tag{3-21}$$

This reduces to

$$S_p^G = \frac{-p}{s + p} \tag{3-22}$$

In Figure 3.25 we show the results of the discussion above. Note that the static sensitivity ($s = 0$) is 1 for the parameter $p$ in the numerator and $-1$ for the parameter $p$ in the denominator.

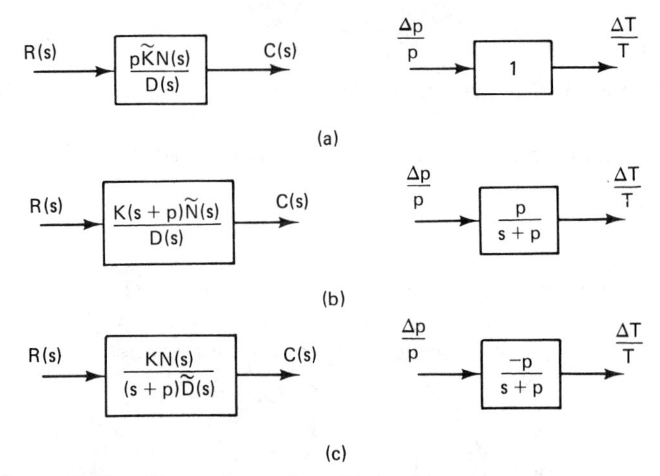


**Figure 3.25.** Open-Loop Sensitivity for:
(a) *p* in gain *K*
(b) *p* in Numerator Dynamics
(c) *p* in Denominator Dynamics

## Some Closed-Loop Sensitivity Expressions: Fixed $H(s)$

We have seen that the closed-loop sensitivity function is given by

$$S_p^T = \frac{p}{G(s)[1 + G(s)H(s)]} \frac{\partial G}{\partial p} \tag{3-23}$$

Note that the corresponding open-loop sensitivity is given by

$$S_p^G = \frac{p}{G(s)} \frac{\partial G}{\partial p}$$

Thus we can use the expression

$$S_p^T = S_p^G \frac{1}{1 + G(s)H(s)} \tag{3-24}$$

The three cases treated before can be dealt with here by use of the expression above. For $K = \tilde{K}p$, we conclude that the closed-loop sensitivity is

$$S_p^T = \frac{1}{1 + G(s)H(s)} \tag{3-25}$$

For

$$N(s) = (s + p)\tilde{N}(s)$$

we conclude that

$$S_p^T = \frac{p}{s + p} \frac{1}{1 + G(s)H(s)} \tag{3-26}$$

For

$$D(s) = (s + p)\tilde{D}(s)$$

we have

$$S_p^T = \frac{-p}{s + p} \frac{1}{1 + G(s)H(s)} \tag{3-27}$$

Figure 3.26 shows in block diagram form the results of the discussion above.

The sensitivity of the closed loop to variations in the parameter $p$ of the feedback function $H(s)$ when $p$ is included as part of the gain can be obtained easily if we assume that

$$H(s) = p\tilde{H}(s)$$

Thus

$$p \frac{\partial H}{\partial p} = H(s)$$

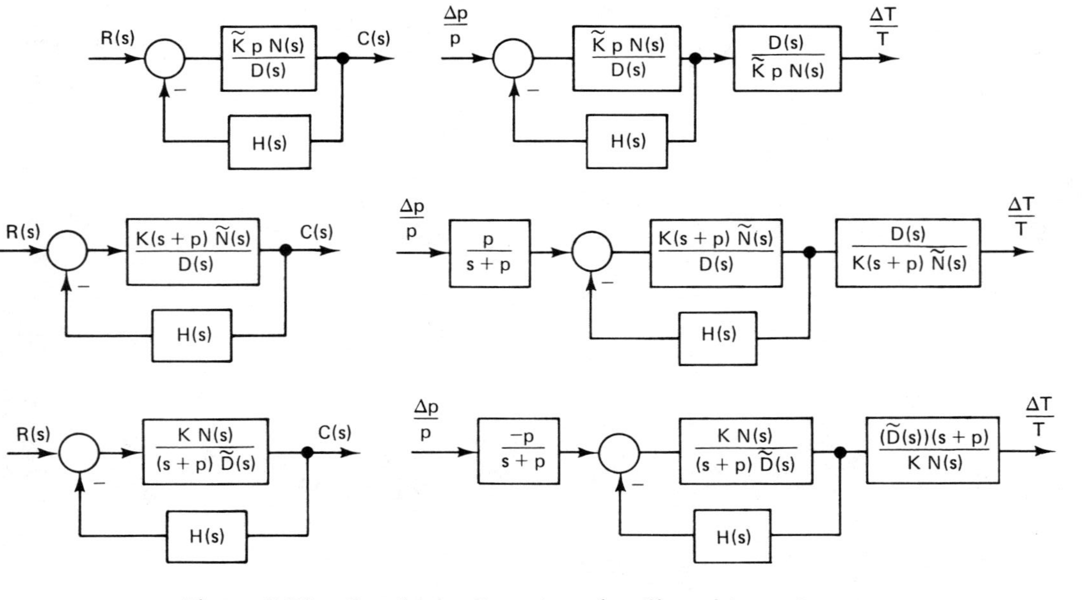

**Figure 3.26.** Sensitivity Functions for Closed-Loop Systems

As a result, the closed-loop sensitivity for variations in $H(s)$ is given by

$$S_p^T = \frac{-G(s)H(s)}{1 + G(s)H(s)} \tag{3-28}$$

Compare this with the corresponding expression if variations in $G(s)$ are considered such that

$$G(s) = p\tilde{G}(s) \tag{3-29}$$

The sensitivity in this case is

$$S_p^T = \frac{1}{1 + G(s)H(s)} \tag{3-30}$$

Note that variations in $H(s)$ result in higher sensitivity. This emphasizes the fact that feedback elements should be carefully selected from components that are insensitive to the environmental conditions.

EXAMPLE 3.2

A unity feedback control system has the transfer function

$$G(s) = \frac{K(s + a)}{s(s + b)(s^2 + cs + d)}$$

The nominal values of the parameters are

$$K = 20 \qquad a = 3 \qquad b = 6$$
$$c = 4 \qquad d = 8$$

Determine the sensitivity of the closed-loop transfer function for changes in $K$, $a$, $b$, $c$, and $d$ taken individually.

SOLUTION

For changes in $K$, we have, by Eq. (3-25),

$$\begin{aligned} S_K^T &= \frac{1}{1 + G(s)H(s)} \\ &= \frac{s(s + b)(s^2 + cs + d)}{K(s + a) + s(s + b)(s^2 + cs + d)} \\ &= \frac{s(s + 6)(s^2 + 4s + 8)}{20(s + 6) + s(s + 6)(s^2 + 4s + 8)} \end{aligned}$$

For changes in $a$, we have, by Eq. (3-26),

$$\begin{aligned} S_a^T &= \frac{a}{s + a}\,\frac{1}{1 + G(s)H(s)} \\ &= \frac{a}{s + a} S_K^T \\ &= \frac{3}{s + 3} S_K^T \end{aligned}$$

For changes in $b$, we have, by Eq. (3-27),

$$S_b^T = \frac{-b}{s+b} \frac{1}{1+G(s)H(s)}$$
$$= \frac{-6}{s+6} S_K^T$$

For changes in $c$ and $d$, we first obtain the open-loop sensitivity using Eq. (3-21):

$$S_p^G = \frac{-p}{D(s)} \frac{\partial D}{\partial p}$$

Let us write

$$D(s) = (s^2 + cs + d)\tilde{D}(s)$$

For variations in $c$, we have

$$\frac{\partial D}{\partial p} = \frac{\partial D}{\partial c} = s\tilde{D}(s)$$

Thus

$$S_c^G = \frac{-cs}{s^2 + cs + d}$$

For variations in $d$, we have

$$\frac{\partial D}{\partial d} = \tilde{D}(s)$$

Thus

$$S_d^G = \frac{-d}{s^2 + cs + d}$$

Thus the closed-loop sensitivity expressions are

$$S_c^T = \frac{-cs}{s^2 + cs + d} S_K^T$$
$$= \frac{-4s}{s^2 + 4s + 8} S_K^T$$
$$S_d^T = \frac{-d}{s^2 + cs + d} S_K^T$$
$$= \frac{-8}{s^2 + 4s + 8} S_K^T$$

# Some Solved Problems

PROBLEM 3A-1

Find the open-loop sensitivity of the systems shown in block diagram form in Figure 3.27 for changes in parameter $p$. Assume first that $p$ is contained in $G_1$, then that $p$ is contained in $G_2$.

SOLUTION

We use the main sensitivity formula

$$S_p^G = \frac{p}{G}\frac{\partial G}{\partial p}$$

For the cascade of elements shown in Figure 3.27a, we have

$$G = G_1G_2$$

Thus, if $p$ is contained in $G_1$, we have

$$\frac{\partial G}{\partial p} = G_2\frac{\partial G_1}{\partial p}$$

As a result,

$$S_{p,G_1}^G = \frac{p}{G_1}\frac{\partial G_1}{\partial p}$$

Similarily, for $p$ contained in $G_2$, we have

$$S_{p,G_2}^G = \frac{p}{G_2}\frac{\partial G_2}{\partial p}$$

For the parallel combination of Figure 3.27b, we have

$$G = G_1 + G_2$$

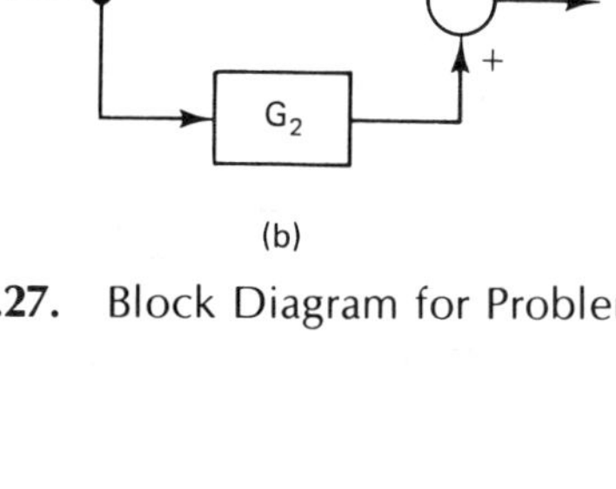

**Figure 3.27.** Block Diagram for Problem 3A-1

Thus, if $p$ is in $G_1$, we have

$$\frac{\partial G}{\partial p} = \frac{\partial G_1}{\partial p}$$

As a result,

$$S^G_{p,G_1} = \frac{p}{G_1 + G_2} \frac{\partial G_1}{\partial p}$$

Similarly, for $p$ in $G_2$, we have

$$S^G_{p,G_2} = \frac{p}{G_1 + G_2} \frac{\partial G_2}{\partial p}$$

PROBLEM 3A-2

Figure 3.28 shows a multiple-loop feedback system in block diagram form. Find expressions for the closed-loop sensitivity to changes in the parameter $p$ for the four cases when $p$ is in $G_1$, $p$ is in $G_2$, $p$ is in $H_1$, and $p$ is in $H_2$.

SOLUTION

Let us consider first the case for which $p$ is included in $G_1$. Denote the reduced feedback loop made of $G_2$ and $H_2$ by $\tilde{G}_2$; we thus have

$$\tilde{G}_2 = \frac{G_2}{1 + G_2 H_2}$$

The open-loop sensitivity for changes in $p$ of $G_1$ is given by

$$S^G_{p,G_1} = \frac{p}{G_1} \frac{\partial G_1}{\partial p}$$

Invoking the closed-loop sensitivity rule, we have

$$S^T_{p,G_1} = S^G_{p,G_1} \frac{1}{1 + G_1 \tilde{G}_2 H_1}$$

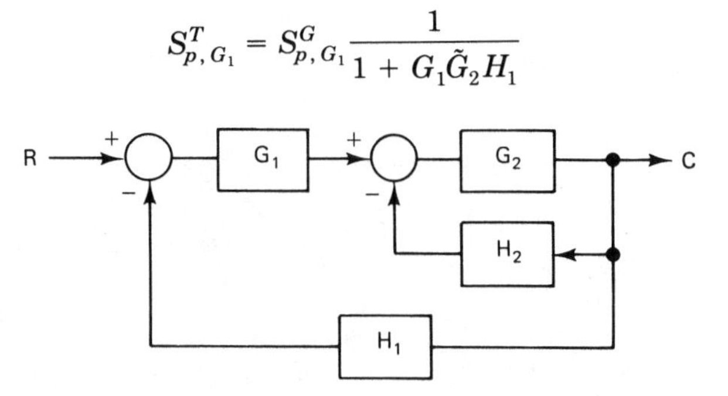


**Figure 3.28.** Block Diagram for Problem 3A-2

Simplifying, we obtain

$$S^T_{p,G_1} = \frac{p}{G_1}\frac{\partial G_1}{\partial p}\frac{1 + G_2H_2}{1 + G_2(H_2 + G_1H_1)}$$

To deal with variations in $p$ included in $G_2$, the simplest means is to use the modified block diagram shown in Figure 3.29. The open-loop sensitivity of $G_1G_2$ is then

$$S^G_{p,G_2} = \frac{p}{G_2}\frac{\partial G_2}{\partial p}$$

The closed-loop sensitivity is thus

$$S^T_{p,G_2} = S^G_{p,G_2}\frac{1}{1 + G_1G_2(H_1 + H_2/G_1)}$$

Simplified, this turns out to be

$$S^T_{p,G_2} = \frac{p}{G_2}\frac{\partial G_2}{\partial p}\frac{1}{1 + G_2(H_2 + G_1H_1)}$$

The situation with $p$ in either $H_1$ or $H_2$ can be handled using the reduced block diagram shown in Figure 3.29. The sensitivity function with respect to variations in $p$ in $H_1$ is

$$S^{\tilde{H}}_{p,H_1} = \frac{p}{H_1 + H_2/G_1}\frac{\partial H_1}{\partial p}$$

The closed-loop sensitivity is then

$$S^T_{p,H_1} = S^{\tilde{H}}_{p,H_1}\frac{-\hat{G}(s)\tilde{H}(s)}{1 + \hat{G}(s)\tilde{H}(s)}$$

where

$$\hat{G}(s) = G_1(s)G_2(s)$$
$$\tilde{H}(s) = H_1(s) + \frac{H_2(s)}{G_1(s)}$$

This reduces to

$$S^T_{p,H_1} = -p\frac{\partial H_1}{\partial p}\frac{G_1(s)G_2(s)}{1 + G_2(s)[H_2(s) + G_1(s)H_1(s)]}$$

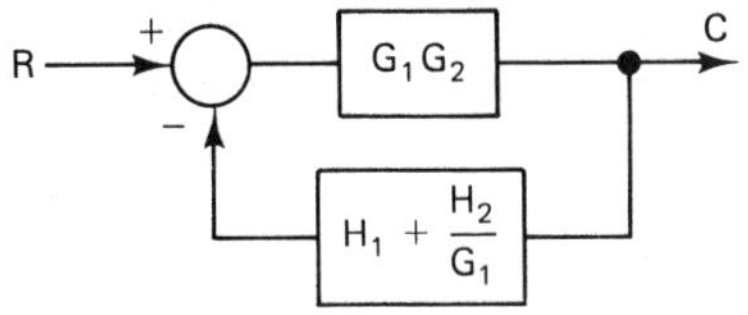

R
+
−
$G_1G_2$
C
$H_1 + \frac{H_2}{G_1}$

**Figure 3.29.** Reduced Block Diagram for Problem 3A-2

A similar argument reveals that

$$S_{p,H_2}^{T} = -p\frac{\partial H_2}{\partial p}\frac{G_2(s)}{1 + G_2(s)[H_2(s) + G_1(s)H_1(s)]}$$

PROBLEM 3A-3

Consider the control system shown in Figure 3.30. It is required to find and compare the open-loop ($h = 0$) and closed-loop system's sensitivity to changes in $K$, $\omega_n$, and $\zeta$ as well as the noise effects.

SOLUTION

The open-loop function is given by

$$G(s) = \frac{K\omega_n^2}{s^2 + 2\zeta\omega_n s + \omega_n^2}$$

The open-loop sensitivity with respect to $K$ is

$$S_K^G = 1$$

To obtain the open-loop sensitivity with respect to $\omega_n$, we have

$$\frac{\partial G}{\partial \omega_n} = \frac{2K\omega_n(s^2 + 2\zeta\omega_n s + \omega_n^2) - K\omega_n^2(2\zeta s + 2\omega_n)}{(s^2 + 2\zeta\omega_n s + \omega_n^2)^2}$$

$$= \frac{2K\omega_n s(s + \zeta\omega_n)}{(s^2 + 2\zeta\omega_n s + \omega_n^2)^2}$$

We now obtain

$$S_{\omega_n}^G = \frac{\omega_n}{K\omega_n^2}\frac{2K\omega_n s(s + \zeta\omega_n)}{s^2 + 2\zeta\omega_n s + \omega_n^2}$$

This reduces to

$$S_{\omega_n}^G = \frac{2s(s + \zeta\omega_n)}{s^2 + 2\zeta\omega_n s + \omega_n^2}$$

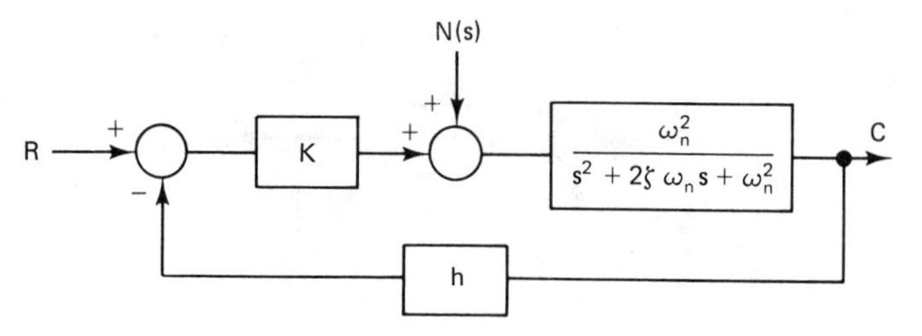


**Figure 3.30.** Block Diagram for Problem 3A-3

To obtain the open-loop sensitivity with respect to $\zeta$, we have

$$\frac{\partial G}{\partial \zeta} = \frac{-2K\omega_n^3 s}{\left(s^2 + 2\zeta\omega_n s + \omega_n^2\right)^2}$$

Thus

$$S_\zeta^G = \frac{-2\zeta\omega_n s}{s^2 + 2\zeta\omega_n s + \omega_n^2}$$

The closed-loop sensitivities are obtained using

$$S_p^T = S_p^G \frac{1}{1 + G(s)H(s)}$$

Thus for our system

$$S_p^T = S_p^G \frac{s^2 + 2\zeta\omega_n s + \omega_n^2}{s^2 + 2\zeta\omega_n s + (1 + Kh)\omega_n^2}$$

Using the obtained open-loop sensitivities, we get

$$S_K^T = \frac{s^2 + 2\zeta\omega_n s + \omega_n^2}{s^2 + 2\zeta\omega_n s + (1 + Kh)\omega_n^2}$$

$$S_{\omega_n}^T = \frac{2s(s + \zeta\omega_n)}{s^2 + 2\zeta\omega_n s + (1 + Kh)\omega_n^2}$$

$$S_\zeta^T = \frac{-2\zeta\omega_n s}{s^2 + 2\zeta\omega_n s + (1 + Kh)\omega_n^2}$$

Note that the closed-loop sensitivities are smaller than the corresponding open-loop sensitivities.

The noise component in the output with feedback is given by

$$C_{N_2}(s) = \frac{\omega_n^2}{s^2 + 2\zeta\omega_n s + (1 + Kh)\omega_n^2} N(s)$$

Without feedback $h = 0$, we get

$$C_{N_1}(s) = \frac{\omega_n^2}{s^2 + 2\zeta\omega_n s + \omega_n^2} N(s)$$

Thus

$$\frac{C_{N_2}(s)}{C_{N_1}(s)} = \frac{s^2 + 2\zeta\omega_n s + \omega_n^2}{s^2 + 2\zeta\omega_n s + (1 + Kh)\omega_n^2} < 1$$

PROBLEM 3A-4

A position-control system is shown in Figure 3.31a. To improve its performance, proportional plus derivative feedback is introduced as shown

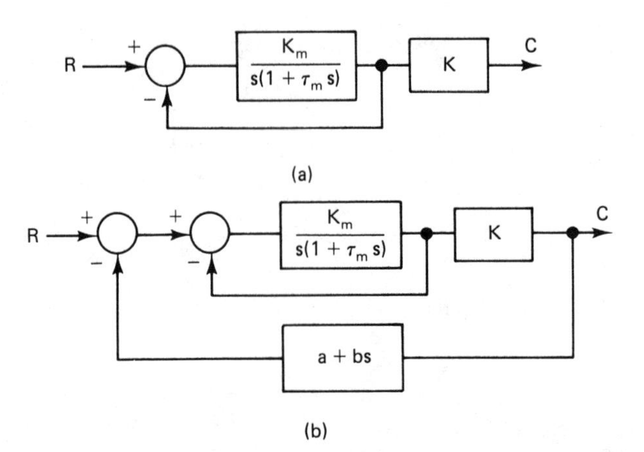


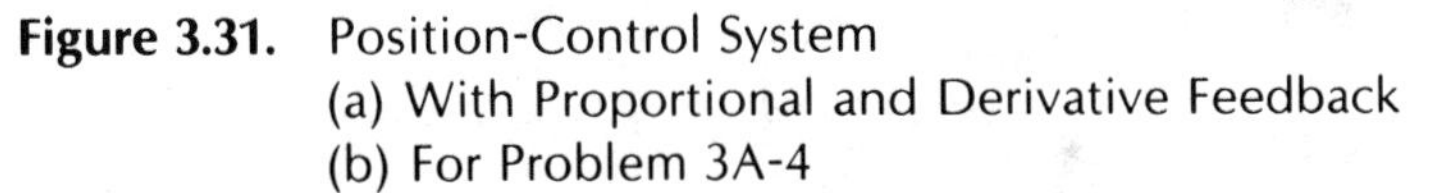

**Figure 3.31.** Position-Control System
(a) With Proportional and Derivative Feedback
(b) For Problem 3A-4

in Figure 3.31b. Compare the sensitivity of both schemes for variations in $K_m$, $\tau_m$, and $K$.

SOLUTION

The simple procedure of reducing the block diagrams to the forms shown in Figure 3.32 is employed. Note that the only difference between the two schemes is in the feedback function $H(s)$. For scheme (a) we have

$$H_a(s) = \frac{1}{K}$$

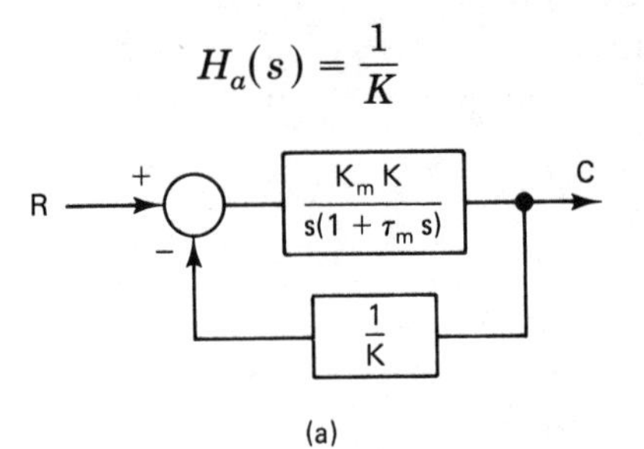


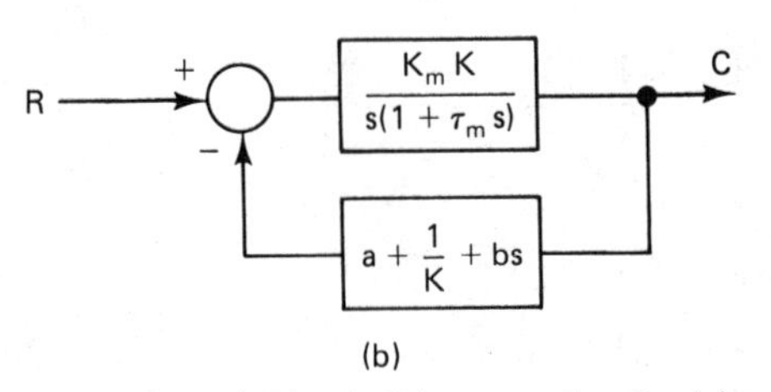

**Figure 3.32.** Reduced Block Diagram for Problem 3A-4

For scheme (b) we have

$$H_b(s) = a + bs + \frac{1}{K}$$

The two schemes have the same open-loop function $G(s)$. Recall that the closed-loop sensitivity function is given by

$$S_p^T = S_p^G \frac{1}{1 + G(s)H(s)}$$

Thus

$$\frac{S_{p_a}^T}{S_{p_b}^T} = \frac{1 + G(s)H_b(s)}{1 + G(s)H_a(s)}$$

But

$$G(s) = \frac{K_m K}{s(1 + \tau_m s)}$$

As a result,

$$\frac{S_{p_a}^T}{S_{p_b}^T} = \frac{K_m K(a + bs + 1/K) + s(1 + \tau_m s)}{K_m K(1/K) + s(1 + \tau_m s)}$$

$$= 1 + \frac{K_m K(a + bs)}{K_m + s(1 + \tau_m s)}$$

Clearly,

$$S_{p_a}^T > S_{p_b}^T$$

The analysis above applies for $p = K_m$ and $p = \tau_m$.

If we desire to find the sensitivity expressions, we note that for $p = K_m$ we have

$$S_{K_m}^G = 1$$

Thus

$$S_{K_{m_a}}^T = \frac{1}{1 + H_a(s)G(s)}$$

$$= \frac{s(1 + \tau_m s)}{K_m + s(1 + \tau_m s)}$$

$$S_{K_{m_b}}^T = \frac{1}{1 + H_b(s)G(s)}$$

$$= \frac{s(1 + \tau_m s)}{K_m K(a + 1/K + bs) + s(1 + \tau_m s)}$$

The open-loop sensitivity of $G(s)$ with respect to $\tau_m$ is found as follows:

$$S_{\tau_m}^{G} = \frac{\tau_m}{G(s)}\frac{\partial G}{\partial \tau_m}$$

$$= \frac{\tau_m}{G(s)}\frac{K_mK}{s}\frac{\partial}{\partial \tau_m}\frac{1}{1+\tau_m s}$$

$$= \frac{K_mK\tau_m}{sG(s)}\frac{-s}{(1+\tau_m s)^2}$$

$$= \frac{-\tau_m s}{1+\tau_m s}$$

Now the closed-loop sensitivities are obtained as

$$S_{\tau_{m_a}}^{T} = \frac{-\tau_m s}{1+\tau_m s}\frac{s(1+\tau_m s)}{s(1+\tau_m s)+K_m}$$

$$= \frac{-\tau_m s^2}{s(1+\tau_m s)+K_m}$$

$$S_{\tau_{m_b}}^{T} = \frac{-\tau_m s}{1+\tau_m s}\frac{s(1+\tau_m s)}{K_mK(a+1/K+bs)+s(1+\tau_m s)}$$

$$= \frac{-\tau_m s^2}{s(1+\tau_m s)+K_m+K_mK(a+bs)}$$

To obtain the sensitivity with respect to $K$, we resort to the reduced forms of Figure 3.33. Clearly, for form (a) we have

$$S_{K_a}^{T} = 1$$

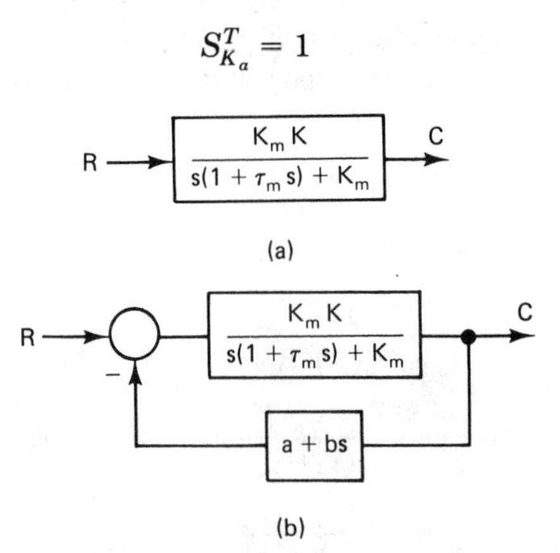


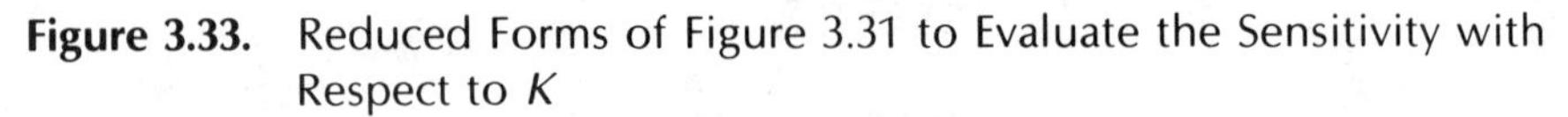

**Figure 3.33.** Reduced Forms of Figure 3.31 to Evaluate the Sensitivity with Respect to $K$

For form (b) we have

$$S_{K_b}^{T} = \cfrac{1}{1 + \cfrac{K_m K(a + bs)}{K_m + s(1 + \tau_m s)}}$$

$$= \frac{K_m + s(1 + \tau_m s)}{K_m[1 + K(a + bs)] + s(1 + \tau_m s)}$$

Again, the same sensitivity ratio is obtained.

# *Problems*

PROBLEM 3B-1

The block diagram of Figure 3.34 shows an automatic braking system used to regulate a class of high-speed trains. The reference input is the distance to stopping point and the actual train position. The gain $K_A$ represents the amplifying stage and $G(s)$ represents an idealization of the train dynamics. The disturbance input is denoted by $N(s)$. Feedback control is employed with

$$H(s) = 1 + h_1 s + h_2 s^2$$

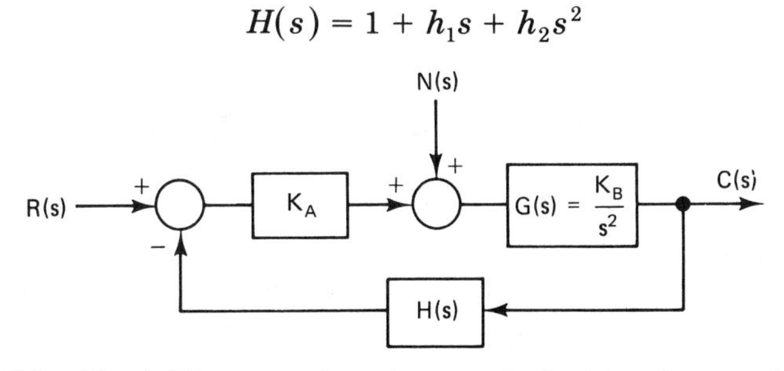


**Figure 3.34.** Block Diagram of an Automatic Braking System for Problem 3B-1

Find the output $C(s)$ in terms of the reference input $R(s)$ and disturbance $N(s)$ and hence show that $K_A$ must be made large to reduce the effect of disturbance in the output.

PROBLEM 3B-2

Assume that $N(s) = 0$ for the system of Problem 3B-1. Show that the closed-loop transfer function can be expressed as

$$\frac{C(s)}{R(s)} = \frac{\omega_n^2}{s^2 + 2\xi\omega_n s + \omega_n^2}$$

where

$$\omega_n^2 = \frac{K_A K_B}{1 + K_A K_B h_2}$$

$$\xi = \tfrac{1}{2} h_1 \omega_n$$

Find $\xi$ and $\omega_n$ for the following nominal parameter values:

$$K_A K_B = 10$$

$$h_1 = h_2 = 0.1$$

PROBLEM 3B-3

A compensating network with transfer function $G_c(s)$ is inserted in the forward loop of the system of Problem 3B-1 and unity feedback is employed as shown in Figure 3.35a. Find $G_c(s)$ so that the closed-loop transfer function remains unchanged [i.e., the block diagram of part (a) is equivalent to that of part (b)].

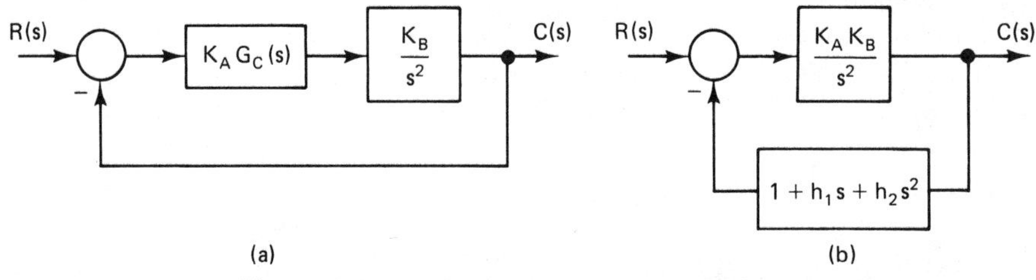


**Figure 3.35.** Block Diagrams for Problem 3B-3

PROBLEM 3B-4

For the system of Problem 3B-2, find the sensitivity of the closed-loop transfer function to changes in the parameters $K_A$, $K_B$, $h_1$, and $h_2$.

PROBLEM 3B-5

Figure 3.36 shows a closed-loop control scheme for an aircraft pitch-rate control mechanism. Find the closed-loop transfer function and its sensitivity to changes in the parameters $M$, $b$, $\xi$, and $\omega_n$. Use the following nominal parameter values:

| | |
|---|---|
| $M = 0.7$ | $\omega_n^2 = 0.65$ |
| $b = 0.008$ | $2\xi\omega_n = 0.0165$ |
| $a = 6.67$ | $h_0 = 1$ |
| $K_1 = 800$ | $h_1 = -0.05$ |
| $K_2 = 0.033$ | |

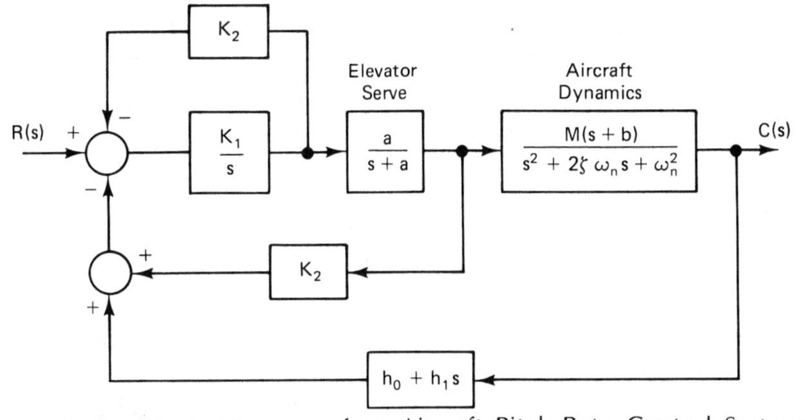


**Figure 3.36.** Block Diagram of an Aircraft Pitch-Rate Control System for Problem 3B-5

PROBLEM 3B-6

A block diagram of a pitch-attitude-control system for a missile is shown in simplified form in Figure 3.37. Find the closed-loop transfer function and its sensitivity to changes in the parameters $K_2$, $a$, $h_1$, and $b$. Use the following nominal values:

$$a = 15 \qquad K_2 = 6.45$$
$$b = 2.14 \qquad h_1 = 0.33$$

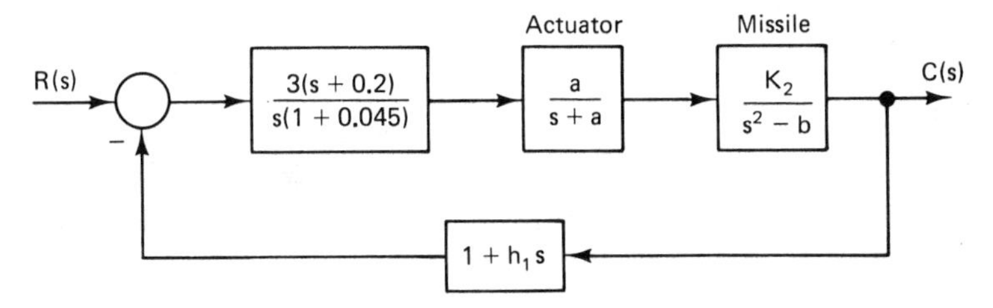


**Figure 3.37.** Block Diagram for Problem 3B-6

PROBLEM 3B-7

An automatic voltage-regulating system is shown in the block diagram of Figure 3.38. Find the expressions for the sensitivity of the closed-loop transfer function to changes in the time constants $\tau_a$, $\tau_g$, and $\tau_v$.

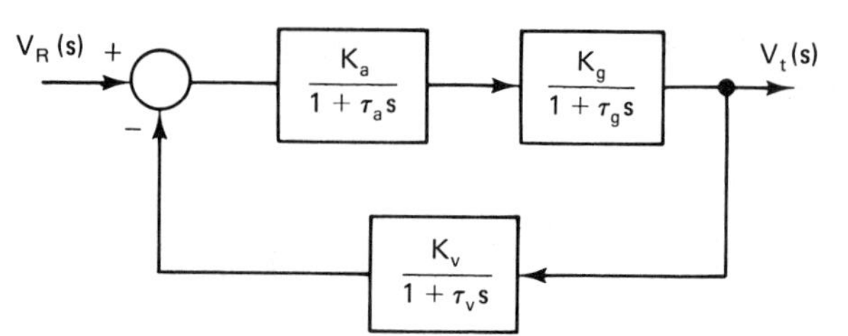


**Figure 3.38.** Block Diagram for Problem 3B-7

PROBLEM 3B-8

Figure 3.39 shows a simplified block diagram for a control system for a large space telescope. Find the closed-loop transfer function and its sensitivity to changes in the gains $K_1$, $K_2$, and $K_3$.

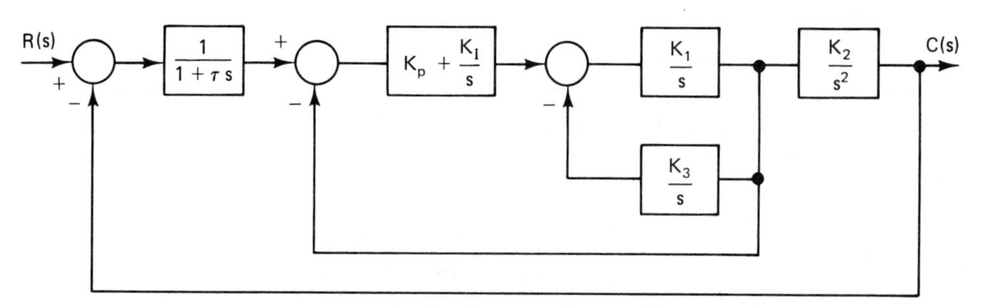


**Figure 3.39.** Block Diagram for Problem 3B-8

PROBLEM 3B-9

The signal flow graph of Figure 3.40 represents the position-control mechanism for a printwheel system. Obtain the system transfer function and its sensitivity to changes in the inductance $L_a$.

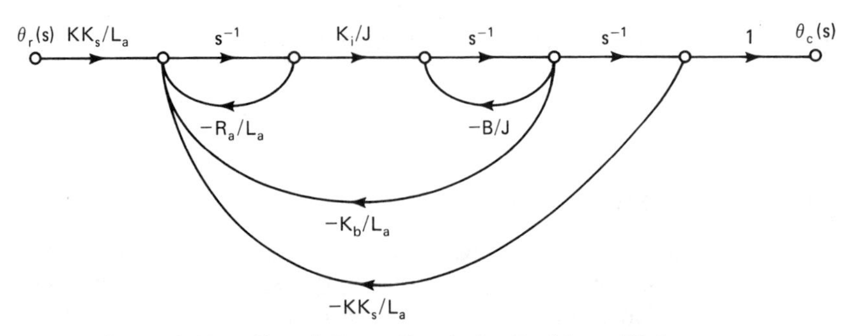


**Figure 3.40.** Signal Flow Graph for Problem 3B-9

PROBLEM 3B-10

A block diagram of a type ST1 potential-source controlled-rectifier exciter is shown in Figure 3.41. Find the closed-loop sensitivity to variations in $K_A$, $\tau_A$, and $\tau_p$.

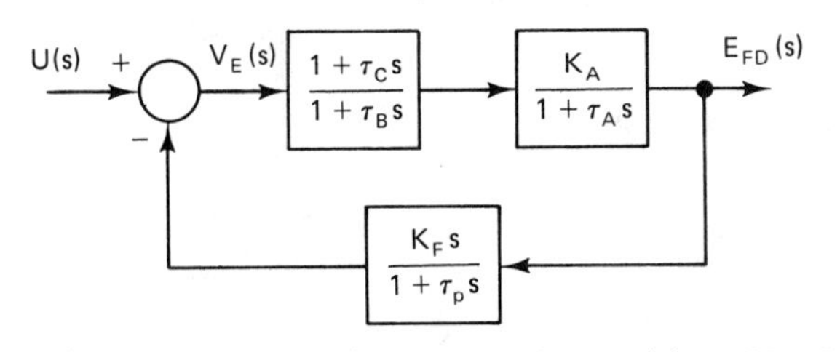


**Figure 3.41.** Block Diagram for Problem 3B-10

PROBLEM 3B-11

A hydrogovernor-turbine model is shown in the block diagram of Figure 3.42. Find the closed-loop transfer function and its sensitivity to changes in $T_w$, $T_d$, and $T_g$.

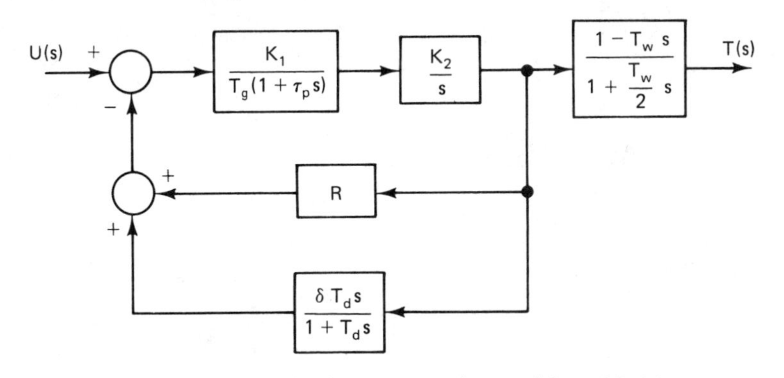


**Figure 3.42.** Block Diagram for Problem 3B-11

# Chapter 4

# STATE-SPACE REPRESENTATION OF CONTROL SYSTEMS

## 4.1 Introduction

An alternative method to describe dynamic systems is the state-space representation. The method provides a means of describing a dynamic system using a compact standard notation. This enables us to apply general control system techniques to various systems without referring to the detailed or concrete nature of the problem. The concept of a *system state* $X$

was introduced by R. E. Kalman in 1963. Kalman's definition is as follows:

> *State.* The state of a system is a mathematical structure consisting of a set of $n$ time-dependent variables $x_1(t), x_2(t), \ldots, x_i(t), \ldots, x_n(t)$, referred to as *state variables*. The system inputs $u_j(t)$ together with the initial conditions of the state variables $x_1(0), x_2(0), \ldots, x_i(0), \ldots, x_n(0)$ are sufficient to uniquely define the system's future response.

We will consider two examples to highlight the concept.

EXAMPLE 4.1

Consider the circuit shown in Figure 4.1. We can write the following relations between the voltage across each element and the current through $i(t)$:

$$v_R(t) = Ri(t) \tag{4-1}$$

$$v_L(t) = L\frac{di}{dt} \tag{4-2}$$

$$v_C(t) = \frac{1}{C}\int i\,dt \tag{4-3}$$

As a result, the applied voltage $v(t)$ is given by

$$v = Ri + L\frac{di}{dt} + \frac{1}{C}\int i\,dt \tag{4-4}$$

Differentiating both sides, we obtain

$$\frac{dv}{dt} = L\frac{d^2i}{dt^2} + R\frac{di}{dt} + \frac{i}{C} \tag{4-5}$$

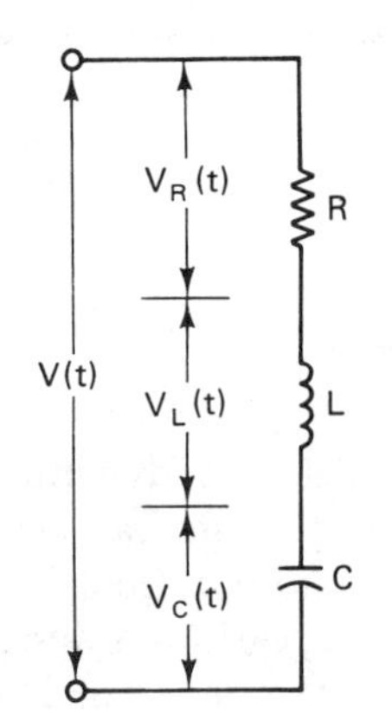


**Figure 4.1.** *RLC* Network

This is a second-order differential equation with $dv/dt$ treated as the input. In terms of the Laplace transform, assuming zero initial conditions, we can write the transfer function as

$$\frac{I(s)}{V(s)} = \frac{Cs}{LCs^2 + RCs + 1} \tag{4-6}$$

Note that the output in this case is taken as the current as shown in Figure 4.2.

We can realize the transfer function of Eq. (4-6) by the block diagrams of Figure 4.2. Note that the second block diagram includes the capacitor voltage $V_C(s)$ as a variable, and in effect what we have done is to replace Eq. (4-6) by the following transfer functions:

$$\frac{I(s)}{V_C(s)} = Cs \tag{4-7}$$

$$\frac{V_C(s)}{V(s)} = \frac{1}{LCs^2 + RCs + 1} \tag{4-8}$$

Equation (4-7) can be rearranged to the form

$$sV_C(s) = \frac{1}{C}I(s) \tag{4-9}$$

We can also write Eq. (4-8) in the form

$$sI(s) = \frac{1}{L}[-V_C(s) - RI(s) + V(s)] \tag{4-10}$$

Employing the Inverse Laplace transform, we can thus write

$$\frac{dv_C}{dt} = \frac{1}{C}i(t) \tag{4-11}$$

$$\frac{di}{dt} = \frac{-1}{L}v_C(t) - \frac{R}{L}i(t) + \frac{1}{L}v(t) \tag{4-12}$$

Let us pause here to consider the result of our development so far. The description of the system behavior has been transformed from that of a second-order differential equation (4-5) to two first-order coupled differential equations (4-11) and (4-12). The variables $v(t)$, $v_C(t)$, and $i(t)$ can be

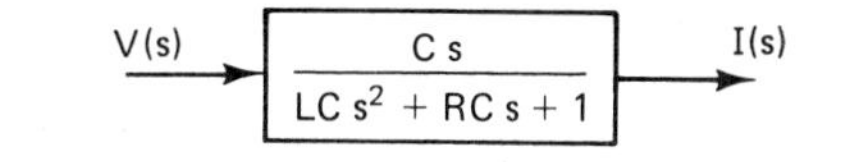


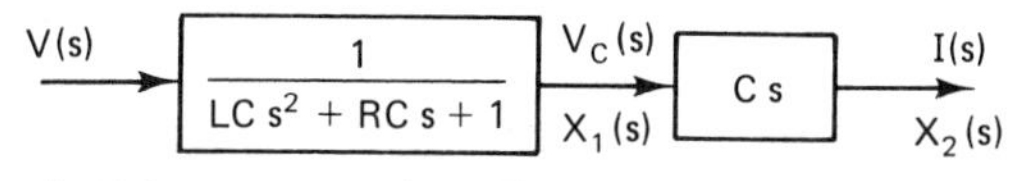

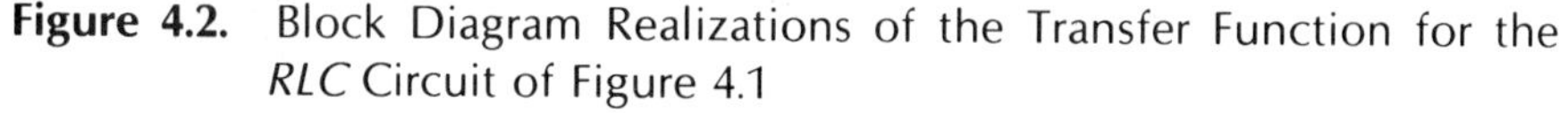


**Figure 4.2.** Block Diagram Realizations of the Transfer Function for the *RLC* Circuit of Figure 4.1

relabeled as $u(t)$, $x_1(t)$, and $x_2(t)$ according to

$$u(t) = v(t) \tag{4-13}$$
$$x_1(t) = v_C(t) \tag{4-14}$$
$$x_2(t) = i(t) \tag{4-15}$$

As a result, we rewrite Eqs. (4-11) and (4-12) as

$$\frac{dx_1}{dt} = \frac{1}{C}x_2 \tag{4-16}$$
$$\frac{dx_2}{dt} = -\frac{1}{L}x_1 - \frac{R}{L}x_2 + \frac{1}{L}u(t) \tag{4-17}$$

The equations above provide the state-space representation of our system. Figure 4.3 shows in block diagram form a representation of the system in accordance with Eqs. (4-16) and (4-17).

Let us observe here that there are two states $x_1$ and $x_2$ for this system. The number of system states is equal to the order of the differential equation describing the system. This can be paraphrased as being equal to the number of initial conditions necessary to obtain a transient response or being equal to the number of energy-storing elements in the system.

The following example illustrates one way of writing the state equation on the basis of the system transfer function.

EXAMPLE 4.2

Obtain a state-space representation of the system described by the transfer function

$$\frac{X(s)}{U(s)} = \frac{1}{s^2 + 4s + 3} \tag{4-18}$$

SOLUTION

In the time domain we can write the differential equation of the system as

$$\frac{d^2x}{dt^2} + 4\frac{dx}{dt} + 3x(t) = u(t) \tag{4-19}$$

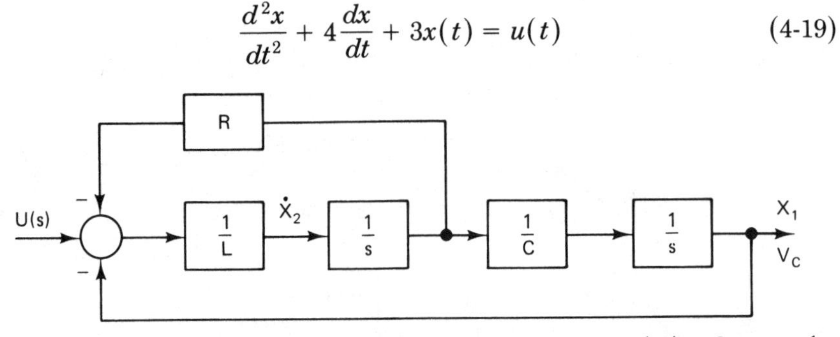


**Figure 4.3.** Equivalent Block Diagram Representation of the System for Example 4.1

We introduce the following state variables:

$$x_1(t) = x(t) \tag{4-20}$$

$$x_2(t) = \dot{x}(t) \tag{4-21}$$

As a result, we have from Eqs. (4-20) and (4-21)

$$\dot{x}_1(t) = x_2(t) \tag{4-22}$$

Moreover, Eqs. (4-19), (4-20), and (4-21) can be combined to give

$$\dot{x}_2(t) = -3x_1(t) - 4x_2(t) + u(t) \tag{4-23}$$

We can write Eqs. (4-22) and (4-23) in vector form as

$$\begin{bmatrix} \dot{x}_1(t) \\ \dot{x}_2(t) \end{bmatrix} = \begin{bmatrix} 0 & 1 \\ -3 & -4 \end{bmatrix} \begin{bmatrix} x_1(t) \\ x_2(t) \end{bmatrix} + \begin{bmatrix} 0 \\ 1 \end{bmatrix} u(t) \tag{4-24}$$

This is described more compactly by

$$\dot{\mathbf{x}}(t) = \mathbf{Ax}(t) + \mathbf{Bu}(t) \tag{4-25}$$

with

$$\mathbf{A} = \begin{bmatrix} 0 & 1 \\ -3 & -4 \end{bmatrix} \tag{4-26}$$

Just in case the reader is not familiar with the vector-matrix algebra used above, the following section details some necessary rules.

# 4.2 Concepts from Linear Algebra

In many engineering problems the concepts of vectors and matrices prove useful in describing process models compactly. To begin with, a *matrix* is an array of elements. The elements of a matrix may be real or complex numbers or functions of time or any other independent variable such as frequency. A *rectangular matrix* with $n$ rows and $m$ columns is referred to as an $n \times m$ matrix. A *square matrix* has an equal number of rows and columns. A general expression for an $n \times m$ matrix $\mathbf{A}$ is

$$\mathbf{A} = \begin{bmatrix} a_{11} & a_{12} & a_{13} & \cdots & a_{1m} \\ a_{21} & a_{22} & a_{23} & \cdots & a_{2m} \\ \vdots & \vdots & \vdots & & \vdots \\ a_{n1} & a_{n2} & a_{n3} & \cdots & a_{nm} \end{bmatrix}$$

or in shorthand,

$$\mathbf{A} = [a_{ij}]_{n \times m}$$

A *column vector* is a matrix with $n$ rows and one column. For example, the column vector $\mathbf{X}$ is written as

$$\mathbf{X} = \begin{bmatrix} x_1 \\ x_2 \\ \vdots \\ x_n \end{bmatrix}$$

The *sum* of two matrices $\mathbf{A}$ and $\mathbf{B}$ of the same dimension $n \times m$ is an $n \times m$ matrix $\mathbf{C}$. Thus

$$\mathbf{C} = \mathbf{A} + \mathbf{B}$$

The element $C_{ij}$ is obtained as the sum of the corresponding elements $A_{ij}$ and $B_{ij}$. Thus

$$C_{ij} = A_{ij} + B_{ij}$$

EXAMPLE 4.3

Assume that

$$\mathbf{A} = \begin{bmatrix} 1 & 2 \\ 8 & 3 \end{bmatrix}$$

$$\mathbf{B} = \begin{bmatrix} 4 & 5 \\ 6 & 7 \end{bmatrix}$$

The sum $\mathbf{C}$ is obtained as

$$\mathbf{C} = \mathbf{A} + \mathbf{B} = \begin{bmatrix} 1 & 2 \\ 8 & 3 \end{bmatrix} + \begin{bmatrix} 4 & 5 \\ 6 & 7 \end{bmatrix} = \begin{bmatrix} 5 & 7 \\ 14 & 10 \end{bmatrix}$$

*Multiplication* of two matrices $\mathbf{A}$ and $\mathbf{B}$ can only be performed if they are dimensionally compatible. Assume that matrix $\mathbf{D}$ is $n \times r$; then $\mathbf{E}$ is dimensionally compatible with $\mathbf{D}$ if the number of rows of $\mathbf{E}$ is equal to the number of columns of $\mathbf{D}$. Thus $\mathbf{E}$ must be $r \times p$. The product $\mathbf{G}$ is written compactly as

$$\mathbf{G}_{(n \times p)} = \mathbf{D}_{(n \times r)} \mathbf{E}_{(r \times p)}$$

In expanded form we have

$$\begin{bmatrix} G_{11} & G_{12} & \cdots & G_{1p} \\ G_{21} & G_{22} & \cdots & G_{2p} \\ \vdots & \vdots & & \vdots \\ G_{n1} & G_{n2} & \cdots & G_{np} \end{bmatrix} = \begin{bmatrix} D_{11} & D_{12} & \cdots & D_{1r} \\ D_{21} & D_{22} & \cdots & D_{2r} \\ \vdots & \vdots & & \vdots \\ D_{n1} & D_{n2} & \cdots & D_{nr} \end{bmatrix} \times \begin{bmatrix} E_{11} & E_{12} & \cdots & E_{1p} \\ E_{21} & E_{22} & \cdots & E_{2p} \\ \vdots & \vdots & & \vdots \\ E_{r1} & E_{r2} & \cdots & E_{rp} \end{bmatrix}$$

with

$$G_{11} = D_{11}E_{11} + D_{12}E_{21} + \cdots + D_{1r}E_{r1}$$
$$G_{12} = D_{11}E_{12} + D_{12}E_{22} + \cdots + D_{1r}E_{r2}$$
$$\vdots$$
$$G_{ij} = \sum_{k=1}^{r} D_{ik}E_{kj} \qquad i = 1,\ldots,n; j = 1,\ldots,p$$

Thus each element of row $i$ of $\mathbf{D}$ is multiplied by the corresponding element of the $j$th column of $\mathbf{E}$ and the products are summed to yield the $ij$th element of $\mathbf{G}$.

Let us pause here and reconsider Eq. (4-24). If the matrix $\mathbf{A}$ is given by

$$\mathbf{A} = \begin{bmatrix} 0 & 1 \\ -3 & -4 \end{bmatrix}$$

Then the vector $\mathbf{x}(t)$ is given by

$$\mathbf{x}(t) = \begin{bmatrix} x_1(t) \\ x_2(t) \end{bmatrix}$$

Thus

$$\mathbf{A}\mathbf{x}(t) = \begin{bmatrix} 0 & 1 \\ -3 & -4 \end{bmatrix} \begin{bmatrix} x_1(t) \\ x_2(t) \end{bmatrix}$$
$$= \begin{bmatrix} x_2(t) \\ -3x_1(t) - 4x_2(t) \end{bmatrix}$$

Note that $\mathbf{A}$ is $2 \times 2$, $\mathbf{x}$ is $2 \times 1$; hence $\mathbf{A}\mathbf{x}(t)$ is $2 \times 1$. The components of $\mathbf{x}$ are functions of time. In the present case the product $[X(t)]\mathbf{A}$ cannot be obtained.

It is possible in some cases to obtain the two products $\mathbf{AB}$ and $\mathbf{BA}$. This happens if $\mathbf{A}$ is $n \times r$ and $\mathbf{B}$ is $r \times n$. In this case $\mathbf{AB}$ is $n \times n$ and $\mathbf{BA}$ is $r \times r$. Clearly, $\mathbf{AB} \neq \mathbf{BA}$, and we say that $\mathbf{A}$ and $\mathbf{B}$ do not *commute*. If $\mathbf{A}$ and $\mathbf{B}$ are both square matrices ($n \times n$), $\mathbf{AB}$ and $\mathbf{BA}$ are both square ($n \times n$). If $\mathbf{AB} = \mathbf{BA}$, we say that $\mathbf{A}$ and $\mathbf{B}$ commute. A number of examples are appropriate here.

EXAMPLE 4.4

Take

$$\mathbf{A} = \begin{bmatrix} 1 & 3 & 2 \\ 5 & 6 & 7 \end{bmatrix}$$
$$\mathbf{B} = \begin{bmatrix} 2 & 3 \\ 1 & 5 \\ 4 & 2 \end{bmatrix}$$

The product **AB** is obtained as

$$\mathbf{AB} = \begin{bmatrix} 1 & 3 & 2 \\ 5 & 6 & 7 \end{bmatrix} \begin{bmatrix} 2 & 3 \\ 1 & 5 \\ 4 & 2 \end{bmatrix}$$

$$= \begin{bmatrix} 13 & 22 \\ 44 & 59 \end{bmatrix}$$

The product **BA** is obtained as

$$\mathbf{BA} = \begin{bmatrix} 2 & 3 \\ 1 & 5 \\ 4 & 2 \end{bmatrix} \begin{bmatrix} 1 & 3 & 2 \\ 5 & 6 & 7 \end{bmatrix} = \begin{bmatrix} 17 & 24 & 25 \\ 26 & 33 & 37 \\ 14 & 24 & 22 \end{bmatrix}$$

Clearly, **AB** is of a different dimension than **BA** and hence cannot be equal. In the following example, both **A** and **B** are square matrices, and hence **AB** and **BA** will be of the same dimension, but $\mathbf{AB} \neq \mathbf{BA}$.

EXAMPLE 4.5

Take

$$\mathbf{A} = \begin{bmatrix} 1 & 2 \\ 3 & 4 \end{bmatrix}$$

$$\mathbf{B} = \begin{bmatrix} 5 & 6 \\ 7 & 8 \end{bmatrix}$$

Here we have

$$\mathbf{AB} = \begin{bmatrix} 1 & 2 \\ 3 & 4 \end{bmatrix} \begin{bmatrix} 5 & 6 \\ 7 & 8 \end{bmatrix} = \begin{bmatrix} 19 & 22 \\ 43 & 50 \end{bmatrix}$$

$$\mathbf{BA} = \begin{bmatrix} 5 & 6 \\ 7 & 8 \end{bmatrix} \begin{bmatrix} 1 & 2 \\ 3 & 4 \end{bmatrix} = \begin{bmatrix} 23 & 34 \\ 31 & 46 \end{bmatrix}$$

Clearly, $\mathbf{AB} \neq \mathbf{BA}$.

EXAMPLE 4.6

Consider the pair of matrices

$$\mathbf{A} = \begin{bmatrix} 1 & 4 \\ 4 & 5 \end{bmatrix}$$

$$\mathbf{B} = \begin{bmatrix} 1 & 5 \\ 5 & 6 \end{bmatrix}$$

$$\mathbf{AB} = \begin{bmatrix} 1 & 4 \\ 4 & 5 \end{bmatrix} \begin{bmatrix} 1 & 5 \\ 5 & 6 \end{bmatrix} = \begin{bmatrix} 21 & 29 \\ 29 & 50 \end{bmatrix}$$

$$\mathbf{BA} = \begin{bmatrix} 1 & 5 \\ 5 & 6 \end{bmatrix} \begin{bmatrix} 1 & 4 \\ 4 & 5 \end{bmatrix} = \begin{bmatrix} 21 & 29 \\ 29 & 50 \end{bmatrix}$$

Clearly, $\mathbf{AB} = \mathbf{BA}$ and thus **A** and **B** commute.

A diagonal matrix **D** is a square matrix with zero elements except for the main diagonal. Thus

$$D_{ij} = 0 \qquad i \neq j$$
$$D_{ii} = D_{ii} \qquad i = 1, \ldots, n$$

A unit matrix **I** is a diagonal matrix with elements of unity along the main diagonal. A $3 \times 3$ unity matrix is written as

$$\mathbf{I} = \begin{bmatrix} 1 & 0 & 0 \\ 0 & 1 & 0 \\ 0 & 0 & 1 \end{bmatrix}$$

Note that for a matrix **A**, premultiplication or postmultiplication by **I** leaves the matrix unchanged:

$$\mathbf{AI} = \mathbf{IA}$$

A matrix whose elements are (differentiable or integrable) time-dependent functions can be differentiated or integrated element by element. Thus

$$\frac{d\mathbf{A}(t)}{dt} = \dot{\mathbf{A}}(t) = \left[\dot{a}_{ij}(t)\right]$$
$$= \begin{bmatrix} \dot{a}_{11} & \dot{a}_{12} & \cdots & \dot{a}_{1n} \\ \dot{a}_{21} & \dot{a}_{22} & \cdots & \dot{a}_{2n} \\ \vdots & \vdots & & \vdots \\ \dot{a}_{n1} & \dot{a}_{n2} & \cdots & \dot{a}_{nn} \end{bmatrix}$$

The derivative of the product of two matrices follows rules of derivatives of scalar products with preservation of order.

$$\frac{d}{dt}[\mathbf{A}(t)\mathbf{B}(t)] = \dot{\mathbf{A}}(t)\mathbf{B}(t) + \mathbf{A}(t)\dot{\mathbf{B}}(t)$$

The integral of a matrix is obtained by integrating each element

$$\int \mathbf{A}(t)\, dt = \left[\int a_{ij}(t)\, dt\right]$$

The Laplace transform $\mathbf{A}(s)$ of a matrix $\mathbf{A}(t)$ is obtained by transforming each element. Thus, $\mathbf{A}(s) = \mathscr{L}\{\mathbf{A}(t)\} = [a_{ij}(s)]$. The inverse Laplace transform of a matrix is similarly obtained as the inverse Laplace transform of the individual elements.

The $k$th power of a matrix **A** denoted by $\mathbf{A}^k$ is defined as the product of **A** and itself $k$ times.

$$\mathbf{A}^k = \mathbf{AAA} \cdots \mathbf{A}$$

In particular, we have

$$\mathbf{A}^2 = \mathbf{A} \cdot \mathbf{A}$$
$$\mathbf{A}^3 = \mathbf{A} \cdot \mathbf{A}^2$$
$$\vdots$$
$$\mathbf{A}^k = \mathbf{A} \cdot \mathbf{A}^{k-1}$$

Consider a square matrix **A** if there is a square matrix **B** such that

$$\mathbf{AB} = \mathbf{BA} = \mathbf{I}$$

Then **B** is the inverse of **A**, denoted by $\mathbf{A}^{-1}$. Thus

$$\mathbf{A} \cdot \mathbf{A}^{-1} = \mathbf{A}^{-1}\mathbf{A} = \mathbf{I}$$

The matrix **A** is invertible (i.e., $\mathbf{A}^{-1}$ exists) if **A** is *nonsingular*, as signified by the condition that the determinant of **A** is not zero:

$$|\mathbf{A}| \neq 0$$

The inverse of the product **BC** exists if both $\mathbf{B}^{-1}$ and $\mathbf{C}^{-1}$ exist. It is easy to prove that

$$(\mathbf{BC})^{-1} = \mathbf{C}^{-1}\mathbf{B}^{-1}$$

For matrices with a small dimension a straightforward procedure for matrix inversion is given by

$$\mathbf{A}^{-1} = \frac{\text{adj}(\mathbf{A})}{|\mathbf{A}|}$$

The *adjoint* of **A** denoted by adj(**A**) is the transpose of the matrix of cofactors of **A** with elements

$$A_{ij} = (-1)^{i+j} M_{ij}$$

The *minors* $M_{ij}$ are determinants of the $(n-1) \times (n-1)$ matrices obtained be deleting the $i$th row and $j$th column from **A**. An example is useful at this point.

EXAMPLE 4.7

Find the inverse of **A**, where

$$\mathbf{A} = \begin{bmatrix} 1 & 2 & 3 \\ 4 & 5 & 6 \\ 7 & 8 & 9 \end{bmatrix}$$

SOLUTION

The determinant of **A** is obtained as

$$\begin{aligned} |\mathbf{A}| &= 1[(5)(9) - (6)(8)] - 2[(4)(9) - (6)(7)] + 3[(4)(8) - (5)(7)] \\ &= 1(45 - 48) - 2(36 - 42) + 3(32 - 35) \\ &= 1(-3) - 2(-6) + 3(-3) \\ &= -3 + 12 - 9 \\ &= 0 \end{aligned}$$

The matrix is singular and an inverse cannot be obtained.

EXAMPLE 4.8

Find the inverse of $\mathbf{A}$, where

$$\mathbf{A} = \begin{bmatrix} 1 & -2 & 3 \\ 4 & 5 & 6 \\ 7 & 8 & 9 \end{bmatrix}$$

SOLUTION

The determinant of $\mathbf{A}$ is obtained as

$$\begin{aligned} |\mathbf{A}| &= 1[(5)(9) - (6)(8)] + 2[(4)(9) - (6)(7)] \\ &\quad + 3[(4)(8) - (5)(7)] \\ &= -3 - 12 - 9 \\ &= -24 \end{aligned}$$

Since $|\mathbf{A}| \neq 0$, we proceed to find the adjoint of $\mathbf{A}$. The matrix of cofactors is obtained as

$$\operatorname{cof}(\mathbf{A}) = \begin{bmatrix} \begin{vmatrix} 5 & 6 \\ 8 & 9 \end{vmatrix} & -\begin{vmatrix} 4 & 6 \\ 7 & 9 \end{vmatrix} & \begin{vmatrix} 4 & 5 \\ 7 & 8 \end{vmatrix} \\ -\begin{vmatrix} -2 & 3 \\ 8 & 9 \end{vmatrix} & \begin{vmatrix} 1 & 3 \\ 7 & 9 \end{vmatrix} & -\begin{vmatrix} 1 & -2 \\ 7 & 8 \end{vmatrix} \\ \begin{vmatrix} -2 & 3 \\ 5 & 6 \end{vmatrix} & -\begin{vmatrix} 1 & 3 \\ 4 & 6 \end{vmatrix} & \begin{vmatrix} 1 & -2 \\ 4 & 5 \end{vmatrix} \end{bmatrix}$$

$$= \begin{bmatrix} -3 & 6 & -3 \\ 42 & -12 & -22 \\ -27 & 6 & 13 \end{bmatrix}$$

Thus transposing $\operatorname{cof}(\mathbf{A})$, we obtain $\operatorname{adj}(\mathbf{A})$ as

$$\operatorname{adj}(\mathbf{A}) = \begin{bmatrix} -3 & 42 & -27 \\ 6 & -12 & 6 \\ -3 & -22 & 13 \end{bmatrix}$$

The inverse of $\mathbf{A}$ is obtained as

$$\mathbf{A}^{-1} = \frac{\operatorname{adj}(\mathbf{A})}{|\mathbf{A}|} = \begin{bmatrix} \dfrac{-3}{-24} & \dfrac{42}{-24} & \dfrac{-27}{-24} \\ \dfrac{6}{-24} & \dfrac{-12}{-24} & \dfrac{6}{-24} \\ \dfrac{-3}{-24} & \dfrac{-22}{-24} & \dfrac{13}{-24} \end{bmatrix}$$

Thus

$$\mathbf{A}^{-1} = \begin{bmatrix} 0.125 & -1.75 & 1.125 \\ -0.25 & 0.5 & -0.25 \\ 0.125 & 0.91667 & -0.54166 \end{bmatrix}$$

We can verify our result by finding the product

$$\mathbf{A}\cdot\mathbf{A}^{-1} = \begin{bmatrix} 1 & -2 & 3 \\ 4 & 5 & 6 \\ 7 & 8 & 9 \end{bmatrix}$$

$$\times \begin{bmatrix} 0.125 & -1.75 & 1.125 \\ -0.25 & 0.5 & -0.25 \\ 0.125 & 0.916667 & -0.54166 \end{bmatrix}$$

$$= \begin{bmatrix} 1 & 0 & 0 \\ 0 & 1 & 0 \\ 0 & 0 & 1 \end{bmatrix} = \mathbf{I}$$

## 4.3 State-Space Formulation

From the discussion of Section 4.1 it can be concluded that the state-space representation of a linear dynamic system is given by the following matrix equation:

$$\dot{\mathbf{x}}(t) = \mathbf{A}\mathbf{x}(t) + \mathbf{B}\mathbf{u}(t) \tag{4-27}$$

In the above, the state vector is denoted by $\mathbf{x}$. In Examples 4.1 and 4.2, the state vector is two-dimensional. The generality of the state formulation follows from the realization that Eq. (4-27) is applicable for an $n$th-order system for which the state vector is $n \times 1$. Clearly, $\dot{\mathbf{x}}(t)$ is the vector of derivatives of the states and is $n \times 1$ in dimension. The system matrix $\mathbf{A}$ is square and is $n \times n$. The input in Eq. (4-27) is denoted by $\mathbf{u}$, and although we dealt only with single input examples for which $\mathbf{u}(t)$ is a scalar, in general $\mathbf{u}(t)$ is an $r \times 1$ vector and thus $\mathbf{B}$ is $n \times r$.

The state equation (4-27) is accompanied by an output equation of the form

$$\mathbf{y}(t) = \mathbf{C}\mathbf{x}(t) + \mathbf{D}\mathbf{u}(t) \tag{4-28}$$

The output vector $\mathbf{y}(t)$ is $m \times 1$, and consequently we can see that $\mathbf{C}$ is $m \times n$ and $\mathbf{D}$ is $m \times r$.

The selection of state variables is an important aspect of the formulation. A straightforward technique selects the dependent variable in the system's differential equation to be the first state, while its first derivative is taken as the second state, and so on. The technique is referred to as the *phase-variable technique*. This is particularly attractive since an interpretation of the resulting states in terms of position, velocity, and acceleration is possible. Example 4.2 was solved using this technique. A pair of higher-order examples are in order now.

EXAMPLE 4.9

Consider the linear dynamic system described by the differential equation

$$5\frac{d^3x}{dt^3} + 3\frac{d^2x}{dt^2} + 2\frac{dx}{dt} + x = 7\sin\omega t$$

Obtain the state-space representation in terms of the phase variables.

SOLUTION

The following state variables are chosen:

$$\begin{aligned} x_1 &= x \\ x_2 &= \dot{x} \\ x_3 &= \ddot{x} \end{aligned}$$

Thus

$$\begin{aligned} \dot{x}_1 &= x_2 \\ \dot{x}_2 &= x_3 \\ \dot{x}_3 &= \frac{-1}{5}x_1 - \frac{2}{5}x_2 - \frac{3}{5}x_3 + \frac{7}{5}\sin\omega t \end{aligned}$$

Let

$$u(t) = \frac{7}{5}\sin\omega t$$

Thus the required state-space form is given by

$$\begin{bmatrix} \dot{x}_1 \\ \dot{x}_2 \\ \dot{x}_3 \end{bmatrix} = \begin{bmatrix} 0 & 1 & 0 \\ 0 & 0 & 1 \\ \frac{-1}{5} & \frac{-2}{5} & \frac{-3}{5} \end{bmatrix} \begin{bmatrix} x_1 \\ x_2 \\ x_3 \end{bmatrix} + \begin{bmatrix} 0 \\ 0 \\ 1 \end{bmatrix} u(t)$$

Clearly,

$$\mathbf{A} = \begin{bmatrix} 0 & 1 & 0 \\ 0 & 0 & 1 \\ -0.2 & -0.4 & -0.6 \end{bmatrix}$$

$$\mathbf{B} = \begin{bmatrix} 0 \\ 0 \\ 1 \end{bmatrix}$$

As a result, we have the form

$$\dot{\mathbf{x}}(t) = \mathbf{A}\mathbf{x}(t) + \mathbf{B}\mathbf{u}(t)$$

EXAMPLE 4.10

Figure 4.4 shows in block diagram form the components of an automatic voltage regulator. Find the state representation of the system using the following state variables:

$$X_1 = V_{TR}$$
$$X_2 = V_V$$
$$X_3 = V_t$$

SOLUTION

From the available transfer functions we can write

$$\frac{V_{TR}}{V_t} = \frac{X_1}{X_3} = \frac{K_V}{1 + sT_V}$$

Thus

$$T_V \dot{x}_1 = -x_1 + K_V x_3$$

or

$$\dot{x}_1 = \frac{-1}{T_V} x_1 + \frac{K_V}{T_V} x_3$$

Also,

$$\frac{V_V}{V_E} = \frac{K_a}{1 + sT_a}$$

We let $u = V_R$; thus

$$\frac{X_2}{u - X_1} = \frac{K_a}{1 + sT_a}$$

Thus

$$\dot{x}_2 = \frac{-K_a}{T_a} x_1 - \frac{x_2}{T_a} + \frac{K_a}{T_a} u$$

Finally,

$$\frac{V_t}{V_V} = \frac{K_g}{1 + sT_g}$$

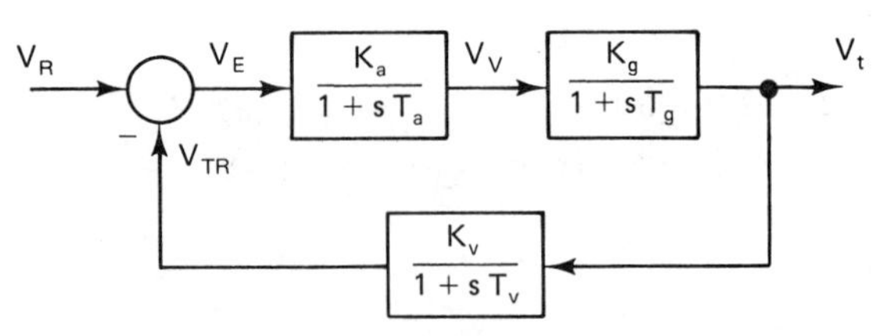


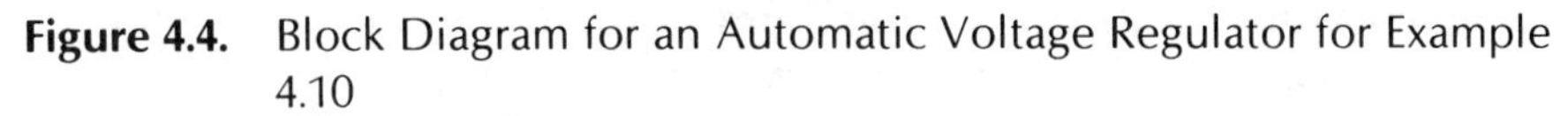
**Figure 4.4.** Block Diagram for an Automatic Voltage Regulator for Example 4.10

or

$$\frac{X_3}{X_2} = \frac{K_g}{1 + sT_g}$$

Thus

$$\dot{x}_3 = \frac{K_g}{T_g}x_2 - \frac{x_3}{T_g}$$

We can thus conclude that the required form is given by

$$\begin{bmatrix} \dot{x}_1 \\ \dot{x}_2 \\ \dot{x}_3 \end{bmatrix} = \begin{bmatrix} \frac{-1}{T_V} & 0 & \frac{K_V}{T_V} \\ \frac{-K_a}{T_a} & \frac{-1}{T_a} & 0 \\ 0 & \frac{K_g}{T_g} & \frac{-1}{T_g} \end{bmatrix} \begin{bmatrix} x_1 \\ x_2 \\ x_3 \end{bmatrix} + \begin{bmatrix} 0 \\ \frac{K_a}{T} \\ 0 \end{bmatrix} u$$

Let us observe here that the phase-variable technique can be used successfully to obtain the state representation for systems with constant numerator transfer functions. The following example illustrates the result of applying the phase-variable method to a system with numerator terms in the transfer function.

EXAMPLE 4.11

Consider the block diagram of Figure 4.5, showing transfer functions of components of a type ST1 potential-source controlled-rectifier exciter. Investigate the phase-variable method's performance in obtaining a state-space formulation using

$$X_1(s) = E_{FD}(s)$$
$$X_2(s) = sE_{FD}(s)$$
$$X_3(s) = s^2E_{FD}(s)$$

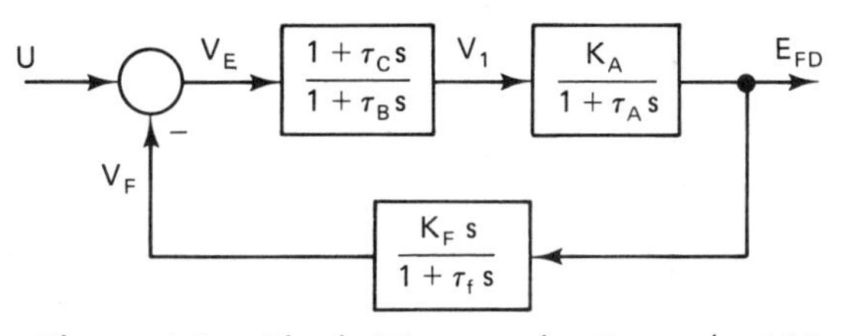


**Figure 4.5.** Block Diagram for Example 4.11

SOLUTION

We can see from the problem statement that

$$\dot{x}_1 = x_2$$
$$\dot{x}_2 = x_3$$

The overall transfer function is obtained as

$$\frac{E_{\mathrm{FD}}(s)}{u(s)} = \frac{K_A(1 + \tau_f s)(1 + \tau_c s)}{K_A K_F s(1 + \tau_c s) + (1 + \tau_A s)(1 + \tau_B s)(1 + \tau_f s)}$$

Thus expanding, we obtain

$$a_3 s^3 E_{\mathrm{FD}}(s) + a_2 s^2 E_{\mathrm{FD}}(s) + a_1 s E_{\mathrm{FD}}(s) + a_0 E_{\mathrm{FD}}(s) = b_2 s^2 u(s) + b_1 s u(s) + b_0 u(s)$$

where

$$\begin{aligned}
a_3 &= \tau_A \tau_B \tau_f \\
a_2 &= \tau_A \tau_B + \tau_f(\tau_A + \tau_B) + \tau_c K_A K_F \\
a_1 &= \tau_A + \tau_B + \tau_f + K_A K_F \\
a_0 &= 1 \\
b_2 &= K_A \tau_f \tau_c \\
b_1 &= K_A(\tau_f + \tau_c) \\
b_0 &= K_A
\end{aligned}$$

Thus

$$a_3 \dot{x}_3 + a_2 x_3 + a_1 x_2 + a_0 x_1 = b_2 \ddot{u} + b_1 \dot{u} + b_0 u$$

The resulting representation is not a bona fide state-space representation, as can be seen from

$$\begin{bmatrix} \dot{x}_1 \\ \dot{x}_2 \\ \dot{x}_3 \end{bmatrix} = \begin{bmatrix} 0 & 1 & 0 \\ 0 & 0 & 1 \\ \dfrac{-a_0}{a_3} & \dfrac{-a_1}{a_3} & \dfrac{-a_2}{a_3} \end{bmatrix} \begin{bmatrix} x_1 \\ x_2 \\ x_3 \end{bmatrix} + \begin{bmatrix} 0 \\ 0 \\ \dfrac{b_0}{a_3} \end{bmatrix} u + \begin{bmatrix} 0 \\ 0 \\ \dfrac{b_1}{a_3} \end{bmatrix} \dot{u} + \begin{bmatrix} 0 \\ 0 \\ \dfrac{b_2}{a_3} \end{bmatrix} \ddot{u}$$

This clearly requires inclusion of the derivatives of the input. The following section discusses possible means of avoiding this problem.

# 4.4 Additional Techniques for Writing the State Equations

To avoid derivatives of the control in the state-space formulation, a number of techniques are available. Two methods are of particular interest here.

## Beck's Method

Consider a system represented by the transfer function

$$\frac{Y(s)}{U(s)} = \frac{\sum_{j=0}^{m} b_j s^j}{\sum_{j=0}^{n} a_j s^j} \tag{4-29}$$

Introducing the variable $X_1(s)$, we can write the transfer function as two equations:

$$Y(s) = \sum_{j=0}^{m} b_j s^j X_1(s) \tag{4-30}$$

$$U(s) = \sum_{j=0}^{n} a_j s^j X_1(s) \tag{4-31}$$

The main feature of Beck's method is the introduction of the state variables $X_j$ defined by

$$X_{j+1}(s) = s^j X_1(s) \qquad j = 0, 1, \ldots, n-1 \tag{4-32}$$

Thus

$$X_j(s) = s^{j-1} X_1(s) \tag{4-33}$$

We can thus conclude that

$$sX_j(s) = X_{j+1}(s) \qquad j = 1, \ldots, n-1 \tag{4-34}$$

Equation (4-30) can thus be written as

$$Y(s) = \sum_{j=0}^{m} b_j X_{j+1}(s) \tag{4-35}$$

Equation (4-31) can be written as

$$a_n s\left[s^{n-1} X_1(s)\right] = U(s) - \sum_{j=0}^{n-1} a_j s^j X_1(s) \tag{4-36}$$

Using Eq. (4-32), we thus have

$$sX_n(s) = \frac{1}{a_n}\left[U(s) - \sum_{j=0}^{n-1} a_j X_{j+1}(s)\right] \tag{4-37}$$

Applying the inverse Laplace transform, we can thus write Eq. (4-34) in the expanded form

$$\dot{x}_1(t) = x_2(t) \tag{4-38}$$

$$\dot{x}_2(t) = x_3(t) \tag{4-39}$$

$$\vdots$$

$$\dot{x}_{n-1}(t) = x_n(t) \tag{4-40}$$

Equation (4-37) provides us with

$$\dot{x}_n(t) = \frac{1}{a_n}\left[-\sum_{j=0}^{n-1} a_j x_{j+1}(t) + u(t)\right] \tag{4-41}$$

The output equation is

$$y(t) = \sum_{j=0}^{m} b_j x_{j+1}(t) \tag{4-42}$$

In vector-matrix form Beck's state-space model is given by

$$\begin{bmatrix} \dot{x}_1(t) \\ \dot{x}_2(t) \\ \vdots \\ \dot{x}_n(t) \end{bmatrix} = \begin{bmatrix} 0 & 1 & 0 & \cdots & 0 \\ 0 & 0 & 1 & \cdots & 0 \\ \vdots & \vdots & \vdots & \vdots & \vdots \\ \dfrac{-a_0}{a_n} & \dfrac{-a_1}{a_n} & \dfrac{-a_2}{a_n} & \cdots & \dfrac{-a_{n-1}}{a_n} \end{bmatrix} \begin{bmatrix} x_1(t) \\ x_2(t) \\ \vdots \\ x_n(t) \end{bmatrix} + \begin{bmatrix} 0 \\ 0 \\ 0 \\ \vdots \\ \dfrac{1}{a_n} \end{bmatrix} u(t) \tag{4-43}$$

This is of the form

$$\dot{\mathbf{x}} = \mathbf{A}\mathbf{x} + \mathbf{B}u \tag{4-44}$$

Note that with this formulation, derivatives of $u$ were avoided.

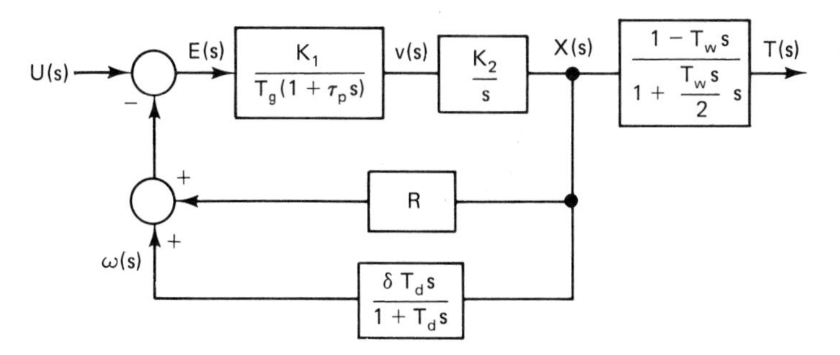


**Figure 4.6.** Hydrogovernor-Turbine Model for Example 4.12

EXAMPLE 4.12

Obtain a state-space representation of the hydrogovernor-turbine model shown in block diagram form in Figure 4.6. Use Beck's method applied to individual blocks in the diagram.

SOLUTION

Consider the transfer function

$$\frac{T(s)}{X(s)} = \frac{1 - T_w s}{1 + (T_w/2)s} \tag{4-45}$$

Applying Beck's idea, we introduce $X_1(s)$ to obtain

$$T(s) = (1 - T_w s)X_1(s) \tag{4-46}$$

$$X(s) = \left(1 + \frac{T_w}{2}s\right)X_1(s) \tag{4-47}$$

Thus the first state $X_1(s)$ is related to the output $T(s)$ by

$$X_1(s) = \frac{T(s)}{1 - T_w s} \tag{4-48}$$

Consider the transfer function

$$\frac{X(s)}{V(s)} = \frac{K_2}{s} = \frac{1}{s/K_2} \tag{4-49}$$

Introduce $X_2(s)$ to obtain

$$\frac{X(s)}{V(s)} = \frac{X_2(s)}{(s/K_2)X_2(s)} \tag{4-50}$$

Thus the second state $X_2(s)$ is

$$X_2(s) = X(s) \tag{4-51}$$

We also have

$$sX_2 = K_2 V(s) \tag{4-52}$$

From Eqs. (4-47) and (4-51), we have

$$\frac{X_1(s)}{X_2(s)} = \frac{1}{1 + (T_w/2)s}$$

Thus

$$x_1 + \frac{T_w}{2}\dot{x}_1 = x_2$$

or

$$\dot{x}_1 = \frac{2}{T_w}(x_2 - x_1)$$

Consider the transfer function

$$\frac{V(s)}{E(s)} = \frac{K_1}{T_g(1 + \tau_p s)} \tag{4-53}$$

Introduce $X_3(s)$ to obtain

$$X_3(s) = V(s) \tag{4-54}$$

$$K_1 E(s) = T_g(1 + \tau_p s) X_3(s) \tag{4-55}$$

From Eqs. (4-52) and (4-54), we conclude that

$$\dot{x}_2 = K_2 x_3 \tag{4-56}$$

Consider the transfer function

$$\frac{\omega(s)}{X(s)} = \frac{\delta T_d s}{1 + T_d s} \tag{4-57}$$

Introduce $X_4(s)$ to obtain

$$X_4(s) = \frac{\omega(s)}{\delta T_d s} \tag{4-58}$$

$$(1 + T_d s) X_4(s) = X(s) \tag{4-59}$$

Recall

$$X_2(s) = X(s)$$

Thus

$$\dot{x}_4 = \frac{1}{T_d}(x_2 - x_4) \tag{4-60}$$

We now have to consider the summing point, at which we have

$$E(s) = U(s) - RX(s) - \omega(s) \tag{4-61}$$

In terms of our state variables, we have

$$\begin{aligned} E(s) &= U(s) - RX_2(s) - \delta T_d s X_4(s) \\ &= U(s) - RX_2(s) - \delta(X_2 - X_4) \\ &= U(s) - (R + \delta)X_2(s) + \delta X_4(s) \end{aligned} \tag{4-62}$$

Thus using Eq. (4-62) in (4-55), we obtain

$$K_1[U(s) - (R + \delta)X_2(s) + \delta X_4(s)] = T_g(1 + \tau_p s)X_3(s)$$

Rearranging, we get

$$\dot{x}_3 = \frac{K_1}{T_g \tau_p}[u - (R + \delta)x_2 + \delta x_4] - \frac{x_3}{\tau_p} \tag{4-63}$$

Our state-space model is thus given by

$$\begin{bmatrix} \dot{x}_1 \\ \dot{x}_2 \\ \dot{x}_3 \\ \dot{x}_4 \end{bmatrix} = \begin{bmatrix} \dfrac{-2}{T_w} & \dfrac{2}{T_w} & 0 & 0 \\ 0 & 0 & K_2 & 0 \\ 0 & \dfrac{-(R + \delta)K_1}{T_g \tau_p} & \dfrac{-1}{\tau_p} & \dfrac{\delta K_1}{\tau_p T_g} \\ 0 & \dfrac{1}{T_d} & 0 & \dfrac{-1}{T_d} \end{bmatrix} \begin{bmatrix} x_1 \\ x_2 \\ x_3 \\ x_4 \end{bmatrix} + \begin{bmatrix} 0 \\ 0 \\ \dfrac{K_1}{\tau_p T_g} \\ 0 \end{bmatrix} u \tag{4-64}$$

## Johnson's Method

Consider the system represented by the transfer function

$$\frac{Y(s)}{U(s)} = \frac{\sum_{j=0}^{n} b_j s^j}{\sum_{j=0}^{n} a_j s^j} \tag{4-65}$$

Cross-multiplying and rearranging, we obtain

$$\sum_{j=0}^{n} s^j\left[b_j U(s) - a_j Y(s)\right] = 0 \tag{4-66}$$

The key to Johnson's method is to add and subtract additional terms in the following fashion for the expanded inverse Laplace-transformed version of Eq. (4-66):

$$\begin{aligned}
&\left[b_0 u(t) - a_0 y(t) - \frac{dx_n}{dt}\right] + \frac{d}{dt}\left[b_1 u(t) - a_1 y(t) + x_n - \frac{dx_{n-1}}{dt}\right] \\
&\quad + \frac{d^2}{dt^2}\left[b_2 u(t) - a_2 y(t) + x_{n-1} - \frac{dx_{n-2}}{dt}\right] \\
&\quad \vdots \\
&\quad + \frac{d^{n-j}}{dt^{n-j}}\left[b_{n-j} u(t) - a_{n-j} y(t) + x_{j+1} - \frac{dx_j}{dt}\right] \\
&\quad \vdots \\
&\quad + \frac{d^{n-1}}{dt^{n-1}}\left[b_{n-1} u(t) - a_{n-1} y(t) + x_2 - \frac{dx_1}{dt}\right] \\
&\quad + \frac{d^n}{dt^n}\left[b_n u(t) - a_n y(t) + x_1(t)\right] = 0
\end{aligned} \tag{4-67}$$

Each term can now be equated to zero to give

$$x_1(t) = a_n y(t) - b_n u(t) \tag{4-68}$$

$$\frac{dx_1}{dt} = x_2 + b_{n-1} u(t) - a_{n-1} y(t) \tag{4-69}$$

$$\frac{dx_2}{dt} = x_3 + b_{n-2} u(t) - a_{n-2} y(t) \tag{4-70}$$

$$\vdots$$

$$\frac{dx_{n-1}}{dt} = x_n + b_1 u(t) - a_1 y(t) \tag{4-71}$$

$$\frac{dx_n}{dt} = b_0 u(t) - a_0 y(t) \tag{4-72}$$

Substituting for $y(t)$ from (4-68), we get

$$\frac{dx_1}{dt} = -\frac{a_{n-1}}{a_n} x_1 + x_2 + \left(b_{n-1} - \frac{b_n a_{n-1}}{a_n}\right) u \tag{4-73}$$

$$\frac{dx_2}{dt} = -\frac{a_{n-2}}{a_n} x_1 + x_3 + \left(b_{n-2} - \frac{b_n a_{n-2}}{a_n}\right) u \tag{4-74}$$

$$\vdots$$

$$\frac{dx_{n-1}}{dt} = -\frac{a_1}{a_n} x_1 + x_n + \left(b_1 - \frac{b_n a_1}{a_n}\right) u \tag{4-75}$$

$$\frac{dx_n}{dt} = -\frac{a_0}{a_n} x_1 + \left(b_0 - \frac{b_n a_0}{a_n}\right) u \tag{4-76}$$

In vector-matrix form we have

$$\begin{bmatrix} \dot{x}_1(t) \\ \dot{x}_2(t) \\ \dot{x}_3(t) \\ \vdots \\ \dot{x}_{n-1}(t) \\ \dot{x}_n(t) \end{bmatrix} = \begin{bmatrix} -\frac{a_{n-1}}{a_n} & 1 & 0 & 0 & \cdots & 0 \\ -\frac{a_{n-2}}{a_n} & 0 & 1 & 0 & \cdots & 0 \\ -\frac{a_{n-3}}{a_n} & 0 & 0 & 1 & \cdots & 0 \\ \vdots & \vdots & \vdots & \vdots & \vdots & \\ -\frac{a_1}{a_n} & 0 & 0 & 0 & \cdots & 1 \\ -\frac{a_0}{a_n} & 0 & 0 & 0 & \cdots & 0 \end{bmatrix} \begin{bmatrix} x_1(t) \\ x_2(t) \\ x_3(t) \\ \vdots \\ x_{n-1}(t) \\ x_n(t) \end{bmatrix} + \begin{bmatrix} b_{n-1} - \frac{b_n a_{n-1}}{a_n} \\ b_{n-2} - \frac{b_n a_{n-2}}{a_n} \\ b_{n-3} - \frac{b_n a_{n-3}}{a_n} \\ \vdots \\ b_0 - \frac{b_n a_0}{a_n} \end{bmatrix} u(t) \tag{4-77}$$

EXAMPLE 4.13

Repeat Example 4.12 by applying Johnson's method to individual blocks.

SOLUTION

The following is a restatement of the transfer function of (4-45).

$$T(s) + \frac{T_w}{2} sT(s) = (1 - T_w s)X(s) \tag{4-78}$$

This can be rearranged to

$$[T(s) - X(s)] + s\left(T_w\left[\frac{T(s)}{2} + X(s)\right]\right) = 0 \tag{4-79}$$

Introduce $X_1(s)$ in Eq. (4-79) to yield

$$[T(s) - X(s) + sX_1(s)] + s\left[T_w\left(\frac{T(s)}{2} + X(s)\right) - X_1(s)\right] = 0 \tag{4-80}$$

As a result, equating the terms in brackets to zero, we get

$$X_1(s) = T_w\left[\frac{T(s)}{2} + X(s)\right] \tag{4-81}$$

$$sX_1(s) = X(s) - T(s) \tag{4-82}$$

Using Eq. (4-81) in (4-82), we get

$$sX_1(s) = \frac{-2}{T_w}X_1(s) + 3X(s)$$

The second transfer function to be considered is

$$\frac{X(s)}{V(s)} = \frac{K_2}{s} \tag{4-83}$$

Thus

$$K_2V(s) - sX(s) = 0 \tag{4-84}$$

Introduce $X_2(s)$:

$$[K_2V(s) - sX_2(s)] - s[X(s) - X_2(s)] = 0 \tag{4-85}$$

Thus

$$X_2(s) = X(s) \tag{4-86}$$

$$sX_2(s) = K_2V(s) \tag{4-87}$$

The third transfer function is that of Eq. (4-53)

$$\frac{V(s)}{E(s)} = \frac{K_1}{T_g(1 + \tau_p s)} \tag{4-88}$$

Thus

$$\left[V(s) - \frac{K_1}{T_g}E(s)\right] + \tau_p s[V(s)] = 0 \tag{4-89}$$

Introduce $X_3(s)$:

$$\left[V(s) - \frac{K_1}{T_g}E(s) + \tau_p sX_3(s)\right] + \tau_p s[V(s) - X_3(s)] = 0 \tag{4-90}$$

Thus

$$X_3(s) = V(s) \tag{4-91}$$

$$sX_3(s) = \frac{1}{\tau_p}\left[\frac{K_1}{T_g}E(s) - V(s)\right] \tag{4-92}$$

The fourth transfer function is that of Eq. (4-57)

$$\frac{\omega(s)}{X(s)} = \frac{\delta T_d s}{1 + T_d s} \tag{4-93}$$

$$\omega(s) + T_d s[\omega(s) - \delta X(s)] = 0 \tag{4-94}$$

Introduce $X_4(s)$ to obtain

$$\omega(s) + sX_4(s) + s\left[T_d(\omega(s) - \delta X(s)) - X_4(s)\right] = 0 \tag{4-95}$$

Thus

$$X_4(s) = T_d\left[\omega(s) - \delta X(s)\right] \tag{4-96}$$

$$sX_4(s) = -\omega(s) \tag{4-97}$$

Recall that from Eq. (4-86)

$$X_2(s) = X(s) \tag{4-98}$$

Thus Eq. (4-96) can be written as:

$$sX_4(s) = -\left[\frac{X_4(s)}{T_d} + \delta X_2(s)\right] \tag{4-99}$$

Now the summing junction yields

$$\begin{aligned} E(s) &= U(s) - RX(s) - \omega(s) \\ &= U(s) - RX_2(s) - \frac{X_4(s)}{T_d} - \delta X_2(s) \end{aligned}$$

$$E(s) = U(s) - (R + \delta)X_2(s) - \frac{X_4(s)}{T_d} \tag{4-100}$$

Thus Eq. (4-92) can be written as

$$sX_3(s) = \frac{1}{\tau_p}\left\{\frac{K_1}{T_g}\left[U(s) - (R + \delta)X_2(s) - \frac{X_4(s)}{T_d}\right] - X_3(s)\right\} \tag{4-101}$$

The representation is almost complete. Collecting relevant equations, it is easy to see that

$$\begin{bmatrix} \dot{x}_1 \\ \dot{x}_2 \\ \dot{x}_3 \\ \dot{x}_4 \end{bmatrix} = \begin{bmatrix} \frac{-2}{T_w} & 3 & 0 & 0 \\ 0 & 0 & K_2 & 0 \\ 0 & \frac{-(R+\delta)K_1}{T_g\tau_p} & \frac{-1}{\tau_p} & \frac{-K_1}{T_gT_d\tau_p} \\ 0 & -\delta & 0 & \frac{-1}{T_d} \end{bmatrix} \times \begin{bmatrix} x_1 \\ x_2 \\ x_3 \\ x_4 \end{bmatrix} + \begin{bmatrix} 0 \\ 0 \\ \frac{K_1}{T_g\tau_p} \\ 0 \end{bmatrix} u \tag{4-102}$$

# 4.5 Free Response and the State Transition Matrix

The previous sections dealt with procedures for obtaining the state equations for linear dynamic systems. In this section we consider the question of determining the solution to the state equations for the case when the input is zero. This case is referred to as the force-free problem for which the dynamics of the system are described by

$$\dot{\mathbf{x}}(t) = \mathbf{A}\mathbf{x}(t) \tag{4-103}$$

There are a number of equivalent ways to obtain the solution to Eq. (4-103). We will introduce two methods for obtaining the required solution $\mathbf{x}(t)$ in terms of the initial state $\mathbf{x}(0)$.

The most direct way to approach Eq. (4-103) is perhaps to employ the Laplace transform method. Equation (4-103) can thus be written in terms of the Laplace operator as

$$s\mathbf{X}(s) - \mathbf{x}(0) = \mathbf{A}\mathbf{x}(s)$$

This can be rearranged to

$$\mathbf{X}(s) = (s\mathbf{I} - \mathbf{A})^{-1}\mathbf{x}(0) \tag{4-104}$$

At this point we should note that Eq. (4-104) is written on the assumption that the matrix $(sI - A)$ is invertible. We will denote the inverse by $\Phi(s)$:

$$\boldsymbol{\Phi}(s) = (s\mathbf{I} - \mathbf{A})^{-1} \tag{4-105}$$

The matrix $\boldsymbol{\Phi}(s)$ is a square $(n \times n)$ matrix with elements that are functions of the Laplace operator $s$. As a result of Eq. (4-105), we can write Eq. (4-104) as

$$\mathbf{X}(s) = \boldsymbol{\Phi}(s)\mathbf{x}(0) \tag{4-106}$$

The inverse Laplace transform is now applied to (4-106) and we can write

$$\mathbf{x}(t) = \boldsymbol{\Phi}(t,0)\mathbf{x}(0) \tag{4-107}$$

In Eq. (4-107) we have

$$\boldsymbol{\Phi}(t,0) = \mathscr{L}^{-1}\{s\mathbf{I} - \mathbf{A}\}^{-1} \tag{4-108}$$

The matrix $\boldsymbol{\Phi}(t,0)$ is called the *state transition matrix*, as this signifies the fact that the state $\mathbf{x}(t)$ at time $t$ is simply a transformation of $\mathbf{x}(0)$ with the matrix $\boldsymbol{\Phi}(t,0)$. An example is taken up at this time.

EXAMPLE 4.14

Find the state transition matrix for the system of Example 4.2.

SOLUTION

To evaluate the state transition matrix associated with **A** of Example 4.2, we use

$$\Phi(s) = (s\mathbf{I} - \mathbf{A})^{-1}$$

Thus

$$\Phi^{-1}(s) = \begin{bmatrix} s & 0 \\ 0 & s \end{bmatrix} - \begin{bmatrix} 0 & 1 \\ -3 & -4 \end{bmatrix}$$
$$= \begin{bmatrix} s & -1 \\ 3 & s+4 \end{bmatrix}$$

As a result,

$$\Phi(s) = \frac{1}{s(s+4)+3}\begin{bmatrix} s+4 & 1 \\ -3 & s \end{bmatrix}$$
$$= \begin{bmatrix} \dfrac{s+4}{s^2+4s+3} & \dfrac{1}{s^2+4s+3} \\ \dfrac{-3}{s^2+4s+3} & \dfrac{s}{s^2+4s+3} \end{bmatrix}$$

Let us assume that

$$\Phi(s) = \begin{bmatrix} \Phi_{11}(s) & \Phi_{12}(s) \\ \Phi_{21}(s) & \Phi_{22}(s) \end{bmatrix}$$

Now

$$\Phi_{11}(s) = \frac{s+4}{(s+3)(s+1)}$$

A partial fraction expansion results in

$$\Phi_{11}(s) = \frac{1.5}{s+1} - \frac{0.5}{s+3}$$

Also, we have

$$\Phi_{12}(s) = \frac{1}{(s+3)(s+1)}$$

Thus using partial fraction expansion we obtain

$$\Phi_{12}(s) = \frac{0.5}{s+1} - \frac{0.5}{s+3}$$

Similarly, we have

$$\Phi_{21}(s) = \frac{-1.5}{s+1} + \frac{1.5}{s+3}$$
$$\Phi_{22}(s) = \frac{1.5}{s+3} - \frac{0.5}{s+1}$$

As a result,

$$\Phi(s) = \begin{bmatrix} \dfrac{1.5}{s+1} - \dfrac{0.5}{s+3} & \dfrac{0.5}{s+1} - \dfrac{0.5}{s+3} \\ \dfrac{-1.5}{s+1} + \dfrac{1.5}{s+3} & \dfrac{1.5}{s+3} - \dfrac{0.5}{s+1} \end{bmatrix}$$

Taking the inverse Laplace transform, we get

$$\Phi(t,0) = \begin{bmatrix} 1.5e^{-t} - 0.5e^{-3t} & 0.5e^{-t} - 0.5e^{-3t} \\ -1.5e^{-t} + 1.5e^{-3t} & 1.5e^{-t} - 0.5e^{-t} \end{bmatrix}$$

This is the required state transition matrix.

From Eqs. (4-107) and (4-103), we can conclude that

$$\dot{\mathbf{x}}(t) = \dot{\Phi}(t,0)\mathbf{x}(0) = A\Phi(t,0)\mathbf{x}(0)$$

As a result, we can see that the state transition matrix satisfies the original equation

$$\dot{\Phi}(t,0) = A\Phi(t,0) \tag{4-109}$$

EXAMPLE 4.15

We can verify Eq. (4-109) for the system of Example 4.14 by direct differentiation.

SOLUTION

$$\dot{\Phi}(t) = \begin{bmatrix} -1.5e^{-t} + 1.5e^{-3t} & -0.5e^{-t} + 1.5e^{-3t} \\ 1.5e^{-t} - 4.5e^{-3t} & -4.5e^{-3t} + 0.5e^{-t} \end{bmatrix}$$

The product of $\mathbf{A}$ and $\mathbf{\Phi}$ is given by

$$\begin{aligned} \mathbf{A}\Phi(t) &= \begin{bmatrix} 0 & 1 \\ -3 & -4 \end{bmatrix} \begin{bmatrix} 1.5e^{-t} - 0.5e^{-3t} & 0.5e^{-t} - 0.5e^{-3t} \\ -1.5e^{-t} + 1.5e^{-3t} & 1.5e^{-3t} - 0.5e^{-t} \end{bmatrix} \\ &= \begin{bmatrix} -1.5e^{-t} + 1.5e^{-3t} & 1.5e^{-3t} - 0.5e^{-t} \\ 1.5e^{-t} - 4.5e^{-3t} & -4.5e^{-3t} + 0.5e^{-t} \end{bmatrix} \end{aligned}$$

Thus the state transition matrix satisfies the original equation

$$\dot{\Phi} = \mathbf{A}\Phi$$

An alternative means of obtaining the free response is to assume the existence of the matrix exponential $e^{-\mathbf{A}t}$ and premultiply both sides of Eq. (4-103) by $e^{-\mathbf{A}t}$ to obtain

$$e^{-\mathbf{A}t}\dot{\mathbf{x}} - e^{-\mathbf{A}t}\mathbf{A}\mathbf{x} = 0$$

As a result,

$$\frac{d}{dt}\left[e^{-\mathbf{A}t}\mathbf{x}(t)\right] = 0$$

Integrating between 0 and $t$, we conclude that

$$e^{-\mathbf{A}t}\mathbf{x}(t) - \mathbf{x}(0) = 0$$

Thus

$$\mathbf{x}(t) = e^{\mathbf{A}t}\mathbf{x}(0) \tag{4-110}$$

Comparing Eqs. (4-107) and (4-110), we can write an alternative expression for the state transition matrix as

$$\Phi(t,0) = e^{\mathbf{A}t} \tag{4-111}$$

Let us observe here that integrating between $t_0$ and $t$ results in

$$e^{-\mathbf{A}t}\mathbf{x}(t) - e^{-\mathbf{A}t_0}\mathbf{x}(t_0) = 0$$

Thus if the initial state at $t = t_0$ is available, we can write

$$\mathbf{x}(t) = e^{\mathbf{A}(t-t_0)}\mathbf{x}(t_0) \tag{4-112}$$

Thus in the terms of the state transition matrix, we have

$$\mathbf{x}(t) = \Phi(t, t_0)\mathbf{x}(t_0) \tag{4-113}$$

where

$$\Phi(t, t_0) = e^{\mathbf{A}(t-t_0)} \tag{4-114}$$

The matrix exponential $e^{\mathbf{A}t}$ can be obtained using an infinite series expansion given by

$$e^{\mathbf{A}t} = I + \sum_{k=1}^{\infty} \frac{\mathbf{A}^k t^k}{k!} \tag{4-115}$$

provided that the series converges. This method is inefficient in general. A simple example will illustrate the procedure.

EXAMPLE 4.16

Find the state transition matrix $\Phi(t,0)$ associated with the matrix

$$\mathbf{A} = \begin{bmatrix} 0 & 0 \\ 0 & 1 \end{bmatrix}$$

SOLUTION

We have

$$sI - \mathbf{A} = \begin{bmatrix} s & 0 \\ 0 & s-1 \end{bmatrix}$$

As a result,

$$(s\mathbf{I} - \mathbf{A})^{-1} = \begin{bmatrix} \frac{1}{s} & 0 \\ 0 & \frac{1}{s-1} \end{bmatrix}$$

We then conclude that

$$\Phi(t,0) = \mathscr{L}^{-1}\{(s\mathbf{I} - \mathbf{A})^{-1}\} = \begin{bmatrix} 1 & 0 \\ 0 & e^t \end{bmatrix}$$

An alternative solution is as follows:

$$\mathbf{A}^2 = \begin{bmatrix} 0 & 0 \\ 0 & 1 \end{bmatrix}\begin{bmatrix} 0 & 0 \\ 0 & 1 \end{bmatrix} = \begin{bmatrix} 0 & 0 \\ 0 & 1 \end{bmatrix} = \mathbf{A}$$

$$\mathbf{A}^3 = \mathbf{A}^2 = \mathbf{A}, \quad \text{etc.}$$

We can use the following expansion:

$$e^{\mathbf{A}t} = \mathbf{I} + \mathbf{A}t + \mathbf{A}^2\frac{t^2}{2!} + \cdots$$

Thus

$$e^{\mathbf{A}t} = \mathbf{I} + \mathbf{A}\left(t + \frac{t^2}{2!} + \cdots\right)$$

or

$$e^{\mathbf{A}t} = \mathbf{I} + \mathbf{A}(e^t - 1)$$

Numerically, we obtain

$$e^{\mathbf{A}t} = \begin{bmatrix} 1 & 0 \\ 0 & 1 \end{bmatrix} + \begin{bmatrix} 0 & 0 \\ 0 & 1 \end{bmatrix}(e^t - 1)$$

$$= \begin{bmatrix} 1 & 0 \\ 0 & e^t \end{bmatrix}$$

The state transition matrix for systems described by a diagonal coefficient matrix $\mathbf{A}$ can be easily obtained as a diagonal matrix whose elements are exponentials of the corresponding elements in $\mathbf{A}$. Consider the matrix $\mathbf{A}$ given by

$$\mathbf{A} = \begin{bmatrix} \lambda_1 & 0 & & & & 0 \\ 0 & \lambda_2 & & & & 0 \\ 0 & 0 & \lambda_3 & & & 0 \\ & & & \ddots & & \\ 0 & 0 & 0 & & \lambda_{n-1} & 0 \\ 0 & 0 & 0 & & 0 & \lambda_n \end{bmatrix}$$

It is clear that

$$\Phi^{-1}(s) = \begin{bmatrix} s-\lambda_1 & 0 & 0 & \cdots & \cdots & 0 \\ 0 & s-\lambda_2 & 0 & \cdots & \cdots & 0 \\ 0 & 0 & s-\lambda_3 & \cdots & \cdots & 0 \\ \vdots & \vdots & 0 & \ddots & & \vdots \\ 0 & 0 & 0 & & s-\lambda_{n-1} & 0 \\ 0 & 0 & 0 & & 0 & s-\lambda_n \end{bmatrix}$$

Thus since the inverse of a diagonal matrix is a diagonal matrix whose elements are the inverse of the original elements, we conclude that

$$\Phi(s) = \begin{bmatrix} \frac{1}{s-\lambda_1} & & & & \\ & \frac{1}{s-\lambda_2} & & & \\ & & \ddots & & \\ & & & \frac{1}{s-\lambda_{n-1}} & \\ & & & & \frac{1}{s-\lambda_n} \end{bmatrix}$$

The inverse Laplace transform gives us

$$\Phi(t,0) = \begin{bmatrix} e^{\lambda_1 t} & & & & \\ & e^{\lambda_2 t} & & & \\ & & \ddots & & \\ & & & e^{\lambda_{n-1} t} & \\ & & & & e^{\lambda_n t} \end{bmatrix}$$

Thus

$$e^{\mathbf{A}t} = \operatorname{diag}(e^{\lambda_i t}) \tag{4-116}$$

for

$$\mathbf{A} = \operatorname{diag}(\lambda_i) \tag{4-117}$$

Of course, the alert reader will realize that the system described above is simply decoupled in the sense that the $i$th component's dynamics are given by

$$\dot{x}_i = \lambda_i x_i$$

The solution is

$$x_i(t) = e^{\lambda_i t} x_i(0)$$

As a result, Eq. (4-116) is obtained. Equation (4-116) is useful in evaluating the state transition matrix of a system once the system equations are transformed into diagonal form as shown in the next section.

## 4.6 State Model Transformations

Consider a system with a state-space representation given by

$$\dot{\mathbf{x}} = \mathbf{A}\mathbf{x} + \mathbf{B}\mathbf{u} \tag{4-118}$$

It is possible to obtain an alternative representation of the system of desired properties by using appropriate transformations. For illustration purposes, we deal with a second-order system; however, our conclusions apply in general.

Assume that instead of using the states $x_1$ and $x_2$ for a second-order system, we desire new states $\tilde{x}_1$ and $\tilde{x}_2$ related to $x_1$ and $x_2$ by

$$\tilde{x}_1 = t_{11}x_1 + t_{12}x_2$$
$$\tilde{x}_2 = t_{21}x_1 + t_{22}x_2$$

In vector form we have

$$\tilde{\mathbf{x}} = \mathbf{T}\mathbf{x} \tag{4-119}$$

where

$$\mathbf{T} = \begin{bmatrix} t_{11} & t_{12} \\ t_{21} & t_{22} \end{bmatrix} \tag{4-120}$$

Differentiating Eq. (4-119), we get

$$\dot{\tilde{\mathbf{x}}} = \mathbf{T}\dot{\mathbf{x}} \tag{4-121}$$

Using Eq. (4-118), we thus have

$$\dot{\tilde{\mathbf{x}}} = \mathbf{T}(\mathbf{A}\mathbf{x} + \mathbf{B}\mathbf{u})$$

Using Eq. (4-119), we get

$$\dot{\tilde{\mathbf{x}}} = \mathbf{T}\mathbf{A}\mathbf{T}^{-1}\tilde{\mathbf{x}} + \mathbf{T}\mathbf{B}\mathbf{u}$$

The equation above is of the form

$$\dot{\tilde{\mathbf{x}}} = \tilde{\mathbf{A}}\tilde{\mathbf{x}} + \tilde{\mathbf{B}}\mathbf{u} \tag{4-122}$$

where

$$\tilde{\mathbf{A}} = \mathbf{T}\mathbf{A}\mathbf{T}^{-1} \tag{4-123}$$

or

$$\tilde{\mathbf{A}}\mathbf{T} = \mathbf{T}\mathbf{A} \tag{4-124}$$

A useful application of the result above is when it is desired to obtain $\tilde{\mathbf{A}}$ in diagonal form to facilitate the calculation of the state transition matrix. An example follows.

EXAMPLE 4.17

For the system of Example 4.2, find a state-space representation for which the coefficient matrix $\mathbf{A}$ is diagonal. Find the relation between the new state variables and the original ones.

SOLUTION

Let us choose $\tilde{\mathbf{A}}$ to be the diagonal matrix:

$$\tilde{\mathbf{A}} = \begin{bmatrix} \lambda_1 & 0 \\ 0 & \lambda_2 \end{bmatrix}$$

We employ

$$\tilde{\mathbf{A}}\mathbf{T} = \mathbf{T}\mathbf{A}$$

Thus

$$\begin{bmatrix} \lambda_1 & 0 \\ 0 & \lambda_2 \end{bmatrix}\begin{bmatrix} t_{11} & t_{12} \\ t_{21} & t_{22} \end{bmatrix} = \begin{bmatrix} t_{11} & t_{12} \\ t_{21} & t_{22} \end{bmatrix}\begin{bmatrix} 0 & 1 \\ -3 & -4 \end{bmatrix}$$

or

$$\begin{bmatrix} \lambda_1 t_{11} & \lambda_1 t_{12} \\ \lambda_2 t_{21} & \lambda_2 t_{22} \end{bmatrix} = \begin{bmatrix} -3t_{12} & t_{11} - 4t_{12} \\ -3t_{22} & t_{21} - 4t_{22} \end{bmatrix}$$

Thus in expanded form we have

$$\lambda_1 t_{11} = -3t_{12}$$
$$\lambda_1 t_{12} = t_{11} - 4t_{12}$$

Combining the relations above, we get

$$\lambda_1^2 + 4\lambda_1 + 3 = 0$$

The solution is

$$\lambda_1 = -1 \quad \text{or} \quad -3$$

Take $\lambda_1 = -1$. Thus

$$t_{11} = 3t_{12}$$

The second row gives

$$\lambda_2 t_{21} = -3t_{22}$$
$$\lambda_2 t_{22} = t_{21} - 4t_{22}$$

Combining the above, we get

$$\lambda_2^2 + 4\lambda_2 + 3 = 0$$

Thus

$$\lambda_2 = -3 \quad \text{or} \quad -1$$

We choose

$$\lambda_2 = -3$$

Note that $\lambda_1$ and $\lambda_2$ occur in pairs. For our choice we get

$$t_{21} = t_{22}$$

Observe that we have only two equations in the four elements $t_{11}$, $t_{12}$, $t_{21}$, and $t_{22}$. Any arbitrary choice of two elements defines the remaining elements. Let us then assume that

$$t_{11} = 3$$
$$t_{22} = 1$$

Thus

$$t_{12} = 1$$
$$t_{21} = 1$$

As a result,

$$\mathbf{T} = \begin{bmatrix} 3 & 1 \\ 1 & 1 \end{bmatrix}$$

The inverse of **T** is given by

$$\mathbf{T}^{-1} = \begin{bmatrix} 0.5 & -0.5 \\ -0.5 & 1.5 \end{bmatrix}$$

Now our diagonal transformation matrix is given by

$$\tilde{\mathbf{A}} = \begin{bmatrix} -1 & 0 \\ 0 & -3 \end{bmatrix}$$

Let us first check to see if indeed $\tilde{\mathbf{A}}$ will be a diagonal matrix. For this we use

$$\tilde{\mathbf{A}} = \mathbf{TAT}^{-1}$$

Thus

$$\begin{aligned} \tilde{\mathbf{A}} &= \begin{bmatrix} 3 & 1 \\ 1 & 1 \end{bmatrix} \begin{bmatrix} 0 & 1 \\ -3 & -4 \end{bmatrix} \begin{bmatrix} 0.5 & -0.5 \\ -0.5 & 1.5 \end{bmatrix} \\ &= \begin{bmatrix} -3 & -1 \\ -3 & -3 \end{bmatrix} \begin{bmatrix} 0.5 & -0.5 \\ -0.5 & 1.5 \end{bmatrix} \\ &= \begin{bmatrix} -1 & 0 \\ 0 & -3 \end{bmatrix} \end{aligned}$$

Using the transformation **T**, the new state variables $\tilde{x}_1$ and $\tilde{x}_2$ are related to the state variables $x_1$ and $x_2$ by

$$\tilde{x}_1 = t_{11}x_1 + t_{12}x_2$$

or

$$\tilde{x}_1 = 3x_1 + x_2$$

and

$$\tilde{x}_2 = t_{21}x_1 + t_{22}x_2$$

or

$$\tilde{x}_2 = x_1 + x_2$$

The inverse relation is given by

$$\mathbf{X} = \mathbf{T}^{-1}\tilde{\mathbf{X}}$$

or

$$\begin{aligned} x_1 &= 0.5\tilde{x}_1 - 0.5\tilde{x}_2 \\ x_2 &= -0.5\tilde{x}_1 + 1.5\tilde{x}_2 \end{aligned}$$

The procedure adopted in this section to obtain the necessary transformations is simple and looks somewhat to be ad hoc. In Section 4.8 we will find a more rigorous basis for finding the diagonalizing transformation. At this time it is appropriate to consider the total response of the dynamic system described by the state equation.

# 4.7 Forced Response

Consider the linear dynamic system described by the state equation

$$\dot{\mathbf{x}}(t) = \mathbf{A}\mathbf{x}(t) + \mathbf{B}\mathbf{u}(t) \tag{4-125}$$

The solution to the state equation provides us with the forced response of the system, giving the variation of $\mathbf{x}(t)$ with time, given an initial state $\mathbf{x}(0)$ and a specified input control $\mathbf{u}(t)$. This case is similar to the free-response case in the sense that we can find the solution using a number of different approaches.

Let us rewrite Eq. (4-125) as

$$\dot{\mathbf{x}}(t) - \mathbf{A}\mathbf{x}(t) = \mathbf{B}\mathbf{u}(t) \tag{4-126}$$

We now premultiply both sides by $e^{-\mathbf{A}t}$, and as a result we write

$$e^{-\mathbf{A}t}\dot{\mathbf{x}}(t) - e^{-\mathbf{A}t}\mathbf{A}\mathbf{x}(t) = e^{-\mathbf{A}t}\mathbf{B}\mathbf{u}(t)$$

The left-hand side can be expressed as

$$\frac{d}{dt}\left[e^{-\mathbf{A}t}\mathbf{x}(t)\right] = e^{-\mathbf{A}t}\mathbf{B}\mathbf{u}(t)$$

Upon integration we conclude that

$$e^{-\mathbf{A}t}\mathbf{x}(t) - e^{-\mathbf{A}t_0}\mathbf{x}(t_0) = \int_{t_0}^{t} e^{-\mathbf{A}\tau}\mathbf{B}\mathbf{u}(\tau)\,d\tau$$

As a result,

$$\mathbf{x}(t) = e^{\mathbf{A}(t-t_0)}\mathbf{x}(t_0) + \int_{t_0}^{t} e^{\mathbf{A}(t-\tau)}\mathbf{B}\mathbf{u}(\tau)\,d\tau \tag{4-127}$$

This provides the required answer, which can be written as

$$\mathbf{x}(t) = \mathbf{\Phi}(t, t_0)\mathbf{x}(t_0) + \int_{t_0}^{t} \mathbf{\Phi}(t, \tau)\mathbf{B}\mathbf{u}(\tau)\,d\tau \tag{4-128}$$

where

$$\mathbf{\Phi}(t, \tau) = e^{\mathbf{A}(t-\tau)} \tag{4-129}$$

It is appropriate here to comment on our result. There are two components to the response, $\mathbf{x}_f(t)$ and $\mathbf{x}_u(t)$.

$$\mathbf{x}(t) = \mathbf{x}_f(t) + \mathbf{x}_u(t) \tag{4-130}$$

The free-response component is given by

$$\mathbf{x}_f(t) = \mathbf{\Phi}(t, t_0)\mathbf{x}(t_0) \tag{4-131}$$

This is simply $\mathbf{x}(t)$ for zero input control. The controlled component $\mathbf{x}_u(t)$ is that due to the control input $\mathbf{u}(t)$ and is given by

$$\mathbf{x}_u(t) = \int_{t_0}^{t} \mathbf{\Phi}(t, \tau)\mathbf{B}\mathbf{u}(\tau)\, d\tau \tag{4-132}$$

We can arrive at Eq. (4-128) using a Laplace-transform-based argument. This follows if we take the Laplace transform of both sides of Eq. (4-125) to obtain

$$s\mathbf{X}(s) - \mathbf{x}(0) = \mathbf{A}\mathbf{X}(s) + \mathbf{B}\mathbf{u}(s) \tag{4-133}$$

Rearranging, we obtain

$$(s\mathbf{I} - \mathbf{A})\mathbf{x}(s) = \mathbf{x}(0) + \mathbf{B}\mathbf{u}(s) \tag{4-134}$$

Recall that

$$\mathbf{\Phi}(s) = (s\mathbf{I} - \mathbf{A})^{-1}$$

Thus

$$\mathbf{X}(s) = \mathbf{\Phi}(s)\mathbf{x}(0) + \mathbf{\Phi}(s)\mathbf{B}\mathbf{u}(s) \tag{4-135}$$

The inverse Laplace transform yields

$$\mathbf{x}(t) = \Phi(t,0)\mathbf{x}(0) + \int_{0}^{t} \mathbf{\Phi}(t, \tau)\mathbf{B}\mathbf{u}(\tau)\, d\tau \tag{4-136}$$

This is precisely Eq. (4-128) with $t_0 = 0$. A pair of examples is appropriate at this time.

EXAMPLE 4.18

Assume that the initial conditions for the system of Example 4.14 are given by

$$x_1(0) = 0$$
$$x_2(0) = 1$$

and that the input is a unit step. Obtain the free response $x_{1_f}(t)$ and $x_{2_f}(t)$ as well as the response to the step input.

SOLUTION

We have seen that the free response is given by

$$\mathbf{x}_f(t) = \mathbf{\Phi}(t,0)\mathbf{x}(0)$$

The free response of the system is thus given by

$$\begin{bmatrix} x_{1_f}(t) \\ x_{2_f}(t) \end{bmatrix} = \begin{bmatrix} 1.5e^{-t} - 0.5e^{-3t} & 0.5e^{-t} - 0.5e^{-3t} \\ -1.5e^{-t} + 1.5e^{-3t} & 1.5e^{-3t} - 0.5e^{-t} \end{bmatrix} \begin{bmatrix} 0 \\ 1 \end{bmatrix}$$

Thus component-wise we have

$$x_{1_f}(t) = 0.5e^{-t} - 0.5e^{-3t}$$
$$x_{2_f}(t) = 1.5e^{-3t} - 0.5e^{-t}$$

With a forcing function $u(t)$, the solution of the state equations is

$$\mathbf{x}(t) = \mathbf{\Phi}(t,0)\mathbf{x}(0) + \int_0^t \mathbf{\Phi}(t,\tau)\mathbf{B}(\tau)u(\tau)\,d\tau$$

The first term has been expressed before. Now the integrand is given by

$$\mathbf{\Phi}(t,\tau)\mathbf{B}(\tau)u(\tau) = \begin{bmatrix} \Phi_{11}(t,\tau) & \Phi_{12}(t,\tau) \\ \Phi_{21}(t,\tau) & \Phi_{22}(t,\tau) \end{bmatrix} \begin{bmatrix} 0 \\ 1 \end{bmatrix} u(\tau)$$
$$= \begin{bmatrix} \Phi_{12}(t,\tau) \\ \Phi_{22}(t,\tau) \end{bmatrix} u(\tau)$$

Thus component-wise we have

$$x_1(t) = x_{1_f}(t) + \int_0^t \Phi_{12}(t,\tau)u(\tau)\,d\tau$$
$$x_2(t) = x_{2_f}(t) + \int_0^t \Phi_{22}(t,\tau)u(\tau)\,d\tau$$

For $u(t) = 1$ we obtain

$$x_1(t) = x_{1_f}(t) + \int_0^t (0.5e^{-(t-\tau)} - 0.5e^{-3(t-\tau)})\,d\tau$$
$$= \tfrac{1}{3}(1 - e^{-3t})$$
$$x_2(t) = x_{2_f}(t) + \int_0^t (1.5e^{-3(t-\tau)} - 0.5e^{-(t-\tau)})\,d\tau$$
$$= e^{-3t}$$

We can verify that

$$\dot{x}_1(t) = x_2(t)$$

as required by the original representation of the system.

EXAMPLE 4.19

Use the transformed system representation obtained in Example 4.17 to solve Example 4.18.

SOLUTION

The state transition matrix for the diagonal matrix $\tilde{\mathbf{A}}$ is given by

$$\tilde{\mathbf{\Phi}}(t,0) = \begin{bmatrix} e^{\lambda_1 t} & 0 \\ 0 & e^{\lambda_2 t} \end{bmatrix}$$

With the matrix $\tilde{\mathbf{A}}$ obtained, the state transition matrix $\tilde{\mathbf{\Phi}}$ is given by

$$\tilde{\mathbf{\Phi}}(t,0) = \begin{bmatrix} e^{-t} & 0 \\ 0 & e^{-3t} \end{bmatrix}$$

The initial states $\tilde{x}_1$ and $\tilde{x}_2$ are

$$\begin{aligned} \tilde{x}_1(0) &= 3x_1(0) + x_2(0) \\ &= (3) \times (0) + 1 = 1 \\ \tilde{x}_2(0) &= x_1(0) + x_2(0) \\ &= 0 + 1 = 1 \end{aligned}$$

Thus the free response of the transformed system is

$$\begin{bmatrix} \tilde{x}_{1_f}(t) \\ \tilde{x}_{2_f}(t) \end{bmatrix} = \begin{bmatrix} e^{-t} & 0 \\ 0 & e^{-3t} \end{bmatrix} \begin{bmatrix} 1 \\ 1 \end{bmatrix}$$

This reduces to

$$\begin{aligned} \tilde{x}_{1_f}(t) &= e^{-t} \\ \tilde{x}_{2_f}(t) &= e^{-3t} \end{aligned}$$

Now to get $x_1(t)$ and $x_2(t)$ we have

$$\begin{aligned} x_{1_f}(t) &= 0.5\tilde{x}_{1_f}(t) - 0.5\tilde{x}_{2_f}(t) \\ &= 0.5e^{-t} - 0.5e^{-3t} \\ x_{2_f}(t) &= -0.5\tilde{x}_{1_f}(t) + 1.5\tilde{x}_{2_f}(t) \\ &= 1.5e^{-3t} - 0.5e^{-t} \end{aligned}$$

This is the same result as obtained before.

For the forced response we have the relation

$$\begin{aligned} \tilde{\mathbf{B}} &= \mathbf{TB} \\ &= \begin{bmatrix} 3 & 1 \\ 1 & 1 \end{bmatrix} \begin{bmatrix} 0 \\ 1 \end{bmatrix} = \begin{bmatrix} 1 \\ 1 \end{bmatrix} \end{aligned}$$

Thus

$$\begin{aligned} \tilde{\mathbf{\Phi}}(t,\tau)\tilde{\mathbf{B}}(\tau)\mathbf{u}(\tau) &= \begin{bmatrix} e^{-(t-\tau)} & 0 \\ 0 & e^{-3(t-\tau)} \end{bmatrix} \begin{bmatrix} 1 \\ 1 \end{bmatrix} u(\tau) \\ &= \begin{bmatrix} e^{-(t-\tau)} \\ e^{-3(t-\tau)} \end{bmatrix} u(\tau) \end{aligned}$$

For $u(t) = 1$,

$$\begin{aligned} \tilde{x}_1(t) &= \tilde{x}_{1_f}(t) + \int_0^t e^{-(t-\tau)}\, d\tau \\ &= e^{-t} + (1 - e^{-t}) \\ &= 1 \\ \tilde{x}_2(t) &= \tilde{x}_{2_f}(t) + \int_0^t e^{-3(t-\tau)}\, d\tau \\ &= e^{-3t} + \left(\tfrac{1}{3}e^{-3(t-\tau)}\right)_0^t \\ &= \tfrac{1}{3}(1 + 2e^{-3t}) \end{aligned}$$

We thus have

$$\begin{aligned} x_1(t) &= 0.5[\tilde{x}_1(t) - \tilde{x}_2(t)] \\ &= 0.5(1 - \tfrac{1}{3} - \tfrac{2}{3}e^{-3t}) \\ &= \tfrac{1}{3}(1 - e^{-3t}) \\ x_2(t) &= -0.5\tilde{x}_1(t) + 1.5\tilde{x}_2(t) \\ &= -0.5 + \tfrac{1}{2}(1 + 2e^{-3t}) \\ &= e^{-3t} \end{aligned}$$

This agrees with the results obtained before.

# *4.8 Characteristic Values and Vectors of a Matrix: Diagonalization*

Consider the vector-matrix equation

$$\mathbf{Y} = \mathbf{AX} \tag{4-137}$$

where $\mathbf{Y}$ and $\mathbf{X}$ are $n \times 1$ column vectors and $\mathbf{A}$ is an $n \times n$ square matrix. This represents a transformation of $\mathbf{X}$ into $\mathbf{Y}$. We would like to find the vector $\mathbf{X}_e$ such that its transformation $\mathbf{Y}_e$ is in the same direction as $\mathbf{X}_e$; thus we would like to find

$$\mathbf{Y}_e = \mathbf{AX}_e \tag{4-138}$$

such that

$$\mathbf{Y}_e = \lambda \mathbf{X}_e \tag{4-139}$$

The condition for this to occur is that there is a solution to

$$\mathbf{AX}_e = \lambda \mathbf{X}_e \tag{4-140}$$

Expanding the equation above results in

$$\begin{aligned} (a_{11} - \lambda)X_{1_e} + a_{12}X_{2_e} + \cdots + a_{1n}X_{n_e} &= 0 \\ a_{21}X_{1_e} + (a_{22} - \lambda)X_{2_e} + \cdots + a_{2n}X_{n_e} &= 0 \\ &\vdots \\ a_{n1}X_{1_e} + a_{n2}X_{2_e} + \cdots + (a_{nn} - \lambda)X_{n_e} &= 0 \end{aligned} \tag{4-141}$$

Thus compactly, we are required to solve

$$(\lambda \mathbf{I} - \mathbf{A})\mathbf{X}_e = \mathbf{0} \tag{4-142}$$

This system has a solution if and only if

$$P(\lambda) = |\lambda \mathbf{I} - \mathbf{A}| = 0 \tag{4-143}$$

The values of $\lambda$ satisfying the above are called the *characteristic values* (or *eigenvalues*) of the matrix **A**. The corresponding solution $\mathbf{X}_e$ is a characteristic vector of the matrix **A**.

The $n$th-order polynomial $P(\lambda)$ is called the *characteristic equation* corresponding to the matrix **A**.

$$P(\lambda) = \lambda^n + a_1\lambda^{n-1} + a_2\lambda^{n-2} + \cdots + a_{n-1}\lambda + a_n$$
$$= 0 \tag{4-144}$$

The roots of $P(\lambda)$ are the characteristic values of **A**.

Some important properties of the characteristic equation are as follows:

1. The characteristic polynomial evaluated at $\lambda = 0$ is related to the determinant of **A**. Since

$$P(0) = |0\mathbf{I} - \mathbf{A}| = (-1)^n|\mathbf{A}| \tag{4-145}$$

Note also that

$$P(0) = a_n$$

2. If $P(\lambda)$ is written in factored form,

$$P(\lambda) = (\lambda - \lambda_1)(\lambda - \lambda_2)\cdots(\lambda - \lambda_n) \tag{4-146}$$

then

$$P(0) = (-1)^n(\lambda_1\lambda_2 \cdots \lambda_n)$$
$$= (-1)^n|\mathbf{A}| \tag{4-147}$$

Thus

$$\lambda_1\lambda_2 \cdots \lambda_n = |\mathbf{A}| \tag{4-148}$$

The product of the characteristic values is equal to the determinant of **A**.

3. The characteristic polynomial can be expanded as

$$P(\lambda) = \lambda^n - (\lambda_1 + \lambda_2 + \cdots + \lambda_n)\lambda^{n-1} + \cdots \tag{4-149}$$

Another way of expanding $P(\lambda)$ is given by

$$P(\lambda) = |\lambda\mathbf{I} - \mathbf{A}| = \lambda^n - (a_{11} + a_{22} + \cdots + a_{nn})\lambda^{n-1} + \cdots \tag{4-150}$$

As a result, we conclude that

$$a_1 = -(\lambda_1 + \lambda_2 + \cdots + \lambda_n)$$
$$= -(a_{11} + a_{22} + \cdots + a_{nn}) \tag{4-151}$$

Thus

$$\sum_{i=1}^{n} \lambda_i = \sum_{i=1}^{n} a_{ii} \tag{4-152}$$

The trace of **A** is defined as

$$T_r = \sum_{i=1}^{n} a_{ii} \tag{4-153}$$

4. Let $A^k$ be the matrix formed by raising **A** to the power of $k$. Thus

$$\mathbf{A}^2 = \mathbf{A} \cdot \mathbf{A}$$
$$\mathbf{A}^3 = \mathbf{A} \cdot \mathbf{A} \cdot \mathbf{A}$$

and so on. The trace of $\mathbf{A}^k$ is denoted $T_r^{(k)}$, we have the following results:

$$a_1 = -T_r^{(1)} \tag{4-154}$$
$$a_2 = -\tfrac{1}{2}\left[a_1 T_r^{(1)} + T_r^{(2)}\right] \tag{4-155}$$
$$a_3 = -\tfrac{1}{3}\left[a_2 T_r^{(1)} + a_1 T_r^{(2)} + T_r^{(3)}\right] \tag{4-156}$$
$$\vdots$$
$$a_n = -\frac{1}{n}\left[a_{n-1} T_r^{(1)} + a_{n-2} T_r^{(2)} + \cdots + a_1 T_r^{(n-1)} + T_r^{(n)}\right] \tag{4-157}$$

This provides a useful recursive formula (called *Bôcher's formula*) for the evaluation of the characteristic equation using a digital computer.

EXAMPLE 4.20

Find the characteristic equation of **A**, where

$$\mathbf{A} = \begin{bmatrix} 3 & -1 & 4 \\ 2 & 2 & 2 \\ 2 & 4 & 0 \end{bmatrix}$$

SOLUTION

Using Bôcher's formula, we have

$$a_1 = -T_r^{(1)} = -(3 + 2 + 0) = -5$$

$$\mathbf{A}^2 = \begin{bmatrix} 3 & -1 & 4 \\ 2 & 2 & 2 \\ 2 & 4 & 0 \end{bmatrix}\begin{bmatrix} 3 & -1 & 4 \\ 2 & 2 & 2 \\ 2 & 4 & 0 \end{bmatrix} = \begin{bmatrix} 15 & 11 & 10 \\ 14 & 10 & 12 \\ 14 & 6 & 16 \end{bmatrix}$$

$$T_r^{(2)} = 41$$

Thus

$$a_2 = -\tfrac{1}{2}\left(a_1 T_r^{(1)} + T_r^{(2)}\right) = -\tfrac{1}{2}\left[(-5)(5) + 41\right] = -8$$

Similarly,

$$\mathbf{A}^3 = \begin{bmatrix} 3 & -1 & 4 \\ 2 & 2 & 2 \\ 2 & 4 & 0 \end{bmatrix} \begin{bmatrix} 15 & 11 & 10 \\ 14 & 10 & 12 \\ 14 & 6 & 16 \end{bmatrix}$$

$$= \begin{bmatrix} 87 & 47 & 82 \\ 86 & 54 & 76 \\ 86 & 62 & 68 \end{bmatrix}$$

$$T_r^{(3)} = 209$$

so that

$$\begin{aligned} a_3 &= -\tfrac{1}{3}\left[a_2 T_r^{(1)} + a_1 T_r^{(2)} + T_r^{(3)}\right] \\ &= -\tfrac{1}{3}[(-8)(5) + (-5)(41) + 209] \\ &= 12 \end{aligned}$$

Thus the characteristic polynomial is

$$P(\lambda) = \lambda^3 - 5\lambda^2 - 8\lambda + 12$$

We can verify the results above as follows:

$$P(\lambda) = |\lambda\mathbf{I} - \mathbf{A}| = \begin{vmatrix} \lambda - 3 & 1 & -4 \\ -2 & \lambda - 2 & -2 \\ -2 & -4 & \lambda \end{vmatrix}$$

$$\begin{aligned} &= (\lambda - 3)[\lambda(\lambda - 2) - 8] - 1(-2\lambda - 4) \\ &\quad -4[8 + 2(\lambda - 2)] \\ &= (\lambda - 3)(\lambda^2 - 2\lambda - 8) + (2\lambda + 4) - (16 + 8\lambda) \end{aligned}$$

$$\begin{aligned} P(\lambda) &= \lambda^3 - 2\lambda^2 - 8\lambda \\ &\quad\;\; - 3\lambda^2 + 6\lambda + 24 \\ &\quad\;\; + 2\lambda + 4 \\ &\quad\;\; - 8\lambda - 16 \\ \hline P(\lambda) &= \lambda^3 - 5\lambda^2 - 8\lambda + 12 \end{aligned}$$

## Characteristic Vectors of a Matrix

Given a matrix $\mathbf{A}$ with characteristic or eigenvalues $\lambda_1, \ldots, \lambda_n$, the eigenvectors of the matrix satisfy the relations

$$\mathbf{A}\mathbf{u}_i = \lambda_i \mathbf{u}_i \tag{4-158}$$

The $\mathbf{u}_i$'s are called *eigenvectors*.

The matrix $\mathbf{U}$ of eigenvectors is nonsingular if the eigenvectors are linearly independent:

$$\mathbf{U} = [\mathbf{u}_1 \quad \mathbf{u}_2 \quad \cdots \quad \mathbf{u}_n] \tag{4-159}$$

Let

$$\mathbf{U}^{-1} = \mathbf{V} = \begin{bmatrix} \mathbf{V}_1^T \\ \mathbf{V}_2^T \\ \vdots \\ \mathbf{V}_n^T \end{bmatrix} \tag{4-160}$$

Since

$$\mathbf{U}^{-1}\mathbf{U} = \mathbf{I}$$

thus in expanded form we have

$$\begin{bmatrix} \mathbf{V}_1^T \\ \mathbf{V}_2^T \\ \vdots \\ \mathbf{V}_n^T \end{bmatrix} [\mathbf{u}_1 \quad \mathbf{u}_2 \quad \cdots \quad \mathbf{u}_n] = \begin{bmatrix} \mathbf{V}_1^T\mathbf{u}_1 & \mathbf{V}_1^T\mathbf{u}_2 & \cdots & \mathbf{V}_1^T\mathbf{u}_n \\ \mathbf{V}_2^T\mathbf{u}_1 & \mathbf{V}_2^T\mathbf{u}_2 & \cdots & \mathbf{V}_2^T\mathbf{u}_n \\ \vdots & & & \\ \mathbf{V}_n^T\mathbf{u}_1 & \mathbf{V}_n^T\mathbf{u}_2 & \cdots & \mathbf{V}_n^T\mathbf{u}_n \end{bmatrix} \tag{4-161}$$

Thus one can conclude that component-wise

$$\mathbf{V}_i^T\mathbf{u}_i = 1 \tag{4-162}$$

$$\mathbf{V}_i^T\mathbf{u}_j = 0 \tag{4-163}$$

## Diagonalization

Consider now the matrix product

$$\tilde{\mathbf{A}} = \mathbf{U}^{-1}\mathbf{A}\mathbf{U} \tag{4-164}$$

Using Eq. (4-160), we have

$$\tilde{\mathbf{A}} = \mathbf{V}\mathbf{A}\mathbf{U} \tag{4-165}$$

In terms of the eigenvectors, we have

$$\tilde{\mathbf{A}} = \mathbf{V}\mathbf{A}[\mathbf{u}_1, \mathbf{u}_2, \ldots, \mathbf{u}_n] \tag{4-166}$$

Multiplying, we get

$$\tilde{\mathbf{A}} = \mathbf{V}[\mathbf{A}\mathbf{u}_1, \mathbf{A}\mathbf{u}_2, \mathbf{A}\mathbf{u}_3, \ldots, \mathbf{A}\mathbf{u}_n] \tag{4-167}$$

Using Eq. (4-158), we get

$$\tilde{\mathbf{A}} = \mathbf{V}[\lambda_1\mathbf{u}_1, \lambda_2\mathbf{u}_2, \lambda_3\mathbf{u}_3, \ldots, \lambda_n\mathbf{u}_n] \tag{4-168}$$

Substituting for $\mathbf{V}$ in partitioned form, we get

$$\tilde{\mathbf{A}} = \begin{bmatrix} \mathbf{V}_1^T \\ \mathbf{V}_2^T \\ \vdots \\ \mathbf{V}_n^T \end{bmatrix} [\lambda_1\mathbf{u}_1, \lambda_2\mathbf{u}_2, \lambda_3\mathbf{u}_3, \ldots, \lambda_n\mathbf{u}_n] \tag{4-169}$$

As a result,

$$\tilde{\mathbf{A}} = \begin{bmatrix} \lambda_1 \mathbf{V}_1^T \mathbf{u}_1, \lambda_2 \mathbf{V}_1^T \mathbf{u}_2, \ldots, \lambda_n \mathbf{V}_1^T \mathbf{u}_n \\ \vdots \\ \lambda_1 \mathbf{V}_n^T \mathbf{u}_1, \lambda_2 \mathbf{V}_n^T \mathbf{u}_2, \ldots, \lambda_n \mathbf{V}_n^T \mathbf{u}_n \end{bmatrix} \tag{4-170}$$

Utilizing Eqs. (4-162) and (4-163), we conclude that

$$\tilde{\mathbf{A}} = \begin{bmatrix} \lambda_1 & 0 & & 0 \\ 0 & \lambda_2 & & \\ \vdots & & \ddots & \\ & & & \lambda_n \end{bmatrix} \tag{4-171}$$

Thus $\tilde{\mathbf{A}}$ is a diagonal matrix whose elements are the eigenvalues of $\mathbf{A}$:

$$\tilde{\mathbf{A}} = \mathbf{U}^{-1}\mathbf{A}\mathbf{U} \tag{4-172}$$

The expression in terms of the transformation $\mathbf{T}$ is

$$\tilde{\mathbf{A}} = \mathbf{T}\mathbf{A}\mathbf{T}^{-1} \tag{4-173}$$

where

$$\mathbf{T}^{-1} = \mathbf{U} \tag{4-174}$$

We have now formally the basis for diagonalizing the state equations. Let us firm up the concepts developed with an example.

EXAMPLE 4.21

Consider the matrix

$$\mathbf{A} = \begin{bmatrix} -2 & 1 \\ 1 & -2 \end{bmatrix}$$

Find the diagonal form using the procedure developed in this section.

SOLUTION

To find the eigenvalues, we have

$$\lambda\mathbf{I} - \mathbf{A} = \begin{bmatrix} \lambda + 2 & -1 \\ -1 & \lambda + 2 \end{bmatrix}$$

Thus, the characteristic polynomial is given by

$$\begin{aligned} P(\lambda) = |\lambda\mathbf{I} - \mathbf{A}| &= (\lambda + 2)^2 - 1 \\ &= \lambda^2 + 4\lambda + 3 \\ &= (\lambda + 1)(\lambda + 3) \end{aligned}$$

Thus the eigenvalues are given by

$$\lambda_1 = -1$$
$$\lambda_2 = -3$$

We compute the eigenvectors as follows:

$$\mathbf{A}\mathbf{u}_1 = \lambda_1 \mathbf{u}_1 \qquad \text{for} \quad \lambda_1 = -1$$

Thus

$$\begin{bmatrix} -2 & 1 \\ 1 & -2 \end{bmatrix} \begin{bmatrix} u_{11} \\ u_{21} \end{bmatrix} = \begin{bmatrix} -u_{11} \\ -u_{21} \end{bmatrix}$$

The first row gives us

$$-2u_{11} + u_{21} = -u_{11}$$

or

$$u_{21} = u_{11}$$

The second row gives us

$$u_{11} - 2u_{21} = -u_{21}$$

or

$$u_{21} = u_{11}$$

Choose

$$u_{21} = u_{11} = 1$$

Thus the first eigenvector is given by

$$\mathbf{u}_1 = \begin{bmatrix} 1 \\ 1 \end{bmatrix}$$

The second eigenvector is obtained as follows:

$$\mathbf{A}\mathbf{u}_2 = \lambda_2 \mathbf{u}_2$$

$$\begin{bmatrix} -2 & 1 \\ 1 & -2 \end{bmatrix} \begin{bmatrix} u_{12} \\ u_{22} \end{bmatrix} = \begin{bmatrix} -3u_{12} \\ -3u_{22} \end{bmatrix}$$

The first row yields

$$-2u_{12} + u_{22} = -3u_{12}$$

or

$$u_{22} = -u_{12}$$

The second row yields

$$u_{12} - 2u_{22} = -3u_{22}$$

or

$$u_{22} = -u_{12}$$

Choose

$$u_{22} = 1 \qquad u_{12} = -1$$

Thus the second eigenvector is given by

$$\mathbf{u}_2 = \begin{bmatrix} -1 \\ 1 \end{bmatrix}$$

Collecting the eigenvectors into a matrix **U**, we have

$$\mathbf{T}^{-1} = [\mathbf{u}_1 \quad \mathbf{u}_2] = \begin{bmatrix} 1 & -1 \\ 1 & 1 \end{bmatrix}$$

Thus

$$\mathbf{T} = \begin{bmatrix} 0.5 & 0.5 \\ -0.5 & 0.5 \end{bmatrix}$$

As a result, the transformed matrix **A** is given by

$$\begin{aligned} \tilde{\mathbf{A}} &= \mathbf{TAT}^{-1} \\ &= \begin{bmatrix} 0.5 & 0.5 \\ -0.5 & 0.5 \end{bmatrix} \begin{bmatrix} -2 & 1 \\ 1 & -2 \end{bmatrix} \begin{bmatrix} 1 & -1 \\ 1 & 1 \end{bmatrix} \\ &= \begin{bmatrix} -0.5 & -0.5 \\ 1.5 & -1.5 \end{bmatrix} \begin{bmatrix} 1 & -1 \\ 1 & 1 \end{bmatrix} = \begin{bmatrix} -1 & 0 \\ 0 & -3 \end{bmatrix} \end{aligned}$$

# 4.9 System Response and Eigenvalues

Consider the force-free system

$$\dot{\mathbf{x}} = \mathbf{Ax} \tag{4-175}$$

Let us apply the transformation

$$\tilde{\mathbf{x}} = \mathbf{Tx} \tag{4-176}$$

Thus we have

$$\dot{\tilde{\mathbf{x}}} = \mathbf{TAT}^{-1}\tilde{\mathbf{x}} \tag{4-177}$$

The transformed system is represented by

$$\dot{\tilde{\mathbf{x}}} = \tilde{\mathbf{A}}\tilde{\mathbf{x}} \tag{4-178}$$

with

$$\tilde{\mathbf{A}} = \mathbf{TAT}^{-1} \tag{4-179}$$

The matrix exponential is thus

$$e^{\tilde{\mathbf{A}}T} = e^{\mathbf{TAT}^{-1}} \tag{4-180}$$

The time-response solution of the transformed system is

$$\tilde{\mathbf{x}}(t) = e^{\tilde{\mathbf{A}}t}\tilde{\mathbf{x}}_0 \tag{4-181}$$

while the solution of the original system is

$$\mathbf{x}(t) = e^{\mathbf{A}t}\mathbf{x}_0 \tag{4-182}$$

Multiply by **T**; then

$$\mathbf{Tx}(t) = \mathbf{T}e^{\mathbf{A}t}\mathbf{x}_0$$
$$\mathbf{Tx}(t) = \mathbf{T}e^{\mathbf{A}t}\mathbf{T}^{-1}\tilde{\mathbf{x}}_0$$

Thus we conclude that

$$\tilde{\mathbf{x}}(t) = \mathbf{T}e^{\mathbf{A}t}\mathbf{T}^{-1}\tilde{\mathbf{x}}_0 \tag{4-183}$$

By comparison

$$e^{\tilde{\mathbf{A}}t} = \mathbf{T}e^{\mathbf{A}t}\mathbf{T}^{-1} \tag{4-184}$$

Thus

$$e^{\tilde{\mathbf{A}}t} = e^{\mathbf{TAT}^{-1}t} = \mathbf{T}e^{\mathbf{A}t}\mathbf{T}^{-1} \tag{4-185}$$

If $\tilde{\mathbf{A}}$ is a diagonal matrix of the form

$$\tilde{\mathbf{A}} = \begin{bmatrix} \lambda_1 & & & 0 \\ & \lambda_2 & & \\ & & \ddots & \\ 0 & & & \lambda_n \end{bmatrix} \tag{4-186}$$

then

$$e^{\tilde{\mathbf{A}}t} = \begin{bmatrix} e^{\lambda_1 t} & & & \\ & e^{\lambda_2 t} & & \\ & & \ddots & \\ & & & e^{\lambda_n t} \end{bmatrix} \tag{4-187}$$

From the above we have a different way to get the state transition matrix:

$$\begin{aligned} \mathbf{e}^{\mathbf{A}t} &= \mathbf{T}^{-1}\mathbf{e}^{\tilde{\mathbf{A}}t}\mathbf{T} \\ &= \mathbf{T}^{-1}\begin{bmatrix} e^{\lambda_1 t} & & & \\ & e^{\lambda_2 t} & & \\ & & \ddots & \\ & & & e^{\lambda_n t} \end{bmatrix}\mathbf{T} \end{aligned} \tag{4-188}$$

Let us define

$$\mathbf{E}_i = \begin{matrix} \\ \\ i \\ \\ \end{matrix}\begin{bmatrix} 0 & & \overset{i}{} & \\ 0 & & & \\ \vdots & & -1 & \\ 0 & & & \end{bmatrix} \tag{4-189}$$

$\mathbf{E}_i$ is a diagonal matrix with zero elements except for the $i$th position. Thus

$$\begin{aligned} e^{\mathbf{A}t} &= \mathbf{T}^{-1}\left(\sum_{i=1}^{n} \mathbf{E}_i e^{\lambda_i t}\right)\mathbf{T} \\ &= \sum_{i=1}^{n} \mathbf{T}^{-1}\mathbf{E}_i\mathbf{T}e^{\lambda_i t} \end{aligned}$$

As a result, we have

$$e^{\mathbf{A}t} = \sum_{i=1}^{n} R_i e^{\lambda_i t} \tag{4-190}$$

where

$$R_i = \mathbf{T}^{-1}\mathbf{E}_i\mathbf{T} \tag{4-191}$$

Thus the solution to the force-free system

$$\dot{\mathbf{x}} = \mathbf{A}\mathbf{x}$$

is given by

$$\mathbf{x}(t) = \sum_{i=1}^{n} R_i e^{\lambda_i t}\mathbf{x}(0) \tag{4-192}$$

The discussion above shows that the free response of a system is equal to the sum of contributions of each eigenvalue $\lambda_i$. This result is important in analyzing the stability of the system, as discussed in Chapter 6.

# *Some Solved Problems*

PROBLEM 4A-1

For the networks shown in Figure 4.7, find the state-space equations.

SOLUTION

For the network of Figure 4.7a, we choose the voltage on the capacitor to be $x_1$ while the current in the inductor is denoted by $x_2$, as shown in Figure 4.8a. Writing Kirchhoff's voltage law for the outer loop, we have

$$e = R_1\left(x_2 + C\frac{dx_1}{dt}\right) + x_1 + R_2C\frac{dx_1}{dt}$$

Thus rearranging, we get

$$\frac{dx_1}{dt} = \frac{-1}{C(R_1 + R_2)}x_1 - \frac{R_1}{C(R_1 + R_2)}x_2 + \frac{e}{C(R_1 + R_2)}$$

The right-hand loop equation is

$$L\frac{dx_2}{dt} = x_1 + R_2C\frac{dx_1}{dt}$$

This can be rewritten as

$$L\frac{dx_2}{dt} = x_1 - \frac{R_2}{R_1 + R_2}x_1 - \frac{R_2R_1}{R_1 + R_2}x_2 + \frac{R_2}{R_1 + R_2}e$$

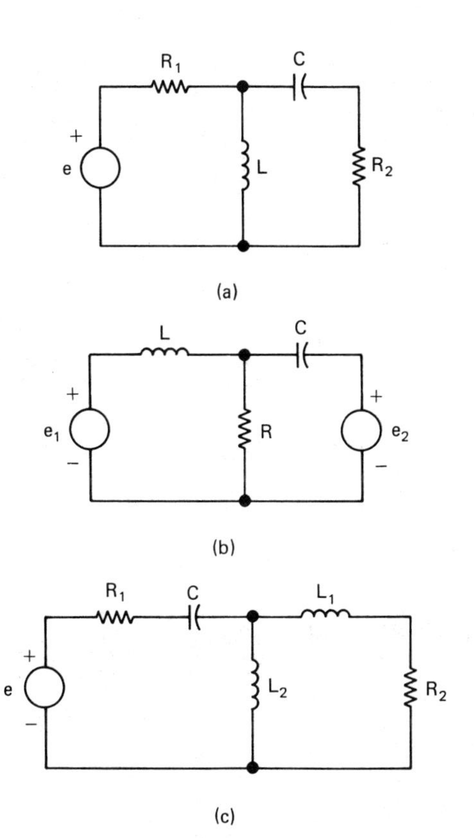


**Figure 4.7.** Networks for Problem 4A-1

Thus we have

$$\frac{dx_2}{dt} = \frac{R_1}{L(R_1 + R_2)}x_1 - \frac{R_1 R_2}{L(R_1 + R_2)}x_2 + \frac{R_2}{L(R_1 + R_2)}e$$

As a result, we conclude that

$$\mathbf{A} = \begin{bmatrix} \dfrac{-1}{C(R_1 + R_2)} & \dfrac{-R_1}{C(R_1 + R_2)} \\ \dfrac{R_1}{L(R_1 + R_2)} & \dfrac{-R_1 R_2}{L(R_1 + R_2)} \end{bmatrix}$$

$$\mathbf{B} = \begin{bmatrix} \dfrac{1}{C(R_1 + R_2)} \\ \dfrac{R_2}{L(R_1 + R_2)} \end{bmatrix}$$

$$\mathbf{x} = \begin{bmatrix} x_1 \\ x_2 \end{bmatrix} \qquad u = e$$

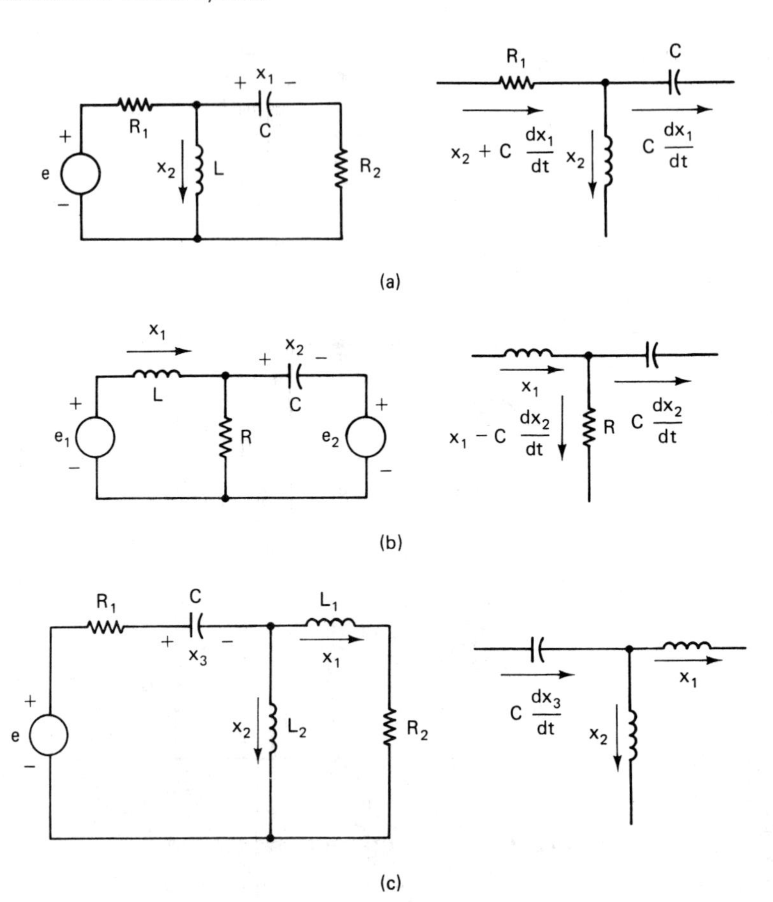


**Figure 4.8.** Identifying State Variables for the Networks of Problem 4A-1

This gives us the state equation in the form

$$\dot{\mathbf{x}} = \mathbf{A}\mathbf{x} + \mathbf{B}u$$

For the network of Figure 4.7b we have the state variables shown in Figure 4.8b. As a result, we have for the outer loop

$$e_1 = L\frac{dx_1}{dt} + x_2 + e_2$$

This is rewritten as

$$\frac{dx_1}{dt} = 0x_1 - \frac{1}{L}x_2 + \frac{1}{L}e_1 - \frac{1}{L}e_2$$

For the right-hand loop, we have

$$\left(x_1 - C\frac{dx_2}{dt}\right)R = x_2 + e_2$$

or

$$\frac{dx_2}{dt} = \frac{x_1}{C} - \frac{x_2}{RC} - \frac{e_2}{RC}$$

In matrix form, we write

$$\begin{bmatrix} \frac{dx_1}{dt} \\ \frac{dx_2}{dt} \end{bmatrix} = \begin{bmatrix} 0 & \frac{-1}{L} \\ \frac{1}{C} & \frac{-1}{RC} \end{bmatrix} \begin{bmatrix} x_1 \\ x_2 \end{bmatrix} + \begin{bmatrix} \frac{1}{L} & \frac{-1}{L} \\ 0 & \frac{-1}{RC} \end{bmatrix} \begin{bmatrix} e_1 \\ e_2 \end{bmatrix}$$

Thus

$$\dot{\mathbf{x}} = \mathbf{A}\mathbf{x} + \mathbf{B}\mathbf{U}$$

The individual matrices are given by

$$\mathbf{A} = \begin{bmatrix} 0 & \frac{-1}{L} \\ \frac{1}{C} & \frac{-1}{RC} \end{bmatrix}$$

$$\mathbf{B} = \begin{bmatrix} \frac{1}{L} & \frac{-1}{L} \\ 0 & \frac{-1}{RC} \end{bmatrix}$$

$$\mathbf{U} = \begin{bmatrix} e_1 \\ e_2 \end{bmatrix}$$

$$\mathbf{x} = \begin{bmatrix} x_1 \\ x_2 \end{bmatrix}$$

For the network of Figure 4.7c, we have the state variables indicated in Figure 4.8c. The sum of the currents at the node $M$ gives

$$\frac{dx_3}{dt} = \frac{1}{C}x_1 + \frac{1}{C}x_2$$

The left-hand loop voltage equation is

$$\begin{aligned} e(t) &= R_1 C\frac{dx_3}{dt} + x_3 + L_2\frac{dx_2}{dt} \\ &= R_1 x_1 + R_1 x_2 + x_3 + L_2\frac{dx_2}{dt} \end{aligned}$$

The outer loop voltage equation is

$$e(t) = R_1(x_1 + x_2) + x_3 + L_1\frac{dx_1}{dt} + R_2 x_1$$

Thus rearranging, we get

$$\frac{dx_1}{dt} = \frac{-1}{L_1}[(R_1 + R_2)x_1 + R_1x_2 + x_3] + \frac{1}{L_1}e(t)$$

$$\frac{dx_2}{dt} = \frac{-1}{L_2}[R_1x_1 + R_1x_2 + x_3 - e(t)]$$

$$\frac{dx_3}{dt} = \frac{1}{C}x_1 + \frac{1}{C}x_2$$

In vector form we have

$$\begin{bmatrix} \dfrac{dx_1}{dt} \\ \dfrac{dx_2}{dt} \\ \dfrac{dx_3}{dt} \end{bmatrix} = \begin{bmatrix} \dfrac{-(R_1 + R_2)}{L_1} & \dfrac{-R_1}{L_1} & \dfrac{-1}{L_1} \\ \dfrac{-R_1}{L_2} & \dfrac{-R_1}{L_2} & \dfrac{-1}{L_2} \\ \dfrac{1}{C} & \dfrac{1}{C} & 0 \end{bmatrix} \begin{bmatrix} x_1 \\ x_2 \\ x_3 \end{bmatrix} + \begin{bmatrix} \dfrac{1}{L_1} \\ \dfrac{1}{L_2} \\ 0 \end{bmatrix} e(t)$$

The result obtained above is the answer required.

PROBLEM 4A-2

For the armature-controlled dc motor and load shown in Figure 4.9, find the state-space model using $\theta$, $\dot{\theta}$, and $i_a$ as the state variables.

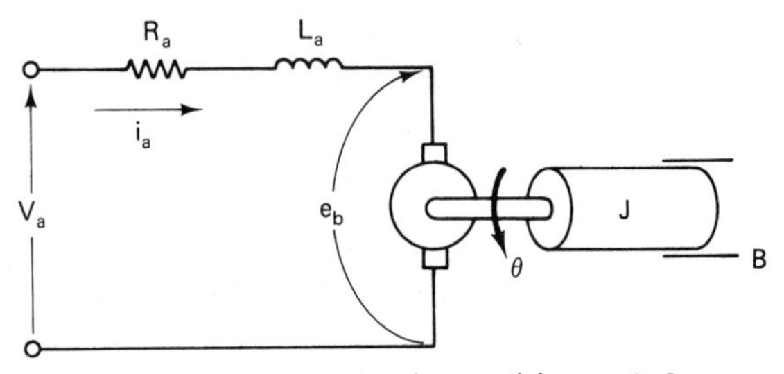


**Figure 4.9.** System for Problem 4A-2

SOLUTION

We can write the following loop equation:

$$V_a - e_b = R_a i_a + L_a \frac{di_a}{dt}$$

The motor's back emf is related to the rotational speed by

$$e_b = K\omega$$

The torque is related to the current in the armature by

$$T = Ki_a$$

In terms of inertia and damping, we have

$$T = J\frac{d^2\theta}{dt^2} + B\frac{d\theta}{dt}$$

Let us designate the state variables as

$$x_1 = \theta$$
$$x_2 = \dot{\theta}$$
$$x_3 = i_a$$

Thus we have

$$\frac{dx_1}{dt} = x_2$$
$$J\frac{dx_2}{dt} + Bx_2 = Kx_3$$
$$V_a - Kx_2 = R_a x_3 + L_a\frac{dx_3}{dt}$$

Rearranging the equations, we have

$$\frac{dx_3}{dt} = -\frac{K}{L_a}x_2 - \frac{R_a}{L_a}x_3 + \frac{V_a}{L_a}$$
$$\frac{dx_2}{dt} = -\frac{B}{J}x_2 + \frac{K}{J}x_3$$
$$\frac{dx_1}{dt} = x_2$$

As a result, we conclude that

$$\mathbf{A} = \begin{bmatrix} 0 & 1 & 0 \\ 0 & \frac{-B}{J} & \frac{K}{J} \\ 0 & \frac{-K}{L_a} & \frac{-R_a}{L_a} \end{bmatrix}$$

$$\mathbf{B} = \begin{bmatrix} 0 \\ 0 \\ \frac{1}{L} \end{bmatrix}$$

where

$$\mathbf{x} = \begin{bmatrix} x_1 \\ x_2 \\ x_3 \end{bmatrix}$$
$$\mathbf{u} = V_a$$

This yields the desired form.

PROBLEM 4A-3

Determine the transition matrix for the following coefficient matrix:

$$\mathbf{A} = \begin{bmatrix} 1 & 2 \\ 4 & 3 \end{bmatrix}$$

SOLUTION

For this matrix we have

$$s\mathbf{I} - \mathbf{A} = \begin{bmatrix} s-1 & -2 \\ -4 & s-3 \end{bmatrix}$$

Thus

$$(s\mathbf{I} - \mathbf{A})^{-1} = \begin{bmatrix} s-3 & 2 \\ 4 & s-1 \end{bmatrix} \frac{1}{\Delta}$$

where $\Delta = (s-5)(s+1)$. Thus

$$\mathbf{\Phi}(s) = (s\mathbf{I} - \mathbf{A})^{-1}$$

or

$$\mathbf{\Phi}(s) = \begin{bmatrix} \Phi_{11}(s) & \Phi_{12}(s) \\ \Phi_{21}(s) & \Phi_{22}(s) \end{bmatrix}$$

Here we have

$$\Phi_{11}(s) = \frac{s-3}{(s-5)(s+1)} = \frac{1}{3}\left(\frac{1}{s-5} + \frac{2}{s+1}\right)$$

$$\Phi_{12}(s) = \frac{2}{(s-5)(s+1)} = \frac{1}{3}\left(\frac{1}{s-5} - \frac{1}{s+1}\right)$$

$$\Phi_{21}(s) = \frac{4}{(s-5)(s+1)} = \frac{2}{3}\left(\frac{1}{s-5} - \frac{1}{s+1}\right)$$

$$\Phi_{22}(s) = \frac{s-1}{(s-5)(s+1)} = \frac{1}{3}\left(\frac{2}{s-5} + \frac{1}{s+1}\right)$$

Using the inverse Laplace transform, we obtain

$$\Phi_{11}(t, t_0) = \tfrac{1}{3}\left(e^{5(t-t_0)} + 2e^{-(t-t_0)}\right)$$

$$\Phi_{12}(t, t_0) = \tfrac{1}{3}\left(e^{5(t,t_0)} - e^{-(t-t_0)}\right)$$

$$\Phi_{21}(t, t_0) = 2\Phi_{12}(t, t_0)$$

$$\Phi_{22}(t, t_0) = \tfrac{1}{3}\left(2e^{5(t-t_0)} + e^{-(t-t_0)}\right)$$

Thus

$$\mathbf{\Phi}(t, t_0) = \begin{bmatrix} \Phi_{11} & \Phi_{12} \\ \Phi_{21} & \Phi_{22} \end{bmatrix}$$

PROBLEM 4A-4

Find the state transition matrix associated with the matrix

$$\mathbf{A} = \begin{bmatrix} 0 & 1 \\ 0 & 1 \end{bmatrix}$$

SOLUTION

The structure of the matrix invites us to investigate the series expansion method. We find that

$$\mathbf{A}^2 = \begin{bmatrix} 0 & 1 \\ 0 & 1 \end{bmatrix}\begin{bmatrix} 0 & 1 \\ 0 & 1 \end{bmatrix} = \begin{bmatrix} 0 & 1 \\ 0 & 1 \end{bmatrix}$$

Thus

$$\mathbf{A} = \mathbf{A}^2 = \mathbf{A}^3 = \cdots$$

As a result,

$$\mathbf{e}^{At} = \mathbf{I} + \mathbf{A}\left(t + \frac{t^2}{2!} + \frac{t^3}{3!} + \cdots\right)$$
$$= \mathbf{I} + \mathbf{A}(e^t - 1)$$

We conclude that

$$\mathbf{e}^{At} = \begin{bmatrix} 1 & e^t - 1 \\ 0 & e^t \end{bmatrix}$$

PROBLEM 4A-5

Consider the system shown in block diagram form in Figure 4.10. It is required to:

(a) Find the state-space representation of the system, using the following state variables.

$$x_1 = x$$
$$x_2 = \dot{x}$$

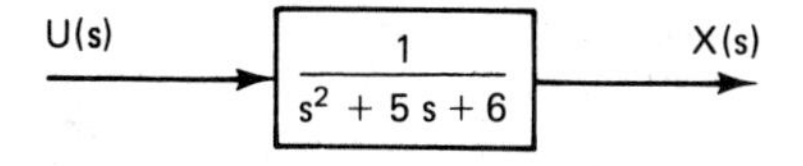


**Figure 4.10.** System for Problem 4A-5

(b) Find the state transition matrix $\Phi(t, 0)$ associated with the representation of part (a).

(c) Verify that for the $\Phi$ obtained in part (b), we have

$$\dot{\Phi} = \mathbf{A}\Phi$$

(d) Find the free (no input) response of the system for

$$x_1(0) = 0$$
$$x_2(0) = 1$$

(e) Find the system's response for the same initial conditions as in part (d) with $u(t) = 1$.

(f) Find a state representation of the system such that the matrix **A** is diagonal.

(g) Verify the results of parts (d) and (e) using the representation of part (f).

SOLUTION

(a) The transfer function of the system is given by

$$\frac{x(s)}{u(s)} = \frac{1}{s^2 + 5s + 6}$$

Thus the differential equation form is given by

$$\ddot{x} + 5\dot{x} + 6x = u$$

Let

$$x_1 = x$$
$$x_2 = \dot{x}$$

Thus

$$\dot{x}_1 = x_2$$
$$\dot{x}_2 = \ddot{x} = -6x_1 - 5x_2 + u$$

As a result,

$$\begin{bmatrix} \dot{x}_1 \\ \dot{x}_2 \end{bmatrix} = \begin{bmatrix} 0 & 1 \\ -6 & -5 \end{bmatrix} \begin{bmatrix} x_1 \\ x_2 \end{bmatrix} + \begin{bmatrix} 0 \\ 1 \end{bmatrix} u$$

Thus the matrix **A** is given by

$$\mathbf{A} = \begin{bmatrix} 0 & 1 \\ -6 & -5 \end{bmatrix}$$

(b) We have

$$\Phi^{-1}(s) = [sI - \mathbf{A}] = \begin{bmatrix} s & -1 \\ 6 & s+5 \end{bmatrix}$$

Thus

$$\Phi(s) = \frac{1}{\Delta} \begin{bmatrix} s+5 & 1 \\ -6 & s \end{bmatrix}$$

The determinant is given by

$$\Delta = s^2 + 5s + 6 = (s+3)(s+2)$$

We now have

$$\phi_{11}(s) = \frac{s+5}{\Delta} = \frac{3}{s+2} - \frac{2}{s+3}$$

$$\phi_{12}(s) = \frac{1}{(s+2)(s+3)} = \frac{1}{s+2} - \frac{1}{s+3}$$

$$\phi_{21}(s) = \frac{-6}{(s+2)(s+3)} = \frac{6}{s+3} - \frac{6}{s+2}$$

$$\phi_{22}(s) = \frac{s}{(s+2)(s+3)} = \frac{3}{s+3} - \frac{2}{s+2}$$

The inverse Laplace transform gives us

$$\boldsymbol{\phi}(t,0) = \begin{bmatrix} 3e^{-2t} - 2e^{-3t} & e^{-2t} - e^{-3t} \\ 6e^{-3t} - 6e^{-2t} & 3e^{-3t} - 2e^{-2t} \end{bmatrix}$$

(c) We have by differentiation

$$\dot{\boldsymbol{\phi}}(t,0) = \begin{bmatrix} 6e^{-3t} - 6e^{-2t} & 3e^{-3t} - 2e^{-2t} \\ 12e^{-2t} - 18e^{-3t} & 4e^{-2t} - 9e^{-3t} \end{bmatrix}$$

We also compute

$$\mathbf{A}\boldsymbol{\phi} = \begin{bmatrix} 0 & 1 \\ -6 & -5 \end{bmatrix} \begin{bmatrix} 3e^{-2t} - 2e^{-3t} & e^{-2t} - e^{-3t} \\ 6e^{-3t} - 6e^{-2t} & 3e^{-3t} - 2e^{-2t} \end{bmatrix}$$

$$= \begin{bmatrix} 6e^{-3t} - 6e^{-2t} & 3e^{-3t} - 2e^{-2t} \\ 12e^{-2t} - 18e^{-3t} & 4e^{-2t} - 9e^{-3t} \end{bmatrix}$$

Thus

$$\mathbf{A}\boldsymbol{\phi} = \dot{\boldsymbol{\phi}}$$

(d) We have for the force-free response,

$$\mathbf{x}(t) = \boldsymbol{\phi}(t,0)\mathbf{x}(0)$$

$$\mathbf{x}(t) = \begin{bmatrix} \phi_{11} & \phi_{12} \\ \phi_{21} & \phi_{22} \end{bmatrix} \begin{bmatrix} x_1(0) \\ x_2(0) \end{bmatrix}$$

Thus with

$$x_1(0) = 0$$
$$x_2(0) = 1$$

we get

$$\mathbf{x}(t) = \begin{bmatrix} \phi_{12} \\ \phi_{22} \end{bmatrix} = \begin{bmatrix} e^{-2t} & -e^{-3t} \\ 3e^{-3t} & -2e^{-2t} \end{bmatrix}$$

(e) With a forcing function we have

$$\mathbf{x}(t) = \boldsymbol{\phi}(t,0)\mathbf{x}(0) + \int_0^t \boldsymbol{\phi}(t,\tau)\mathbf{B}(\tau)u(\tau)\,d\tau$$

Now the integrand is given by

$$\boldsymbol{\phi}(t,\tau)\mathbf{B}(\tau)u(\tau) = \begin{bmatrix} \phi_{11} & \phi_{12} \\ \phi_{21} & \phi_{22} \end{bmatrix}\begin{bmatrix} 0 \\ 1 \end{bmatrix} u(\tau)$$

$$= \begin{bmatrix} \phi_{12}(t,\tau) \\ \phi_{22}(t,\tau) \end{bmatrix} u(\tau)$$

$$= \begin{bmatrix} e^{-2(t-\tau)} & -e^{-3(t-\tau)} \\ 3e^{-3(t-\tau)} & -2e^{2(t-\tau)} \end{bmatrix}$$

Thus

$$\int_0^t \boldsymbol{\phi}(t,\tau)\mathbf{B}(\tau)u(\tau)\,d\tau = \begin{bmatrix} e^{-2t}\left(\int_0^t e^{2\tau}\,d\tau\right) - e^{-3t}\left(\int_0^t e^{3\tau}\,d\tau\right) \\ 3e^{-3t}\int_0^t e^{3\tau}\,d\tau - 2e^{-2t}\int_0^t e^{2\tau}\,d\tau \end{bmatrix}$$

$$= \begin{bmatrix} \frac{1}{6} - 0.5e^{-2t} + \frac{1}{3}e^{-3t} \\ e^{-2t} - e^{-3t} \end{bmatrix}$$

As a result,

$$x_1(t) = (e^{-2t} - e^{-3t}) + \tfrac{1}{6} - 0.5e^{-2t} + \tfrac{1}{3}e^{-3t}$$
$$= \tfrac{1}{6} + 0.5e^{-2t} - \tfrac{2}{3}e^{-3t}$$
$$x_2(t) = (3e^{-3t} - 2e^{-2t}) + e^{-2t} - e^{-3t}$$
$$= 2e^{-3t} - e^{-2t}$$

(f) Let

$$\hat{\mathbf{x}} = \mathbf{T}\mathbf{x}$$

Differentiating, we get

$$\dot{\hat{\mathbf{x}}} = \mathbf{T}\dot{\mathbf{x}} = \mathbf{TAx} + \mathbf{TBu}$$

or

$$\dot{\hat{\mathbf{x}}} = \mathbf{TAT}^{-1}\hat{\mathbf{x}} + \mathbf{TBu}$$

Thus

$$\dot{\hat{\mathbf{x}}} = \tilde{\mathbf{A}}\hat{\mathbf{x}} + \tilde{\mathbf{B}}\mathbf{u}$$

where

$$\tilde{\mathbf{A}} = \mathbf{TAT}^{-1}$$

or

$$\tilde{\mathbf{A}}\mathbf{T} = \mathbf{TA}$$

$$\begin{bmatrix} a_{11} & 0 \\ 0 & a_{22} \end{bmatrix}\begin{bmatrix} t_{11} & t_{12} \\ t_{21} & t_{22} \end{bmatrix} = \begin{bmatrix} t_{11} & t_{12} \\ t_{21} & t_{22} \end{bmatrix}\begin{bmatrix} 0 & 1 \\ -6 & -5 \end{bmatrix}$$

Expanding, we have

$$a_{11}t_{11} = -6t_{12} \tag{1}$$
$$a_{11}t_{12} = t_{11} - 5t_{12} \tag{2}$$
$$a_{22}t_{21} = -6t_{22} \tag{3}$$
$$a_{22}t_{22} = t_{21} - 5t_{22} \tag{4}$$

Substitution of Eq. (2) in (1) gives

$$a_{11}(a_{11}t_{12} + 5t_{12}) + 6t_{12} = 0$$

or

$$a_{11}^2 + 5a_{11} + 6 = 0$$
$$(a_{11} + 3)(a_{11} + 2) = 0$$

Thus we take $a_{11} = -2$. Substitution of Eq. (4) in (3) gives

$$a_{22}(a_{22}t_{22} + 5t_{22}) = -6t_{22}$$
$$(a_{22} + 3)(a_{22} + 2) = 0$$

Take

$$a_{22} = -3$$

Thus

$$-2t_{11} = -6t_{12} \qquad t_{11} = 3t_{12}$$
$$-3t_{21} = -6t_{22} \qquad t_{21} = 2t_{22}$$

Take $t_{12} = 1$ to obtain $t_{11} = 3$; similarly, $t_{22} = 2$ to obtain $t_{21} = 4$. Thus the transform matrix $\mathbf{T}$ is obtained as

$$\mathbf{T} = \begin{bmatrix} 3 & 1 \\ 4 & 2 \end{bmatrix}$$

The inverse of $\mathbf{T}$ is obtained as

$$\mathbf{T}^{-1} = \begin{bmatrix} 1 & \dfrac{-1}{2} \\ -2 & \dfrac{3}{2} \end{bmatrix}$$

Since

$$\mathbf{x} = \mathbf{T}^{-1}\hat{\mathbf{x}}$$

we conclude that

$$x_1 = \hat{x}_1 - 0.5\hat{x}_2$$
$$x_2 = -2\hat{x}_1 + 1.5\hat{x}_2$$

(f) We have

$$\tilde{\mathbf{A}} = \begin{bmatrix} -2 & 0 \\ 0 & -3 \end{bmatrix}$$

The associated state transition matrix is thus given by

$$\tilde{\boldsymbol{\phi}} = \begin{bmatrix} e^{-2t} & 0 \\ 0 & e^{-3t} \end{bmatrix}$$

(g) For part (d) we have

$$\hat{x}_1 = 3x_1 + x_2$$
$$\hat{x}_2 = 4x_1 + 2x_2$$

Thus the initial conditions are

$$\hat{x}_1(0) = 3(0) + 1 = 1$$
$$\hat{x}_2(0) = 4(0) + 2 = 2$$

Thus the free response is obtained as

$$\hat{\mathbf{x}}_f(t) = \begin{bmatrix} e^{-2t} & 0 \\ 0 & e^{-3t} \end{bmatrix} \begin{bmatrix} 1 \\ 2 \end{bmatrix} = \begin{bmatrix} e^{-2t} \\ 2e^{-3t} \end{bmatrix}$$

Expanding, we get

$$\hat{x}_{1_f}(t) = e^{-2t}$$
$$\hat{x}_{2_f}(t) = 2e^{-3t}$$

Thus for the original system form, we have

$$x_{1_f}(t) = e^{-2t} - e^{-3t}$$
$$x_{2_f}(t) = -2e^{-2t} + 3e^{-3t}$$

For part (e)

$$\hat{\mathbf{x}}(t) = \hat{\mathbf{x}}_f(t) + \int_0^t \tilde{\boldsymbol{\phi}}(t,\tau)\tilde{\mathbf{B}}(\tau)u(\tau)\,d\tau$$

$$\tilde{\boldsymbol{\phi}}(t,\tau)\tilde{\mathbf{B}}(\tau)u(\tau) = \begin{bmatrix} e^{-2(t-\tau)} & 0 \\ 0 & e^{-3(t-\tau)} \end{bmatrix} \begin{bmatrix} 3 & 1 \\ 4 & 2 \end{bmatrix} \begin{bmatrix} 0 \\ 1 \end{bmatrix} u(\tau)$$

$$= \begin{bmatrix} e^{-2(t-\tau)} \\ 2e^{-3(t-\tau)} \end{bmatrix} u(\tau)$$

$$\int_0^t \tilde{\boldsymbol{\phi}}(t,\tau)\tilde{\mathbf{B}}(\tau)u(\tau)\,d\tau = \begin{bmatrix} e^{-2t}\left(\dfrac{e^{2t}}{2}\right)\Big|_0^t \\ 2e^{-3t}\left(\dfrac{e^{3\tau}}{3}\right)\Big|_0^t \end{bmatrix}$$

$$= \begin{bmatrix} 0.5e^{-2t}(e^{2t} - 1) \\ \dfrac{2e^{-3t}}{3}(e^{3t} - 1) \end{bmatrix}$$

Thus

$$\hat{x}_1(t) = e^{-2t} + 0.5(1 - e^{-2t})$$
$$= 0.5 + 0.5e^{-2t}$$
$$\hat{x}_2(t) = 2e^{-3t} + \tfrac{2}{3}(1 - e^{-3t})$$
$$= \tfrac{2}{3} + \tfrac{4}{3}e^{-3t}$$

Thus

$$\begin{aligned}
x_1(t) &= \hat{x}_1(t) - 0.5\hat{x}_2(t) \\
&= 0.5 + 0.5e^{-2t} - \tfrac{1}{3} - \tfrac{2}{3}e^{-3t} \\
&= \tfrac{1}{6} + \tfrac{1}{2}e^{-2t} - \tfrac{2}{3}e^{-3t} \\
x_2(t) &= -2\hat{x}_1(t) + 1.5\hat{x}_2(t) \\
&= -1 - e^{-2t} + 1 + 2e^{-3t} \\
&= 2e^{-3t} - e^{-2t}
\end{aligned}$$

Clearly, we obtain the same results.

# *Problems*

PROBLEM 4B-1

Consider the network of Figure 4.11 and choose the following state variables:

$$X_1 = V_{c_1}$$
$$X_2 = V_{c_2}$$

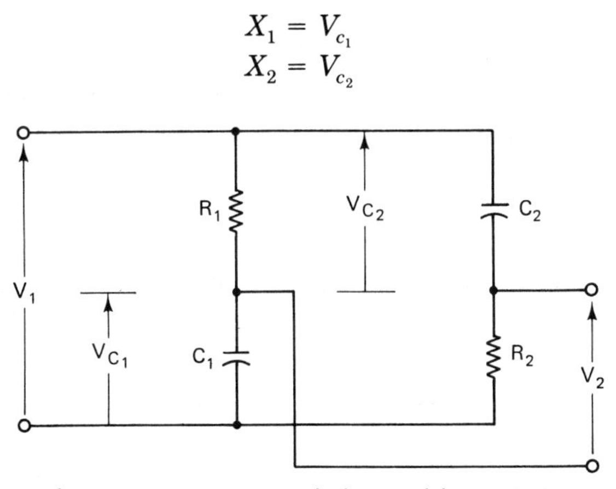


**Figure 4.11.** Network for Problem 4B-1

Let the input voltage $V_1$ be denoted by $u$ and the output voltage $V_2$ denoted by $y$. Show that the state representation of the network with the above assumptions is given by

$$\begin{bmatrix} \dot{x}_1 \\ \dot{x}_2 \end{bmatrix} = \begin{bmatrix} \dfrac{-1}{R_1C_1} & 0 \\ 0 & \dfrac{-1}{R_1C_1} \end{bmatrix} \begin{bmatrix} x_1 \\ x_2 \end{bmatrix} + \begin{bmatrix} \dfrac{1}{R_1C_1} \\ \dfrac{1}{R_2C_2} \end{bmatrix} u$$

$$y = [-1 \quad -1] \begin{bmatrix} x_1 \\ x_2 \end{bmatrix} + u$$

PROBLEM 4B-2

For the system shown in Figure 4.12, assume that the state variables are given by

$$x_1 = i_f$$
$$x_2 = e_2$$

Find the state equations and hence calculate the state transition matrix for

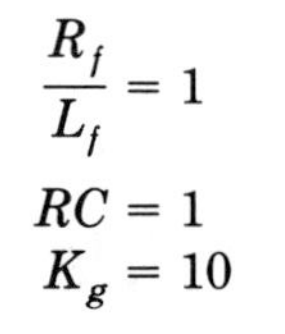

$$\frac{R_f}{L_f} = 1$$
$$RC = 1$$
$$K_g = 10$$

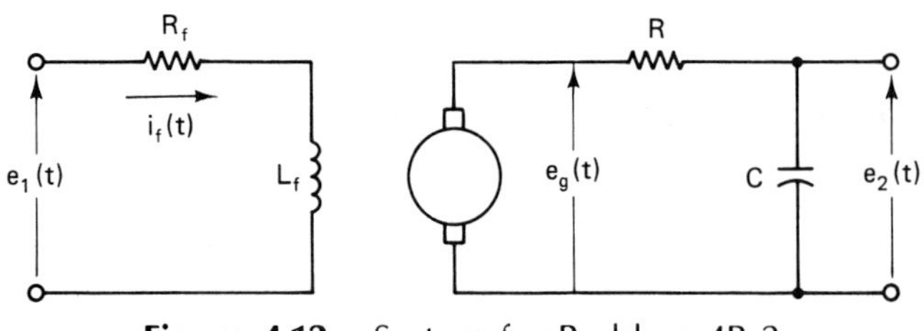


**Figure 4.12.** System for Problem 4B-2

PROBLEM 4B-3

Obtain a state-space representation of the twin-T network shown in Figure 4.13.

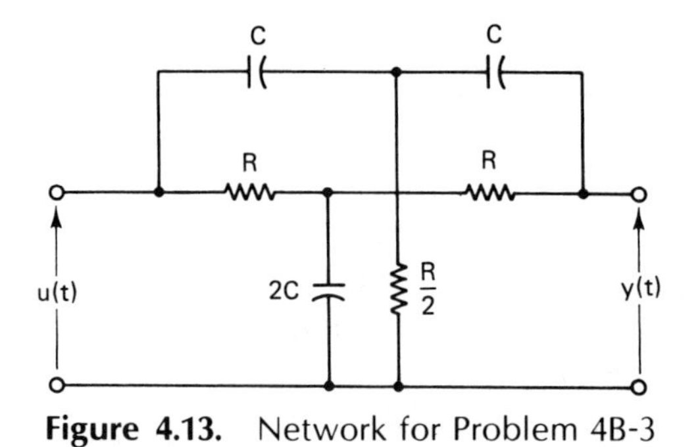


**Figure 4.13.** Network for Problem 4B-3

PROBLEM 4B-4

Find the state-space representation of the ladder network of Problem 2B-6 using the capacitor voltages as state variables.

PROBLEM 4B-5

Find the state-space representation for the network of Problem 2B-9 using the capacitor voltages as state variables.

PROBLEM 4B-6

Consider the automatic braking system of Problem 3B-1. Choose the following state variables:

$$x_1(t) = c(t)$$
$$x_2(t) = \dot{c}(t)$$

There are two inputs, which are denoted by $u_1$ and $u_2$ according to

$$u_1(t) = r(t)$$
$$u_2(t) = n(t)$$

Obtain the state-space representation of the system.

PROBLEM 4B-7

Find a state-space representation of the aircraft pitch-rate control mechanism of Problem 3B-5 using the closed-loop transfer function and assuming that the output $c(t)$ and its derivatives are the state variables.

PROBLEM 4B-8

Use Beck's method to find a state-space representation for the system of Problem 3B-5.

PROBLEM 4B-9

Find a state-space representation for the pitch-altitude-control mechanism of Problem 3B-6 using the closed-loop transfer function and assuming that the output $c(t)$ and its derivatives are the state variables.

PROBLEM 4B-10

Use Beck's method to find a state-space representation for the system of Problem 3B-6.

PROBLEM 4B-11

Repeat Problem 4B-10 using Johnson's method.

PROBLEM 4B-12

Find a state-space representation for the large space telescope of Problem 3B-8 using a method of your choice.

PROBLEM 4B-13

Obtain a state-space representation of the system described by the transfer function

$$\frac{X(s)}{U(s)} = \frac{1}{s^2 + 7s + 12}$$

PROBLEM 4B-14

Obtain a state-space representation of the system described by the transfer function

$$\frac{X(s)}{U(s)} = \frac{2(s+5)}{s^2 + 7s + 12}$$

PROBLEM 4B-15

Consider the system described by

$$\begin{bmatrix} \dot{x}_1 \\ \dot{x}_2 \end{bmatrix} = \begin{bmatrix} -7 & -1 \\ 3 & -1 \end{bmatrix} \begin{bmatrix} x_1 \\ x_2 \end{bmatrix} + \begin{bmatrix} 1 \\ 2 \end{bmatrix} u$$

$$y = x_1 + 3x_2$$

Find the transfer function of the system $Y(s)/U(s)$.

PROBLEM 4B-16

Find the state transition matrix associated with the coefficient matrix given by

$$\mathbf{A} = \begin{bmatrix} 0 & 1 & 0 \\ -3 & -4 & 0 \\ 0 & 0 & -2 \end{bmatrix}$$

PROBLEM 4B-17

Find the state transition matrix for the system of Problem 4B-13.

PROBLEM 4B-18

Find the state transition matrix for the system of Problem 4B-14.

PROBLEM 4B-19

Find the state transition matrix for the system of Example 4.9.

PROBLEM 4B-20

The system of Example 4.9 is subject to the following initial conditions:

$$x_1(0) = 0 \qquad x_2(0) = 1 \qquad x_3(0) = 0$$

Find the total response $x(t)$ using results of Problem 4B-19 for the input $u(t) = \frac{7}{5} \sin \omega t$.

PROBLEM 4B-21

Find the force-free response of the system of Example 4.10 assuming the following parameter values:

$$\begin{aligned} T_v &= 0.2 \qquad & K_v &= 5 \\ T_a &= 0.1 \qquad & K_a &= 2 \\ T_g &= 0.2 \qquad & K_g &= 10 \end{aligned}$$

Assume the following initial conditions:

$$\begin{aligned} x_1(0) &= 1 \\ x_2(0) &= 0.5 \\ x_3(0) &= 1.2 \end{aligned}$$

PROBLEM 4B-22

For the system of Problem 4B-3, find a state-space representation for which the coefficient matrix **A** is diagonal. Find the relation between the new state variables and the original ones.

PROBLEM 4B-23

Repeat Problem 4B-22 for the system of Problem 4B-14.

PROBLEM 4B-24

Assume that the input $u(t)$ to the system of Problem 4B-14 is a unit step. Find the output $x(t)$ assuming zero initial conditions.

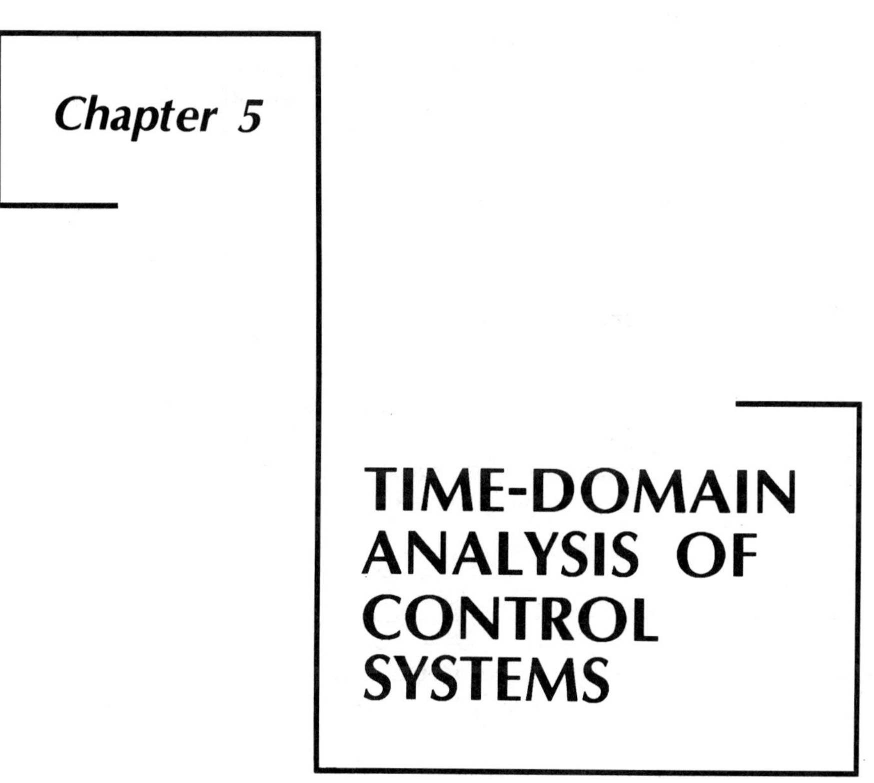


## 5.1 Introduction

The previous chapters dealt with control systems by using mathematical tools such as the Laplace transform and state-space methods in conjunction with physical laws to arrive at models either in the transfer-function form or the state-space form. The modeling process enables us to classify systems according to their common mathematical features. System order (first, second, and higher) is a feature that is immediately recognizable from the discussions of the previous chapters.

The present chapter deals with the response of control systems in the time domain. Salient performance characteristics of systems will be discussed. We start by an analysis of the error signal for a number of standard test signals. This leads to a natural classification of systems according to the number of poles at the origin, which is denoted by system type. The time response of control systems due to standard test signal input is treated for systems with order from one to four.

# 5.2 Test Signals

To compare the performance characteristics of control systems, a number of standard test signals are used. This makes the analysis process more mathematically tractable. Design specifications are commonly given in terms of response to the following test signals.

## Step Input Function

An instantaneous change in the reference input signal at time $t = 0$ from zero to a value $R_0$, which is held constant for all $t > 0$ defines a step function of magnitude $R_0$.

$$\begin{aligned} r(t) &= R_0 \qquad t > 0 \\ &= 0 \qquad t < 0 \end{aligned} \tag{5-1}$$

The step function is shown in Figure 5.1a. The Laplace transform of the step $r(t)$ is given by

$$R(s) = \frac{R_0}{s} \tag{5-2}$$

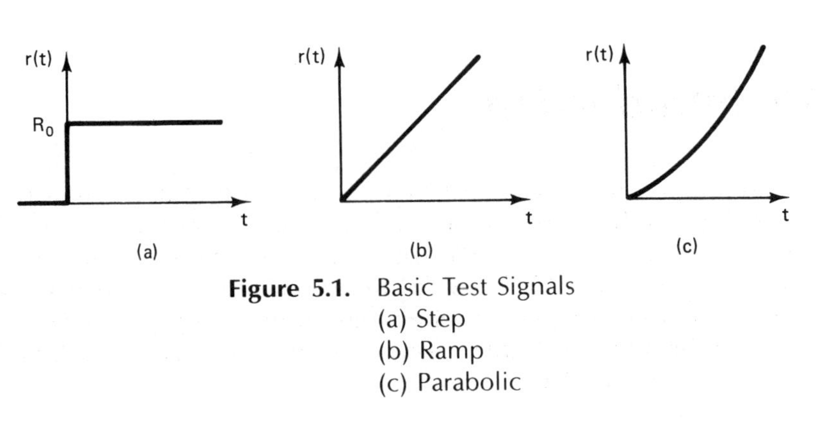


**Figure 5.1.** Basic Test Signals
(a) Step
(b) Ramp
(c) Parabolic

## Ramp Input Function

A ramp function has a constant rate of change with respect to time as shown in Figure 5.1b.

$$r(t) = At \qquad t \geq 0$$
$$= 0 \qquad t < 0 \tag{5-3}$$

The Laplace transform of a ramp function is given by

$$R(s) = \frac{A}{s^2} \tag{5-4}$$

## Parabolic Input Function

A parabolic function has the form

$$r(t) = At^2 \qquad t \geq 0$$
$$= 0 \qquad t < 0 \tag{5-5}$$

The Laplace transform of a parabolic function is given by

$$R(s) = \frac{2A}{s^3} \tag{5-6}$$

A parabolic function is faster than a ramp and is the highest-order practically used test signal.

# *5.3 Error Series*

Let us consider the closed-loop system shown in Figure 5.2. The Laplace transform of the error signal is given by

$$E(s) = R(s)F(s) \tag{5-7}$$

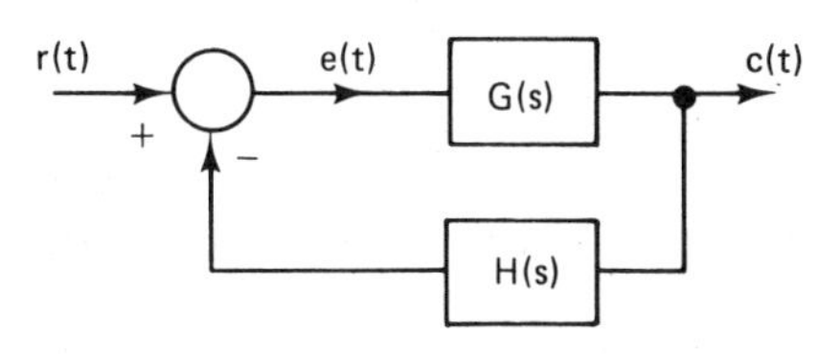


**Figure 5.2.** Closed-Loop System

where the error transfer function $F(s)$ is defined by

$$F(s) = \frac{1}{1 + G(s)H(s)} \tag{5-8}$$

The inverse Laplace transform of $E(s)$ provides us with the error signal $e(t)$ as a convolution integral

$$e(t) = \int_0^t f(\tau) r(t - \tau)\, d\tau \tag{5-9}$$

The lower integration limit is zero since $r$ is zero for negative time. The inverse Laplace transform of $F(s)$ is denoted by $f(\tau)$.

The function $r(t - \tau)$ can be expanded in a Taylor series provided that derivatives of $r(t)$ exist as

$$r(t - \tau) = r(t) + \sum_{k=1}^{\infty} \frac{(-\tau)^k}{k!} \frac{d^{(k)} r(t)}{dt^{(k)}} \tag{5-10}$$

As a result, the error signal can be written as

$$e(t) = \int_0^t f(\tau) \left[ r(t) + \sum_{k=1}^{\infty} \frac{(-\tau)^k}{k!} \frac{d^{(k)} r(t)}{dt^{(k)}} \right] d\tau \tag{5-11}$$

Let us define the generalized error coefficient $C_k(t)$ by

$$C_k(t) = (-1)^k \int_0^t \tau^k f(\tau)\, d\tau \qquad k = 0, 1, 2, \ldots \tag{5-12}$$

Thus the error signal of Eq. (5-11) is written as

$$e(t) = \sum_{k=0}^{\infty} \frac{C_k(t)}{k!} \frac{d^{(k)} r(t)}{dt^{(k)}} \tag{5-13}$$

The steady-state error $e_{ss}$ is obtained as the limit of the error as time goes to infinity:

$$e_{ss} = \lim_{t \to \infty} e(t)$$

In terms of the steady-state error as a function of time, we have

$$e_{ss} = \lim_{t \to \infty} e_s(t) \tag{5-14}$$

where $e_s(t)$ is the steady-state part of $e(t)$ and is given by

$$e_s(t) = \sum_{k=0}^{\infty} \frac{C_k}{k!} \frac{d^{(k)} r_s(t)}{dt^{(k)}} \tag{5-15}$$

Here $r_s(t)$ denotes the steady-state part of $r(t)$ and the $C_k$ are given by

$$C_k = \lim_{t \to \infty} C_k(t) = (-1)^k \int_0^{\infty} \tau^k f(\tau)\, d\tau \tag{5-16}$$

We can simplify the amount of work required to find the steady-state generalized error coefficients by using the following results. The functions $F(s)$ and $f(\tau)$ are related by the Laplace transform relation

$$F(s) = \int_0^\infty f(\tau)e^{-\tau s}\,d\tau \tag{5-17}$$

Taking the limits of both sides as $s$ tends to zero gives

$$\begin{aligned}\lim_{s\to 0} F(s) &= \lim_{s\to 0}\int_0^\infty f(\tau)e^{-\tau s}\,d\tau \\ &= \int_0^\infty f(\tau)\,d\tau \end{aligned}\tag{5-18}$$

As a result,

$$C_0 = \lim_{s\to 0} F(s) \tag{5-19}$$

Taking the derivative of Eq. (5-17) with respect to $s$, we obtain

$$\begin{aligned}\lim_{s\to 0}\frac{dF(s)}{ds} &= \lim_{s\to 0}\int_0^\infty -\tau f(\tau)e^{-\tau s}\,d\tau \\ &= -\int_0^\infty \tau f(\tau)\,d\tau \end{aligned}\tag{5-20}$$

Thus

$$C_1 = \lim_{s\to 0}\frac{dF(s)}{ds} \tag{5-21}$$

In a similar manner we can conclude that

$$C_k = \lim_{s\to 0}\frac{d^{(k)}F(s)}{ds^{(k)}} \tag{5-22}$$

We can arrive at the same results utilizing a Maclaurin series expansion of the function $F(s)$, which yields

$$E_s(s) = F(0)R(s) + \left.\frac{dF(s)}{ds}\right|_{s=0} sR(s) + \left.\frac{d^2F(s)}{ds^2}\right|_{s=0}\frac{s^2}{2!}R(s) + \cdots \tag{5-23}$$

Thus

$$E_s(s) = \sum_{k=0}^{\infty}\frac{C_k}{k!}s^kR(s) \tag{5-24}$$

where the $C_k$ values are defined by Eqs. (5-19) through (5-22).

To illustrate the application of the error series, we consider the following example.

EXAMPLE 5.1

Consider a closed-loop system with the following particulars:

$$G(s) = \frac{K}{s + \alpha}$$
$$H(s) = b$$

Thus using Eq. (5-8), we obtain

$$F(s) = \frac{s + \alpha}{s + \beta}$$

where we define

$$\beta = \alpha + Kb$$

The error coefficients are obtained as

$$C_0 = \lim_{s \to 0} F(s) = \frac{\alpha}{\beta}$$
$$C_1 = \lim_{s \to 0} \frac{dF(s)}{ds} = \frac{Kb}{\beta^2}$$
$$C_2 = \lim_{s \to 0} \frac{d^2F(s)}{ds^2} = \frac{-2Kb}{\beta^3}$$
$$\vdots$$

The error series is given by the expression

$$e_s(t) = C_0 r_s(t) + C_1 \frac{dr_s(t)}{dt} + \frac{C_2}{2} \frac{d^2 r_s(t)}{dt^2} + \cdots$$

Using the expressions for $C_0$, $C_1$, and $C_2$ arrived at for this system, we get

$$e_s(t) = \frac{\alpha}{\beta} r_s(t) + \frac{Kb}{\beta^2} \frac{dr_s(t)}{dt} - \frac{2Kb}{\beta^3} \frac{d^2 r_s(t)}{dt^2} + \cdots$$

Assuming now that the input signal is a unit step function, we have

$$r_s(t) = 1 \qquad 0 \le t$$

All derivatives of $r_s$ are zero; consequently, we have

$$e_s(t) = \frac{\alpha}{\beta} = \frac{\alpha}{\alpha + Kb}$$

Thus the steady-state error for this system is a constant for a unit step input.

If the input is a unit ramp function, then

$$r_s(t) = t$$
$$\frac{dr_s(t)}{dt} = 1$$

All higher-order derivatives of $r_s(t)$ are zero. As a result,

$$e_s(t) = \frac{\alpha}{\beta}t + \frac{Kb}{\beta^2}$$

or

$$e_s(t) = \frac{\alpha}{\beta}\left[t + \left(\frac{1}{\alpha} - \frac{1}{\beta}\right)\right]$$

We can conclude that the steady-state error of the system for a ramp input increases linearly with time. In the limit as $t$ approaches infinity, the steady-state error is infinite.

For a unit parabolic input $r_s(t) = t^2/2$, we can show that for this system

$$e_s(t) = \frac{\alpha}{\beta}\left[\frac{t^2}{2} + \left(\frac{1}{\alpha} - \frac{1}{\beta}\right)\left(t - \frac{2}{\beta}\right)\right]$$

The error increases with the square of the time elapsed.

EXAMPLE 5.2

Consider the system with unity feedback and a forward transfer function of the form

$$G(s) = \frac{\omega_n^2}{s(s + 2\zeta\omega_n)}$$

Thus

$$1 + G(s)H(s) = \frac{s^2 + 2\zeta\omega_n s + \omega_n^2}{s(s + 2\zeta\omega_n)}$$

We then have, by Eq. (5-8),

$$F(s) = \frac{s^2 + 2\zeta\omega_n s}{s^2 + 2\zeta\omega_n s + \omega_n^2}$$

As a result,

$$C_0 = \lim_{s\to 0} F(s) = 0$$

The derivative of $F(s)$ is obtained as

$$\frac{dF(s)}{ds} = \frac{2\omega_n^2(s + \zeta\omega_n)}{\left(s^2 + 2\zeta\omega_n s + \omega_n^2\right)^2}$$

As a result,

$$C_1 = \lim_{s\to 0} \frac{dF(s)}{ds} = \frac{2\zeta}{\omega_n}$$

The second derivative of $F(s)$ is obtained as

$$\frac{d^2F(s)}{ds^2} = \frac{2\omega_n^2\left[\left(s^2 + 2\zeta\omega_n s + \omega_n^2\right) - 4(s + \zeta\omega_n)^2\right]}{\left(s^2 + 2\zeta\omega_n s + \omega_n^2\right)^3}$$

As a result,

$$C_2 = \lim_{s\to 0} \frac{d^2F(s)}{ds} = \frac{2(1 - 4\zeta^2)}{\omega_n^2}$$

# 5.4 Steady-State Error Constants

The steady-state error of a control system depends on the input function $r(t)$ as is clear from the generalized error series. A conventional means of describing the steady-state error is in terms of standard test signals such as the step function, ramp function, and the parabolic function. We will examine the error measures obtained using results of generalized error series.

## Steady-State Step Error Constant

Consider the Laplace transform of the steady-state error $E_s(s)$ for a unit step input. Clearly, we have

$$E_s(s) = C_0 \frac{1}{s} \tag{5-25}$$

The steady-state step error constant $K_p$ is defined as the ratio of the steady-state output to steady-state error. Thus

$$K_p = \frac{c_{ss}(t)}{e_{ss}(t)} \tag{5-26}$$

Using the final-value theorem we conclude that

$$K_p = \frac{\lim_{s\to 0} s\left[R(s) - E_s(s)\right]}{\lim_{s\to 0} sE_s(s)}$$

$$= \frac{1 - C_0}{C_0} \tag{5-27}$$

As a result,

$$K_p = \frac{1}{C_0} - 1 \tag{5-28}$$

$$E_s(s) = \frac{1}{1 + K_p}\frac{1}{s} \tag{5-29}$$

Thus the steady-state error is a step function for a unit step function input

$$e_s(t) = \frac{1}{1 + K_p} \tag{5-30}$$

An alternative expression for $e_s(t)$, in terms of the open-loop transfer function, is given by

$$e_s(t) = \lim_{s \to 0} \frac{sR(s)}{1 + G(s)H(s)} \tag{5-31}$$

For a unit step input we conclude that

$$e_s(t) = \frac{1}{1 + \lim_{s \to 0} G(s)H(s)} \tag{5-32}$$

Comparison of Eqs. (5-30) and (5-32) provides us with the conventional definition of the step error constant:

$$K_p = \lim_{s \to 0} G(s)H(s) \tag{5-33}$$

EXAMPLE 5.3

Again consider the forward loop transfer function of Example 5.1 expressed as

$$G(s) = \frac{K}{s + \alpha}$$

but this time assume unity feedback. Hence

$$K_p = \lim_{s \to 0} \frac{K}{s + \alpha} = \frac{K}{\alpha}$$

Using the result of Example 5.1, we have

$$C_0 = \frac{\alpha}{\beta}$$

Thus an alternative way of finding $K_p$ is as follows:

$$K_p = \frac{\beta}{\alpha} - 1 = \frac{\beta - \alpha}{\alpha} = \frac{Kb}{\alpha}$$

Since $b = 1$ in the present case, the two results agree.

## Steady-State Ramp Error Constant

Consider the Laplace transform of the steady-state error $E_s(s)$ for a unit ramp input.

$$R(s) = \frac{1}{s^2}$$

Thus

$$E_s(s) = C_0 \frac{1}{s^2} + C_1 \frac{1}{s} \tag{5-34}$$

Assuming that

$$C_0 = 0 \tag{5-35}$$

the conventional steady-state ramp (velocity) error constant is defined as the ratio of the steady-state derivative of the output to the steady-state error; thus

$$K_v = \frac{\left.\frac{d}{dt} c(t)\right|_{\text{ss}}}{e_{\text{ss}}(t)} = \frac{\lim\limits_{t \to \infty} \dot{c}(t)}{\lim\limits_{t \to \infty} e(t)} \tag{5-36}$$

The output is given in terms of $r(t)$ and $e(t)$ as

$$c(t) = r(t) - e(t) \tag{5-37}$$

Thus

$$\dot{c}(t) = \dot{r}(t) - \dot{e}(t) \tag{5-38}$$

The final-value theorem gives us

$$\begin{aligned} \lim_{t \to \infty} \dot{c}(t) &= \lim_{s \to 0} s[sR(s) - C_1] \\ &= 1 \\ \lim_{t \to \infty} e(t) &= \lim_{s \to 0} sC_1 \frac{1}{s} = C_1 \end{aligned}$$

As a result we conclude that when $C_0 = 0$, we have

$$K_v = \frac{1}{C_1} \tag{5-39}$$

The steady-state error is thus

$$E_s(s) = \frac{1}{K_v} \frac{1}{s}$$

The inverse Laplace transform gives

$$e_s(t) = \frac{1}{K_v} \tag{5-40}$$

An alternative expression for $e_s(t)$ in terms of the open-loop transfer function is given by

$$e_s(t) = \lim_{s \to 0} \frac{sR(s)}{1 + G(s)H(s)} \tag{5-41}$$

For a unit ramp input we conclude that

$$e_s(t) = \frac{1}{\lim_{s \to 0} sG(s)H(s)} \tag{5-42}$$

Thus we obtain the conventional expression for the ramp error constant given by

$$K_v = \lim_{s \to 0} sG(s)H(s) \tag{5-43}$$

Let us emphasize here that the relation between $K_v$ and $C_1$ just described is valid only for systems where $C_0 = 0$. The value of $K_v$ can still be obtained using Eq. (5-43). The following two examples illustrate the concepts developed so far.

EXAMPLE 5.4

Consider the closed-loop system of Examples 5.1 and 5.3. Clearly, for finite $K$, $C_0$ is not zero, and the value of $C_1$ was found to be

$$C_1 = \frac{Kb}{\beta^2}$$

Using Eq. (5-43), we find that

$$K_v = \lim_{s \to 0} s \frac{K}{s + \alpha} = 0$$

Clearly, $K_v \neq 1/C_1$

EXAMPLE 5.5

Consider the system of Example 5.2; here we have found that

$$C_0 = 0$$

$$C_1 = \frac{2\zeta}{\omega_n}$$

Using Eq. (5-43), we obtain

$$K_v = \lim_{s \to 0} s \frac{\omega_n^2}{s(s + 2\zeta\omega_n)} = \frac{\omega_n}{2\zeta}$$

Thus the following relation is true for this system:

$$C_1 = \frac{1}{K_v}$$

## Steady-State Parabolic Error Constant

This time we assume a parabolic input function for which

$$R(s) = \frac{2}{s^3} \tag{5-44}$$

Assuming that the coefficients $C_0$ and $C_1$ are zero,

$$C_0 = C_1 = 0 \tag{5-45}$$

we have

$$E_s(s) = C_2 \frac{1}{s} \tag{5-46}$$

The conventional steady-state parabolic (acceleration) error constant $K_a$ is defined as the ratio of the steady-state second derivative of the output to the steady-state error; thus

$$K_a = \frac{\left.\dfrac{d^2}{dt^2}c(t)\right|_{\text{ss}}}{e_{\text{ss}}(t)} \tag{5-47}$$

Using arguments similar to those adopted for $K_v$, we can conclude that

$$K_a = \frac{1}{C_2} \tag{5-48}$$

Also, we can assert that

$$K_a = \lim_{s \to 0} s^2 G(s)H(s) \tag{5-49}$$

# 5.5 System Types

Consider a system with the loop transfer function given by

$$G(s)H(s) = \frac{K_m N(s)}{s^m D(s)} \tag{5-50}$$

The polynomials $N(s)$ and $D(s)$ are assumed to be of the form

$$N(s) = 1 + b_1 s + b_2 s^2 + \cdots \tag{5-51}$$

$$D(s) = 1 + a_1 s + a_2 s^2 + \cdots \tag{5-52}$$

$K_m$ is the open-loop gain. Note that we have an $m$th-order pole at $s = 0$. The type of a feedback control system refers to the order ($m$) of the pole at $s = 0$. Thus if $m = 0$, the system is referred to as a *type* 0 *system*, for $m = 1$ it is called a *type* 1 *system*, and so on. The steady-state error of the system in response to input functions depends on the system type.

The error coefficients $C_0$, $C_1$, and $C_2$ as well as the error constants $K_p$, $K_v$, and $K_a$ provide measures of the steady-state error as established earlier. For the system of type $m$, we have

$$F(s) = \frac{s^m D(s)}{s^m D(s) + K_m N(s)} \tag{5-53}$$

The error coefficients are given by

$$C_0 = \lim_{s \to 0} F(s) \tag{5-54}$$

$$C_1 = \lim_{s \to 0} \frac{dF(s)}{ds} \tag{5-55}$$

$$C_2 = \lim_{s \to 0} \frac{d^2F}{ds^2} \tag{5-56}$$

The error constants are given by

$$K_p = \lim_{s \to 0} G(s)H(s) \tag{5-57}$$

$$K_v = \lim_{s \to 0} sG(s)H(s) \tag{5-58}$$

$$K_a = \lim_{s \to 0} s^2 G(s)H(s) \tag{5-59}$$

## Type 0 System

For a type 0 system, we have $m = 0$; thus

$$G(s) = \frac{K_m N(s)}{D(s)} \tag{5-60}$$

We can immediately assert that

$$K_p = K_m \tag{5-61}$$

$$K_v = 0 \tag{5-62}$$

$$K_a = 0 \tag{5-63}$$

The system of Examples 5.1, 5.3, and 5.4 is a type 0 system.

For this system type we have

$$F(s) = \frac{D(s)}{D(s) + K_m N(s)} \tag{5-64}$$

We obtain

$$C_0 = \lim_{s \to 0} F(s) = \frac{1}{1 + K_m} \tag{5-65}$$

Note that

$$\lim_{s \to 0} N(s) = 1$$
$$\lim_{s \to 0} D(s) = 1$$

The derivative of $F(s)$ is given by

$$\frac{dF}{ds} = \frac{K_m(ND' - DN')}{(D + K_m N)^2}$$

Note that the derivatives are given by

$$N' = b_1 + 2b_2 s + \cdots$$
$$D' = a_1 + 2a_2 s + \cdots$$

As a result,

$$\lim_{s \to 0} \frac{dF}{ds} = \frac{K_m(a_1 - b_1)}{(1 + K_m)^2}$$

Thus

$$C_1 = \frac{K_m(a_1 - b_1)}{(1 + K_m)^2} \tag{5-66}$$

The second derivative of $F(s)$ is given by

$$\frac{d^2F}{ds^2} = \frac{K_m[(ND'' - DN'')(D + K_m N) - 2(ND' - DN')(D' + K_m N')]}{(D + K_m N)^3}$$

Note that

$$N'' = 2b_2 + 6b_3 s + \cdots$$
$$D'' = 2a_2 + 6a_3 s + \cdots$$

Thus

$$\lim_{s \to 0} \frac{d^2F}{ds^2} = \frac{2K_m[(a_2 - b_2)(1 + K_m) - (a_1 - b_1)(a_1 + K_m b_1)]}{(1 + K_m)^3}$$

As a result,

$$C_2 = \frac{2K_m}{(1 + K_m)^3}[(1 + K_m)(a_2 - b_2) - (a_1 + b_1 K_m)(a_1 - b_1)] \tag{5-67}$$

Let us consider the system of Example 5.1 to find $C_0$, $C_1$, and $C_2$ using the formulas just derived.

EXAMPLE 5.6

Consider the system of Example 5.1, with $b = 1$, the open-loop function can be written in the form

$$G(s)H(s) = \frac{K/\alpha}{1 + (s/\alpha)}$$

Thus

$$K_m = \frac{K}{\alpha}$$
$$a_1 = \frac{1}{\alpha}$$
$$b_1 = b_2 = a_2 = 0$$

We find $C_0$ as

$$C_0 = \frac{1}{1 + K_m} = \frac{1}{1 + K/\alpha} = \frac{\alpha}{\alpha + K}$$

This is the same result as obtained in Example 5.1. We next find $C_1$ as

$$C_1 = \frac{K_m(a_1 - b_1)}{(1 + K_m)^2} = \frac{(K/\alpha)(1/\alpha)}{(1 + K/\alpha)^2} = \frac{K}{(\alpha + K)^2}$$

Again this confirms the results of Example 5.1. Finally, we calculate $C_2$ as

$$C_2 = \frac{2(K/\alpha)}{(1 + K/\alpha)^3}\left(-\frac{1}{\alpha^2}\right) = \frac{-2K}{(\alpha + K)^3}$$

This is exactly the same result as obtained before.

The steady-state error of a type 0 system subject to a unit step input is given by a constant value

$$e_s(t) = \frac{1}{1 + K_m}$$

When the system is subject to unit ramp input, we have

$$e_s(t) = \frac{1}{1 + K_m} t + \frac{K_m(a_1 - b_1)}{(1 + K_m)^2}$$

Thus the error increases with time and its final value approaches infinity, or

$$e_{ss} \to \infty$$

The result above can also be obtained from the observation that with

$$K_v = 0$$

$$e_{ss} = \frac{1}{K_v} \to \infty$$

Thus a type 0 system cannot track a ramp input.

## Type 1 System

For a type 1 system we have $m = 1$; thus

$$G(s) = \frac{K_m N(s)}{sD(s)} \tag{5-68}$$

We can assert that

$$K_p = \infty \tag{5-69}$$

$$K_v = K_m \tag{5-70}$$

$$K_a = 0 \tag{5-71}$$

For this system type we have

$$F(s) = \frac{sD(s)}{sD(s) + K_m N(s)} \tag{5-72}$$

As a result,

$$C_0 = \lim_{s \to 0} F(s) = 0 \tag{5-73}$$

The derivative of $F(s)$ is obtained as

$$\frac{dF}{ds} = \frac{K_m[N(D + sD') - sDN']}{[sD(s) + K_m N(s)]^2}$$

As a result,

$$C_1 = \lim_{s \to 0} \frac{dF}{ds} = \frac{1}{K_m} \tag{5-74}$$

Note that the condition $C_0 = 0$ is satisfied, and hence for a type 1 system,

$$K_v = \frac{1}{C_1} \tag{5-75}$$

To obtain the second derivative of $F(s)$, we set

$$X(s) = sD(s) + K_m N(s)$$
$$Q(s) = N(D + sD') - sDN'$$

Thus

$$\frac{dF}{ds} = \frac{K_m Q}{X^2}$$

As a result,

$$\frac{d^2F}{ds^2} = \frac{K_m(XQ' - 2X'Q)}{X^3}$$

Note that

$$X' = D + sD' + K_m N'$$
$$Q' = N(2D' + sD'') - sDN''$$
$$\lim_{s \to 0} X(s) = K_m$$
$$\lim_{s \to 0} X'(s) = 1 + K_m b_1$$
$$\lim_{s \to 0} Q(s) = 1$$
$$\lim_{s \to 0} Q'(s) = 2a_1$$

Thus

$$C_2 = \lim_{s \to 0} \frac{d^2F}{ds^2}$$
$$= \frac{K_m[K_m(2a_1) - 2(1 + K_m b_1)]}{K_m^3}$$

or in final form

$$C_2 = \frac{2[K_m(a_1 - b_1) - 1]}{K_m^2} \tag{5-76}$$

Let us now take an example.

EXAMPLE 5.7

The system of Example 5.2 is a type 1 system with

$$K_m = \frac{\omega_n}{2\zeta}$$
$$a_1 = \frac{1}{2\zeta\omega_n}$$
$$b_1 = 0$$

Thus

$$C_0 = 0$$

$$C_1 = \frac{1}{K_m} = \frac{2\zeta}{\omega_n}$$

$$C_2 = \frac{2\left(\dfrac{\omega_n}{2\zeta}\dfrac{1}{2\zeta\omega_n} - 1\right)}{(\omega_n/2\zeta)^2}$$

This reduces to

$$C_2 = \frac{2(1 - 4\zeta^2)}{\omega_n^2}$$

These results confirm our earlier conclusions.

The steady-state error of a type 1 system due to a unit step input is zero, since $C_0 = 0$:

$$e_s(t) = 0$$

For a unit ramp input the steady-state error is a constant:

$$e_s(t) = \frac{1}{K_m}$$

For a unit parabolic input the steady-state error increases with time since $C_1$ and $C_2$ exist.

## Type 2 System

For a type 2 system, we have

$$G(s) = \frac{K_m N(s)}{s^2 D(s)} \tag{5-77}$$

We can assert that

$$K_p = \infty \tag{5-78}$$

$$K_v = \infty \tag{5-79}$$

$$K_a = K_m \tag{5-80}$$

For this system we have

$$F(s) = \frac{s^2 D(s)}{s^2 D(s) + K_m N(s)} \tag{5-81}$$

As a result,

$$C_0 = \lim_{s \to 0} F(s) = 0 \tag{5-82}$$

TABLE 5.1
ERROR COEFFICIENTS AND CONSTANTS FOR THREE SYSTEM TYPES

| | TYPE 0 | TYPE 1 | TYPE 2 |
|---|---|---|---|
| $C_0$ | $\frac{1}{1+K_m}$ | 0 | 0 |
| $C_1$ | $\frac{K_m(a_1 - b_1)}{(1+K_m)^2}$ | $\frac{1}{K_m}$ | 0 |
| $C_2$ | Eq. (5-67) | Eq. (5-76) | $\frac{1}{K_m}$ |
| $K_p$ | $K_m$ | $\infty$ | $\infty$ |
| $K_v$ | 0 | $K_m$ | $\infty$ |
| $K_a$ | 0 | 0 | $K_m$ |

The derivative of $F(s)$ is obtained as

$$\frac{dF}{ds} = \frac{K_m s\left[N(2D + sD') - sDN'\right]}{\left(s^2D + K_mN\right)^2}$$

Consequently,

$$C_1 = \lim_{s\to 0} \frac{dF}{ds} = 0 \tag{5-83}$$

Since $C_0$ and $C_1$ are zero, we conclude according to the relation between the error coefficients and error constants that

$$C_2 = \frac{1}{K_m} \tag{5-84}$$

This can be verified by taking the second derivative of $F$ and the limit as $s \to 0$.

Table 5.1 summarizes the values of the error coefficients and parameters for systems type 0, 1, and 2.

# 5.6 Time Response of First- and Second-Order Systems

The previous sections considered the problem of characterizing the steady-state error of a closed-loop control system in terms of an error series as well as error constants. Consideration was given to classification of systems according to the number of poles at the origin, which gives rise to system type. We are now interested in studying the response of systems of the first and second order to the standard test signals of Section 5.2.

## First-Order Systems

A first-order system is characterized by the transfer function

$$\frac{C(s)}{R(s)} = \frac{K}{1 + Ts} \tag{5-85}$$

The unit impulse response is obtained by setting

$$R(s) = 1$$

As a result,

$$C(s) = \frac{K}{1 + Ts}$$

Alternatively,

$$C(s) = \frac{K}{T} \frac{1}{s + 1/T} \tag{5-86}$$

The inverse Laplace transform yields

$$c(t) = \frac{K}{T} e^{-t/T} \tag{5-87}$$

Note that

$$c(0) = \frac{K}{T} \tag{5-88}$$

$$\lim_{t \to \infty} c(t) = 0 \tag{5-89}$$

The unit impulse response of a first-order system is sketched in Figure 5.3.

To obtain the unit step response, we set

$$R(s) = \frac{1}{s}$$

Thus

$$C(s) = \frac{K}{s(1 + Ts)} \tag{5-90}$$

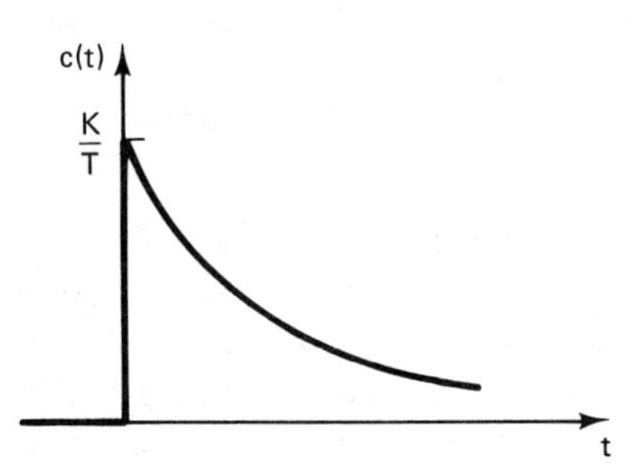

**Figure 5.3.** Unit Impulse Response of a First-Order System

Performing a partial fraction expansion, we obtain

$$C(s) = K\left(\frac{1}{s} - \frac{1}{s + 1/T}\right) \tag{5-91}$$

The inverse Laplace transform provides us with the required unit step response as

$$c(t) = K(1 - e^{-t/T}) \tag{5-92}$$

Note that

$$c(0) = 0$$
$$\lim_{t\to\infty} c(t) = K \tag{5-93}$$

A sketch of the unit step response of a first-order system is shown in Figure 5.4. After $T$ seconds, the response is 0.632 times its steady-state value. $T$ is the system time constant. From a practical point of view the system reaches a steady state after four to five time constants.

Assume now that the system input is a unit ramp; thus

$$R(s) = \frac{1}{s^2}$$

As a result,

$$C(s) = \frac{K}{s^2(1 + Ts)} \tag{5-94}$$

Performing a partial fraction expansion, we obtain

$$C(s) = K\left(\frac{1}{s^2} - \frac{T}{s} + \frac{T^2}{1 + Ts}\right) \tag{5-95}$$

The inverse Laplace transform yields

$$c(t) = K(t - T + Te^{-t/T}) \tag{5-96}$$

The error is given by

$$e(t) = r(t) - c(t) = t(1 - K) + KT(1 - e^{-t/T}) \tag{5-97}$$

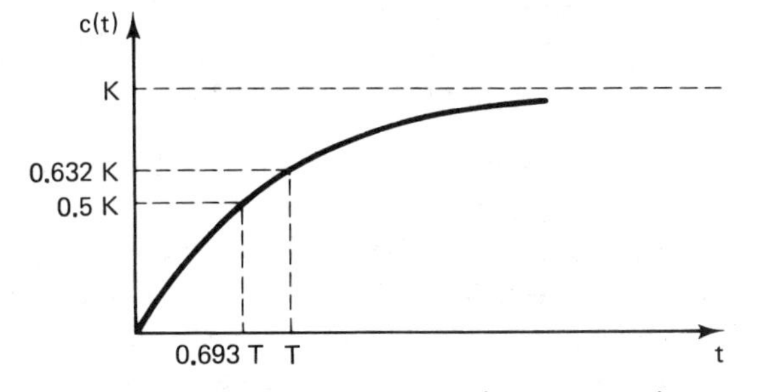


**Figure 5.4.** Unit Step Response of a First-Order System

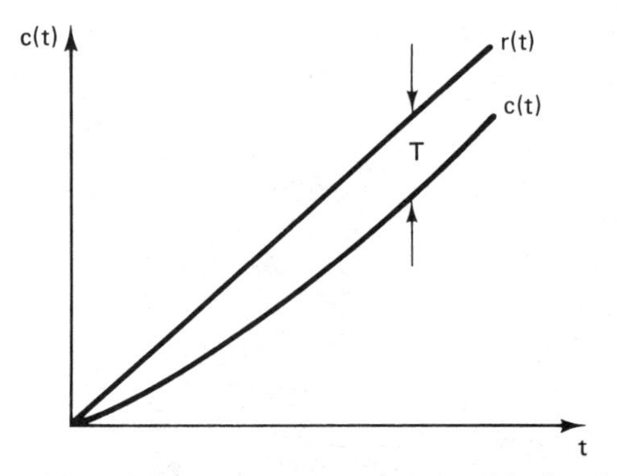


**Figure 5.5.** Ramp Response of a First-Order System

The steady-state value of $e(t)$ for $K = 1$ is given by

$$\begin{aligned} e_{ss} &= \lim_{t \to \infty} e(t) \\ &= T \end{aligned} \tag{5-98}$$

It is thus clear that a first-order system cannot track a ramp input in the sense that a steady error equal to the system time constant will result as shown in Figure 5.5.

## Second-Order Systems

Consider a control system with the closed-loop transfer function given by

$$\frac{C(s)}{R(s)} = \frac{\omega_n^2}{s^2 + 2\zeta\omega_n s + \omega_n^2} \tag{5-99}$$

Assume that the denominator is factored as

$$s^2 + 2\zeta\omega_n s + \omega_n^2 = (s - r_1)(s - r_2) \tag{5-100}$$

The nature of the roots depends on the value of $\zeta$, the damping factor, as is evident from the expression

$$r_{1,2} = \left(-\zeta \pm \sqrt{\zeta^2 - 1}\right)\omega_n \tag{5-101}$$

Clearly, the roots $r_1$ and $r_2$ are real valued for $\zeta \geq 1$ and, in this case, the system is called overdamped. For $0 \leq \zeta < 1$, the roots are complex conjugates and the system is called underdamped. When $\zeta = 0$, the system is undamped. The system is negatively damped for $\zeta < 0$ and the roots will be in the right-half of the $s$-plane. We will study the impulse response and the step response of the system for the various cases of damping discussed above.

### Impulse Response

The Laplace transform of the output $C(s)$ for a unit impulse input $R(s) = 1$ is given by

$$C(s) = \frac{\omega_n^2}{r_1 - r_2}\left(\frac{1}{s - r_1} - \frac{1}{s - r_2}\right) \tag{5-102}$$

This applies for $r_1 \neq r_2$, that is, for $\zeta \neq 1$. Substituting the values of $r_1$ and $r_2$, and performing the inverse Laplace transformation, we get

$$c(t) = \frac{\omega_n e^{-\zeta\omega_n t}}{2\sqrt{\zeta^2 - 1}}\left[\exp\left(\sqrt{\zeta^2 - 1}\,\omega_n t\right) - \exp\left(-\sqrt{\zeta^2 - 1}\,\omega_n t\right)\right] \tag{5-103}$$

For the overdamped case, $\zeta > 1$ and the expression $\sqrt{\zeta^2 - 1}$ is real valued. Recall that

$$\sinh\theta = \frac{e^{\theta} - e^{-\theta}}{2} \tag{5-104}$$

As a result, we may write the impulse response for an overdamped second-order system as

$$c(t) = \frac{\omega_n}{\sqrt{\zeta^2 - 1}} e^{-\zeta\omega_n t} \sinh\left(\sqrt{\zeta^2 - 1}\,\omega_n t\right) \tag{5-105}$$

It is noted here that $c(t) \geq 0$ for all $t$.

The case when $\zeta = 1$ is handled differently. Here we have $r_1 = r_2$, and consequently

$$C(s) = \frac{\omega_n^2}{(s + \omega_n)^2} \tag{5-106}$$

The inverse Laplace transform for this case is given by

$$c(t) = \omega_n^2 t e^{-\omega_n t} \tag{5-107}$$

The system is said to be critically damped. Again we note that $c(t) \geq 0$, for all $t$, as shown in Figure 5.6.

The maximum value of $c(t)$ for the overdamped case is obtained by setting the derivative of $c(t)$ with respect to time to zero. This occurs for

$$\sqrt{\zeta^2 - 1}\cosh\left(\sqrt{\zeta^2 - 1}\,\omega_n t\right) - \zeta \sinh\left(\sqrt{\zeta^2 - 1}\,\omega_n t\right) = 0 \tag{5-108}$$

This is satisfied for the time $t$ for the first overshoot which is defined by

$$t_p = \frac{1}{\omega_n\sqrt{\zeta^2 - 1}} \tanh^{-1}\frac{\sqrt{\zeta^2 - 1}}{\zeta} \tag{5-109}$$

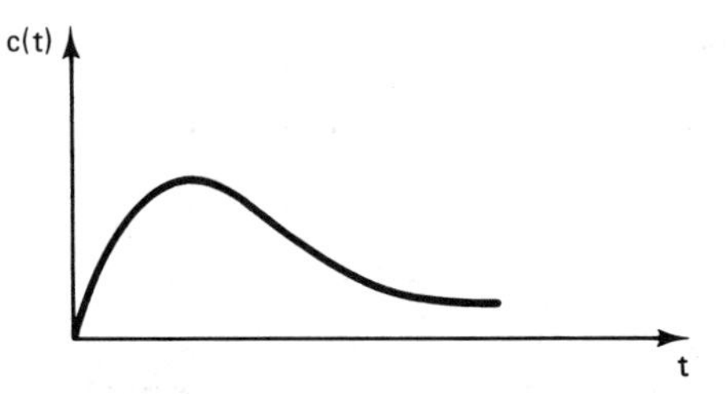


**Figure 5.6.** Impulse Response of a Critically Damped Second-Order System

Alternatively,

$$t_p = \frac{1}{\omega_n\sqrt{\zeta^2 - 1}}\cosh^{-1}\zeta$$

or

$$t_p = \frac{1}{\omega_n\sqrt{\zeta^2 - 1}}\ln\left(\zeta + \sqrt{\zeta^2 - 1}\right) \tag{5-110}$$

As a result, the maximum value of $c(t)$ is given by

$$c_{\max} = \omega_n\left(\zeta + \sqrt{\zeta^2 - 1}\right)^{-\zeta/\sqrt{\zeta^2-1}} \tag{5-111}$$

Figure 5.7 shows typical impulse response curves for overdamped second-order systems.

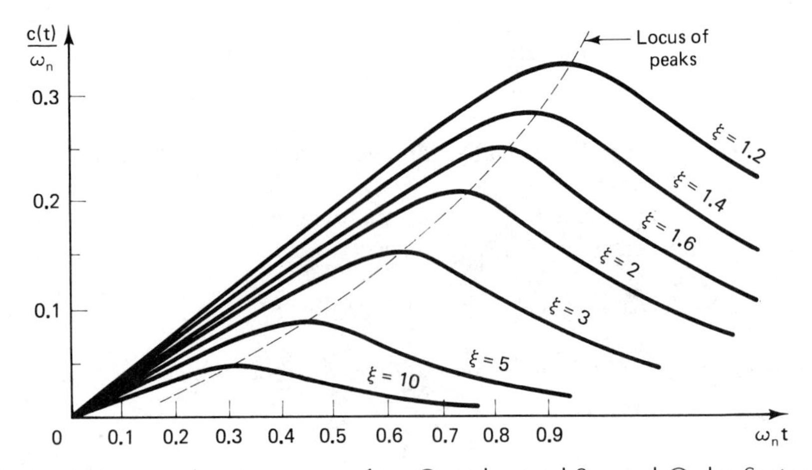


**Figure 5.7.** Impulse Response of an Overdamped Second-Order System

For the critically damped case, we have the maximum value of $c(t)$ occurring at

$$t_p = \frac{1}{\omega_n} \tag{5-112}$$

The maximum value is

$$\begin{aligned} c_{\max} &= \omega_n e^{-1} \\ &= 0.3679\omega_n \end{aligned} \tag{5-113}$$

We now turn our attention to the underdamped case for which $0 < \zeta < 1$. Here we find that $\sqrt{\zeta^2 - 1}$ is an imaginary number. As a result, the roots are

$$r_1 = \sigma + j\omega_d \tag{5-114}$$

$$r_2 = \sigma - j\omega_d \tag{5-115}$$

where we define

$$\sigma = -\zeta\omega_n \tag{5-116}$$

$$\omega_d = \omega_n\sqrt{1 - \zeta^2} \tag{5-117}$$

The radian frequency $\omega_d$ is called the *damped natural radian frequency.* The impulse response is obtained as

$$c(t) = \frac{\omega_n^2}{\omega_d} e^{+\sigma t} \sin \omega_d t \tag{5-118}$$

To obtain the maximum value of $c(t)$, commonly denoted as the *overshoot*, we set the derivative of $c(t)$ with respect to $t$ to zero. This gives us

$$\sigma \sin \omega_d t + \omega_d \cos \omega_d t = 0 \tag{5-119}$$

As a result,

$$t_p = \frac{1}{\omega_d} \tan^{-1} \frac{\sqrt{1 - \zeta^2}}{\zeta} \tag{5-120}$$

The maximum overshoot is

$$c_{\max} = \omega_n \exp\left( \frac{-\zeta}{\sqrt{1 - \zeta^2}} \tan^{-1} \frac{\sqrt{1 - \zeta^2}}{\zeta} \right) \tag{5-121}$$

Figure 5.8 shows a typical unit impulse response for an underdamped second-order system, sometimes called a free response. Note that the output has both positive and negative values, depending on the time instant as opposed to the case of overdamped and critically damped systems, where the output is always positive.

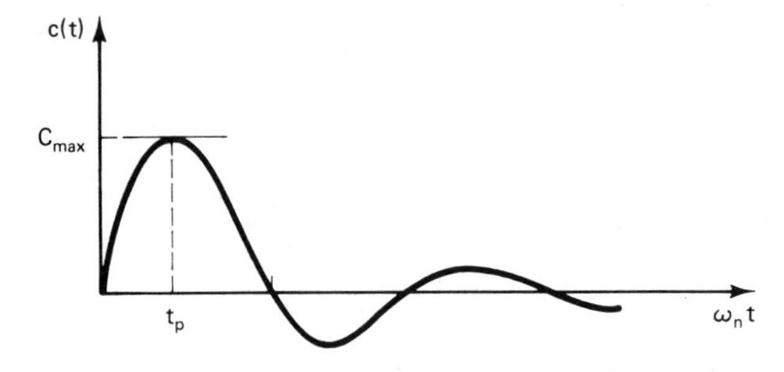


**Figure 5.8.** Unit Impulse Response of an Underdamped Second-Order System

### Step Response

Consider now the case of a unit step input to the second-order system. Here the Laplace transform of the output is given by

$$C(s) = \frac{\omega_n^2}{s(s - r_1)(s - r_2)} \tag{5-122}$$

Performing a partial fraction expansion, we obtain

$$C(s) = \frac{1}{s} + \frac{1}{2\sqrt{\zeta^2 - 1}}\left(\frac{1}{-\zeta + \sqrt{\zeta^2 - 1}}\,\frac{1}{s - r_1} + \frac{1}{\zeta + \sqrt{\zeta^2 - 1}}\,\frac{1}{s - r_2}\right) \tag{5-123}$$

The inverse Laplace transform gives us the step response, as

$$c(t) = 1 + \frac{\omega_n}{2\sqrt{\zeta^2 - 1}}\left(\frac{e^{r_1 t}}{r_1} - \frac{e^{r_2 t}}{r_2}\right) \tag{5-124}$$

where

$$r_1 = \left(-\zeta + \sqrt{\zeta^2 - 1}\right)\omega_n \tag{5-125}$$

$$r_2 = \left(-\zeta - \sqrt{\zeta^2 - 1}\right)\omega_n \tag{5-126}$$

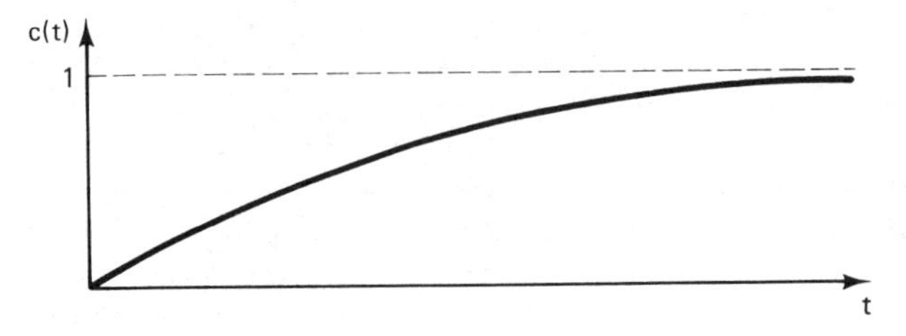

**Figure 5.9.** Step Response of an Overdamped Second-Order System

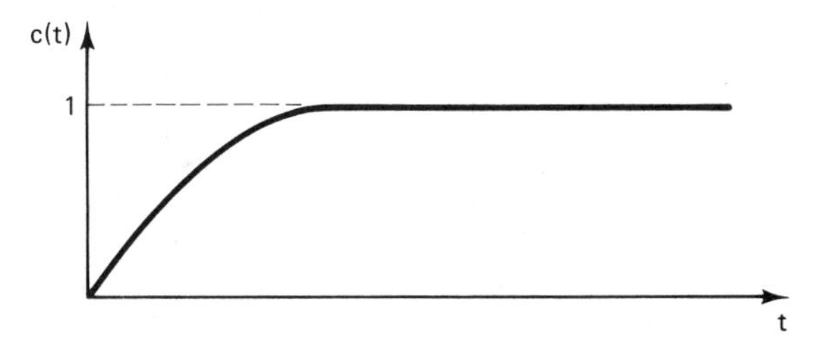


**Figure 5.10.** Step Response of a Critically Damped Second-Order System

This is the response for the overdamped case $\zeta \geq 1$, and a typical time response is plotted in Figure 5.9.

The case for which $\zeta = 1$ (i.e., for a critically damped system) yields a time response given by

$$c(t) = 1 - (1 + \omega_n t)e^{-\omega_n t} \tag{5-127}$$

The response is shown in Figure 5.10.

We turn our attention now to the important case of an underdamped system for which $0 < \zeta < 1$. The time response in this case is given by

$$c(t) = 1 + \frac{e^{-\zeta\omega_n t}}{\sqrt{1 - \zeta^2}} \sin\left(\omega_n\sqrt{1 - \zeta^2}\, t - \alpha\right) \tag{5-128}$$

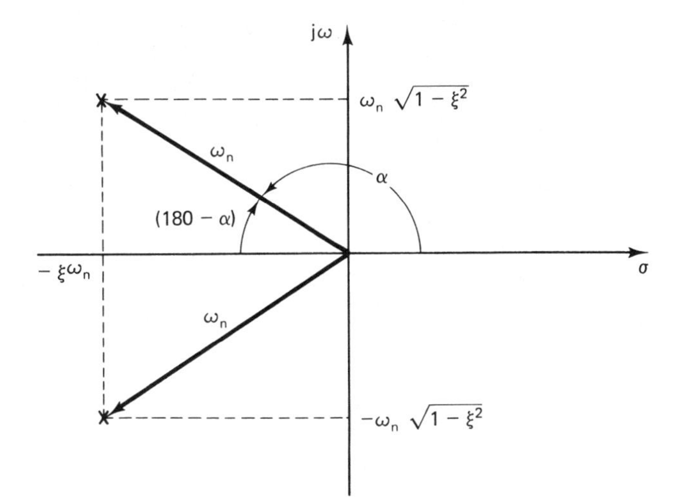


**Figure 5.11.** Pole Configuration in the *s*-Plane for an Underdamped Second-Order System

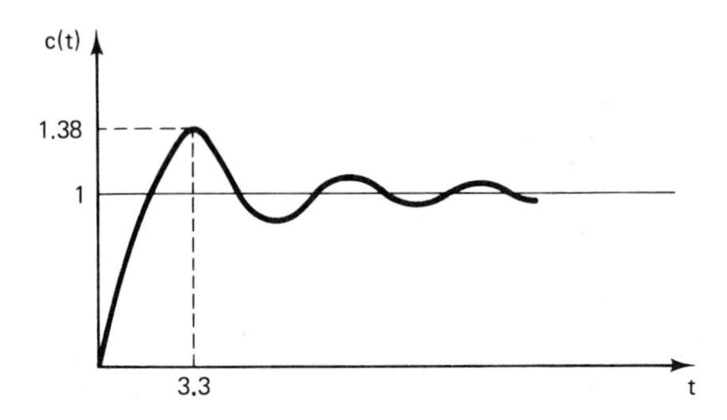


**Figure 5.12.** Step Response of an Underdamped Second-Order System

where the angle $\alpha$ is as defined in Figure 5.11.

$$\cos\alpha = -\zeta$$
$$\sin\alpha = \sqrt{1-\zeta^2}$$

A typical time response is shown in Figure 5.12.

Inspection of Eq. (5-128) and Figure 5.12 reveals that the step response of the underdamped second-order system is oscillatory with a frequency $\omega_d$ that varies with $\zeta$ as

$$\omega_d = \sqrt{1-\zeta^2}\,\omega_n \tag{5-129}$$

Figure 5.13 shows the step response of underdamped, damped and over-damped second-order systems. If the damping ratio $\zeta$ is equal to zero, the response is undamped and oscillations continue indefinitely. The time response in this case is given by

$$c(t) = 1 - \cos\omega_n t \tag{5-130}$$

The time to the first overshoot and the value of output at that time are of interest. These are obtained by setting

$$\frac{dc}{dt} = 0$$

As a result,

$$0 = \frac{d}{dt}\left[e^{-\zeta\omega_n t}\sin\left(\omega_n\sqrt{1-\zeta^2}\,t-\alpha\right)\right]$$

This yields

$$-\zeta\omega_n e^{-\zeta\omega_n t}\sin\left(\omega_n\sqrt{1-\zeta^2}\,t-\alpha\right) + \omega_n\sqrt{1-\zeta^2}\,e^{-\zeta\omega_n t}\cos\left(\omega_n\sqrt{1-\zeta^2}\,t-\alpha\right) = 0 \tag{5-131}$$

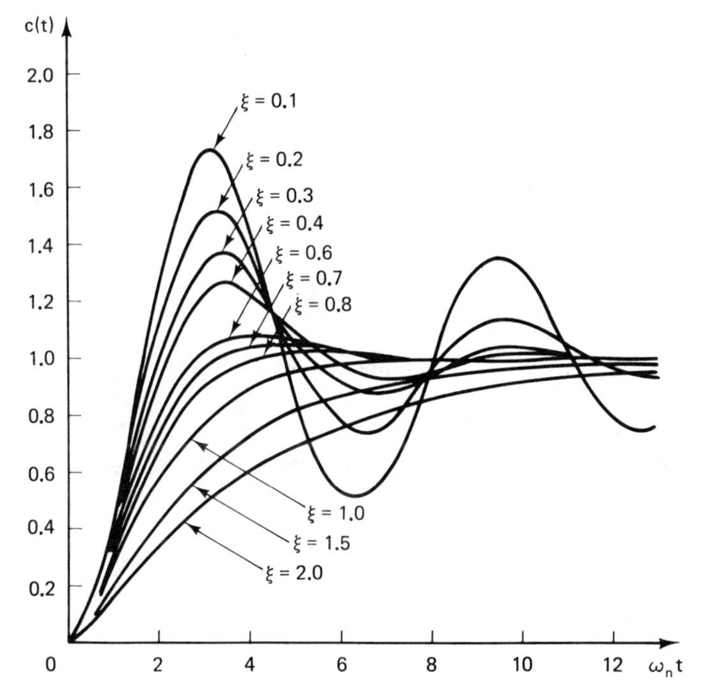


**Figure 5.13.** Step Response of a Second-Order System for Various Values of $\zeta$

Thus the condition for maximum is

$$\cos\alpha\sin\left(\omega_n\sqrt{1-\zeta^2}\,t-\alpha\right)+\sin\alpha\cos\left(\omega_n\sqrt{1-\zeta^2}\,t-\alpha\right)=0 \quad (5\text{-}132)$$

This reduces to

$$\sin\left(\omega_n\sqrt{1-\zeta^2}\,t\right)=0$$

This is satisfied for

$$\omega_n\sqrt{1-\zeta^2}\,t=0,\pi,2\pi,\ldots \quad (5\text{-}133)$$

Thus the time to first overshoot is given by

$$t_p=\frac{\pi}{\omega_n\sqrt{1-\zeta^2}} \quad (5\text{-}134)$$

The peak value of the output is

$$c_p=c(t_p)=1+\frac{e^{-\zeta\pi/\sqrt{1-\zeta^2}}}{\sqrt{1-\zeta^2}}\sin(\pi-\alpha) \quad (5\text{-}135)$$

This reduces to

$$c_p = 1 + M_p \tag{5-136}$$

where

$$M_p = e^{-\zeta\pi/\sqrt{1-\zeta^2}} \tag{5-137}$$

This is usually expressed as a percentage of the input. We thus define

$$\text{maximum percent overshoot} = \exp\left(-\frac{\zeta\pi}{\sqrt{1-\zeta^2}}\right) \times 100$$

The variation of the maximum percent overshoot with the damping ratio is shown in Figure 5.14.

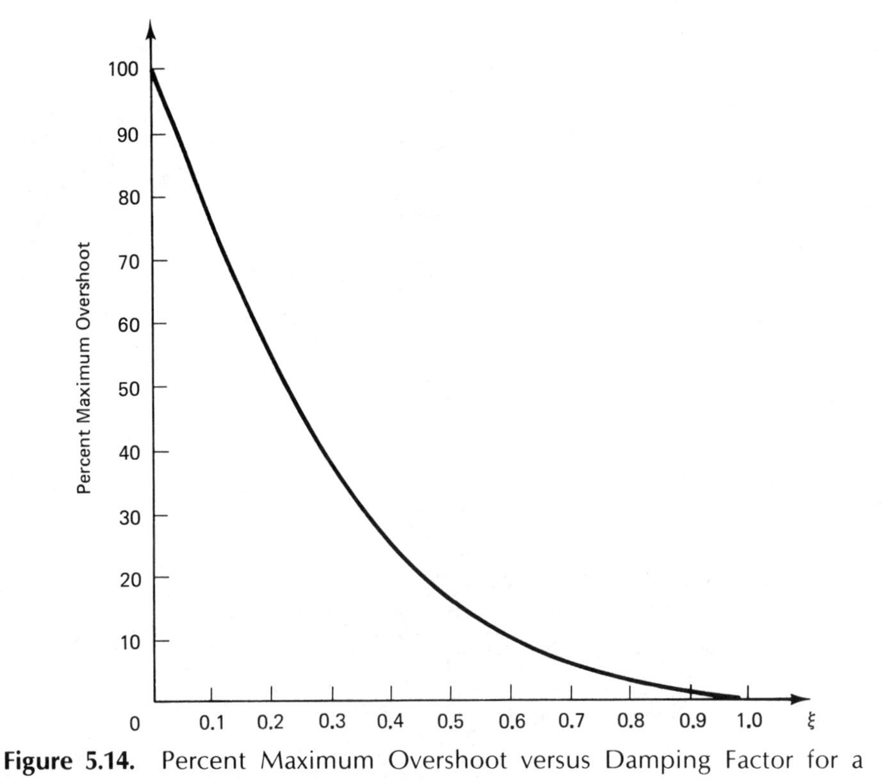


**Figure 5.14.** Percent Maximum Overshoot versus Damping Factor for a Second-Order System

# 5.7 Time-Response Specifications

In practice, the required performance characteristics of a control system are given in terms of quantities derived from the transient response of a second-order underdamped system to unit step input. The following form of the output time response based on Eq. (5-128) is used to derive the quantities of interest.

$$c(t) - 1 = -e^{-\zeta\omega_n t}\left(\cos\omega_d t + \frac{\zeta}{\sqrt{1-\zeta^2}}\sin\omega_d t\right) \tag{5-138}$$

where

$$\omega_d = \omega_n\sqrt{1-\zeta^2} \tag{5-139}$$

1. *Delay time* $t_d$. This is the required time for the step response to attain half of its steady-state value. Thus $t_d$ is obtained as the solution to

$$0.5 = e^{-\zeta\omega_n t_d}\left(\cos\omega_d t_d + \frac{\zeta}{\sqrt{1-\zeta^2}}\sin\omega_d t_d\right) \tag{5-140}$$

2. *Rise time* $t_r$. This is the time required for the response to rise from 10 to 90%, 5 to 95%, or 0 to 100% of its final value. For underdamped systems, 0 to 100% rise time is normally used. For overdamped systems, 10 to 90% is commonly used.

   We will take $t_r$ to occur for $c(t_r) = 1$. Thus

$$0 = \frac{e^{-\zeta\omega_n t_r}}{\sqrt{1-\zeta^2}}\sin\left(\omega_n\sqrt{1-\zeta^2}\,t_r - \alpha\right)$$

   This is satisfied for

$$\omega_n\sqrt{1-\zeta^2}\,t_r - \alpha = n\pi \qquad n = 0,1,\ldots$$

   As a result we conclude that

$$t_r = \frac{\alpha}{\omega_d} \tag{5-141}$$

   where

$$\cos\alpha = -\zeta \tag{5-142}$$

3. *Peak time* $t_p$. This is the time required for the response to reach the first peak of the overshoot. As obtained before, we have

$$t_p = \frac{\pi}{\omega_d} \tag{5-143}$$

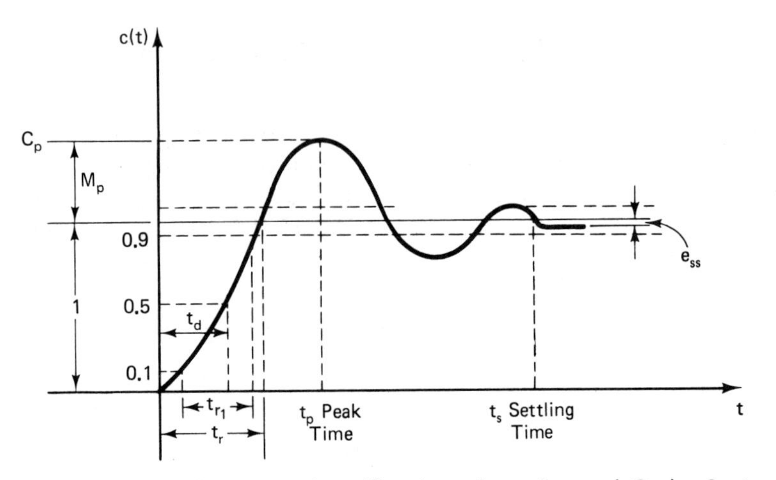


**Figure 5.15.** Performance Specifications for a Second-Order System

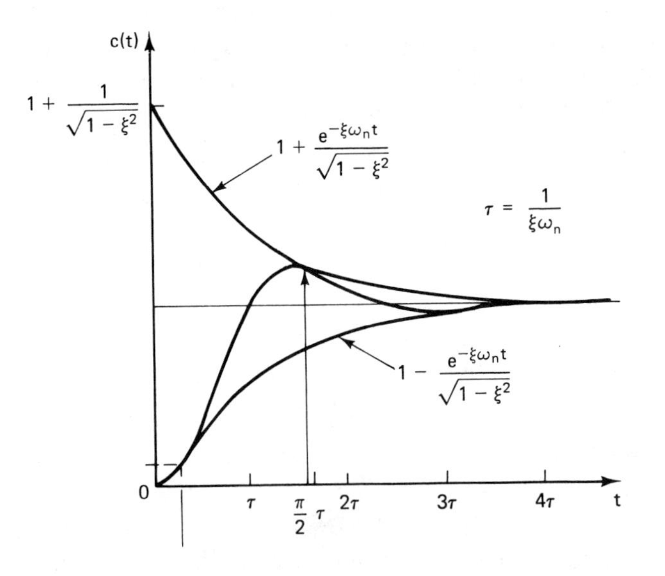


**Figure 5.16.** Defining Settling Time

4. *Maximum overshoot* $M_p$. This is the peak or maximum value of the time response. As before, we have

$$M_p = e^{-(\zeta/\sqrt{1-\zeta^2})\pi} \tag{5-144}$$

Figure 5.15 shows the various specification quantities discussed here.

5. *Settling time* $t_s$. This is the time required for the response to reach and stay within a range about the final value of size specified by an absolute percentage of the final value (usually 5% or 2%). The curves $(1 \pm e^{-\zeta\omega_n t/\sqrt{1-\zeta^2}})$ are the envelope curves of the transient response and have a time constant

$$\tau = \frac{1}{\zeta\omega_n} \tag{5-145}$$

This is shown in Figure 5.16.

The speed of decay depends on the value of $\tau$. Settling times for different $\zeta$ are shown in Figure 5.17. For the 2% criterion

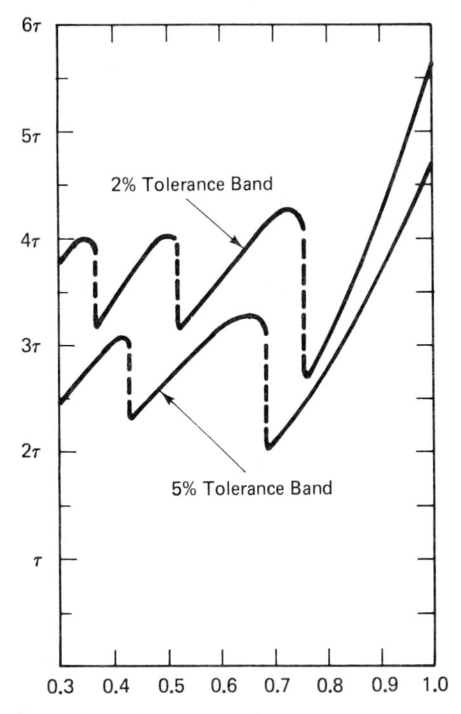


**Figure 5.17.** Settling Time ($t_s$) versus $\zeta$ Curves for a Second-Order System

we can use

$$t_s = 4\tau \tag{5-146}$$

For the 5% criterion, we use

$$t_s = 3\tau$$

Figure 5.17 shows the variation of the settling time with the value of $\zeta$ for the two criteria considered.

EXAMPLE 5.8

Find the transfer function of a second-order system with time to peak of $t_p = \pi/12$ seconds, and a maximum overshoot $M_p = 0.095$.

SOLUTION

We have

$$t_p = \frac{\pi}{\omega_d}$$

For

$$t_p = \frac{\pi}{12}$$

we get

$$\omega_d = 12$$

Thus

$$\omega_n\sqrt{1-\zeta^2} = 12$$

We also have

$$M_p = 0.095 = e^{-(\zeta/\sqrt{1-\zeta^2})\pi}$$

Thus

$$\zeta = 0.600$$

As a result,

$$\omega_n = 15$$

The transfer function is thus given by

$$\frac{C(s)}{R(s)} = \frac{225}{s^2 + 18s + 225}$$

6. *Logarithmic decrement* $\delta$. The exponentially damped sinusoid of the step response exhibits oscillations whose amplitude decays as a geometric series. At $t = t_p$ we have

$$M_p = e^{-\sigma\pi/\omega_d} \tag{5-147}$$

After one oscillation, at

$$t = t_p + \frac{2\pi}{\omega_d} = \frac{3\pi}{\omega_d}$$

the amplitude is equal to

$$M_{p_2} = e^{-3\sigma\pi/\omega_d}$$

The ratio of amplitudes is thus

$$\frac{M_p}{M_{p_2}} = e^{2\zeta\pi/\sqrt{1-\zeta^2}} \tag{5-148}$$

The logarithmic decrement $\delta$ is defined as the logarithm of the amplitude ratio:

$$\delta = \ln e^{2\zeta\pi/\sqrt{1-\zeta^2}}$$

Thus

$$\delta = \frac{2\zeta\pi}{\sqrt{1-\zeta^2}} \tag{5-149}$$

It is a function of $\zeta$ only.

EXAMPLE 5.9

Consider the system shown in block diagram form in Figure 5.18. Assume that

$$\zeta = 0.707$$
$$\omega_n = 10 \text{ rad / sec}$$

Calculate:

(a) The delay time.
(b) The rise time.
(c) The overshoot $M_p$.
(d) The settling time.

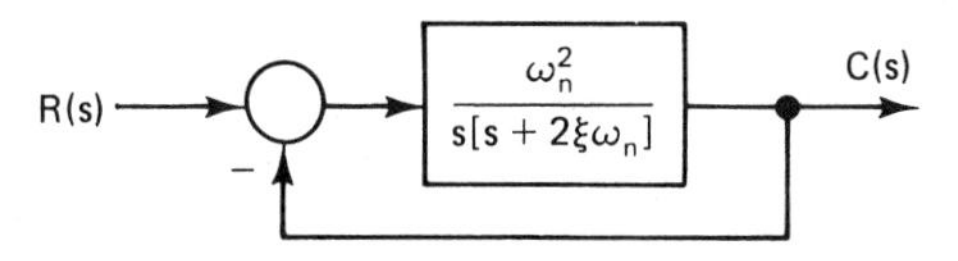

**Figure 5.18.** System Block Diagram for Example 5.9

SOLUTION

The damped frequency $\omega_d$ is calculated as

$$\omega_d = \omega_n\sqrt{1-\zeta^2} = 7.07 \text{ rad / sec}$$
$$\sigma = -\zeta\omega_n = -7.07 \text{ rad / sec}$$

(a) The delay time is obtained as the solution to

$$0.5 = e^{-7.07t_d}(\cos 7.07t_d + \sin 7.07t_d)$$

The above reduces to requiring the solution of

$$e^x - 2\sqrt{2}\sin(x + 45°) = 0$$

where

$$x = 7.07t_d$$

We obtain the solution in $x$ iteratively as

$$x = 1.0135$$

Thus

$$t_d = 143.35 \times 10^{-3} \text{ sec}$$

(b) The rise time is given by

$$t_r = \frac{\alpha}{\omega_d}$$

since

$$\cos\alpha = -0.707$$
$$\alpha = 2.356 \text{ rad}$$

Thus

$$t_r = 0.333 \text{ sec}$$

(c) The overshoot $M_p$ is given by

$$M_p = e^{-(\zeta/\sqrt{1-\zeta^2})\pi}$$

This is calculated to be

$$M_p = 0.0432$$

or 4.32% overshoot.

(d) The settling time $t_s$ is

$$t_s = \frac{4}{\zeta\omega_n} = \frac{4}{7.07} = 0.566 \text{ sec}$$

# 5.8 Higher-Order Systems

The concepts applied in Section 5.6 to obtain the time response of first-order and second-order systems can be used also with higher-order systems. In this section we deal with two specific systems of higher order to illustrate the procedure.

## Third-Order System

Consider a system with transfer function

$$\frac{C(s)}{R(s)} = \frac{K}{(s+\alpha)(s^2 + 2\zeta\omega_n s + \omega_n^2)} \tag{5-150}$$

We can perform a partial fraction expansion as follows:

$$\frac{C(s)}{R(s)} = K\left(\frac{A}{s+\alpha} + \frac{B + Es}{s^2 + 2\zeta\omega_n s + \omega_n^2}\right) \tag{5-151}$$

To evaluate the constants, $A$, $B$, and $E$, we have

$$A(s^2 + 2\zeta\omega_n s + \omega_n^2) + (B + Es)(s + \alpha) = 1 \tag{5-152}$$

Equating the coefficients of $s^2$ on both sides of Eq. (5-152), we conclude that

$$A + E = 0$$

Substituting $s = -\alpha$, we get

$$A = \frac{1}{(\omega_n^2 + \alpha^2) - 2\zeta\omega_n\alpha}$$

Thus

$$E = \frac{1}{2\zeta\omega_n\alpha - (\omega_n^2 + \alpha^2)}$$

since $E = -A$. The coefficient of the absolute term on both sides of Eq. (5-152) yields

$$A\omega_n^2 + \alpha B = 1$$

Thus

$$B = \frac{1 - \omega_n^2 A}{\alpha}$$

This reduces to

$$B = \frac{\alpha - 2\zeta\omega_n}{(\omega_n^2 + \alpha^2) - 2\zeta\omega_n\alpha}$$

or in terms of $A$,

$$B = A(\alpha - 2\zeta\omega_n)$$

As a result, the transfer function in partial fraction form is given by

$$\frac{C(s)}{R(s)} = KA\left[\frac{1}{s+\alpha} + \frac{(\alpha - 2\zeta\omega_n) - s}{s^2 + 2\zeta\omega_n s + \omega_n^2}\right] \tag{5-153}$$

### Impulse Response

If the system is subject to a unit impulse input, then

$$R(s) = 1$$

As a result,

$$C(s) = KA\left[\frac{1}{s+\alpha} + \frac{(\alpha - 2\zeta\omega_n) - s}{s^2 + 2\zeta\omega_n s + \omega_n^2}\right] \tag{5-154}$$

Let us define

$$C_A(s) = KA\frac{1}{s+\alpha} \tag{5-155}$$

$$C_B(s) = KA\frac{(\alpha - 2\zeta\omega_n) - s}{s^2 + 2\zeta\omega_n s + \omega_n^2} \tag{5-156}$$

Thus

$$C(s) = C_A(s) + C_B(s) \tag{5-157}$$

The impulse response is thus

$$c(t) = c_A(t) + c_B(t) \tag{5-158}$$

Let us recall that

$$\mathscr{L}^{-1}\left\{\frac{1}{s+\alpha}\right\} = e^{-\alpha t}$$

Thus

$$c_A(t) = KA\,e^{-\alpha t} \tag{5-159}$$

We also recall that

$$\mathscr{L}^{-1}\left\{\frac{1}{s^2 + 2\zeta\omega_n s + \omega_n^2}\right\} = \frac{e^{\sigma t}}{\omega_d}\sin\omega_d t$$

$$\mathscr{L}^{-1}\left\{\frac{s}{s^2 + 2\zeta\omega_n s + \omega_n^2}\right\} = \frac{e^{\sigma t}}{\omega_d}(\sigma\sin\omega_d t + \omega_d\cos\omega_d t)$$

where

$$\omega_d = \omega_n\sqrt{1-\zeta^2}$$

$$\sigma = -\zeta\omega_n$$

As a result,

$$c_B(t) = \frac{KAe^{\sigma t}}{\omega_d}[(\alpha - \zeta\omega_n)\sin\omega_d t - \omega_d\cos\omega_d t] \tag{5-160}$$

We need the following result:

$$D^2 = (\alpha - \zeta\omega_n)^2 + \omega_d^2$$

$$= \alpha^2 - 2\zeta\alpha\omega_n + \omega_n^2 = \frac{1}{A} \tag{5-161}$$

Thus we may write $c_B(t)$ as

$$c_B(t) = \frac{KAD\,e^{\sigma t}}{\omega_d}\left(\frac{\alpha - \zeta\omega_n}{D}\sin\omega_d t - \frac{\omega_d}{D}\cos\omega_d t\right) \tag{5-162}$$

Define the angle $\phi$ by

$$\cos\phi = \frac{\alpha - \zeta\omega_n}{D} \tag{5-163}$$

$$\sin\phi = \frac{\omega_d}{D} \tag{5-164}$$

Thus

$$c_B(t) = \frac{KADe^{\sigma t}}{\omega_d}\sin(\omega_d t - \phi) \tag{5-165}$$

But since

$$D = \frac{1}{\sqrt{A}} \tag{5-166}$$

we get

$$c_B(t) = \frac{K\sqrt{A}}{\omega_d}e^{\sigma t}\sin(\omega_d t - \phi) \tag{5-167}$$

We now have the impulse response of the third-order system in the following form:

$$c(t) = KA\left[e^{-\alpha t} + \frac{e^{-\zeta\omega_n t}}{\omega_d\sqrt{A}}\sin(\omega_d t - \phi)\right] \tag{5-168}$$

where

$$A = \frac{1}{\omega_n^2 + \alpha^2 - 2\zeta\omega_n\alpha} \tag{5-169}$$

$$\cos\phi = (\alpha - \zeta\omega_n)\sqrt{A} \tag{5-170}$$

$$\sin\phi = \omega_d\sqrt{A} \tag{5-171}$$

An alternative form of the impulse response is given by

$$c(t) = \frac{K}{\omega_n^2(\beta^2 - 2\zeta\beta + 1)}\left[e^{-\alpha t} + \frac{\sqrt{\beta^2 - 2\zeta\beta + 1}}{\sqrt{1 - \zeta^2}}e^{-\zeta\omega_n t}\sin(\omega_d t - \phi)\right] \tag{5-172}$$

where

$$\beta = \frac{\alpha}{\omega_n} \tag{5-173}$$

### Step Response

If the system is subject to a unit step input, then

$$R(s) = \frac{1}{s}$$

As a result,

$$C(s) = KA\left[\frac{1}{s(s+\alpha)} + \frac{\alpha - 2\zeta\omega_n}{s(s^2 + 2\zeta\omega_n s + \omega_n^2)} - \frac{1}{s^2 + 2\zeta\omega_n s + \omega_n^2}\right] \tag{5-174}$$

Performing partial fraction expansions on the first two terms, and collecting terms, the above reduces to

$$C(s) = \frac{K}{\alpha\omega_n^2}\left[\frac{1}{s} - \frac{1}{1 + \beta^2 - 2\zeta\beta}\frac{1}{s+\alpha} - \frac{\alpha(1 + 2\beta\zeta - 4\zeta^2)}{1 + \beta^2 - 2\zeta\beta}\frac{1}{s^2 + 2\zeta\omega_n s + \omega_n^2} - \frac{\beta(\beta - 2\zeta)}{1 + \beta^2 - 2\beta\zeta}\frac{s}{s^2 + 2\zeta\omega_n s + \omega_n^2}\right] \tag{5-175}$$

where

$$\beta = \frac{\alpha}{\omega_n} \tag{5-176}$$

The inverse Laplace transform of $C(s)$ is given by

$$c(t) = \frac{K}{\alpha\omega_n^2}\left[1 - \frac{e^{-\alpha t}}{(\beta^2 - 2\zeta\beta + 1)} + \frac{\beta e^{-\zeta\omega_n t}}{\sqrt{1-\zeta^2}\sqrt{\beta^2 - 2\zeta\beta + 1}}\sin(\omega_d t - \psi)\right] \tag{5-177}$$

where

$$\sin\psi = \frac{(2\zeta - \beta)\sqrt{1-\zeta^2}}{\sqrt{\beta^2 - 2\zeta\beta + 1}}$$

$$\cos\psi = \frac{1 + \beta\zeta - 2\zeta^2}{\sqrt{\beta^2 - 2\zeta\beta + 1}}$$

An alternative expression for $\psi$ is given by

$$\psi = \tan^{-1}\frac{\sqrt{1-\zeta^2}}{-\zeta} + \tan^{-1}\frac{\sqrt{1-\zeta^2}}{\beta - \zeta} \tag{5-178}$$

## Fourth-Order System

Consider a system with the transfer function given by

$$\frac{C(s)}{R(s)} = \frac{K}{(s^2 + 2\zeta_A\omega_A s + \omega_A^2)(s^2 + 2\zeta_B\omega_B s + \omega_B^2)} \tag{5-179}$$

Performing a partial fraction expansion, we have

$$\frac{C(s)}{R(s)} = K\left(\frac{a + bs}{s^2 + 2\zeta_A\omega_A s + \omega_A^2} + \frac{e + ds}{s^2 + 2\zeta_B\omega_B s + \omega_B^2}\right) \tag{5-180}$$

The constants $a$, $b$, $e$, and $d$ are obtained from

$$(a + bs)(s^2 + 2\zeta_B\omega_B s + \omega_B^2) + (e + ds)(s^2 + 2\zeta_A\omega_A s + \omega_A^2) = 1 \tag{5-181}$$

Equating the coefficients of $s^3$ on both sides, we then have

$$b + d = 0$$

Equating the coefficients of the absolute term, we have

$$a\omega_B^2 + e\omega_A^2 = 1$$

Assume that the polynomial $(s^2 + 2\zeta_A\omega_A s + \omega_A^2)$ has roots $r_1$ and $r_2$. Thus in rectangular form

$$r_1 = \sigma_A + j\omega_{dA} \tag{5-182}$$

$$r_2 = \sigma_A - j\omega_{dA} \tag{5-183}$$

where

$$\sigma_A = -\zeta_A\omega_A \tag{5-184}$$

$$\omega_{dA} = \omega_A\sqrt{1 - \zeta_A^2} \tag{5-185}$$

In polar form

$$r_1 = \omega_A\angle\theta_A \tag{5-186}$$

$$r_2 = \omega_A\angle -\theta_A \tag{5-187}$$

with

$$\cos\theta_A = -\zeta_A \tag{5-188}$$

Substituting $s = r_1$ in Eq. (5-181), we get

$$(a + br_1)(r_1^2 + 2\zeta_B\omega_B r_1 + \omega_B^2) = 1$$

Substituting $s = r_2$ in Eq. (5-181), we get

$$(a + br_2)(r_2^2 + 2\zeta_B\omega_B r_2 + \omega_B^2) = 1$$

As a result, we get

$$a = \frac{1}{r_2 - r_1}\left(\frac{r_2}{r_1^2 + 2\zeta_B\omega_B r_1 + \omega_B^2} - \frac{r_1}{r_2^2 + 2\zeta_B\omega_B r_2 + \omega_B^2}\right) \tag{5-189}$$

$$b = \frac{1}{r_1 - r_2}\left(\frac{1}{r_1^2 + 2\zeta_B\omega_B r_1 + \omega_B^2} - \frac{1}{r_2^2 + 2\zeta_B\omega_B r_2 + \omega_B^2}\right) \tag{5-190}$$

This reduces to

$$a = \frac{r_1^2 + r_1 r_2 + r_2^2 + 2\zeta_B\omega_B(r_1 + r_2) + \omega_B^2}{D} \tag{5-191}$$

$$b = \frac{-2\zeta_B\omega_B - (r_1 + r_2)}{D} \tag{5-192}$$

where

$$D = \left(r_1^2 + 2\zeta_B\omega_B r_1 + \omega_B^2\right)\left(r_2^2 + 2\zeta_B\omega_B r_2 + \omega_B^2\right) \tag{5-193}$$

The denominator $D$ can be reduced using the properties of $r_1$ and $r_2$ as follows. Using the polar form, we get

$$D = \left(\omega_A^2 e^{j2\theta_A} + 2\zeta_B\omega_A\omega_B e^{j\theta_A} + \omega_B^2\right)\left(\omega_A^2 e^{-j2\theta_A} + 2\zeta_B\omega_A\omega_B e^{-j\theta_A} + \omega_B^2\right) \tag{5-194}$$

Observe that the quantity in the second set of parentheses is the complex conjugate to that in the first set of parentheses; thus

$$\begin{aligned} D &= |\omega_A^2 e^{j2\theta_A} + 2\zeta_B\omega_A\omega_B e^{j\theta_A} + \omega_B^2|^2 \\ &= \left(\omega_A^2 \cos 2\theta_A + 2\zeta_B\omega_A\omega_B \cos\theta_A + \omega_B^2\right)^2 \\ &\quad + \left(\omega_A^2 \sin 2\theta_A + 2\zeta_B\omega_A\omega_B \sin\theta_B\right)^2 \end{aligned} \tag{5-195}$$

Thus after some manipulation we get

$$D = \left(\omega_A^2 - \omega_B^2\right)^2 + 4\omega_A\omega_B(\zeta_A\omega_A - \zeta_B\omega_B)(\zeta_A\omega_B - \zeta_B\omega_A) \tag{5-196}$$

We have the following relations:

$$r_1 + r_2 = 2\sigma_A = -2\zeta_A\omega_A \tag{5-197}$$

$$r_1^2 + r_1 r_2 + r_2^2 = \omega_A^2(4\zeta_A^2 - 1) \tag{5-198}$$

As a result, we obtain modified forms of Eq. (5-191) and (5-192) given by

$$a = \frac{\left(\omega_B^2 - \omega_A^2\right) + 4\zeta_A\omega_A(\zeta_A\omega_A - \zeta_B\omega_B)}{D} \tag{5-199}$$

$$b = \frac{2(\zeta_A\omega_A - \zeta_B\omega_B)}{D} \tag{5-200}$$

where

$$D = \left(\omega_A^2 - \omega_B^2\right)^2 + 4\omega_A\omega_B(\zeta_A\omega_A - \zeta_B\omega_B)(\zeta_A\omega_B - \zeta_B\omega_A) \tag{5-201}$$

To complete the partial fraction coefficients, we have

$$e = \frac{1 - a\omega_B^2}{\omega_A^2} \tag{5-202}$$

$$d = -b \tag{5-203}$$

**Impulse Response**

Recall that

$$\mathscr{L}^{-1}\left\{\frac{1}{s^2 + 2\zeta_A\omega_A s + \omega_A^2}\right\} = \frac{e^{\sigma_A t}}{\omega_{dA}}\sin(\omega_{dA}t) \tag{5-204}$$

We also have

$$\mathscr{L}^{-1}\left\{\frac{s}{s^2 + 2\zeta_A\omega_A s + \omega_A^2}\right\} = \frac{1}{\omega_{dA}}\left[\sigma_A e^{\sigma_A t}\sin(\omega_{dA}t) + \omega_{dA}e^{\sigma_A t}\cos(\omega_{dA}t)\right]$$

$$= \frac{e^{\sigma_A t}}{\omega_{dA}}\left[\sigma_A\sin(\omega_{dA}t) + \omega_{dA}\cos(\omega_{dA}t)\right] \tag{5-205}$$

Thus

$$\mathscr{L}^{-1}\left\{\frac{a + bs}{s^2 + 2\zeta_A\omega_A s + \omega_A^2}\right\} = \frac{e^{\sigma_A t}}{\omega_{dA}}\left[(a + b\sigma_A)\sin\omega_{dA}t + b\omega_{dA}\cos\omega_{dA}t\right] \tag{5-206}$$

From the expressions for $a$, $b$, $\sigma_A$, and $\omega_{dA}$ we can conclude that

$$a + b\sigma_A = \frac{\left(\omega_B^2 - \omega_A^2\right) + 2\zeta_A\omega_A(\zeta_A\omega_A - \zeta_B\omega_B)}{D}$$

$$b\omega_{dA} = \frac{2\omega_A(\zeta_A\omega_A - \zeta_B\omega_B)\sqrt{1 - \zeta_A^2}}{D}$$

We can also show that

$$(a + b\sigma_A)^2 + (b\omega_{dA})^2 = \frac{1}{D}$$

Thus we define an angle $\phi_A$ such that

$$\cos\phi_A = (a + b\sigma_A)\sqrt{D} \tag{5-207}$$

$$\sin\phi_A = b\omega_{dA}\sqrt{D} \tag{5-208}$$

As a result, we have

$$\mathscr{L}^{-1}\left\{\frac{a + bs}{s^2 + 2\zeta_A\omega_A s + \omega_A^2}\right\} = \frac{e^{\sigma_A t}}{\omega_{dA}\sqrt{D}}\sin\left(\omega_{d_A}t + \phi_A\right) \tag{5-209}$$

We consider next

$$\mathscr{L}^{-1}\left\{\frac{e + ds}{s^2 + 2\zeta_B\omega_B s + \omega_B^2}\right\} = \frac{e^{\sigma_B t}}{\omega_{d_B}}\left[(e + d\sigma_B)\sin\omega_{d_B}t + d\omega_{d_B}\cos\omega_{d_B}t\right] \tag{5-210}$$

Using the properties of $e$, $d$, $\sigma_B$, and $\omega_{d_B}$, we can conclude that

$$e + d\sigma_B = \frac{\left(\omega_A^2 - \omega_B^2\right) - 2\zeta_B\omega_B(\zeta_A\omega_A - \zeta_B\omega_B)}{D} \tag{5-211}$$

$$d\omega_{d_B} = \frac{-2\omega_B\sqrt{1 - \zeta_B^2}\,(\zeta_A\omega_A - \zeta_B\omega_B)}{D} \tag{5-212}$$

We can also show that

$$(e + d\sigma_B)^2 + (d\omega_{d_B})^2 = \frac{1}{D} \tag{5-213}$$

Thus we can define that angle $\phi_B$ such that

$$\cos\phi_B = (e + d\sigma_B)\sqrt{D} \tag{5-214}$$

$$\sin\phi_B = b\omega_{d_B}\sqrt{D} \tag{5-215}$$

As a result, we have

$$\mathscr{L}^{-1}\left\{\frac{e + ds}{s^2 + 2\zeta_B\omega_B s + \omega_B^2}\right\} = \frac{e^{\sigma_B t}}{\omega_{d_B}\sqrt{D}} \sin\left(\omega_{d_B} t + \phi_B\right) \tag{5-216}$$

This completes the derivation of the ingredients of the impulse response of the system, as in this case $R(s) = 1$.

To summarize, the impulse response of the fourth-order system is given by

$$c(t) = \frac{K}{\sqrt{D}}\left[c_A(t) + c_B(t)\right] \tag{5-217}$$

where $c_A(t)$ and $c_B(t)$ are the responses due to poles of polynomials $A$ and $B$. These are given by

$$c_A(t) = \frac{e^{-\zeta_A\omega_A t}}{\omega_A\sqrt{1 - \zeta_A^2}} \sin\left(\sqrt{1 - \zeta_A^2}\,\omega_A t + \phi_A\right) \tag{5-218}$$

$$c_B(t) = \frac{e^{-\zeta_B\omega_B t}}{\omega_B\sqrt{1 - \zeta_B^2}} \sin\left(\sqrt{1 - \zeta_B^2}\,\omega_B t + \phi_B\right) \tag{5-219}$$

where

$$\cos\phi_A = \frac{\left(\omega_B^2 - \omega_A^2\right) + 2\zeta_A\omega_A(\zeta_A\omega_A - \zeta_B\omega_B)}{\sqrt{D}} \tag{5-220}$$

$$\sin\phi_A = \frac{2\omega_A\sqrt{1 - \zeta_A^2}\,(\zeta_A\omega_A - \zeta_B\omega_B)}{\sqrt{D}} \tag{5-221}$$

$$\cos\phi_B = \frac{\left(\omega_A^2 - \omega_B^2\right) + 2\zeta_B\omega_B(\zeta_B\omega_B - \zeta_A\omega_A)}{\sqrt{D}} \tag{5-222}$$

$$\sin\phi_B = \frac{2\omega_B\sqrt{1 - \zeta_B^2}\,(\zeta_B\omega_B - \zeta_A\omega_A)}{\sqrt{D}} \tag{5-223}$$

$$D = \left(\omega_A^2 - \omega_B^2\right)^2 + 4\omega_A\omega_B(\zeta_A\omega_A - \zeta_B\omega_B)(\zeta_A\omega_B - \zeta_B\omega_A) \tag{5-224}$$

# *Some Solved Problems*

PROBLEM 5A-1

Consider a unity feedback control system with the open-loop transfer function of

$$G(s) = \frac{K}{s(s^2 + 2\zeta\omega_n s + \omega_n^2)}$$

Find the error coefficients $C_0$, $C_1$, and $C_2$.

SOLUTION

The function $F(s)$ is given by

$$F(s) = \frac{sD}{sD + K}$$

where

$$D = s^2 + 2\zeta\omega_n s + \omega_n^2$$

Clearly,

$$C_0 = \lim_{s \to 0} F(s) = 0$$

The first derivative of $F(s)$ is given by

$$\frac{dF(s)}{ds} = \frac{K(3s^2 + 4\zeta\omega_n s + \omega_n^2)}{(sD + K)^2}$$

Thus

$$C_1 = \lim_{s \to 0} \frac{dF}{ds} = \frac{\omega_n^2}{K}$$

The second derivative is

$$\frac{d^2F}{ds^2} = \frac{2K\left[(sD + K)(3s + 2\zeta\omega_n) - (3s^2 + 4\zeta\omega_n s + \omega_n^2)^2\right]}{(sD + K)^3}$$

Thus

$$C_2 = \frac{2\omega_n}{K^2}(2K\zeta - \omega_n^3)$$

PROBLEM 5A-2

Consider a unity feedback control system with the open-loop transfer function of

$$G(s) = \frac{K}{s^n}$$

where $n$ is an integer. Find the error coefficients $C_0$, $C_1$, and $C_2$.

SOLUTION

The function $F(s)$ is given by

$$F(s) = \frac{s^n}{K + s^n}$$

As a result,

$$C_0 = \lim_{s \to 0} F(s) = 0$$

The derivative of $F(s)$ is given by

$$\frac{dF(s)}{ds} = \frac{Kns^{n-1}}{(K + s^n)^2}$$

Thus we have

$$C_1 = \lim_{s \to 0} \frac{Kns^{n-1}}{(K + s^n)^2}$$

The limits provide us with

$$\begin{aligned} C_1 &= \frac{1}{K} \qquad n = 1 \\ &= 0 \qquad n > 1 \end{aligned}$$

The second derivative of $F(s)$ is obtained as

$$\frac{d^2F(s)}{ds^2} = \frac{Kns^{n-2}[K(n-1) - (n+1)s^n]}{(K + s^n)^3}$$

Thus

$$C_2 = \lim_{s \to 0} \frac{d^2F(s)}{ds^2}$$

The limits provide us with

$$\begin{aligned} C_2 &= \frac{-2}{K^2} \qquad n = 1 \\ C_2 &= \frac{2}{K} \qquad n = 2 \\ C_2 &= 0 \qquad n > 2 \end{aligned}$$

PROBLEM 5A-3

Consider a unity feedback control system with the following open-loop transfer function:

$$G(s) = \frac{K(s + a)}{s + b}$$

Find the error coefficients $C_0$, $C_1$, and $C_2$.

SOLUTION

The function $F(s)$ is obtained as

$$F(s) = \frac{s+b}{(1+K)s+(b+Ka)}$$

As a result,

$$C_0 = \lim_{s\to 0} F(s) = \frac{b}{b+Ka}$$

The derivative of $F(s)$ is obtained as

$$\frac{dF(s)}{ds} = \frac{K(a-b)}{[(1+K)s+(b+Ka)]^2}$$

As a result,

$$C_1 = \lim_{s\to 0} \frac{dF(s)}{ds} = \frac{K(a-b)}{(b+Ka)^2}$$

The second derivative of $F(s)$ is obtained as

$$\frac{d^2F(s)}{ds^2} = \frac{2K(b-a)(1+K)}{[(1+K)s+(b+Ka)]^3}$$

Thus

$$C_2 = \frac{2K(b-a)(1+K)}{(b+Ka)^3}$$

# Problems

PROBLEM 5B-1

Consider the automatic braking system of Problem 3B-1 with $N = 0$. Find the error series coefficients $C_0$, $C_1$, and $C_2$, as well as the steady-state step, ramp, and parabolic error constants $K_p$, $K_v$, and $K_a$. What is the type of this system?

PROBLEM 5B-2

For the system of Problem 3B-5, reduce the block diagram to the canonical (standard) form of Figure 5.2 and hence determine the system type as well as the error series coefficients and the error constants.

PROBLEM 5B-3

For Problem 3B-6, determine the type of the system as well as the error series coefficients and error constants.

PROBLEM 5B-4

Repeat Problem 5B-3 for the system of Problem 3B-7.

PROBLEM 5B-5

Repeat Problem 5B-2 for the system of Problem 3B-8.

PROBLEM 5B-6

Repeat Problem 5B-3 for the system of Problem 3B-10.

PROBLEM 5B-7

Repeat Problem 5B-2 for the system of Problem 3B-11.

PROBLEM 5B-8

A first-order system is represented by Eq. (5-85). The unit step response of the system is known to reach a steady-state value of 5. At $t = 1$ sec, the response is $c(1) = 2$. Find the gain $K$ and the time constant $\tau$ for this system.

PROBLEM 5B-9

A first-order system is represented by the transfer function

$$\frac{C(s)}{R(s)} = \frac{1}{s+2}$$

The input function $r(t)$ consists of two impulses as shown in Figure 5.19. Find and sketch the output $c(t)$.

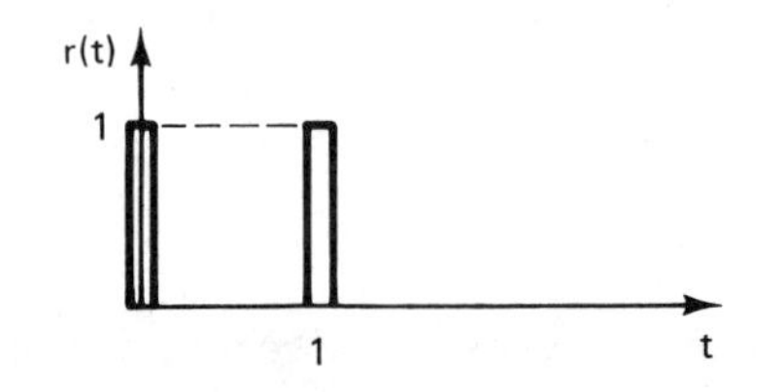


**Figure 5.19.** Input Function for the System of Problem 5B-9

PROBLEM 5B-10

Find the transfer function of a second-order system with time to peak of $t_p = 0.44$ sec and a maximum overshoot of $M_p = 0.0456$.

PROBLEM 5B-11

Consider the system shown in block diagram form in Figure 5.18. Assume that

$$\xi = 0.8$$
$$\omega_n = 12 \text{ rad / sec}$$

Calculate:

(a) The delay time.
(b) The rise time.
(c) The overshoot $M_p$.
(d) The settling time.

PROBLEM 5B-12

Consider the second-order system represented by the canonical form

$$\frac{C(s)}{R(s)} = \frac{\omega_n^2}{s^2 + 2\xi\omega_n s + \omega_n^2}$$

For an underdamped system, it is possible to achieve a response that does not exhibit any oscillations by the appropriate choice of the input function $r(t)$. Figure 5.20 shows the input function and the desired output. The input consists of a step input of magnitude $A$ applied at $t = 0$. At time $t_1$, a delayed step of magnitude $B$ is applied. Show that for a given settled response, $c(t) = H$ for $t > t_1$, the values of $t_1$, $A$, and $B$ are given by

$$t_1 = \frac{\pi}{\omega_d}$$
$$A = \frac{H}{1 + M_p}$$
$$B = AM_p$$

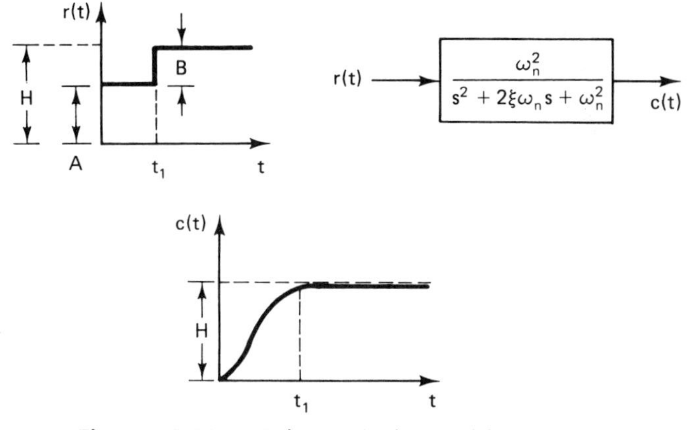


Figure 5.20. Schematic for Problem 5B-12

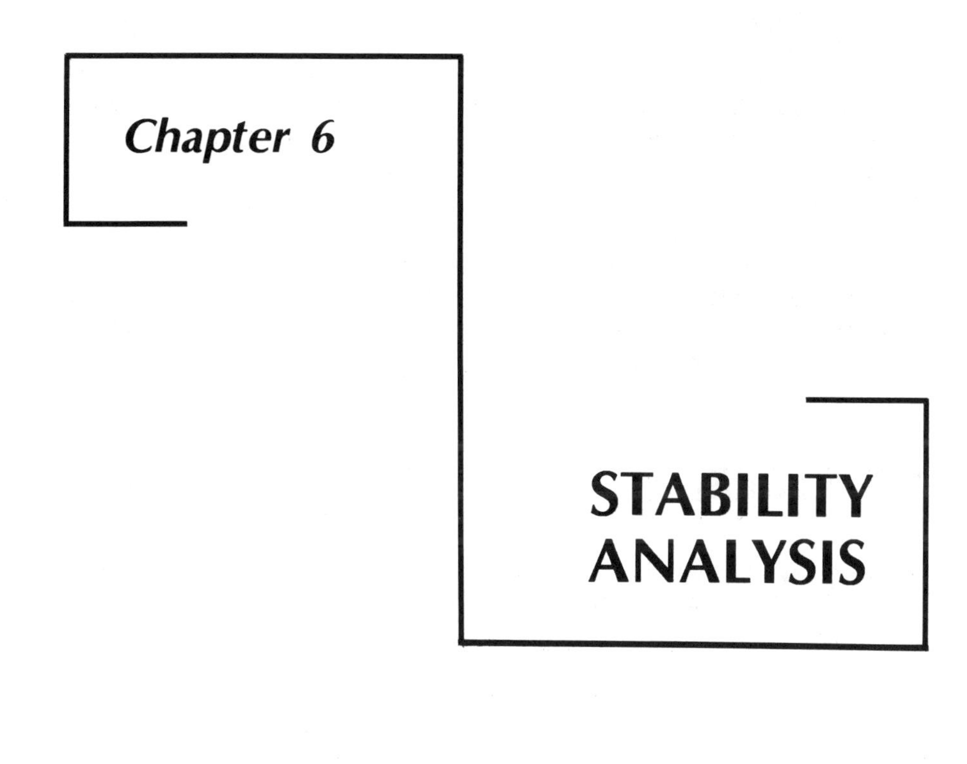


## 6.1 Introduction

This chapter is devoted to one of the most important (if not the most important) aspects of control systems analysis and design, generally referred to as *system stability*. Simply stated, a system is stable if for every bounded input, the output is bounded. This is referred to as *bounded input–bounded output* (BIBO) *stability*. It is clear from this definition that a design that results in an unstable system is not only useless but can lead to hazardous if not disastrous consequences. An unstable system will respond to a bounded input with an output that grows with time in magnitude to levels that can cause serious damage to life, property, and other tangibles. To ascertain system stability is the first task the system designer must undertake.

The systems treated in this book are those described as linear time invariant. They are characterized by ordinary linear differential equations with constant coefficients. Our development so far has dealt with system representation and time-response analysis. The systems discussed can be described either using the transfer function approach of Chapter 2 or the state-space representation of Chapter 4. In both cases we were able to find the details of the time response of the system considered for a given input function. It is thus clear that we can evaluate the stability of the system with the tools developed in previous chapters. The intention of this chapter is to present further tools that enable us to evaluate system stability without having to evaluate the total response of the system.

## 6.2 Stability in the s-Domain

Consider the closed-loop control system with forward transfer function $G(s)$ and feedback function $H(s)$; the Laplace transform of the output $C(s)$ in response to an input $R(s)$ is given by

$$C(s) = \frac{G(s)R(s)}{1 + G(s)H(s)}$$

The inverse Laplace transform introduced in Chapter 2 can be used to evaluate the time response $c(t)$ following a partial fraction expansion as illustrated in both Chapters 2 and 5. Since we are interested in checking the stability of the system in the BIBO sense, a natural question arises concerning the nature of the required input signal $r(t)$. It turns out that the impulse function is sufficient as a test signal. This is shown in the following.

Let $T(s)$ denote the closed-loop transfer function

$$T(s) = \frac{C(s)}{R(s)} = \frac{G(s)}{1 + G(s)H(s)}$$

The transform of the output $C(s)$ is thus

$$C(s) = T(s)R(s)$$

The convolution integral Eq. (2-17) provides us with an expression for the time response $c(t)$ given by

$$c(t) = \int_0^{\infty} T(\tau)r(t-\tau)\,d\tau$$

The magnitude of $c(t)$ is

$$|c(t)| = \left|\int_0^{\infty} T(\tau)r(t-\tau)\,d\tau\right|$$

We now invoke the following inequality:

$$\left|\int f(s)g(s)\,ds\right| \leq \int |f(s)|\,|g(s)|ds$$

As a result,

$$|c(t)| \leq \int_0^\infty |T(\tau)|\,|r(t-\tau)|d\tau$$

Let us recall our initial premise of a bounded input, stated mathematically as

$$|r(t-\tau)| \leq R_A$$

Thus

$$|c(t)| \leq R_A \int_0^\infty |T(\tau)|d\tau$$

The integral in the right-hand side of the inequality above is simply the impulse response of the system. Thus we can see that if

$$\int_0^\infty |T(\tau)|d\tau \leq R_B$$

then

$$|c(t)| \leq R_A R_B$$

What we have just proved is that the output $c(t)$ is bounded provided that the system's impulse response is bounded. It is thus enough to consider the case with $R(s) = 1$ in asserting system stability.

The Laplace transform of the output $C(s)$ for an impulse input is given by

$$C(s) = \frac{G(s)}{1 + G(s)H(s)}$$

Assume that $G(s)$ is given in terms of numerator and denominator functions as

$$G(s) = \frac{N_G(s)}{D_G(s)}$$

Similarly,

$$H(s) = \frac{N_H(s)}{D_H(s)}$$

Thus

$$C(s) = \frac{N_G(s)D_H(s)}{D_G(s)D_H(s) + N_G(s)N_H(s)}$$

Let us define

$$N_C(s) = N_G(s)D_H(s)$$
$$D_C(s) = D_G(s)D_H(s) + N_G(s)N_H(s)$$

As a result, we have

$$C(s) = \frac{N_C(s)}{D_C(s)}$$

The denominator $D_C(s)$ is factored according to

$$D_C(s) = (s - \lambda_1)(s - \lambda_2)\cdots(s - \lambda_n)$$

The parameters $\lambda_1, \lambda_2, \ldots, \lambda_n$ are the roots of

$$D_C(s) = 0$$

A partial fraction expansion will yield

$$C(s) = \frac{A_1}{s - \lambda_1} + \frac{A_2}{s - \lambda_2} + \cdots + \frac{A_n}{s - \lambda_n}$$

The time-domain expression for the output is clearly given by

$$c(t) = A_1 e^{\lambda_1 t} + A_2 e^{\lambda_2 t} + \cdots + A_n e^{\lambda_n t}$$

Let us note here that some of the roots may be complex but they occur as complex-conjugate pairs.

The output $c(t)$ is unbounded if any of the roots $\lambda_i$ has a positive real part. On the other hand, if all roots $\lambda_i$ have negative real parts, then the output $c(t)$ is bounded and the system is stable.

Let us consider the function $P(s)$, defined by

$$P(s) = 1 + G(s)H(s)$$

Using the expressions for $G$ and $H$ in terms of numerator and denominator functions, we have

$$P(s) = \frac{D_G(s)D_H(s) + N_G(s)N_H(s)}{D_G(s)D_H(s)}$$
$$= \frac{D_C(s)}{D_G(s)D_H(s)}$$

It is clear that $P(s) = 0$ implies that $D_C(s) = 0$. Stated differently, the roots of $P(s)$ and $D_C(s)$ are identical. The function $P(s)$ is referred to as the *characteristic equation* of the system.

We can now state a major $s$-domain stability result. The system is stable if all roots of the characteristic equation $P(s)$ have negative real parts. The roots of the characteristic equation are also the poles of the closed-loop transfer function.

Figure 6.1 summarizes the relation between impulse response and root location in the $s$-plane for roots in the (stable) left half of the $s$-plane. The

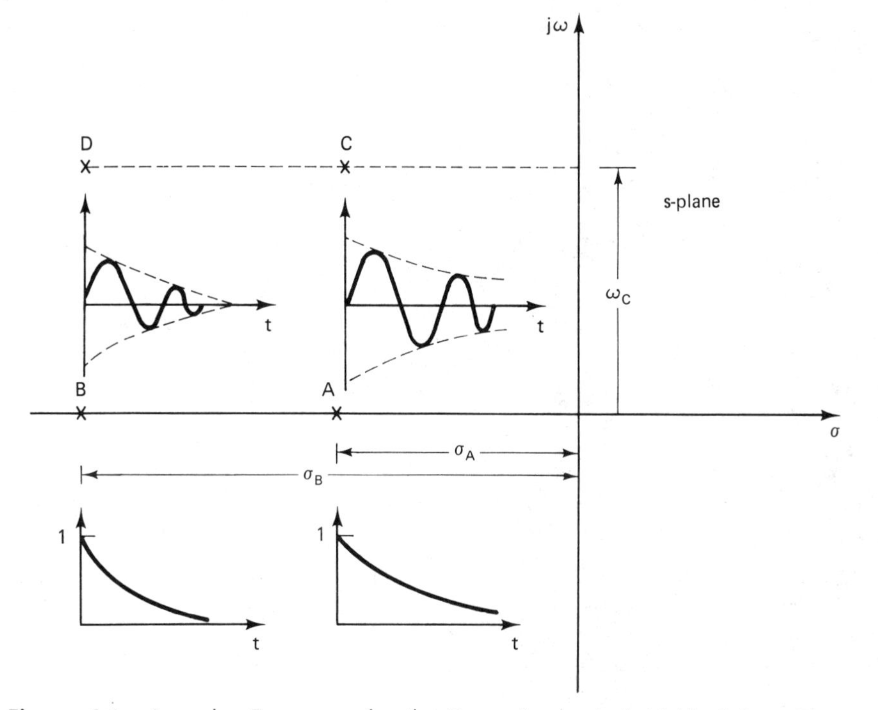


**Figure 6.1.** Impulse Response for the Roots in the Left Half of the s-Plane

response for both roots $A$ and $B$ on the real axis at $s_A = -\sigma_A$ and $s_B = -\sigma_B$ is a decaying exponential. Note, however, that the rate of decay for root $B$ is faster than that for root $A$. Root $C$ is given by

$$s_C = -\sigma_A + j\omega_C$$

It is conjugate to the root $\tilde{C}$, which is not shown but is given by

$$s_{\tilde{C}} = -\sigma_A - j\omega_C$$

The pair of roots correspond to an underdamped second-order polynomial and the impulse response is a damped sinusoid. Root $D$ and its partner $\tilde{D}$ also give rise to a damped sinusoid as shown. Note again that the rate of decay is faster for $D$ than for $C$.

The impulse response for roots on the imaginary axis and in the right half of the $s$-plane is shown in Figure 6.2. Instability is signified by the increase of the output as time is increased for the poles $G$ and $H$. On the imaginary axis the system is critically stable.

From the previous discussion it appears that we need to solve for the roots of the characteristic equation to verify the stability of our linear time-invariant system. It turns out, however, that since we are interested

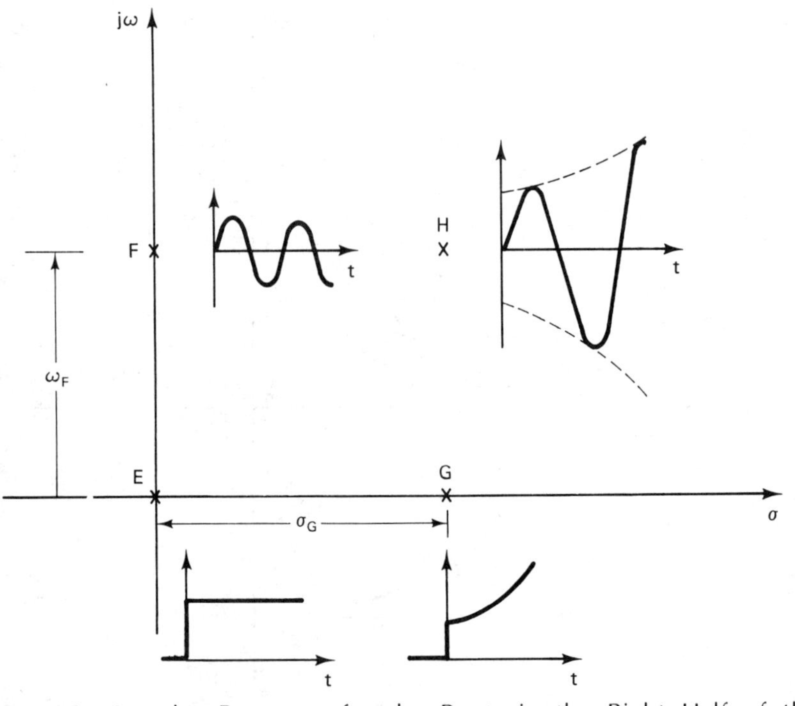


**Figure 6.2.** Impulse Response for the Roots in the Right Half of the *s*-Plane and on the Imaginary Axis

only in the sign of the real part of the roots, efficient tests are available to detect the presence of roots that cause instability. This is the subject of our next section.

## 6.3 Routh–Hurwitz and Related Criteria

We have concluded in the preceding section that the stability of the closed-loop system can be determined from the location of the roots of the characteristic equation given by

$$P(s) = 1 + G(s)H(s) = 0$$

We can express the characteristic equation as an $n$th-order polynomial given by

$$P(s) = s^n + a_1 s^{n-1} + a_2 s^{n-2} + \cdots + a_{n-1} s + a_n = 0$$

We have seen that if all the roots of $P(s)$ have negative real parts, then the system is stable. To detect the presence of roots with positive real parts we can utilize some fundamental results that make solving for the roots unnecessary.

## Preliminary Test

A useful test on the polynomial $P(s)$ relies on the fact that $P(s)$ can be factored as

$$P(s) = \left[(s - \sigma_1^2) + \omega_1^2\right] \cdots \left[(s - \sigma_{m-1})^2 + \omega_{m-1}^2\right]$$
$$(s - \lambda_{m+1}) \cdots (s - \lambda_n)$$

Note that we assume that the quadratic terms occur for odd indices $i = 1, 3, \ldots, m - 1$ for bookkeeping purposes.

For a stable system, we require that

$$\sigma_i < 0 \qquad i = 1, 3, 5, \ldots, m - 1$$
$$\lambda_i < 0 \qquad i = m + 1, \ldots, n$$

As a result, we require that all coefficients of the polynomial $P(s)$ be present and be all positive. This is a direct result of the following property:

$$a_i = (-1)^i \sum \text{(products of roots taken } i \text{ at a time)}$$

A first step in evaluating system stability is then to inspect for missing coefficients and for changes in sign of the coefficients. This is a necessary but insufficient condition. This means that if $P(s)$ passes this test, there is no guarantee that $P(s)$ is stable. This can be seen from inspection of

$$P(s) = s^3 + 0.5s^2 + 3.5s + 4$$

This polynomial passes the test but has two complex-conjugate roots in the right half of the $s$-plane as seen from

$$P(s) = (s + 1)(s^2 - 0.5s + 4)$$

On the other hand, if $P(s)$ fails this preliminary test, the system is unstable.

To test $P(s)$ further, we have a fundamental result in the Hurwitz criterion.

## Hurwitz Criterion

The Hurwitz criterion proceeds with the construction of the Hurwitz determinants $\Delta_i$ for $i = 1, 2, \ldots, n$, where $n$ is the order of $P(s)$. The determinants are given by

$$\Delta_1 = a_1$$

$$\Delta_2 = \begin{vmatrix} a_1 & a_3 \\ 1 & a_2 \end{vmatrix}$$

$$\Delta_3 = \begin{vmatrix} a_1 & a_3 & a_5 \\ 1 & a_2 & a_4 \\ 0 & a_1 & a_3 \end{vmatrix}$$

$$\vdots$$

$$\Delta_n = \begin{vmatrix} a_1 & a_3 & a_5 & \cdots & 0 \\ 1 & a_2 & a_4 & & \\ 0 & a_1 & a_3 & & \\ & 1 & a_2 & & \\ & & a_1 & & \\ \cdots & \cdots & \cdots & \cdots & \cdots \\ & & & a_n & 0 \\ & & & a_{n-1} & 0 \\ & & & a_{n-2} & a_n \end{vmatrix}$$

The Hurwitz criterion states that $P(s)$ is stable if all $\Delta_i > 0$. Note that we need to evaluate all the determinants, not a very pleasant chore indeed. A reasonable improvement results if we consider the Lienard–Chipart theorem.

## Lienard–Chipart Theorem

The Lienard–Chipart theorem states that if all coefficients of $P(s)$ are positive, then $P(s)$ is a stable polynomial (i.e., has no roots in the right-half of the $s$-plane) if and only if

1. For odd values of $n$, $\Delta_i > 0$ for $i = 2, 4, 6, \ldots, n - 1$.
2. For even values of $n$, $\Delta_i > 0$ for $i = 3, 5, 7, \ldots, n - 1$.

Thus it is sufficient to check the odd determinants for an even-order polynomial, and vice versa. The theorem essentially reduces the computational burden from that corresponding to the Hurwitz criterion. Before we present the considerably simpler Routh–Hurwitz criterion, we take our first numerical example of this chapter.

EXAMPLE 6.1

Consider the characteristic polynomial

$$P(s) = s^4 + 2s^3 + 3s^2 + 4s + 5$$

All coefficients are positive and we may thus proceed to form the Hurwitz determinants

$$\Delta_1 = 2 > 0$$

$$\Delta_2 = \begin{vmatrix} 2 & 4 \\ 1 & 3 \end{vmatrix} = 2$$

$$\Delta_3 = \begin{vmatrix} 2 & 4 & 0 \\ 1 & 3 & 5 \\ 0 & 2 & 4 \end{vmatrix} = -12$$

$$\Delta_4 = \begin{vmatrix} 2 & 4 & 0 & 0 \\ 1 & 3 & 5 & 0 \\ 0 & 2 & 4 & 0 \\ 0 & 1 & 3 & 5 \end{vmatrix} = -60$$

The system is clearly unstable in accordance with Hurwitz criterion. Note that the Lienard–Chipart test would have enabled us to conclude instability on the basis of the sign of $\Delta_3$.

We are now ready to state the Routh–Hurwitz criterion. As will be seen, this elegant test involves considerably less computational effort than do the preceding tests.

## Routh–Hurwitz Criterion

This criterion provides the basis for determining the location of roots of a polynomial in $s$ with real coefficients with respect to the left half and right half of the $s$-plane, without having to solve for the roots. Consider again the $n$th-order polynomial

$$P(s) = s^n + a_1 s^{n-1} + a_2 s^{n-2} + a_3 s^{n-3} + \cdots + a_{n-1} s + a_n = 0$$

As we stated earlier, the following two necessary conditions must be satisfied:

1. All the coefficients $a_i$ have the same sign.
2. None of the coefficients is zero.

The Routh–Hurwitz criterion provides a necessary and sufficient condition for all roots of $P(s)$ to lie in the left half of the $s$-plane, from information contained in the Routh array. The array contains $(n + 1)$ rows. The first two rows contain the coefficients of the polynomial in the following manner

$$\begin{array}{r|cccccc} \text{Row 1} & 1 & a_2 & a_4 & a_6 & a_8 & \cdots \\ \text{Row 2} & a_1 & a_3 & a_5 & a_7 & a_9 & \cdots \end{array}$$

The third, fourth, and remaining rows are calculated sequentially. The complete array takes on the form

| | | | | | |
|---|---|---|---|---|---|
| Row 1 | $s^7$ | 1 | $a_2$ | $a_4$ | $a_6$ |
| Row 2 | $s^6$ | $a_1$ | $a_3$ | $a_5$ | $a_7$ |
| Row 3 | $s^5$ | $b_1$ | $b_3$ | $b_5$ | |
| Row 4 | $s^4$ | $c_1$ | $c_3$ | $c_5$ | |
| Row 5 | $s^3$ | $d_1$ | $d_3$ | | |
| Row 6 | $s^2$ | $e_1$ | $e_3$ | | |
| Row 7 | $s^1$ | $f_1$ | | | |
| Row 8 | $s^0$ | $g_1$ | | | |

The above assumes that $n = 7$. The elements of row 3 are obtained from the contents of rows 1 and 2 according to

$$b_1 = \frac{-1}{a_1}\begin{vmatrix} 1 & a_2 \\ a_1 & a_3 \end{vmatrix}$$

$$b_3 = \frac{-1}{a_1}\begin{vmatrix} 1 & a_4 \\ a_1 & a_5 \end{vmatrix}$$

$$b_5 = \frac{-1}{a_1}\begin{vmatrix} 1 & a_6 \\ a_1 & a_7 \end{vmatrix}$$

The elements of row 4 are obtained from the contents of rows 2 and 3 as

$$c_1 = \frac{-1}{b_1}\begin{vmatrix} a_1 & a_3 \\ b_1 & b_3 \end{vmatrix}$$

$$c_3 = \frac{-1}{b_1}\begin{vmatrix} a_1 & a_5 \\ b_1 & b_5 \end{vmatrix}$$

$$c_5 = \frac{-1}{b_1}\begin{vmatrix} a_1 & a_7 \\ b_1 & 0 \end{vmatrix}$$

The other rows are formed in a similar fashion.

As soon as we have constructed the Routh array, the next step is to examine the contents of the left most column. Conclusions on root location can be drawn as follows:

1. If all elements of the left column of the Routh array are of the same sign, the roots of the polynomial are all in the left-hand side of the $s$-plane.
2. If there are changes of sign in the elements of the left column, the number of sign changes indicates the number of roots with positive real parts.

The relation between the Routh array and the Hurwitz determinants can be seen from the following:

$$
\begin{array}{c|c}
s^7 & 1 \\
s^6 & a_1 = \Delta_1 \\
s^5 & b_1 = \dfrac{\Delta_2}{\Delta_1} \\
s^4 & c_1 = \dfrac{\Delta_3}{\Delta_2} \\
s^3 & d_1 = \dfrac{\Delta_4}{\Delta_3} \\
s^2 & e_1 = \dfrac{\Delta_5}{\Delta_4} \\
s^1 & f_1 = \dfrac{\Delta_6}{\Delta_5} \\
s^0 & g_1 = \dfrac{\Delta_7}{\Delta_6}
\end{array}
$$

We can thus assert that if all the Hurwitz determinants are positive, the elements in the first column would also be of the same sign.

It is in order to take a couple of examples at this stage.

EXAMPLE 6.2

Consider the characteristic polynomial given by

$$P(s) = s^4 + 22s^3 + 164s^2 + 458s + 315$$

The first two rows are given by

$$
\begin{array}{c|ccc}
s^4 & 1 & 164 & 315 \\
s^3 & 22 & 458 &
\end{array}
$$

The elements of the third row are

$$b_1 = \frac{-1}{22}\begin{vmatrix} 1 & 164 \\ 22 & 458 \end{vmatrix} = 143.18$$

$$b_3 = \frac{-1}{22}\begin{vmatrix} 1 & 315 \\ 22 & 0 \end{vmatrix} = 315$$

We thus have

$$
\begin{array}{c|ccc}
s^4 & 1 & 164 & 315 \\
s^3 & 22 & 458 & \\
s^2 & 143.18 & 315 &
\end{array}
$$

The element of the fourth row is

$$c_1 = \frac{-1}{143.18}\begin{vmatrix} 22 & 458 \\ 143.18 & 315 \end{vmatrix} = 409.6$$

Our array now looks as follows:

| | | | |
|---|---|---|---|
| $s^4$ | 1 | 164 | 315 |
| $s^3$ | 22 | 458 | 0 |
| $s^2$ | 143.18 | 315 | 0 |
| $s^1$ | 409.6 | 0 | 0 |
| $s^0$ | 315 | | |

The last row was added in as it is simple to obtain.

It is clear that the polynomial does not have any roots in the right half of the $s$-plane. This is clear from the factorization.

$$P(s) = (s+1)(s+5)(s+7)(s+9) = 0$$

Our next example will be of the sixth order and will turn out to have unstable roots.

EXAMPLE 6.3

Consider the characteristic polynomial

$$P(s) = s^6 + 4s^5 + 11s^4 + 53s^3 + 102s^2 + 150s + 100$$

The first two rows of the Routh array are given by

| | | | | |
|---|---|---|---|---|
| $s^6$ | 1 | 11 | 102 | 100 |
| $s^5$ | 4 | 53 | 150 | |

The elements of the third row are calculated as

$$b_1 = \frac{-1}{4}\begin{vmatrix} 1 & 11 \\ 4 & 53 \end{vmatrix} = -2.25$$

$$b_3 = \frac{-1}{4}\begin{vmatrix} 1 & 102 \\ 4 & 150 \end{vmatrix} = 64.5$$

$$b_5 = \frac{-1}{4}\begin{vmatrix} 1 & 100 \\ 4 & 0 \end{vmatrix} = 100$$

Thus the first three rows of the Routh array are

| | | | | |
|---|---|---|---|---|
| $s^6$ | 1 | 11 | 102 | 100 |
| $s^5$ | 4 | 53 | 150 | |
| $s^4$ | −2.25 | 64.5 | 100 | |

We note that $b_1$ is negative, indicating instability.

If we like to proceed, we calculate the fourth row as

$$c_1 = \frac{-1}{-2.25}\begin{vmatrix} 4 & 53 \\ -2.25 & 64.5 \end{vmatrix} = 167.67$$

$$c_3 = \frac{-1}{-2.25}\begin{vmatrix} 4 & 150 \\ -2.25 & 100 \end{vmatrix} = 327.78$$

Thus we have the first four rows as

$$\begin{array}{c|cccc} s^6 & 1 & 11 & 102 & 100 \\ s^5 & 4 & 53 & 150 & \\ s^4 & -2.25 & 64.5 & 100 & \\ s^3 & 167.67 & 327.78 & & \end{array}$$

The elements of the fifth row are

$$d_1 = \frac{-1}{167.67}\begin{vmatrix} -2.25 & 64.5 \\ 167.67 & 327.78 \end{vmatrix} = 68.9$$

$$d_3 = \frac{-1}{167.67}\begin{vmatrix} -2.25 & 100 \\ 167.67 & 0 \end{vmatrix} = 100$$

The five rows are now listed as

$$\begin{array}{c|cccc} s^6 & 1 & 11 & 102 & 100 \\ s^5 & 4 & 53 & 150 & \\ s^4 & -2.25 & 64.5 & 100 & \\ s^3 & 167.67 & 327.78 & & \\ s^2 & 68.9 & 100 & & \end{array}$$

The element of the sixth row is

$$e_1 = \frac{-1}{68.9}\begin{vmatrix} 167.67 & 327.78 \\ 68.9 & 100 \end{vmatrix} = 84.43$$

We can now list the full array as

$$\begin{array}{c|cccc} s^6 & 1 & 11 & 102 & 100 \\ s^5 & 4 & 53 & 150 & \\ s^4 & -2.25 & 64.5 & 100 & \\ s^3 & 167.67 & 327.78 & & \\ s^2 & 68.9 & 100 & & \\ s^1 & 84.43 & & & \\ s^0 & 100 & & & \end{array}$$

There are two changes in the sign of elements of the first column, indicating that there are two roots in the right half of the $s$-plane. The system is clearly unstable.

# 6.4 Special Cases

The application of the Routh–Hurwitz criterion to the polynomials of Examples 6.2 and 6.3 presented us with no difficulties in setting up the arrays and consequently in investigating the stability of the systems concerned. In certain instances complications arise. Two special cases are considered here.

## Case 1: A Zero Element in the First Column

In this case the first element in a row is zero but not all other elements of the row are zero. We will show by way of an example this case.

EXAMPLE 6.4

Consider the polynomial given by

$$P(s) = s^4 + s^3 + 2s^2 + 2s + 5 = 0$$

The first three rows are given by

$$\begin{array}{c|ccc} s^4 & 1 & 2 & 5 \\ s^3 & 1 & 2 & \\ s^2 & 0 & 5 & \end{array}$$

The third row has a zero as the first element. We thus have to handle the situation in a manner that enables us to proceed.

There is a number of ways to circumvent this difficulty. Three methods are discussed here:

1. Replace the null element with a small number $\varepsilon$ and proceed with the construction of the array. Let $\varepsilon$ tend to zero to arrive at the conclusions.
2. Multiply the original polynomial by a factor $(s + \alpha)$, where $\alpha$ is a positive number that introduces an additional negative root. For simplicity use $(s + 1)$. Form the array for the new polynomial.
3. Substitute $s = 1/x$ in the original polynomial and proceed with the new polynomial.

We now illustrate the first method for Example 6.4. The full array is thus given in terms of $\varepsilon$ as

$$\begin{array}{c|ccc} s^4 & 1 & 2 & 5 \\ s^3 & 1 & 2 & \\ s^2 & \varepsilon & 5 & \\ s^1 & \dfrac{2\varepsilon - 5}{\varepsilon} & & \\ s^0 & 5 & & \end{array}$$

In the limit $\varepsilon \to 0$ the term $(2\varepsilon - 5)/\varepsilon$ is negative and hence there are two changes in sign of the first column. This indicates instability.

Let us use the second method, suggesting multiplication of $P(s)$ by $(s + 1)$ to get $\tilde{P}(s)$.

$$\begin{aligned} \tilde{P}(s) &= (s + 1)(s^4 + s^3 + 2s^2 + 2s + 5) \\ &= s^5 + 2s^4 + 3s^3 + 4s^2 + 7s + 5 \end{aligned}$$

We set up our Routh array as follows:

$$\begin{array}{c|ccc} s^5 & 1 & 3 & 7 \\ s^4 & 2 & 4 & 5 \\ s^3 & 1 & 4.5 & \\ s^2 & -5 & 5 & \\ s^1 & 5.5 & 0 & \\ s^0 & 5 & & \end{array}$$

There are two changes in sign, leading to the same conclusions as above.

The third method suggests substituting $s = 1/x$. We thus consider

$$\begin{aligned} \hat{P}(s) &= P(s)|_{s=1/x} \\ &= \frac{1}{x^4} + \frac{1}{x^3} + \frac{2}{x^2} + 2\frac{1}{x} + 5 = 0 \end{aligned}$$

Multiply by $x^4$ to obtain

$$\tilde{\tilde{P}}(x) = 5x^4 + 2x^3 + 2x^2 + x + 1 = 0$$

The Routh array is thus

$$\begin{array}{c|ccc} x^4 & 5 & 2 & 1 \\ x^3 & 2 & 1 & \\ x^2 & -0.5 & 1 & \\ x^1 & 5 & & \\ x^0 & 1 & & \end{array}$$

Again we get two changes in sign of the left-hand column elements to yield the same conclusion.

Having illustrated the first case, we turn our attention to the second case, which turns out to be a little bit more challenging.

## Case 2: All Elements of One Row Are Zero

The second special case arises when we encounter a row full of zeros on our way to finish the Routh's array. This encounter is a symptom of the existence of one or more of the following situations:

1. Pair of real roots, with opposite signs but equal magnitude
2. Pairs of roots on the imaginary axis
3. Pairs of complex-conjugate roots in symmetry about the origin of the $s$-plane.

Before embarking on remedying this situation, it is perhaps instructive to justify our proposals. First take the polynomial

$$\begin{aligned} P_1(s) &= (s^2 - 4)(s + 1)(s + 8) \\ &= s^4 + 9s^3 + 4s^2 - 36s - 32 \end{aligned}$$

The Routh array is constructed as

$$\begin{array}{c|ccc} s^4 & 1 & 4 & -32 \\ s^3 & 9 & -36 & \\ s^2 & 8 & -32 & \\ s^1 & 0 & 0 & \end{array}$$

The premature termination is due to the presence of roots $s = \pm 2$.

For the second situation, we take the polynomial

$$\begin{aligned} P_2(s) &= (s^2 + 4)(s + 1) \\ &= s^3 + s^2 + 4s + 4 \end{aligned}$$

The Routh array is constructed as

$$\begin{array}{c|cc} s^3 & 1 & 4 \\ s^2 & 1 & 4 \\ s^1 & 0 & 0 \end{array}$$

The premature termination is due to the roots $s = \pm j2$.

To illustrate the rise of the third situation, let us take

$$\begin{aligned} P_3(s) &= (s^4 + 4)(s + 3)(s^2 + 2s + 3) \\ &= s^7 + 5s^6 + 9s^5 + 9s^4 + 4s^3 + 20s^2 + 36s + 36 \end{aligned}$$

Note that we have roots at $s = \pm 1 \pm j1$. The Routh array is given by

$$\begin{array}{lcccc}
s^7 & 1 & 9 & 4 & 36 \\
s^6 & 5 & 9 & 20 & 36 \\
s^5 & \frac{36}{5} & 0 & \frac{144}{5} & \\
s^4 & 9 & 0 & 36 & \\
s^3 & 0 & 0 & &
\end{array}$$

Again we encounter the premature termination.

In order to proceed, we find the so-called *auxiliary equation*, which results from the row immediately above the row with zero terms. The auxiliary equation is denoted by $A(s)$. For the polynomial $P_1(s)$, we have the $s^2$ row to form $A_1(s)$ as

$$A_1(s) = 8s^2 - 32 = 0$$

or

$$A_1(s) = s^2 - 4 = 0$$

Note that $A_1(s)$ is a common divisor of $P_1(s)$.

For the polynomial $P_2(s)$, we have the $s^2$ row to form $A_2(s)$ as

$$A_2(s) = s^2 + 4$$

Again $A_2(s)$ is a common divisor of $P_2(s)$.

For the polynomial $P_3(s)$, we have the $s^4$ row to yield

$$A_3(s) = 9s^4 + 0s^2 + 36 = 0$$

Thus

$$A_3(s) = s^4 + 4 = 0$$

Again, $A_3(s)$ is a common divisor of $P_3(s)$.

The next step is to differentiate $A(s)$ with respect to $s$ to obtain a polynomial whose coefficients are used to replace the row with zero elements. For our three examples we have

$$\frac{dA_1(s)}{ds} = 2s$$

$$\frac{dA_2(s)}{ds} = 2s$$

$$\frac{dA_3(s)}{ds} = 4s^3$$

We will show our conclusions for $P_1(s)$. The Routh array turns out to be

$$\begin{array}{c|ccc}
s^4 & 1 & 4 & -32 \\
s^3 & 9 & -36 & \\
s^2 & 8 & -32 & \\
s^1 & 2 & \longleftarrow & \qquad \dfrac{dA_1}{ds} = 0 \\
s^0 & -32 & &
\end{array}$$

The one sign change indicates that one root is unstable.

For $P_2(s)$, the Routh array is given by

$$\begin{array}{c|cc}
s^3 & 1 & 4 \\
s^2 & 1 & 4 \\
s^1 & 2 & \longleftarrow \qquad \dfrac{dA_2}{ds} = 0 \\
s^0 & 4 &
\end{array}$$

No sign changes indicating stability.

Finally, for $P_3(s)$ we have the partial Routh array given by

$$\begin{array}{c|cccc}
s^7 & 1 & 9 & 4 & 36 \\
s^6 & 5 & 9 & 20 & 36 \\
s^5 & \dfrac{36}{5} & 0 & \dfrac{144}{5} & \\
s^4 & 9 & 0 & 36 & \\
s^3 & 4 & 0 \longleftarrow & & \qquad \dfrac{dA_3}{ds} = 0 \\
s^2 & 0 & 36 & &
\end{array}$$

We encounter once again the zero in the left column and have to use the $\varepsilon$ procedure of case 1. We will write down the elements of the array starting from $s^4$

$$\begin{array}{c|ccc}
s^4 & 9 & 0 & 36 \\
s^3 & 4 & 0 & \\
s^2 & \varepsilon & 36 & \\
s^1 & \dfrac{-4 \times 36}{\varepsilon} & 0 & \\
s^0 & 36 & &
\end{array}$$

Clearly, there are two sign changes in the left column and the system is unstable.

# 6.5 Application of the Routh–Hurwitz Criterion

Our discussion so far has centered on determining whether or not a given characteristic equation has roots in the right half of the $s$-plane, using the Routh–Hurwitz criterion. The practical application of the criterion is concerned with the design of a control system where ranges of certain parameters have to be obtained ensuring a stable operation. We illustrate the concepts involved using a number of examples.

EXAMPLE 6.5

The characteristic equation of a closed-loop control system is given by

$$P(s) = s(s^2 + 1.4s + 1) + K = 0$$

The Routh array is given by

$$\begin{array}{c|cc} s^3 & 1 & 1 \\ s^2 & 1.4 & K \\ s^1 & \dfrac{1.4 - K}{1.4} & \\ s^0 & K & \end{array}$$

The condition on $K$ to retain stability is clearly given by

$$0 < K < 1.4$$

EXAMPLE 6.6

The open-loop transfer function for the closed-loop system shown in Figure 6.3 is given by

$$G(s) = \frac{K}{s} G_D(s)$$

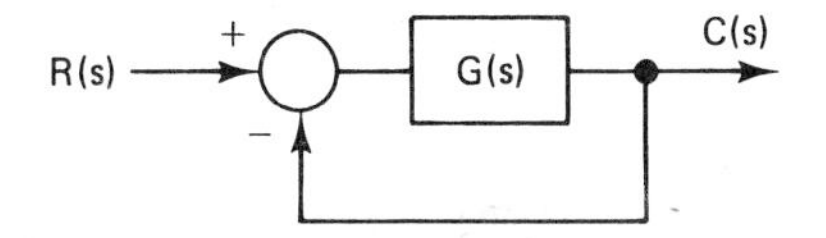


**Figure 6.3.** Block Diagram for Example 6.6

The transfer function $G_D(s)$ is an approximation for a delay, and is given by

$$G_D(s) = \frac{[1 - (sT/4)]^2}{[1 + (sT/4)]^2}$$

Investigate the stability of the system using the Routh–Hurwitz criterion.

SOLUTION

The characteristic equation for this system is given by

$$s\left[1 + \left(\frac{sT}{4}\right)\right]^2 + K\left[1 - \left(\frac{sT}{4}\right)\right]^2 = 0$$

A few simplification steps result in

$$P(s) = s^3 + \left(K + \frac{8}{T}\right)s^2 + \frac{8}{T^2}(2 - KT)s + \frac{16K}{T^2} = 0$$

We now set up the Routh array

$$\begin{array}{c|cc} s^3 & 1 & \frac{8}{T^2}(2 - KT) \\ s^2 & K + \frac{8}{T} & \frac{16K}{T^2} \\ s^1 & b_1 & 0 \\ s^0 & \frac{16K}{T^2} & \end{array}$$

The value of $b_1$ is given by

$$b_1 = \frac{8(2 - KT)}{T^2} - \frac{16K}{T(8 + KT)} > 0$$

For stability $K$ and $T$ should be positive. In addition, $b_1$ should be positive. This can be reduced to requiring that

$$16 - 8KT - K^2T^2 > 0$$

Solving the quadratic, we get

$$(r_1 - KT)(r_2 + KT) > 0$$

The roots $r_1$ and $r_2$ are given by

$$r_1 = 4(\sqrt{2} - 1)$$
$$r_2 = 4(\sqrt{2} + 1)$$

Since $(r_2 + KT)$ is positive, we require that

$$r_1 - KT > 0$$

Thus for stability

$$KT < r_1$$

Numerically,

$$KT < 4(\sqrt{2} - 1) \simeq 1.656$$

It is clear that the longer the delay $T$, the lower is the limit on the gain $K$ for stability as shown graphically in Figure 6.4.

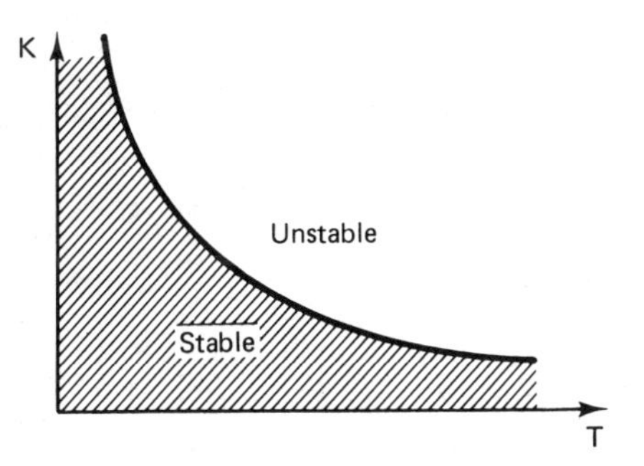


**Figure 6.4.** Stability Boundary in the $K-T$ Plane for Example 6.6

EXAMPLE 6.7

Consider the multiple-loop control system shown in Figure 6.5. Use the Routh–Hurwitz criterion to find the range of $K$ for the minor loop to be stable. Find the ranges of $K_c$ for a stable closed-loop system for the maximum allowed value of $K$.

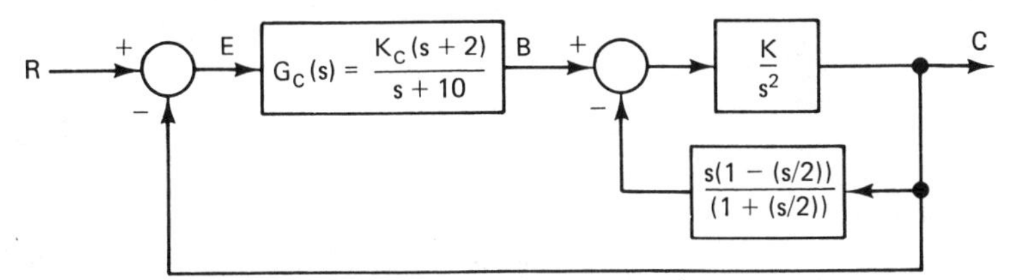


**Figure 6.5.** Block Diagram for Example 6.7

SOLUTION

The transfer function of the minor loop is given by

$$\frac{C(s)}{B(s)} = \frac{K/s^2}{1 + \dfrac{K[1-(s/2)]}{s[1+(s/2)]}} = \frac{K(2+s)}{s^3 + (2-K)s^2 + 2Ks}$$

The characteristic equation is thus

$$P_m(s) = s[s^2 + (2-K)s + 2K] = 0$$

The Routh array for the minor loop is

$$\begin{array}{c|cc} s^2 & 1 & 2K \\ s^1 & 2-K & \\ s^0 & 2K & \end{array}$$

For stability we require that

$$2 - K > 0 \qquad \text{and} \qquad K > 0$$

Thus the range of $K$ is

$$0 < K < 2$$

The closed-loop transfer function is given by

$$\frac{C(s)}{R(s)} = \frac{\dfrac{KK_c(2+s)^2}{s(s+10)[s^2+(2-K)s+2K]}}{1 + \dfrac{KK_c(2+s)^2}{s(s+10)[s^2+(2-K)s+2K]}}$$

Simplifying, we obtain

$$\frac{C(s)}{R(s)} = \frac{KK_c(2+s)^2}{s(s+10)[s^2+(2-K)s+2K] + KK_c(2+s)^2}$$

As a result, the closed-loop characteristic equation is given by

$$P_M(s) = s(s+10)[s^2+(2-K)s+2K] + KK_c(2+s)^2$$

Expanding and collecting terms, we obtain

$$P_M(s) = s^4 + (12-K)s^3 + (20 - 8K + KK_c)s^2 + (20K + 4KK_c)s + 4KK_c = 0$$

The Routh array is obtained as

$$\begin{array}{c|ccc} s^4 & 1 & 20 - 8K + KK_c & 4KK_c \\ s^3 & 12 - K & 20K + 4KK_c & 0 \\ s^2 & b_1 & 4KK_c & \\ s^1 & c_1 & 0 & \\ s^0 & 4KK_c & & \end{array}$$

The term $b_1$ is given by

$$b_1 = (20 - 8K + KK_c) - \frac{4K(K_c + 5)}{12 - K}$$

$$= \frac{240 - 136K + 8K^2 + 8KK_c - K^2K_c}{12 - K}$$

This reduces to

$$b_1 = \frac{8(15 - K)(2 - K) + KK_c(8 - K)}{12 - K}$$

Note that with $0 < K < 2$, we have $b_1 > 0$ for $K_c > 0$. The term $c_1$ is given by

$$c_1 = 4K\left[5 + K_c\left(1 - \frac{12 - K}{b_1}\right)\right]$$

Assume that we fix $K$ at the limit dictated by the stability requirements for the minor loop. Thus

$$K = 2$$

The corresponding value of $b_1$ is obtained as

$$b_1 = 1.2K_c$$

The value of $c_1$ is calculated as

$$c_1 = 8(K_c - 3.33)$$

Thus for the overall system to be stable we have the requirement

$$K_c > 3.33$$

To conclude this section, an example that is less involved is taken to show the utility of the Routh–Hurwitz criterion as a design tool.

EXAMPLE 6.8

Consider the system shown in Figure 6.6. Find the region in the $K$–$K_1$ plane for the system to be stable using the Routh–Hurwitz criterion.

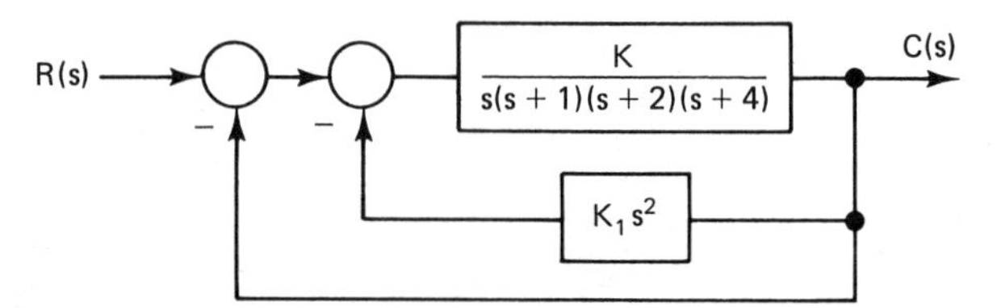


**Figure 6.6.** Block Diagram for the System of Example 6.8

SOLUTION

It is not a difficult task to show that the characteristic polynomial of the system is given by

$$P(s) = s^4 + 7s^3 + (14 + KK_1)s^2 + 8s + K = 0$$

The Routh array is thus given by

$$\begin{array}{c|ccc} s^4 & 1 & 14 + KK_1 & K \\ s^3 & 7 & 8 & \\ s^2 & 12.86 + KK_1 & K & \\ s^1 & 8 - \dfrac{7K}{12.86 + KK_1} & 0 & \\ s^0 & K & & \end{array}$$

It is clear that for stability we require that

$$8 - \frac{7K}{12.86 + KK_1} > 0$$

The inequality is written as

$$102.86 + K(8K_1 - 7) > 0$$

For $(8K_1 - 7) > 0$, the inequality requires that

$$K > \frac{102.86}{7 - 8K_1}$$

For $(8K_1 - 7) < 0$, the inequality requires that

$$K < \frac{102.86}{7 - 8K_1}$$

The region in the $K$–$K_1$ plane where the system characteristic equation has no roots in the right half of the $s$-plane is shown in Figure 6.7. Any combinations of $K$ and $K_1$ in this region will provide for a stable operation.

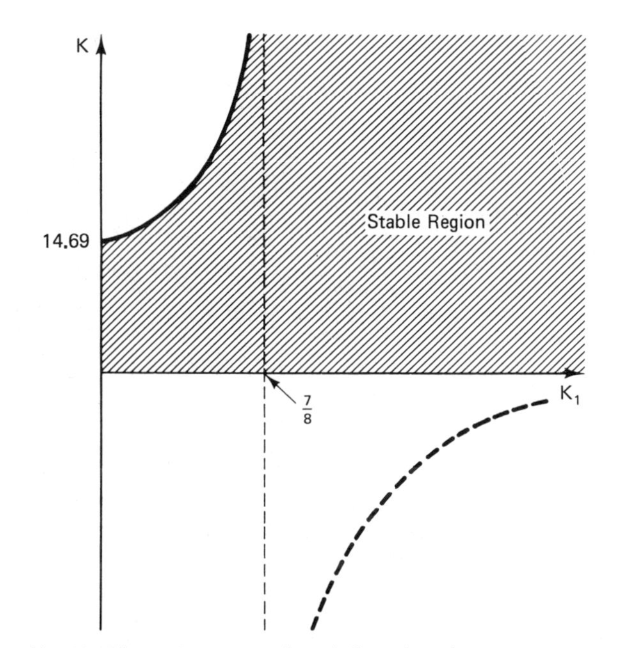


**Figure 6.7.** $K - K_1$ Plane Region of Stability for the System of Example 6.8

# *Some Solved Problems*

PROBLEM 6A-1

Use the Routh–Hurwitz criterion to test the polynomial $P(s)$ for unstable roots. Obtain the roots of $P(s)$ on the basis of the Routh array.

$$P(s) = s^4 + 2s^3 + 11s^2 + 18s + 18$$

SOLUTION

The Routh array is constructed as follows:

$$\begin{array}{c|ccc} s^4 & 1 & 11 & 18 \\ s^3 & 2 & 18 & \\ s^2 & 2 & 18 & \end{array}$$

There is premature termination at the $s^1$ row. The auxiliary equation is

$$A(s) = 2s^2 + 18 = 0$$

Dividing through 2, we get

$$A(s) = s^2 + 9$$

Now taking the derivative, we obtain

$$\frac{dA}{ds} = 2s$$

We can now finish the Routh array:

$$\begin{array}{c|ccc} s^4 & 1 & 11 & 18 \\ s^3 & 2 & 18 & \\ s^2 & 2 & 18 & \\ s^1 & 2 & 0 & \\ s^0 & 18 & & \end{array}$$

There are no sign changes in the first column and we can assert that there are no unstable roots.

The auxiliary equation is a common divisor. We thus perform the following long division:

$$\frac{P(s)}{A(s)} = \frac{s^4 + 2s^3 + 11s^2 + 18s + 18}{s^2 + 9}$$

The result is

$$\frac{P(s)}{A(s)} = s^2 + 2s + 2$$

From the above,

$$P(s) = (s^2 + 9)(s^2 + 2s + 2) = 0$$

The roots are obtained as

$$s = \pm j3$$
$$s = -1 \pm j1$$

PROBLEM 6A-2

Use the Routh–Hurwitz criterion to test for the presence of roots in the right half of $s$-plane for the polynomial

$$P(s) = s^6 + s^5 + 3s^4 + 3s^3 + 3s^2 + 2s + 1$$

SOLUTION

The Routh array is constructed as

$$\begin{array}{c|cccc}
s^6 & 1 & 3 & 3 & 1 \\
s^5 & 1 & 3 & 2 & \\
s^4 & \varepsilon & 1 & 1 & \\
s^3 & \dfrac{3\varepsilon - 1}{\varepsilon} & \dfrac{2\varepsilon - 1}{\varepsilon} & 0 & \\
s^2 & \dfrac{2\varepsilon^2 - 4\varepsilon + 1}{1 - 3\varepsilon} & 1 & & \\
s^1 & \dfrac{\varepsilon(4\varepsilon - 1)}{2\varepsilon^2 - 4\varepsilon + 1} & 0 & & \\
s^0 & 1 & & &
\end{array}$$

Note that a zero was encountered at the $s^4$ first column and the $\varepsilon$ method was invoked. In the limit as $\varepsilon \to 0$ we get

$$\begin{array}{c|cccc}
s^6 & 1 & 3 & 3 & 1 \\
s^5 & 1 & 3 & 2 & \\
s^4 & 0 & 1 & 1 & \\
s^3 & -\infty & -\infty & 0 & \\
s^2 & 1 & 1 & & \\
s^1 & 0 & 0 & &
\end{array}$$

The zero row at $s^1$ indicates that we need the auxiliary equation

$$A(s) = s^2 + 1$$

Its derivative is

$$\frac{dA}{ds} = 2s$$

Thus the proper array is

$$\begin{array}{c|cccc}
s^6 & 1 & 3 & 3 & 1 \\
s^5 & 1 & 3 & 2 & \\
s^4 & 0 & 1 & 1 & \\
s^3 & -\infty & -\infty & 0 & \\
s^2 & 1 & 1 & & \\
s^1 & 2 & 0 & & \\
s^0 & 1 & & &
\end{array}$$

There are two sign changes in the first column indicating two roots in the right half of the $s$-plane.

The auxiliary equation shows us that $s^2 + 1$ is a divisor. Performing the long division, we obtain

$$P(s) = (s^2 + 1)(s^4 + s^3 + 2s^2 + 2s + 1)$$

We can perform a check on our result by testing the polynomial

$$\tilde{P}(s) = s^4 + s^3 + 2s^2 + 2s + 1$$

The Routh array is obtained as

| | | | |
|---|---|---|---|
| $s^4$ | 1 | 2 | 1 |
| $s^3$ | 1 | 2 | |
| $s^2$ | $\varepsilon$ | 1 | |
| $s^1$ | $\dfrac{2\varepsilon - 1}{\varepsilon}$ | | |
| $s^0$ | 1 | | |

In the limit we have two changes in sign and we have the same conclusion.

If we use the substitution $x = 1/s$, we get

$$\hat{Q}(s) = x^4 + 2x^3 + 2x^2 + x + 1$$

The Routh array is given by

| | | | |
|---|---|---|---|
| $x^4$ | 1 | 2 | 1 |
| $x^3$ | 2 | 1 | |
| $x^2$ | 1.5 | 1 | |
| $x^1$ | $-\dfrac{1}{3}$ | | |
| $x^0$ | 1 | | |

This confirms our conclusions.

## PROBLEM 6A-3

We have learned that a first step in checking the stability of a polynomial $P(s)$ is to check for missing coefficients and changes in their sign. An additional test can be carried out by forming the two polynomials

$$Q_0(x) = \frac{1}{2}[P(s) + P(-s)]\Big|_{s^2=x}$$

$$Q_1(x) = \frac{1}{2s}[P(s) - P(-s)]\Big|_{s^2=x}$$

If both $Q_0(x)$ and $Q_1(x)$ have negative real zeros, the original polynomial $P(s)$ is stable. This can represent a reasonable reduction in computing effort.

Apply this test to the polynomial

$$P(s) = s^8 + 9s^7 + 35s^6 + 77s^5 + 105s^4 + 91s^3 + 49s^2 + 15s + 3000$$

SOLUTION

We have by the definitions

$$Q_0(s) = s^8 + 35s^6 + 105s^4 + 49s^2 + 3000$$
$$Q_1(s) = 9s^6 + 77s^4 + 91s^2 + 15$$

Substituting $s^2 = x$, we obtain

$$Q_0(x) = x^4 + 35x^3 + 105x^2 + 49x + 3000$$
$$Q_1(x) = 9x^3 + 77x^2 + 91x + 15$$

We cannot apply the test directly. Therefore, we split $Q_0(x)$ again:

$$R_0(y) = \frac{1}{2}[Q_0(x) + Q_0(-x)]\Big|_{x^2=y}$$
$$R_0(y) = y^2 + 105y + 3000$$

The roots of $R_0(y)$ are not real:

$$r = \frac{-105 \pm \sqrt{(105)^2 - 12{,}000}}{2}$$

As a result, we conclude that $Q_0(x)$ and hence $P(s)$ have roots in the right half of the $s$-plane.

For this problem we can show that $Q_0(s)$ has roots in the right half of the $s$-plane using the Routh array.

The Routh array for $Q_0(x)$ is given by

| | | | |
|---|---|---|---|
| $x^4$ | 1 | 105 | 3000 |
| $x^3$ | 35 | 49 | |
| $x^2$ | 103.6 | 3000 | |
| $x^1$ | −964.51 | | |
| $x^0$ | 3000 | | |

It is clear that $Q_0(x)$ is not a stable polynomial and hence $P(s)$ includes roots in the right half of the $s$-plane.

PROBLEM 6A-4

To remedy special case 1, that involves a zero entry of a first column in a row, the text suggests replacing $s$ by $1/x$ and proceeding with $\tilde{P}(x) = x^n P(1/x)$. This procedure may run into the same snag. An example polynomial for which this last statement is true is given by

$$P(s) = s^6 + 3s^5 + 2s^4 + 6s^3 + 3s^2 + 6s + 3$$

Verify the statements made and apply the $\varepsilon$ method to resolve the question of stability of $P(s)$.

SOLUTION

The new polynomial is given by

$$\tilde{P}(x) = 3x^6 + 6x^5 + 3x^4 + 6x^3 + 2x^2 + 3x + 1$$

The partial Routh array is given by

$$\begin{array}{c|cccc} x^6 & 3 & 3 & 2 & 1 \\ x^5 & 6 & 6 & 3 & \\ x^4 & 0 & 0.5 & 1 & \end{array}$$

Thus this approach will not resolve the question directly.

Applying the Routh criterion directly with $\varepsilon$ inserted as soon as we encounter a zero, we have

$$\begin{array}{c|cccc} s^6 & 1 & 2 & 3 & 3 \\ s^5 & 3 & 6 & 6 & \\ s^4 & \varepsilon & 1 & 3 & \\ s^3 & \dfrac{6\varepsilon - 3}{\varepsilon} & \dfrac{6\varepsilon - 9}{\varepsilon} & & \\ s^2 & \dfrac{-2\varepsilon^2 + 5\varepsilon - 1}{2\varepsilon - 1} & 3 & & \\ s^1 & \dfrac{-4\varepsilon^3 + 4\varepsilon^2 - 5\varepsilon}{-2\varepsilon^2 + 5\varepsilon - 1} & 0 & & \\ s^0 & 3 & & & \end{array}$$

In the limit we have for $\varepsilon \to 0$,

$$\begin{array}{c|cccc} s^6 & 1 & 2 & 3 & 3 \\ s^5 & 3 & 6 & 6 & \\ s^4 & 0 & 1 & 3 & \\ s^3 & -\infty & -\infty & & \\ s^2 & 1 & 3 & & \\ s^1 & 0 & 0 & & \\ s^0 & 3 & & & \end{array}$$

There are two roots in the right half of the $s$-plane.

PROBLEM 6A-5

The following tenth-order polynomial has received a lot of attention in the control systems literature.

$$P(s) = s^{10} + s^9 + 2s^8 + 2s^7 + s^6 + 2s^5 + 6s^4 + 7s^3 + 10s^2 + 6s + 4$$

The methods described in this text work for this polynomial. Verify this statement by applying the Routh–Hurwitz criterion.

SOLUTION

The first three rows of the Routh array are given by

| | | | | | | |
|---|---|---|---|---|---|---|
| $s^{10}$ | 1 | 2 | 1 | 6 | 10 | 4 |
| $s^9$ | 1 | 2 | 2 | 7 | 6 | |
| $s^8$ | 0 | $-1$ | $-1$ | 4 | 4 | |

We have to invoke the $\varepsilon$ method. The continuation of the array is given by

| | | | | | |
|---|---|---|---|---|---|
| $s^9$ | 1 | 2 | 2 | 7 | 6 |
| $s^8$ | $\varepsilon$ | $-1$ | $-1$ | 4 | 4 |
| $s^7$ | $2+\frac{1}{\varepsilon}$ | $2+\frac{1}{\varepsilon}$ | $7-\frac{4}{\varepsilon}$ | $6-\frac{4}{\varepsilon}$ | |
| $s^6$ | $d_1$ | $d_2$ | $d_3$ | 4 | |
| $s^5$ | $e_1$ | $e_2$ | $e_3$ | | |
| $s^4$ | $f_1$ | $f_2$ | 4 | | |
| $s^3$ | $g_1$ | $g_2$ | | | |

The elements of the $s^6$ row are given by

$$d_1 = -(1+\varepsilon)$$

$$d_2 = \frac{-(7\varepsilon^2 - 2\varepsilon + 1)}{2\varepsilon + 1}$$

$$d_3 = \frac{-6\varepsilon^2 + 12\varepsilon + 4}{2\varepsilon + 1}$$

Note that in the limit $d_1$ and $d_2$ become negative. We proceed with the elements of the $s^5$ row to obtain

$$e_1 = \frac{5(1-\varepsilon)}{1+\varepsilon}$$

$$e_2 = \frac{\varepsilon + 15}{\varepsilon + 1}$$

$$e_3 = \frac{6\varepsilon + 10}{\varepsilon + 1}$$

The elements of the $s^4$ row are

$$f_1 = \frac{37\varepsilon^3 - 12\varepsilon^2 + 61\varepsilon + 10}{5(1-\varepsilon)(1+2\varepsilon)}$$

$$f_2 = \frac{42\varepsilon^3 - 52\varepsilon^2 + 76\varepsilon + 30}{5(1-\varepsilon)(1+2\varepsilon)}$$

The elements of the $s^3$ row are

$$g_1 = \frac{247\varepsilon^4 + 73\varepsilon^3 + 521\varepsilon^2 + 695\varepsilon}{(\varepsilon + 1)(37\varepsilon^3 - 12\varepsilon^2 + 61\varepsilon + 10)}$$

$$g_2 = \frac{2(111\varepsilon^4 + 49\varepsilon^3 + 273\varepsilon^2 + 335\varepsilon)}{(\varepsilon + 1)(37\varepsilon^3 - 12\varepsilon^2 + 61\varepsilon + 10)}$$

Note now that $g_1$ and $g_2$ tend to zero in the limit.

The Routh array in the limit, down to the $s^3$ row, is thus given by

$$\begin{array}{c|cccccc}
s^{10} & 1 & 2 & 1 & 6 & 10 & 4 \\
s^9 & 1 & 2 & 2 & 7 & 6 & \\
s^8 & 0 & -1 & -1 & 4 & 4 & \\
s^7 & \infty & \infty & -\infty & -\infty & & \\
s^6 & -1 & -1 & 4 & 4 & & \\
s^5 & 5 & 15 & 10 & & & \\
s^4 & 2 & 6 & 4 & & & \\
s^3 & 0 & 0 & & & &
\end{array}$$

It is now incumbent on us to use the auxiliary equation

$$A(s) = 2s^4 + 6s^2 + 4$$

The derivative is

$$\frac{dA}{ds} = 8s^3 + 12s$$

We can now continue with the Routh array, which is given below:

$$\begin{array}{c|ccc}
s^5 & 5 & 15 & 10 \\
s^4 & 2 & 6 & 4 \\
s^3 & 8 & 12 & \\
s^2 & 3 & 4 & \\
s^1 & \frac{4}{3} & & \\
s^0 & 4 & &
\end{array}$$

It is clear that two changes in sign occur and thus for $P(s)$ there are two roots in the right half of the $s$-plane.

# Problems

PROBLEM 6B-1

Use the Hurwitz, Lienard–Chipart, and Routh–Hurwitz criteria to determine the stability of the following polynomials.

(a) $$P(s) = s^3 + 20s^2 + 113s + 154$$

(b) $$P(s) = s^4 + 21s^3 + 133s^2 + 267s + 154$$

(c) $$P(s) = s^3 + 25s^2 + 187s + 363$$

(d) $$P(s) = s^4 + 5s^3 + 5s^2 + 10s + 24$$

PROBLEM 6B-2

Use the Routh–Hurwitz stability criterion to determine if the feedback control system shown in Figure 6.8 is stable for the following transfer functions.

(a) $$G(s) = \frac{10}{s(s^2 + 6s + 18)}$$

(b) $$G(s) = \frac{2s + 1}{s^2(100s^2 + 500s + 25)}$$

(c) $$G(s) = \frac{0.1(s + 3)}{s(s + 0.2)(s + 0.6)(s + 4)}$$

(d) $$G(s) = \frac{40}{s(s + 2)(s + 3)}$$

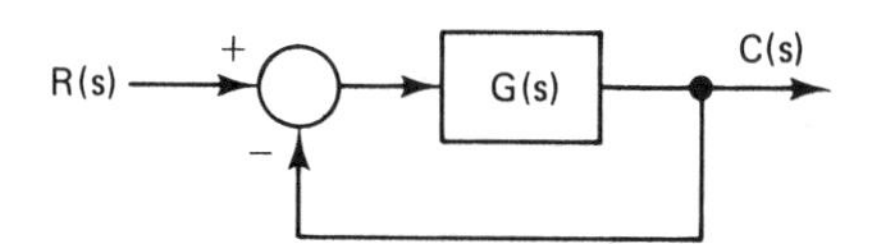


**Figure 6.8.** Block Diagram for Problem 6B-2

PROBLEM 6B-3

Use the Routh–Hurwitz criterion to check the stability of the system of Problem 3B-1 for positive values of $K_A$, $K_B$, $h_1$, and $h_2$.

PROBLEM 6B-4

Use the Routh–Hurwitz criterion to check the stability of the system of Problem 3B-5 for the given nominal parameter values.

PROBLEM 6B-5

Use the Routh–Hurwitz criterion to establish the range of the parameter $M$ required for the system of Problem 3B-5 to be stable. Assume that all other parameters are kept at their nominal value.

PROBLEM 6B-6

Repeat Problem 6B-5 for the parameter $a$. Assume that all other parameters, including $M$, are kept at their nominal value.

PROBLEM 6B-7

Repeat Problem 6B-5 for the parameter $\omega_n$. Assume that all other parameters, including $M$ and $a$, are kept at their nominal value.

PROBLEM 6B-8

Use the Routh–Hurwitz criterion to check the stability of the system of Problem 3B-6 for the parameter values indicated.

PROBLEM 6B-9

Use the Routh–Hurwitz criterion to establish the range of $K_2$ for the system of Problem 3B-6 to be stable with $a$, $b$, and $h_1$ kept at their nominal values.

PROBLEM 6B-10

Repeat Problem 6B-9 to establish the range of the parameter $a$ for stability of the system of Problem 3B-6.

PROBLEM 6B-11

Repeat Problem 6B-9 to establish the range of $h_1$ for stability of the system of Problem 3B-6.

PROBLEM 6B-12

Apply the method suggested in Problem 6A-3 to the polynomial.

$$P(s) = 2s^4 + s^3 + 3s^2 + 5s + 10$$

PROBLEM 6B-13

Repeat Problem 6B-12 for the polynomial

$$P(s) = s^6 + 10s^5 + 300s^4 + 72s^3 + 73s^2 + 38s + 8$$

(*Hint*: Cardan's formula for cubic equations states that for $a_3x^3 + a_2x^2 + a_1x + a_0$ to have all three real zeros, the condition $a_1^2 - 3a_2a_0 \geq 0$ should be satisfied.)

PROBLEM 6B-14

Repeat Problem 6B-12 for the polynomial of Problem 6A-4.

# Chapter 7

# ROOT-LOCUS METHOD

## 7.1 Introduction

The root-locus method was first presented by W. R. Evans in a five-page article, "Graphical Analysis of Control Systems," published in Volume 67 of the *Transactions* of the American Institute of Electrical Engineers in 1948. The importance of the method is due to the fact that it provides a simple correlation between system parameters and the essential features of the system response. The method is simple and this appeals to control systems designers, for it provides them with a tool to evaluate the effects of changing a parameter on the performance characteristics of the system.

The root-locus method is based on the relation between the poles and zeros of the closed-loop transfer function and those of the open-loop transfer function, given by the characteristic equation. The method establishes the locus of the roots of the characteristic equation as a parameter is varied. To illustrate the features of the method, we begin with an introductory example.

## 7.2 Introductory Example

Consider the system shown in Figure 7.1. The closed-loop transfer function is given by

$$\frac{C(s)}{R(s)} = \frac{K}{s^2 + (bK + 1/\tau)s + aK} \tag{7-1}$$

The characteristic equation of this system is given by

$$P(s) = s^2 + \left(bK + \frac{1}{\tau}\right)s + aK = 0 \tag{7-2}$$

The roots of $P(s)$ are given by

$$s = \frac{1}{2}\left[-\left(bK + \frac{1}{\tau}\right) \pm \sqrt{\left(bK + \frac{1}{\tau}\right)^2 - 4aK}\right] \tag{7-3}$$

The location of the roots depends on the parameters $K$, $a$, $b$, and $\tau$. As a result, the relative stability of the system depends on the choice of these

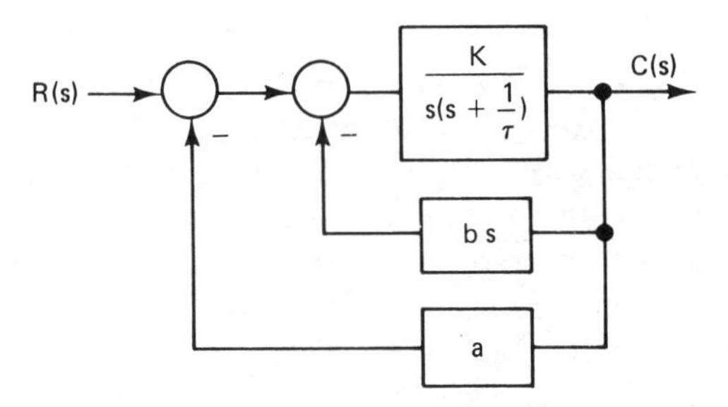


**Figure 7.1.** System for the Introductory Example

parameters. A root locus with respect to a parameter is the locus of the roots of the characteristic equation as the parameter is varied from zero to infinity.

To be specific, let us consider the four cases where one parameter is allowed to vary while the other three remain fixed for the system under consideration.

## Case 1

To begin with, let us assume that

$$b = \frac{1}{\tau} = a = 1$$

In this case, the characteristic equation is

$$P(s) = s^2 + (1 + K)s + K = 0 \tag{7-4}$$

The two roots are thus given by

$$s_1 = -K$$
$$s_2 = -1$$

It is clear that the location of the first root depends on the value of $K$. For

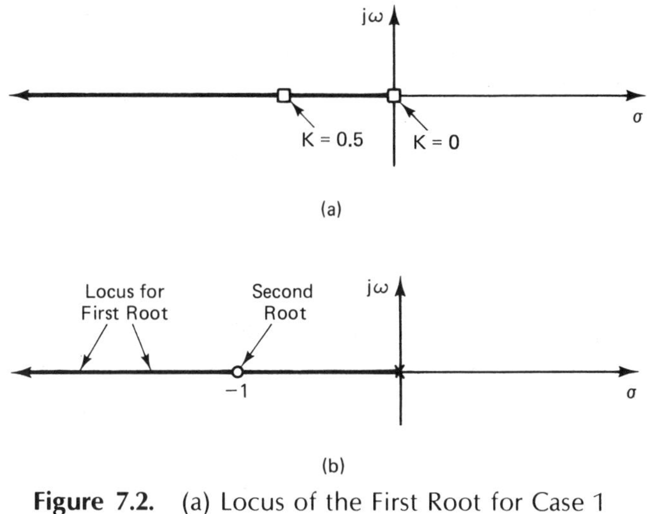


**Figure 7.2.** (a) Locus of the First Root for Case 1
(b) Complete Root Locus for Case 1

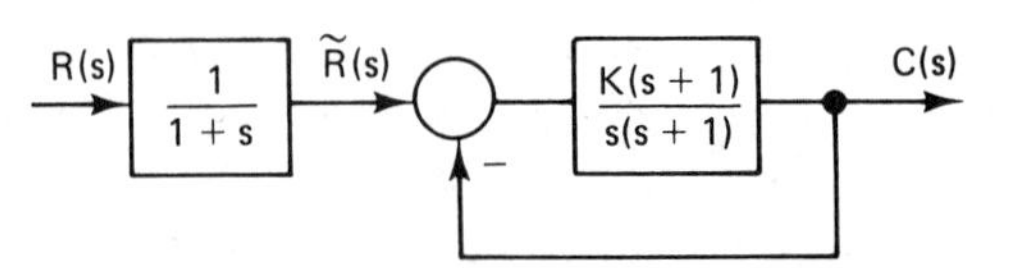


**Figure 7.3.** Reduced Block Diagram for Case 1

example,

$$K = 0 \qquad s_1 = 0$$
$$K = 0.5 \qquad s_1 = -0.5$$
$$K \to \infty \qquad s_1 \to -\infty$$

The locus of the first root in the $s$-plane is as shown in Figure 7.2a. The second root is independent of the choice of $K$. The complete root locus with respect to $K$ is shown in Figure 7.2b.

Let us note here that for this case a reduction of the block diagram of Figure 7.1 results in the block diagram shown in Figure 7.3. Clearly, the reduced system has the characteristic equation given by (7-4).

## Case 2

The second case involves varying the parameter $a$. Let us assume that

$$K = b = \frac{1}{\tau} = 1$$

Thus the characteristic equation of the system for this case is given by

$$P(s) = s^2 + 2s + a \tag{7-5}$$

The root locus locations are thus

$$s_{1,2} = -1 \pm \sqrt{1 - a}$$

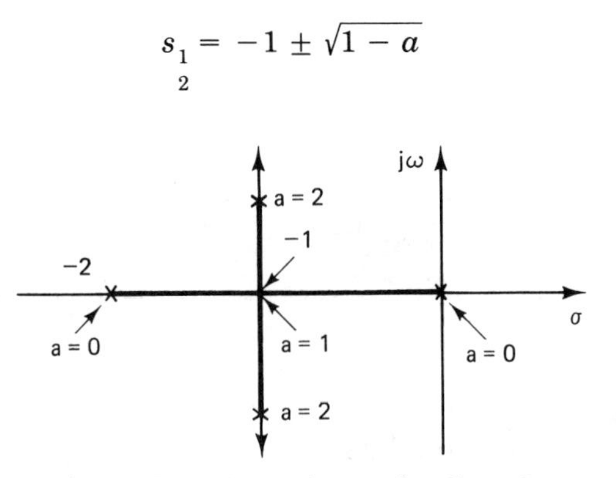


**Figure 7.4.** Root Locus for Case 2

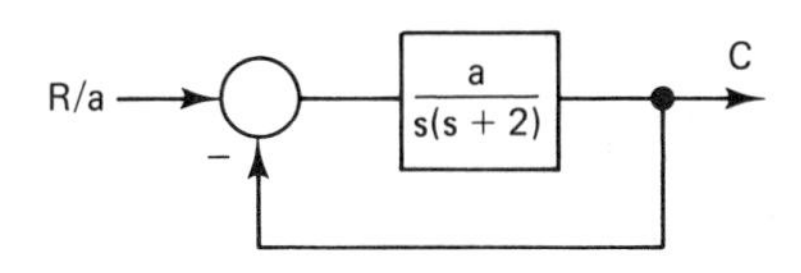


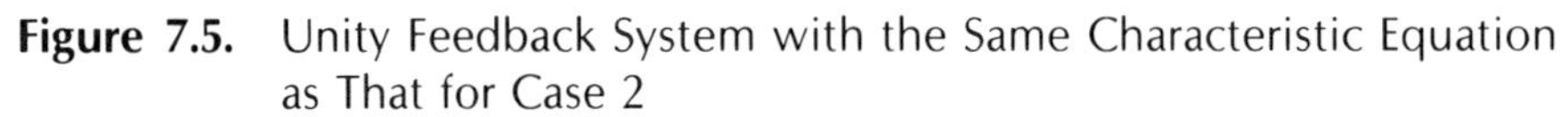

**Figure 7.5.** Unity Feedback System with the Same Characteristic Equation as That for Case 2

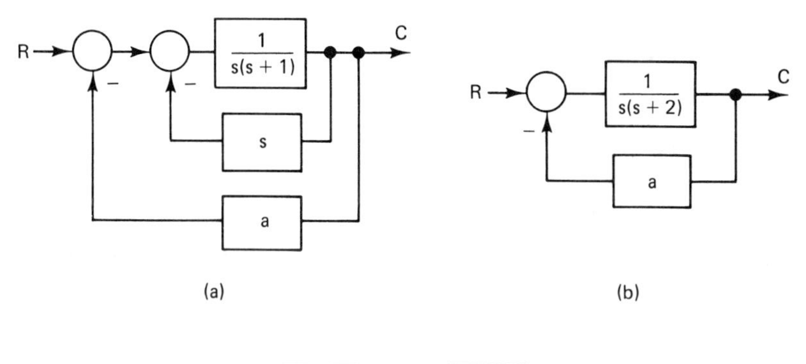


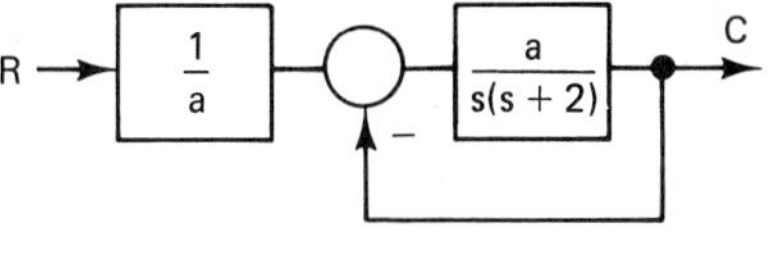


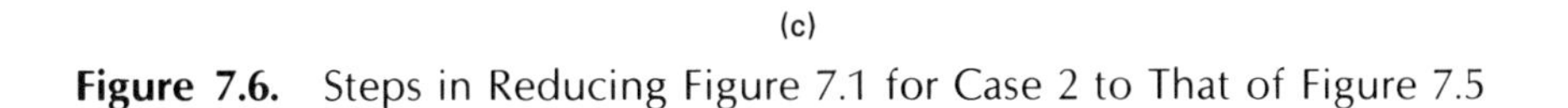

**Figure 7.6.** Steps in Reducing Figure 7.1 for Case 2 to That of Figure 7.5

The values of the roots are given by

$$
\begin{array}{lll}
a = 0 & s_1 = 0 & s_2 = -2 \\
a = 1 & s_1 = -1 & s_2 = -1 \\
a = 2 & s_1 = -1 + j & s_2 = -1 - j
\end{array}
$$

Therefore, the root loci as $a$ varies from zero to infinity are as shown in Figure 7.4. Figure 7.5 shows a block diagram of a system with a unity feedback characteristic polynomial given by Eq. (7-5). The steps in reducing the block diagram of Figure 7.1 for case 2 to that of Figure 7.5 are shown in Figure 7.6.

## Case 3

The third case involves varying the parameter $b$. Let us assume that

$$K = a = \frac{1}{\tau} = 1$$

Thus the characteristic equation is

$$P(s) = s^2 + (1 + b)s + 1 \tag{7-6}$$

The characteristic roots are

$$s_{\substack{1\\2}} = \frac{-(1 + b) \pm \sqrt{b^2 + 2b - 3}}{2} \tag{7-7}$$

Let us examine the nature of the roots:

$$b = 0 \qquad s_{\substack{1\\2}} = -0.5 \pm j0.866 = e^{\pm j120°}$$

$$b = 1 \qquad s_{\substack{1\\2}} = -1$$

$$b = 2 \qquad s_1 = -0.38 \quad s_2 = -2.61$$

$$b \to \infty \qquad s_1 \to -\infty \quad s_2 \to 0$$

It is clear that for $0 < b < 1$, the roots are complex conjugates. The $s$-plane relation between the real and imaginary parts gives the equation of the locus. This is obtained by letting

$$s_1 = -\sigma + j\omega$$

Thus from Eq. (7-7) we get

$$\sigma = 0.5(1 + b)$$

$$\omega = 0.5\sqrt{3 - 2b - b^2}$$

This yields

$$b = 2\sigma - 1$$

and

$$4\omega^2 = 3 - 2b - b^2$$

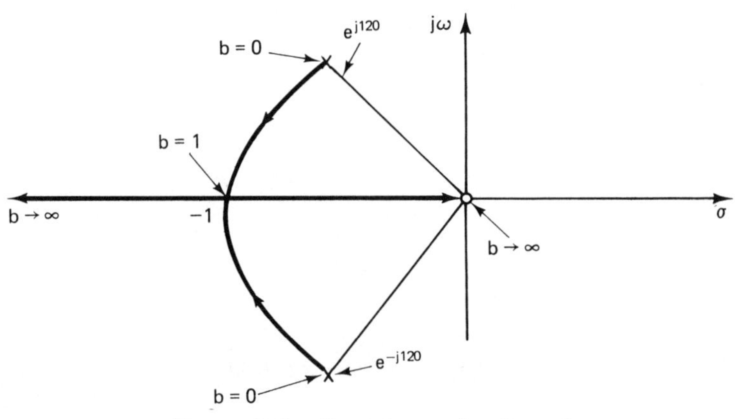


**Figure 7.7.** Root Locus for Case 3

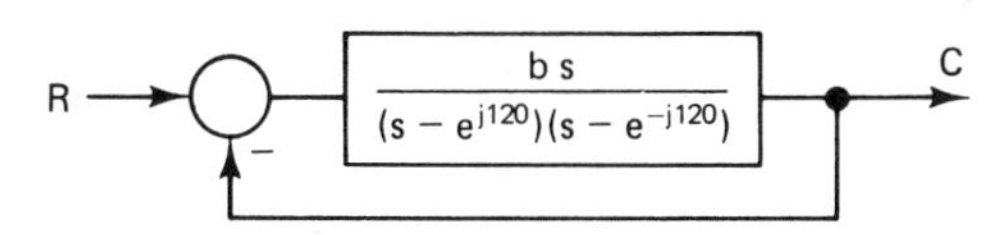


**Figure 7.8.** Unity Feedback System Producing the Characteristic Equation for Case 3

Eliminating $b$, we get

$$4\omega^2 = 3 - 2(2\sigma - 1) - (2\sigma - 1)^2$$

This reduces to

$$\sigma^2 + \omega^2 = 1$$

This is the equation of a circle with the center at the origin and radius of 1. The root locus is shown in Figure 7.7.

Again the unity feedback system shown in block diagram form in Figure 7.8 produces the same characteristic equation for this case.

## Case 4

The fourth case involves varying $1/\tau$, where

$$K = a = b = 1$$

The characteristic equation is

$$P(s) = s^2 + \left(1 + \frac{1}{\tau}\right)s + 1$$

This is clearly similar to the previous case.

The main point to be emphasized here is that root loci are constructed for variations of a parameter. It is more convenient to construct the loci if the parameter is the gain of the open-loop transfer function in a feedback system as shown in Figure 7.9. Observe now that the gain in the system shown in Figure 7.3 is $K$, while that in Figure 7.5 is $a$, and in Figure 7.8 the gain is $b$. Note that we have unity feedback in each case.

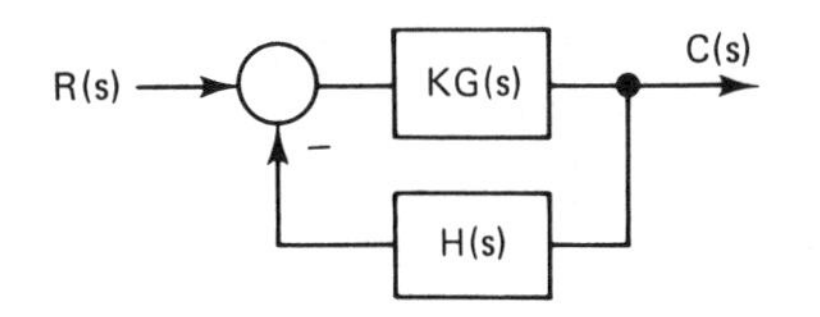


**Figure 7.9.** Closed-Loop System

# 7.3 Properties of the Root Locus

Having discussed the basic underlying concepts defining the root-locus plots for a relatively simple system, we now turn our attention to a discussion of the general properties of the root locus. Our discussion deals with a general form of the system transfer function. The discussion leads to useful rules for the construction of the root locus.

The characteristic equation of the system shown in Figure 7.9 is

$$P(s) = 1 + KG(s)\,H(s) = 0 \tag{7-8}$$

It is clear that the roots of the characteristic equation satisfy the two conditions

$$K\,|G(s)H(s)| = 1 \tag{7-9}$$

$$\arg[G(s)H(s)] = 180° \pm n(360°) \tag{7-10}$$

Let us introduce

$$\tilde{G}(s) = G(s)H(s) \tag{7-11}$$

We further assume that

$$\tilde{G}(s) = \frac{\prod_{i=1}^{M}(s - z_i)}{\prod_{k=1}^{N}(s - p_k)} \tag{7-12}$$

where $p_i$ are the poles of $\tilde{G}$ and $z_i$ are the zeros of $\tilde{G}$. The characteristic equation can thus be written as

$$\prod_{k=1}^{N}(s - p_k) + K\prod_{i=1}^{M}(s - z_i) = 0 \tag{7-13}$$

It is perhaps appropriate at this juncture to comment on terms commonly encountered in system studies. The *poles* of a transfer function such as that described by Eq. (7-12) are the values of $s$ that make $\tilde{G}(s) = \infty$. It is thus clear that according to this definition, the values $s = p_i$ specify the poles of $\tilde{G}(s)$ of Eq. (7-12). The *zeros* of a transfer function are the values of $s$ that make $\tilde{G}(s) = 0$. As a result of this definition, the values $s = z_i$ specify the zeros of $\tilde{G}(s)$ of Eq. (7-12). It is worth noting that poles and zeros are roots of the denominator and numerator polynomials of the transfer function respectively. The term "root" in control systems terminology is reserved for polynomials such as the characteristic equation. The term "poles of a system" implies certain time-response and frequency-domain characteristics and is preferred to "roots of a system."

If we consider $K = 0$, we find that the roots of the characteristic equation occur for

$$\prod_{k=1}^{N} (s - p_k) = 0 \tag{7-14}$$

Thus the poles of $\tilde{G}(s)$ provide the starting points for the locus. On the other hand, as $K \to \infty$, we have the characteristic equation reducing to

$$\prod_{i=1}^{M} (s - z_i) = 0 \tag{7-15}$$

As a result, we conclude that the zeros of $\tilde{G}(s)$ provide the termination points of the locus. We can now state the first rule of root-locus construction.

*Rule 1:* The root locus starts at the poles of the loop function $\tilde{G}(s)$ and terminates at the zeros of $\tilde{G}(s)$.

Examination of Figures 7.7 and 7.8 reveals that this rule is satisfied for case 3 of Section 7.2.

Considering the angle condition applied on the real axis, given by Eq. (7-10),

$$\arg[\tilde{G}(s)] = \pm 180° \pm n(360°) \tag{7-16}$$

we can conclude that segments of the locus on the real axis exist only to the left of an odd number of poles and zeros of $\tilde{G}(s)$. This is shown in Figure 7.10. We thus have rule 2:

*Rule 2:* The locus segments on the real axis always lie in a section to the left of an odd number of poles and zeros.

The third rule concerns the number of separate segments of the locus. The number of poles is equal to the number of separate loci, since each locus starts at a pole of the loop function. Thus we have

*Rule 3:* The number of separate segments of the locus is equal to the number of loop function poles.

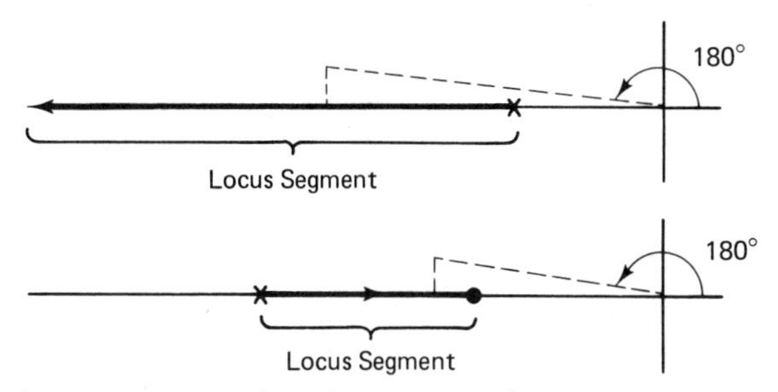


**Figure 7.10.** Rule 2 for Root-Locus Construction

Complex roots give rise to sections of the locus not on the real axis. Since these roots occur as conjugate pairs, we have

*Rule 4:* The root locus must be symmetrical with respect to the real axis.

The root loci start at the poles and terminate at the zeros according to rule 1. The number of poles $N$ can exceed the number of finite zeros $M$. In this case $(N\text{-}M)$ sections must terminate at infinity along asymptotic lines, whose center $\sigma_A$ and angle $\phi_A$ are given in

*Rule 5:* The number of locus sections terminating at infinity is equal to the excess of poles over zeros.

The center of the asymptotes $\sigma_A$ on the real axis is given by

$$\sigma_A = \frac{\sum_{k=1}^{N} p_k - \sum_{i=1}^{M} z_i}{N - M} \tag{7-17}$$

The angles of the asymptotes are

$$\phi_A = \frac{(2q + 1)180°}{N - M} \qquad q = 0, 1, 2, \ldots, N - M - 1 \tag{7-18}$$

An example is appropriate at this time to illustrate the foregoing rules.

EXAMPLE 7.1

Consider the loop function given by

$$\tilde{G}(s) = \frac{(s + 3)(s + 4)}{s(s + 1)(s + 2)(s + 5)(s + 6)}$$

We have two zeros at

$$z_1 = -3$$
$$z_2 = -4$$

There are five poles located at

$$p_1 = 0$$
$$p_2 = -1$$
$$p_3 = -2$$
$$p_4 = -5$$
$$p_5 = -6$$

According to rule 1, five segments of the root locus start at the five poles with two segments terminating at the two zeros. The locus segments on the real axis according to rule 2 are between $p_1$ and $p_2$, $p_3$ and $z_1$, $z_2$ and $p_4$, and $p_5$ and minus infinity. This situation is shown in Figure 7.11.

According to rule 3, we should have five separate segments of the locus. Note that so far we have identified two segments on the real axis that start at a pole and terminate at a finite zero. According to rule 5, we should have three segments (excess of poles over zeros) terminating at infinity. The center of the asymptotes is given by

$$\sigma_A = \frac{(0 - 1 - 2 - 5 - 6) - (-3 - 4)}{5 - 2} = -2.33$$

The angles of the asymptotes are

$$\phi_A = 60°, 180°, \quad \text{and} \quad 300°$$

The construction of the root locus for this example system will be continued following our study of next rule.

## Breakaway Rule

Segments of the locus on the real axis exist according to rule 2 starting at a pole ($K = 0$) or terminating at a zero ($K \to \infty$) according to rule 1. Consider the case when a segment exists between two poles $p_1$ and $p_2$ as shown in Figure 7.11. The gain $K$ increases as we move to the left of $p_1$ and to the right of $p_2$ on the real axis. There is a point in the interval ($p_1$ to $p_2$) where $K$ is a maximum at which the locus breaks away from the real axis. We thus have

> *Rule 6:* Breakaway points are points on the real axis where two or more branches of the root locus depart from the real axis.

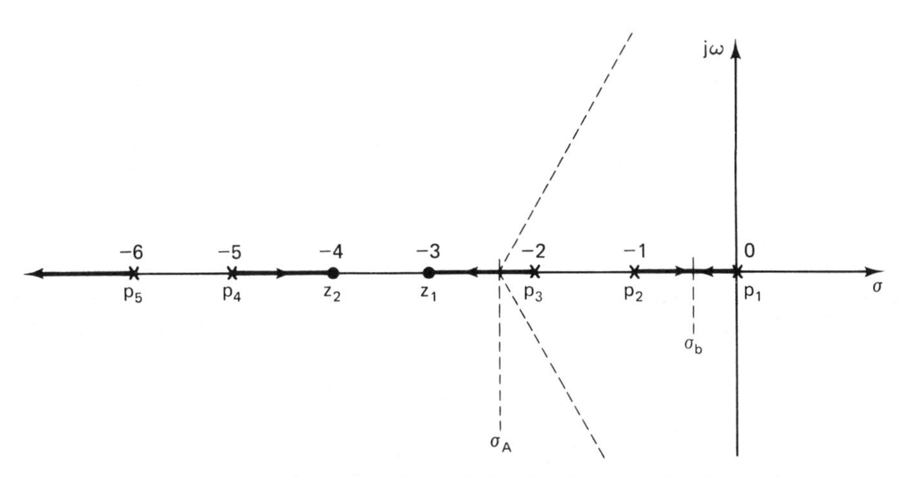


**Figure 7.11.** Partial Realization of the Root Locus for Example 7.1

Their location $\sigma_b$ is obtained by maximizing $K(\sigma)$, where the characteristic equation for $s = \sigma$ is written as

$$1 + K\tilde{G}(\sigma) = 0$$

Thus

$$K = \frac{-1}{\tilde{G}(\sigma)} \tag{7-19}$$

The maximum value of $K$ is obtained at

$$\frac{\partial K}{\partial \sigma} = 0 \tag{7-20}$$

The resulting equation is solved for $\sigma_b$.

EXAMPLE 7.1 (continued)

For our loop function, we can write

$$-K = \frac{s(s+1)(s+2)(s+5)(s+6)}{s^2 + 7s + 12}$$

Let $s = \sigma$, and expand to obtain

$$-K = \frac{\sigma^5 + 14\sigma^4 + 65\sigma^3 + 112\sigma^2 + 60\sigma}{\sigma^2 + 7\sigma + 12}$$

Condition (7-20) is applied by taking the derivative of $K$ with respect to $\sigma$ and setting the result to zero. As a result, we obtain

$$(5\sigma^4 + 56\sigma^3 + 195\sigma^2 + 224\sigma + 60)(\sigma^2 + 7\sigma + 12) - (\sigma^5 + 14\sigma^4 + 65\sigma^3 + 112\sigma^2 + 60\sigma)(2\sigma + 7) = 0$$

After some algebra we get

$$3\sigma^6 + 56\sigma^5 + 419\sigma^4 + 1582\sigma^3 + 3064\sigma^2 + 2688\sigma + 720 = 0$$

There are six roots to the equation above. The roots are determined iteratively. A search in the neighborhood between 0 and $-1$ where we anticipate our breakaway point yields a root of interest at

$$\sigma_b = -0.45385$$

Note that there are three other real roots for the sixth-order equation, at $-1.6425$, $-3.4577$, and $-5.45$, which do not belong to the root locus. A moment's reflection shows this last statement to be true, as we have established that segments of the root locus on the real axis for this example are in the intervals $(0, -1)$, $(-2, -3)$, $(-4, -5)$, and $(-6, -\infty)$. The three additional roots do not lie in these intervals.

To continue on we will need the following rule.

## Rule on Intersection with the Imaginary Axis

Segments on the loci can exist in the right half of the $s$-plane. This signifies instability. We have the seventh rule, which enables us to determine the intersection of the locus with the imaginary axis. Note that once the locus enters the right half of the $s$-plane, the system is unstable. We thus have:

> *Rule 7:* The Routh–Hurwitz criterion provides the points at which the loci cross the imaginary axis.

EXAMPLE 7.1 (continued)

The characteristic equation is

$$s(s+1)(s+2)(s+5)(s+6)+K(s+3)(s+4)=0$$

or

$$s^5+14s^4+65s^3+(112+K)s^2+(60+7K)s+12K=0$$

The first two rows of the Routh array are

$$\begin{array}{c|ccc} s^5 & 1 & 65 & 60+7K \\ s^4 & 14 & (112+K) & 12K \end{array}$$

The third row is

$$\begin{array}{c|ccc} s^3 & \dfrac{798-K}{14} & \dfrac{840+86K}{14} & 0 \end{array}$$

It is clear that there is a critical $K$ beyond which the system is unstable. This occurs for $K \geq 798$. This is only the first condition, we should continue on.

The fourth row is

$$\begin{array}{c|cc} s^2 & \dfrac{77{,}616-518K-K^2}{798-K} & 12K \end{array}$$

For stability we require that

$$\frac{77{,}616-518K-K^2}{798-K} \geq 0$$

or

$$\frac{(121.39-K)(639.39+K)}{798-K} \geq 0$$

Since we have $K \leq 798$ from the third row's requirement, we conclude that

$$K \leq 121.39$$

is a requirement for stability.

The fifth row is

$$s^1 \mid A$$

where

$$A = \frac{\dfrac{77{,}616 - 518K - K^2}{798 - K}\,\dfrac{840 + 86K}{14} - 12K\dfrac{798 - K}{14}}{\dfrac{77{,}616 - 518K - K^2}{798 - K}}$$

$$= \frac{(77{,}616 - 518K - K^2)(840 + 86K) - 12K(798 - K)^2}{14(77{,}616 - 518K - K^2)}$$

$$A = \frac{7(K - 29.02876)(K^2 + 296.74K + 22{,}917.96)}{(K + 639.39)(K - 121.39)}$$

With $K \leq 121.39$, we see that for stability as dictated by $A \geq 0$, we should

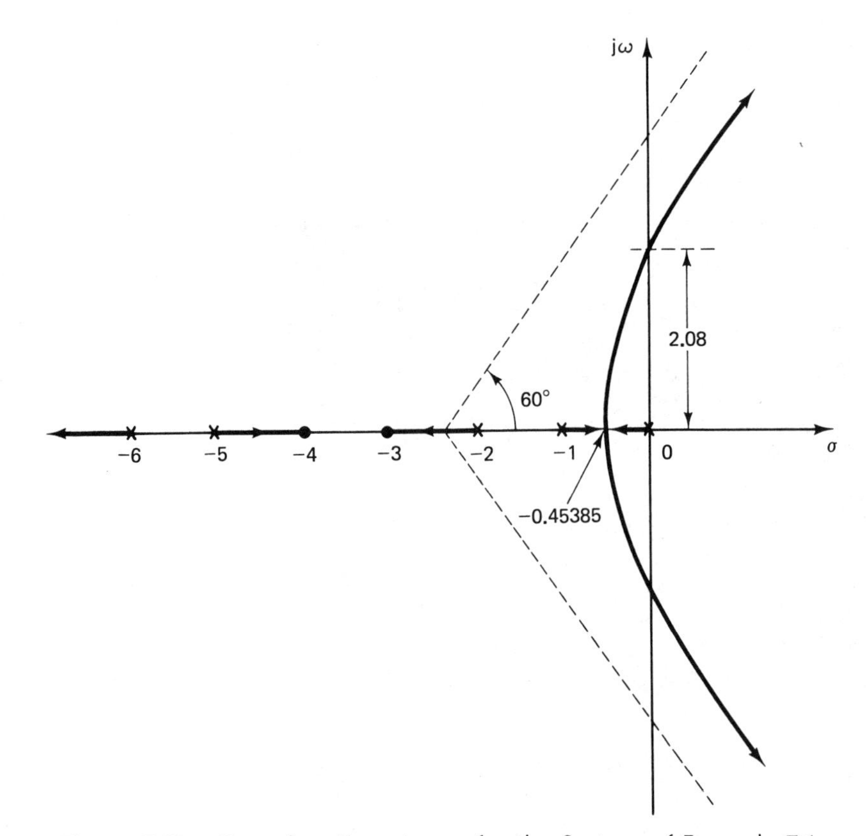


**Figure 7.12.** Complete Root Locus for the System of Example 7.1

require that

$$K \leq 29.02876 \simeq 29$$

As a conclusion,

$$K_{\text{crit}} \simeq 29$$

The frequency of oscillation for critical stability provides us with the intersection with the imaginary axis. This is obtained by solving the characteristic equation associated with the fourth row:

$$\frac{77{,}616 - 518K - K^2}{798 - K} s^2 + 12K = 0$$

With the value of $K_{\text{crit}}$ obtained, we can conclude from the auxiliary equation associated with the $s^2$ row that

$$\omega_{\text{osc}} = 2.08$$

The complete root locus is shown in Figure 7.12.

## Rule on Departure Angle

When complex poles are encountered in the open-loop function $\tilde{G}(s)$, the locus will start at that complex pole in a direction determined by the departure angle given in the following rule.

*Rule 8:* The departure angle from a complex pole is given by

$$\theta_D = 180° + \arg[\tilde{G}'(s)] \qquad (7\text{-}21)$$

where arg $[\tilde{G}']$ is the phase angle of $\tilde{G}$ computed at the complex pole but dropping the contribution of that particular pole.

Note that $\theta_D$ is measured as usual in the counterclockwise direction from the horizontal ($\sigma$ axis) in the $s$-plane. An example will illustrate the application of this rule.

EXAMPLE 7.2

Consider a system with loop transfer function given by

$$\tilde{G}(s) = \frac{1}{s(s^2 + 2s + 3)}$$

The poles are at

$$p_1 = 0$$
$$p_2 = -1 + j\sqrt{2}$$
$$p_3 = -1 - j\sqrt{2}$$

Clearly, there is a segment on the real axis extending from the origin to minus infinity, as shown in Figure 7.13. Since there are no zeros, we conclude that the segments approach infinity along asymptotic lines that intersect at

$$\sigma_A = \frac{-2}{3}$$

The angles are

$$\phi_A = 60^\circ, 180^\circ, \text{ and } 300^\circ$$

Two locus segments will depart from the complex poles. Consider the pole $p_2$,

$$\tilde{G}(p_2) = \frac{1}{p_2(p_2 - p_3)(p_2 - p_2)}$$

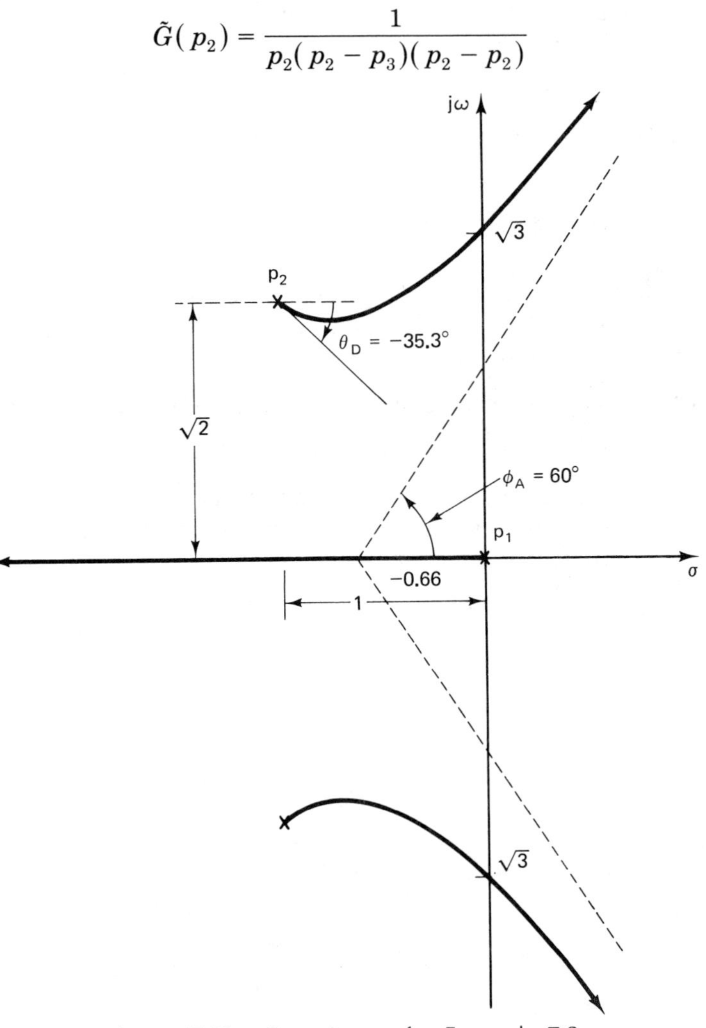


**Figure 7.13.** Root Locus for Example 7.2

Dropping the contribution of $p_2$, we get

$$\tilde{G}'(p_2) = \frac{1}{p_2(p_2 - p_3)}$$

$$\tilde{G}'(p_2) = \frac{1}{(-1 + j\sqrt{2})(j2\sqrt{2})}$$

$$= 0.204\underline{/-215.26^\circ}$$

Thus

$$\arg\left[\tilde{G}'(p_2)\right] = -215.26^\circ$$

As a result, the departure angle is

$$\theta_D = 180 - 215.26^\circ = -35.26^\circ$$

This is shown in Figure 7.13.

To complete the root locus, we need to find the intersection with the imaginary axis. The characteristic equation is given by

$$s(s^2 + 2s + 3) + K = 0$$

or

$$s^3 + 2s^2 + 3s + K = 0$$

The Routh array is thus constructed as

$$\begin{array}{c|cc} s^3 & 1 & 3 \\ s^2 & 2 & K \\ s^1 & 3 - \dfrac{K}{2} & 0 \\ s^0 & K & \end{array}$$

Thus for stability we require that

$$3 - \frac{K}{2} > 0$$

Thus

$$K \leq 6$$

The characteristic equation for the $s^2$ row is

$$2s^2 + K_{\text{crit}} = 0$$

or

$$2s^2 + 6 = 0$$

Thus we get the intersection with the imaginary axis as

$$s = \pm j\sqrt{3}$$

It is interesting to note that we can obtain the intersection with the imaginary axis by solving the characteristic equation for $s = j\omega$. In this case we have

$$(j\omega)^3 + 2(j\omega)^2 + 3(j\omega) + K = 0$$

or

$$-j\omega^3 - 2\omega^2 + 3j\omega + K = 0$$

Separating real and imaginary parts, we get

$$-\omega^3 + 3\omega = 0$$

and

$$-2\omega^2 + K = 0$$

Thus

$$\omega_{\text{osc}} = \sqrt{3}$$
$$K_{\text{crit}} = 6$$

For systems with higher order, the Routh method is preferred.

## Rule on Locus Calibration

With the basic features of the root locus determined using the foregoing rules, a sketch of the locus is easily obtained. The precise location of points on the locus and the associated gain $K$ can be obtained by simply using the characteristic equation

$$1 + K\tilde{G}(\sigma + j\omega) = 0$$

Note that we have replaced $s$ by $(\sigma + j\omega)$ in writing the characteristic equation. The equation above is complex and we can replace it by two equations in terms of the real and imaginary parts. The two equations provide us with the values of two unknowns, such as $K$ and $\omega$, if $\sigma$ is specified, $K$ and $\sigma$ if $\omega$ is specified, and $\sigma$ and $\omega$ if $K$ is specified. A few examples will illustrate the procedures involved.

EXAMPLE 7.3

Consider the system of Example 7.2. It is required to find the gain $K$ and the radian frequency $\omega$ for $\sigma = -0.5$.

SOLUTION

The characteristic equation for this system is given by

$$1 + \frac{K}{s(s^2 + 2s + 3)} = 0$$

Substituting

$$s = -0.5 + j\omega$$

we obtain after some manipulation

$$K = 1.125 + 0.5\omega^2 + j\omega(\omega^2 - 1.75)$$

The imaginary part gives

$$\omega^2 - 1.75 = 0$$

Thus

$$\omega = \pm\sqrt{1.75} = 1.3228$$

The real part gives

$$K = 1.125 + 0.5(1.75) = 2$$

EXAMPLE 7.4

Find the gain $K$ and $\sigma$ on the root locus of Example 7.2 for $\omega = 1.5$.

SOLUTION

Substitute

$$s = \sigma + j1.5$$

in the characteristic equation to obtain

$$(\sigma + j1.5)\left[(\sigma + j1.5)^2 + 2(\sigma + j1.5) + 3\right] + K = 0$$

Separating real and imaginary parts after some manipulation we obtain

$$\sigma(\sigma^2 + 2\sigma + 0.75) - 4.5(\sigma + 1) + K = 0 \qquad \text{(real)}$$

$$1.5(\sigma^2 + 2\sigma + 0.75) + 3\sigma(\sigma + 1) = 0 \qquad \text{(imaginary)}$$

Solving the imaginary equation for $\sigma$, we obtain a solution on the locus as

$$\sigma = -0.2257$$

Thus we use the real equation to calculate $K$ as

$$K = 3.56$$

# 7.4 Root Locus for Systems with Time Delay

Let us consider the problem of constructing the root locus for systems where a time delay of $\tau$ seconds is involved in the loop function. Figure 7.14 shows in block diagram form two systems involving delay in the forward path and

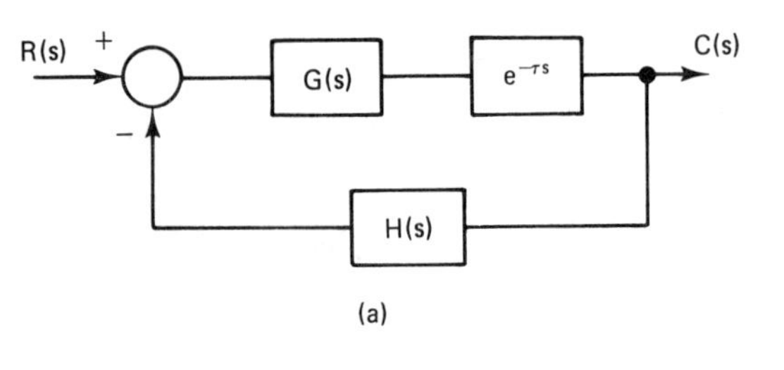


(a)

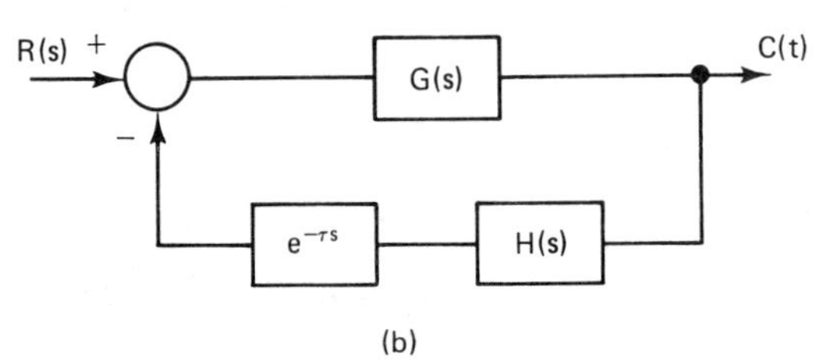


(b)

**Figure 7.14.** Closed-Loop Systems with Time Delays Involved

the feedback path. In either case, the characteristic equation is given by

$$P(s) = 1 + KG(s)H(s)e^{-\tau s} = 0 \tag{7-22}$$

Using the definitions of Eqs. (7-11) and (7-12), we can write the characteristic equation as

$$P(s) = 1 + K\tilde{G}(s)e^{-\tau s} = 0 \tag{7-23}$$

Alternatively,

$$\prod_{k=1}^{N}(s - p_k) + Ke^{-\tau s}\prod_{i=1}^{M}(s - z_i) = 0 \tag{7-24}$$

or

$$e^{\tau s}\prod_{k=1}^{N}(s - p_k) + K\prod_{i=1}^{M}(s - z_i) = 0 \tag{7-25}$$

The presence of the exponential in the characteristic equation necessitates some modifications to the rules for constructing the root locus, which will be examined here.

Let us consider the case when $K = 0$. Here we obtain from Eq. (7-25), with $s = \sigma + j\omega$,

$$e^{\tau s}\prod_{k=1}^{N}(s - p_k) = 0$$

This is satisfied for $s = p_k(k = 1,\ldots,N)$ and $\sigma \to -\infty$. As a result, we have the first part of rule 1, given by

*Rule 1(a):* The root locus for the system with time delay starts at the poles of the loop function $\tilde{G}(s)$ and $\sigma \to -\infty$.

If we consider the case when $K \to \infty$, then Eq. (7-24) can be written as

$$e^{-\tau s} \prod_{i=1}^{M} (s - z_i) = 0$$

This is satisfied for $s = z_i$ $(i = 1, \ldots, M)$ and $\sigma \to \infty$. Thus the second part of rule 1 is obtained as:

*Rule 1(b):* The root locus for the system with time delay terminates at the zeros of the loop function $\tilde{G}(s)$ and $\sigma \to \infty$.

For points in the $s$-plane to lie on the locus, Eq. (7-23) must be satisfied. If $s$ is represented by the rectangular form

$$s = \sigma + j\omega$$

then we have

$$1 + Ke^{-\tau\sigma}\tilde{G}(s)e^{-j\omega\tau} = 0 \tag{7-26}$$

Thus we have the magnitude condition

$$K = \frac{e^{\tau\sigma}}{|\tilde{G}(s)|} \tag{7-27}$$

and the angle condition

$$\arg[\tilde{G}(s)] = \pm 180°(1 + 2n) + \omega\tau \qquad n = 0, 1, \ldots \tag{7-28}$$

On the real axis, $\omega = 0$; thus the last term in Eq. (7-28) does not exist. In this case Eq. (7-16) applies, and we can conclude that rule 2 of Section 7.3 applies.

*Rule 2:* The locus segments on the real axis always lie in a section to the left of an odd number of poles and zeros of $\tilde{G}(s)$.

We know that the number of roots of an $N$th-order polynomial is $N$. Examination of Eq. (7-24) for $\tau = 0$ reveals that there will be $N$ roots (equal to number of poles, assuming that there are more poles than zeros) and hence $N$ separate segments of the locus. For $\tau \neq 0$, the situation is different. Recall that the exponential function can be expanded as the infinite series

$$e^{-\tau s} = \sum_{k=0}^{\infty} \frac{(-\tau s)^k}{k!} \tag{7-29}$$

Thus Eq. (7-24) has an infinite number of roots, and consequently there are an infinite number of root-locus segments. We thus have:

*Rule 3:* The number of separate segments of the locus is infinite.

The rule on symmetry of the locus about the real axis remains unchanged in view of the infinite series (7-29), which has real coefficients. We thus have complex-conjugate pairs of roots.

*Rule 4:* The root locus must be symmetrical with respect to the real axis.

As we have an infinite number of segments of the locus, we thus have an infinite number of asymptotes. Now asymptotic lines occur for $s \to \infty$; thus according to rule 1, the asymptotes are at $\sigma \to \infty$ for $K \to \infty$, and $\sigma \to -\infty$ for $K = 0$.

For $s \to \infty$, the contribution of the finite poles and zeros may be neglected and thus for the asymptotes

$$\lim_{s\to\infty} \tilde{G}(s) = \lim_{s\to\infty} \left[ \prod_{i=1}^{M} (s - z_i) \Big/ \prod_{k=1}^{N} (s - p_k) \right]$$
$$= \lim_{s\to\infty} (s)^{-(N-M)}$$

Consider the asymptotes for $K = 0$, $\sigma \to -\infty$; we have

$$\lim_{s\to\infty} \arg[\tilde{G}(s)] = -(N - M)(180°)$$

If $(N - M)$ is an odd integer, we thus have

$$\lim_{s\to\infty} \arg[\tilde{G}(s)] = \pm 180°$$

On the other hand, for $(N - M) =$ even integer

$$\lim_{s\to\infty} \arg[\tilde{G}(s)] = 0°$$

Thus Eq. (7-28) yields for $(N - M)$ odd

$$\pm(180°) = \pm(180°)(1 + 2n) + \omega\tau$$

This gives the values of $\omega$, the intersection of the asymptotes with the imaginary axis, as

$$\omega = \frac{\pm 2n(\pi)}{\tau} \qquad n = 0, 1, \ldots$$

For $(N - M)$ even, we get from Eq. (7-28),

$$0 = \pm 180°(1 + 2n) + \omega\tau$$

Thus

$$\omega = \frac{\pm(1 + 2n)(\pi)}{\tau} \qquad n = 0, 1, \ldots$$

Consider the asymptotes for $K \to \infty$, $\sigma \to \infty$; we have by a similar argument, for $(N - M)$ odd or even,

$$\omega = \frac{\pm(1 + 2n)(\pi)}{\tau}$$

we can thus write rule 5 as follows:

*Rule 5:* The number of asymptotes of the root locus is infinite and they are parallel to the real axis of the $s$-plane. The intersections of the asymptotes with the imaginary axis are given by

$$\omega = \frac{q\pi}{\tau}$$

where $q$ depends on $K$ and $(N - M)$ as follows:

$$\begin{array}{lll} K = 0, & (N-M) \text{ odd:} & q = \text{even integers} \\ & (N-M) \text{ even:} & q = \text{odd integers} \\ K \to \infty, & (N-M) \text{ odd:} & q = \text{odd integers} \\ & (N-M) \text{ even:} & q = \text{odd integers} \end{array}$$

The breakaway rule in our situation is similar to the one obtained in Section 7.3. Thus

*Rule 6:* The location of the breakaway points is determined by maximizing

$$K = \frac{-e^{\tau\sigma}}{\tilde{G}(\sigma)}$$

This requires

$$\frac{\partial K}{\partial \sigma} = 0$$

The rule on the intersection with the imaginary axis based on the Routh–Hurwitz criterion is no longer valid since $e^{-Ts}$ is not an algebraic expression. Instead, we have

*Rule 7:* The intersections of the root locus with the imaginary axis can be obtained by solving the complex equation

$$1 + K\tilde{G}(j\omega)e^{-j\omega\tau} = 0$$

for $K$ and $\omega$.

The rule on departure angles remains unaltered. Thus we have

*Rule 8:* The departure angle from a complex pole is given by

$$\theta_D = 180° + \arg\left[\tilde{G}'(s)e^{-\tau s}\right]$$

where the argument is taken by dropping the contribution of the complex pole already dealt with.

Locus calibration is done in a manner similar to that discussed in Section 7.3. Here we have to solve

$$1 + K\tilde{G}(\sigma + j\omega)e^{-\tau(\sigma + j\omega)} = 0$$

for $s = \sigma + j\omega$. Alternatively, we must solve

$$1 + K\tilde{G}(re^{j\theta})e^{-\tau r(\cos\theta + j\sin\theta)} = 0 \tag{7-30}$$

where $s$ is expressed in the polar form such that

$$s = re^{j\theta} = r\angle\theta$$

It is appropriate at this time to consider a number of examples.

EXAMPLE 7.5

Consider a unity feedback control system with forward transfer function given by

$$G(s) = \frac{e^{-\tau s}}{s^2}$$

The characteristic equation is

$$s^2 + Ke^{-\tau s} = 0$$

The number of finite poles is $N = 2$, and there are no zeros; thus the locus starts at the double pole $s = 0$ and $\sigma \to -\infty$ and terminates at $\sigma \to \infty$, according to rule 1. According to rule 2, we do not have any segments on the real axis other than at the origin $s = 0$. Rules 3 and 4 apply here as well. Since $N - M$ is even, then according to rule 5, we can conclude that our asymptotes intersect the imaginary axis at

$$K = 0: \qquad \omega = \frac{\pi}{\tau}, \frac{3\pi}{\tau}, \frac{5\pi}{\tau}, \ldots$$

$$K \to \infty: \qquad \omega = \frac{\pi}{\tau}, \frac{3\pi}{\tau}, \frac{5\pi}{\tau}, \ldots$$

On the imaginary axis, $s = j\omega$; thus

$$-\omega^2 + Ke^{-j\omega\tau} = 0$$

or

$$-\omega^2 + K(\cos\omega\tau - j\sin\omega\tau) = 0$$

The imaginary part yields

$$\sin\omega\tau = 0$$

The real part gives

$$K = \frac{\omega^2}{\cos\omega\tau}$$

For $\omega\tau = 0, 2\pi, 4\pi, \ldots$, we have $\sin\omega\tau = 0$ and $\cos\omega\tau = 1$. Thus

$$K = \omega^2$$

For $\omega\tau = \pi, 3\pi, \ldots$, we have $\sin\omega\tau = 0$ and $\cos\omega\tau = 1$. Thus

$$K = -\omega^2$$

Thus, for positive $K$, we have the intersection of the locus with the imaginary axis at

$$\omega = \frac{2\tilde{q}\pi}{\tau} \qquad \tilde{q} = 0, 1, 2, \ldots$$

The corresponding gain is

$$K = \frac{4\pi^2}{\tau^2}\tilde{q}^2$$

or

$$K = \frac{39.478}{\tau^2}\tilde{q}^2$$

Figure 7.15 shows the root locus for positive $\omega$ and positive $K$.

EXAMPLE 7.6

Consider the transfer function

$$G(s) = \frac{e^{-\tau s}}{(s+2)(s+4)}$$

Assume that $\tau = 1$ seconds for simplicity. The steps in developing the root

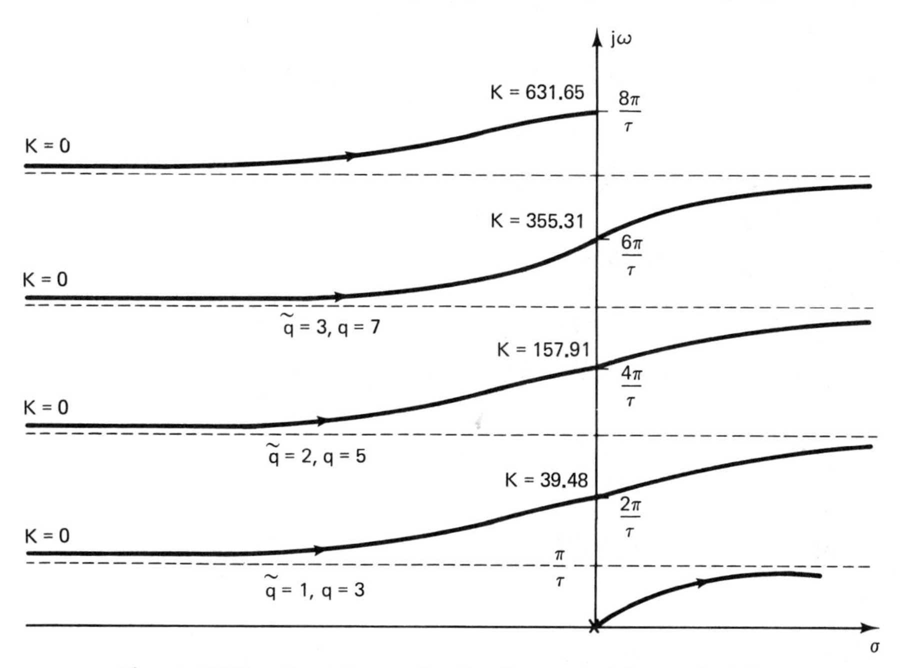


**Figure 7.15.** Root Locus for the System of Example 7.5

locus are similar to those for Example 7.5. Our poles in this case are at

$$p_1 = -2 \qquad \text{and} \qquad p_2 = -4$$

On the imaginary axis we have $s = j\omega$; thus the characteristic equation is

$$(2 + j\omega)(4 + j\omega) + Ke^{-j\omega\tau} = 0$$

This reduces to

$$(8 - \omega^2) + j6\omega + K(\cos \omega\tau - j\sin \omega\tau) = 0$$

The real part yields

$$K\cos \omega\tau + 8 - \omega^2 = 0$$

As a result, the intersection with the imaginary axis satisfies

$$\tan \omega\tau = \frac{6\omega}{\omega^2 - 8}$$

Clearly, there is an infinite number of intersections. We are interested in the first solution, which is obtained iteratively at

$$\omega = 1.925$$

The value of the gain $K$ for this value is

$$K = \frac{6\omega}{\sin \omega\tau} = 12.31$$

The location of the breakaway points is obtained by maximizing

$$K = -(\sigma + 2)(\sigma + 4)e^{\tau\sigma}$$

Thus we obtain

$$\frac{\partial K}{\partial \sigma} = (\sigma + 2)e^{\tau\sigma} + (\sigma + 4)e^{\tau\sigma} + (\sigma + 2)(\sigma + 4)\tau e^{\tau\sigma} = 0$$

As a result, for $\tau = 1$ we get

$$\sigma^2 + 8\sigma + 14 = 0$$

Thus

$$\sigma_b = \frac{-8 \pm \sqrt{64 - 56}}{2} = -4 \pm \sqrt{2}$$

Since the breakaway point must lie between $-2$ and $-4$, we deduce that

$$\sigma_b = -4 + \sqrt{2} = -2.58$$

The root locus for this example is shown in Figure 7.16.

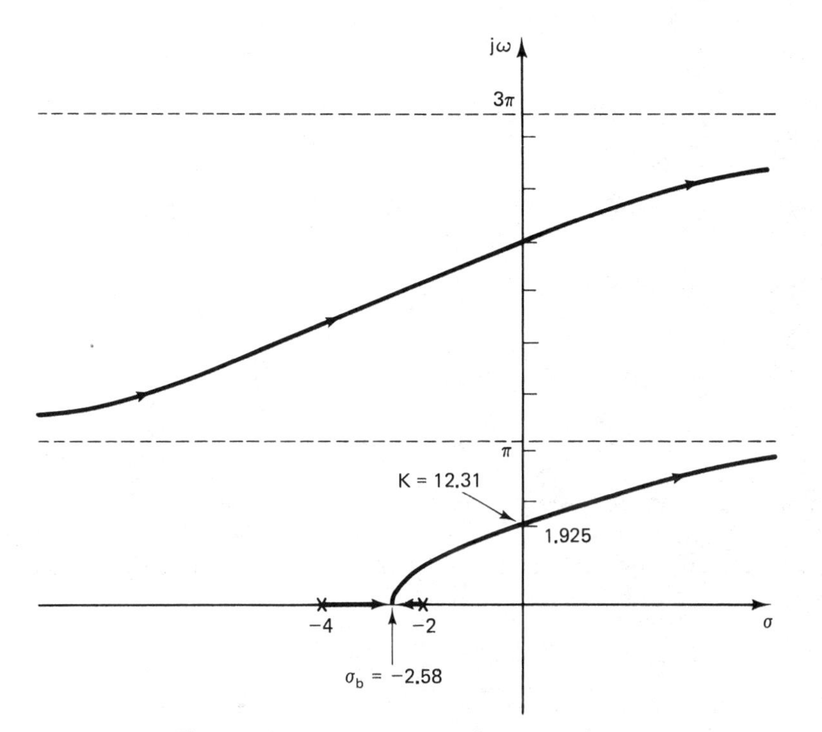


**Figure 7.16.** Root Locus for Example 7.6

# *7.5 Transient Response from the Root Locus*

Once the basic features of the root locus for a control system have been established, the designer can obtain useful information about the system's performance. The locus determines the location of the closed-loop poles of the system for a given value of the gain $K$. Knowledge of these poles is then utilized to determine the transient response of the system to various input functions. It should be emphasized that an exact determination of the poles still requires solving the complex-valued characteristic equation, which is an iterative process. The root-locus sketch, however, provides us with guidelines as to the nature and approximate location of the closed-loop poles. In the next example we focus our attention on the performance of the system when the gain $K$ results in the closed-loop poles being restricted to the real axis.

EXAMPLE 7.7

Consider the unity feedback control system with loop transfer function given by

$$\tilde{G}(s) = \frac{1}{(s+4)(s+8)(s+12)(s+16)}$$

There are four poles at

$$\begin{aligned} p_1 &= -4 \\ p_2 &= -8 \\ p_3 &= -12 \\ p_4 &= -16 \end{aligned}$$

There are four asymptotes intersecting at $-10$ with angles $45°$, $135°$, $225°$, and $-45°$. Application of the breakaway rule results in finding two breakaway points at $\sigma_{b_1} = -5.5$ and at $\sigma_b = -14.47$. The gain at the breakaway points is found to be

$$K_b = 256$$

The Routh criterion is used to find the critical value of $K$ for stability as

$$K_{\text{crit}} = 32256$$

The intersection with the imaginary axis occurs at

$$\omega_{\text{osc}} = 8.94$$

The root locus is shown in Figure 7.17.

Consider the case when the gain $K$ is given by

$$K = 200$$

From inspection of the root locus, we can assert that the characteristic equation will have four real roots, since $K < K_b$. We can also assert that the roots are in the following neighborhoods:

Root 1: in neighborhood $(-4$ to $-5.5)$
Root 2: in neighborhood $(-5.5$ to $-8)$
Root 3: in neighborhood $(-12$ to $-14.47)$
Root 4: in neighborhood $(-14.47$ to $-16)$

Thus the root locus provides us with rough information about the root location. To be exact we have to solve the characteristic equation for this case iteratively:

$$s^4 + 40s^3 + 560s^2 + 3200s + 6344 = 0$$

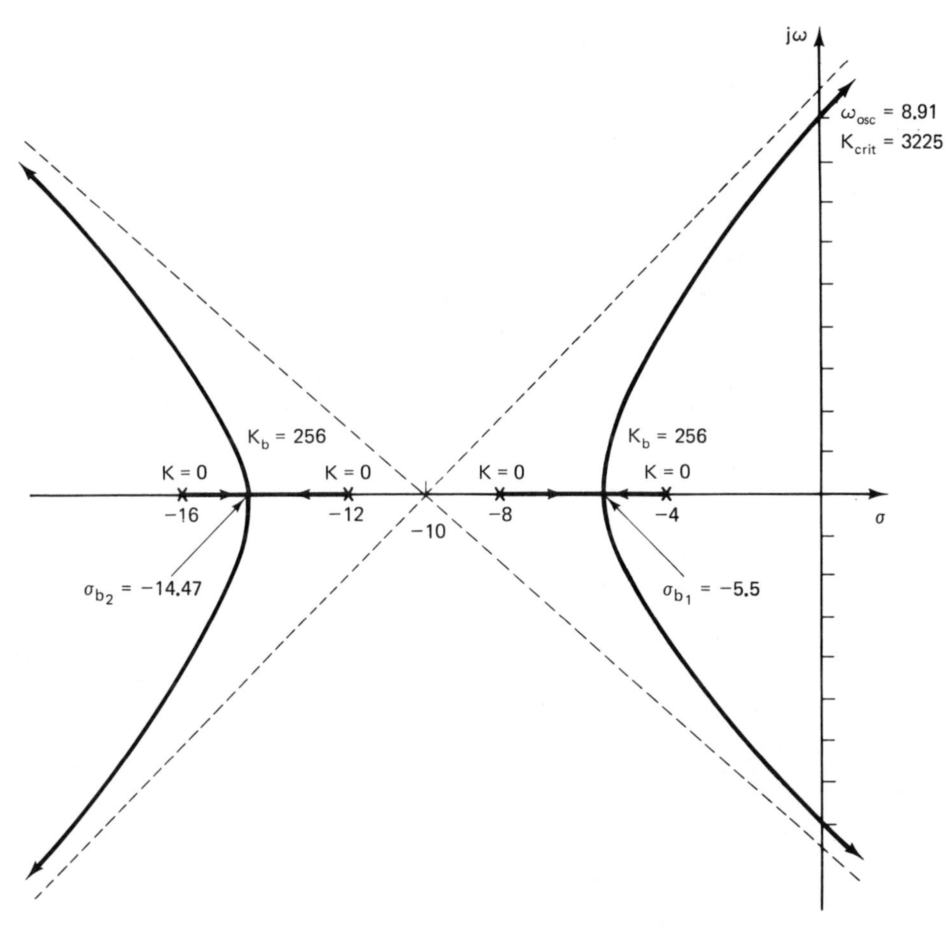


**Figure 7.17.** Root Locus for Example 7.7

The roots are found to be

$$r_1 = -4.76$$
$$r_2 = -6.46$$
$$r_3 = -13.54$$
$$r_4 = -15.24$$

We can thus conclude that the closed-loop transfer function is given by

$$\frac{C(s)}{R(s)} = \frac{K}{(s - r_1)(s - r_2)(s - r_3)(s - r_4)}$$

Performing a partial fraction expansion, we have

$$\frac{C(s)}{R(s)} = K\left(\frac{A_1}{s - r_1} + \frac{A_2}{s - r_2} + \frac{A_3}{s - r_3} + \frac{A_4}{s - r_4}\right)$$

$$A_1 = \left[(r_1 - r_2)(r_1 - r_3)(r_1 - r_4)\right]^{-1} = 0.006373$$

$$A_2 = \left[(r_2 - r_1)(r_2 - r_3)(r_2 - r_4)\right]^{-1} = -0.009443$$

$$A_3 = \left[(r_3 - r_1)(r_3 - r_2)(r_3 - r_4)\right]^{-1} = 0.009443$$

$$A_4 = \left[(r_4 - r_1)(r_4 - r_2)(r_4 - r_3)\right]^{-1} = -0.006373$$

Thus

$$\frac{C(s)}{R(s)} = 200\left(\frac{0.006373}{s + 4.76} - \frac{0.009443}{s + 6.46} + \frac{0.009443}{s + 13.54} - \frac{0.006373}{s + 15.24}\right)$$

The impulse response is obtained for

$$R(s) = 1$$

Taking the inverse Laplace transform, we get

$$c(t) = 200\left[0.006373\left(e^{-4.76t} - e^{-15.24t}\right) + 0.009443\left(e^{-13.54t} - e^{-6.46t}\right)\right]$$

Note that the transient terms due to the roots of $r_3 = -13.54$ and $r_4 = -15.24$ decay in a much shorter time than the time of transient terms due to the dominant roots, $r_1 = -4.76$ and $r_2 = -6.46$.

Let us observe here that for $256 < K < 32{,}256$ in Example 7.7, the closed-loop poles are two pairs of complex-conjugate roots. The characteristic equation can thus be factored as the product of two second-order polynomials.

$$P(s) = P_A(s)P_B(s)$$

where

$$P_A(s) = s^2 + 2\zeta_A\omega_{n_A}s + \omega_{n_A}^2$$

$$P_B(s) = s^2 + 2\zeta_B\omega_{n_B}s + \omega_{n_B}^2$$

The roots of polynomial $P_A$ are given by

$$r_1 = \sigma_A + j\omega_{d_A}$$

$$r_2 = \sigma_A - j\omega_{d_A}$$

where

$$\sigma_A = -\zeta_A\omega_{n_A}$$

$$\omega_{d_A} = \omega_{n_A}\sqrt{1 - \zeta_A^2}$$

Similarly, the roots of polynomial $P_B$ are given by

$$r_3 = \sigma_B + j\omega_{d_B}$$

$$r_4 = \sigma_B - j\omega_{d_B}$$

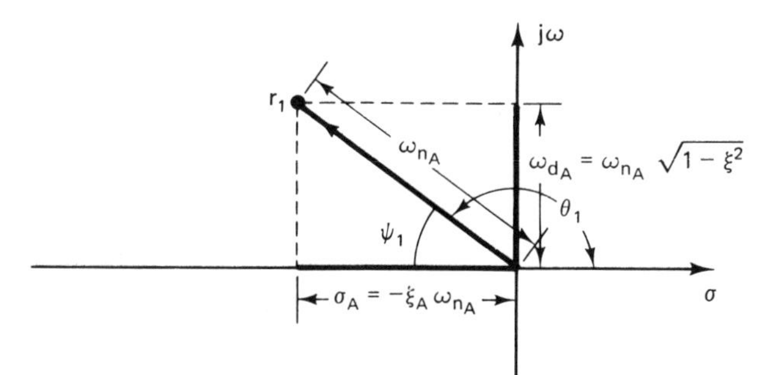


**Figure 7.18.** Geometry in the *s*-Plane for a Complex Root of the Characteristic Equation

where

$$\sigma_B = -\zeta_B \omega_{n_B}$$

$$\omega_{d_B} = \omega_{n_B}\sqrt{1 - \zeta_B^2}$$

Note that the damping factors $\zeta_A$ and $\zeta_B$ are less than 1 for an underdamped system.

The root $r_1$ is shown in the *s*-plane of Figure 7.18. In complex form we have

$$r_1 = |r_1| \angle \theta_1$$

From the geometry of the figure we have

$$|r_1| = \omega_{n_A}$$

$$\theta_1 = 180° - \psi_1$$

$$\cos \psi_1 = \zeta_A$$

It is thus evident that if $\zeta_A$ is specified, the direction of the vector $r_1$ is determined. The exact location of $r_1$ is at the intersection of the vector with the root locus. This will also enable us to determine $K$ and the other roots.

EXAMPLE 7.8

Assume that it is desired to have a damping ratio $\zeta_A$ for the system of Example 7.7 given by

$$\zeta_A = 0.8$$

Determine the necessary gain $K$, find all the roots of the characteristic equation, and determine the impulse response of the system.

SOLUTION

As we are given $\zeta_A$, we can determine the angle $\theta_1$ as

$$\theta_1 = 180° - \cos^{-1} 0.8 = 143.13°$$

The characteristic equation is given by

$$P(s) = s^4 + 40s^3 + 560s^2 + 3200s + 6144 + K = 0$$

Substituting $s = \omega_{n_A} \angle \theta_1$, we get

$$\omega_{n_A}^4 \angle 4\theta_1 + 40\omega_{n_A}^3 \angle 3\theta_1 + 560\omega_{n_A}^2 \angle 2\theta_1 + 3200\omega_{n_A} \angle \theta_1 + 6144 + K = 0$$

Separating real and imaginary parts, we have first, for the real part,

$$\omega_{n_A}^4 \cos 4\theta_1 + 40\omega_{n_A}^3 \cos 3\theta_1 + 560\omega_{n_A}^2 \cos 2\theta_1 + 3200\omega_{n_A} \cos \theta_1 + 6144 + K = 0$$

The imaginary part gives

$$\omega_{n_A}^4 \sin 4\theta_1 + 40\omega_{n_A}^3 \sin 3\theta_1 + 560\omega_{n_A}^2 \sin 2\theta_1 + 3200\omega_{n_A} \sin \theta_1 = 0$$

This reduces to a third-order equation,

$$-\omega_{n_A}^3(0.5376) + 37.44\omega_{n_A}^2 - 537.6\omega_{n_A} + 1920.00 = 0$$

There is a solution at

$$\omega_{n_A} = 5.536$$

As a result, we have by substitution in the equation for the real part the required value of the gain obtained as

$$K = 1625.8$$

The second-order polynomial $P_A(s)$ is thus given by

$$\begin{aligned} P_A(s) &= s^2 + 2\zeta_A\omega_{n_A}s + \omega_{n_A}^2 \\ &= s^2 + 2(0.8)(5.536)s + (5.536)^2 \\ &= s^2 + 8.858s + 30.65 \end{aligned}$$

The characteristic equation with the obtained value of $K$ is given by

$$P(s) = s^4 + 40s^3 + 560s^2 + 3200s + 7769.8$$

We can now obtain the polynomial $P_B(s)$ as

$$\begin{aligned} P_B(s) &= \frac{P(s)}{P_A(s)} \\ &= s^2 + 31.1417s + 253.49 \end{aligned}$$

Thus

$$\begin{aligned} \omega_{n_B} &= \sqrt{253.49} = 15.92 \\ \zeta_B &= 0.978 \end{aligned}$$

We can now obtain

$$\begin{aligned} \sigma_B &= -\zeta_B\omega_{n_B} = -15.57 \\ \omega_{d_B} &= \omega_{n_B}\sqrt{1 - \zeta_B^2} = 3.32 \end{aligned}$$

The closed-loop transfer function is given by

$$\frac{C(s)}{R(s)} = \frac{K}{P(s)}$$

In factored form, we have

$$\frac{C(s)}{R(s)} = \frac{K}{P_A(s)P_B(s)}$$

Performing a partial fraction expansion, we obtain

$$\frac{C(s)}{R(s)} = K\left(\frac{a + bs}{s^2 + 2\zeta_A\omega_{n_A}s + \omega_{n_A}^2} + \frac{e + ds}{s^2 + 2\zeta_B\omega_{n_B}s + \omega_{n_B}^2}\right)$$

This is of the form discussed in Chapter 5. We thus compute the following parameters:

$$\begin{aligned}
D &= \left(\omega_A^2 - \omega_B^2\right)^2 + 4\omega_A\omega_B(\zeta_A\omega_A - \zeta_B\omega_B)(\zeta_A\omega_B - \zeta_B\omega_A) \\
&= \left[(5.536)^2 - (15.92)^2\right]^2 + 4(5.536)(15.92) \\
&\quad \times [(0.8)(5.536) - (0.978)(15.92)][(0.8)(15.92) - (0.978)(5.536)] \\
&= 20{,}889.62 \\
\cos\phi_A &= \frac{\left(\omega_B^2 - \omega_A^2\right) + 2\zeta_A\omega_A(\zeta_A\omega_A - \zeta_B\omega_B)}{\sqrt{D}} \\
&= 0.859 \\
\sin\phi_A &= \frac{2\omega_A\sqrt{1 - \zeta_A^2}\,(\zeta_A\omega_A - \zeta_B\omega_B)}{\sqrt{D}} \\
&= -0.512
\end{aligned}$$

Thus

$$\begin{aligned}
\phi_A &= -30.8° \\
\cos\phi_B &= \frac{\left(\omega_A^2 - \omega_B^2\right) + 2\zeta_B\omega_B(\zeta_B\omega_B - \zeta_A\omega_A)}{\sqrt{D}} \\
&= 0.859 \\
\sin\phi_B &= \frac{2\omega_B\sqrt{1 - \zeta_B^2}\,(\zeta_B\omega_B - \zeta_A\omega_A)}{\sqrt{D}} \\
&= 0.512
\end{aligned}$$

Thus

$$\phi_B = 30.8°$$

We can write the complete impulse response as

$$c(t) = \frac{K}{\sqrt{D}}[c_A(t) + c_B(t)]$$

$$c(t) = 11.249[c_A(t) + c_B(t)]$$

with

$$c_A(t) = 0.301e^{-4.429t}\sin(3.3218t - 30.8°)$$
$$c_B(t) = 0.301e^{-15.57t}\sin(3.3218t + 30.8°)$$

Figure 7.19 shows the root locus of Example 7.7 labeled to illustrate the required steps for this example.

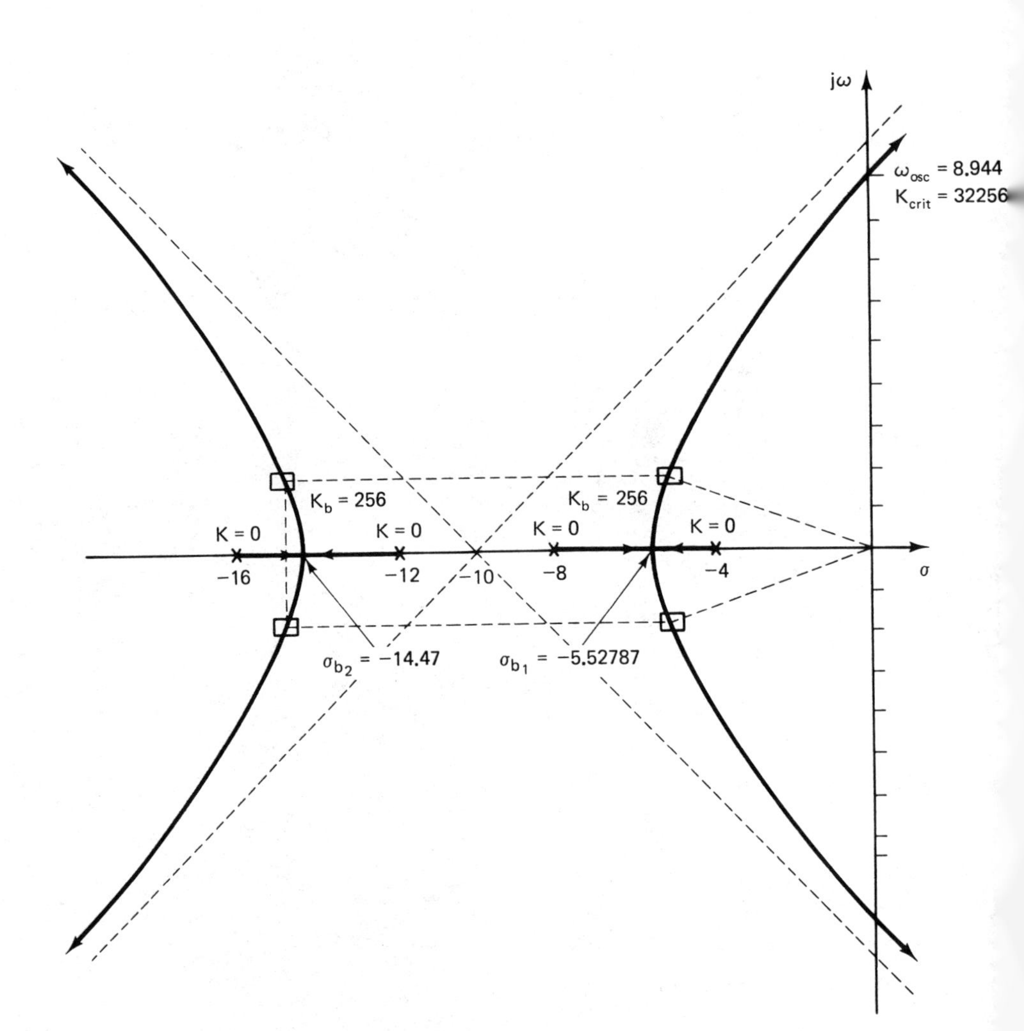


**Figure 7.19.** Root Locus for Example 7.7 with the Details Required for Example 7.8

# *Some Solved Problems*

## PROBLEM 7A-1

It is required to sketch the root locus for a system with unity feedback and a forward transfer function given by

$$G(s) = \frac{K}{s(s+a)(s+b)}$$

for $K$ varying from zero to infinity, with

$$a = 1 \qquad \text{and} \qquad b = 5$$

## SOLUTION

With $K$ as the varying parameter, a straightforward application of the rules results in the required locus. We have

$$G(s) = \frac{K}{s(s+a)(s+b)}$$

There are three poles at

$$p_1 = 0 \qquad p_2 = -a \qquad p_3 = -b$$
$$\qquad\qquad\quad = -1 \qquad\quad = -5$$

There are no finite zeros. We thus have three separate loci which terminate at infinity. The center of asymptotes is at

$$\sigma_A = \frac{-(a+b)}{3} = -2$$

The angles of the asymptotes are

$$\begin{aligned} \phi_A &= (2q+1)60^\circ \\ &= 60^\circ, 180^\circ, 300^\circ \end{aligned}$$

We apply the Routh–Hurwitz criterion to find the intersection with the imaginary axis:

$$\begin{aligned} s(s+1)(s+5) + K &= 0 \\ s^3 + 6s^2 + 5s + K &= 0 \end{aligned}$$

The Routh array is thus given by

$$\begin{array}{c|cc} s^3 & 1 & 5 \\ s^2 & 6 & K \\ s^1 & \dfrac{30-K}{6} & 0 \\ s^0 & K & 0 \end{array}$$

The value of $K$ at the intersection with the imaginary axis is obtained from the third row as

$$K = 30$$

The roots at the intersection with the imaginary axis are obtained from

$$6s^2 + 30 = 0$$

or

$$s = \pm j\sqrt{5}$$

Alternatively, we solve the characteristic equation with $K = 30$:

$$\begin{aligned} s^3 + 6s^2 + 5s + 30 &= 0 \\ (s + 6)(s^2 + 5) &= 0 \end{aligned}$$

Thus the roots are

$$\begin{aligned} s_{\substack{1\\2}} &= \pm j\sqrt{5} \\ s_3 &= -6 \end{aligned}$$

We now obtain the breakaway point:

$$\begin{aligned} K &= \frac{-1}{G(\sigma)} \\ &= -(\sigma)(\sigma + 1)(\sigma + 5) = -(\sigma^3 + 6\sigma^2 + 5\sigma + 30) \end{aligned}$$

Differentiating, we obtain

$$\begin{aligned} \frac{\partial K}{\partial \sigma} &= -(3\sigma^2 + 12\sigma + 5) = 0 \\ \sigma_b &= \frac{-12 \pm \sqrt{144 - 60}}{6} = \frac{-12 \pm 9.17}{6} \\ &= -0.47 \quad \text{or} \quad -3.53 \end{aligned}$$

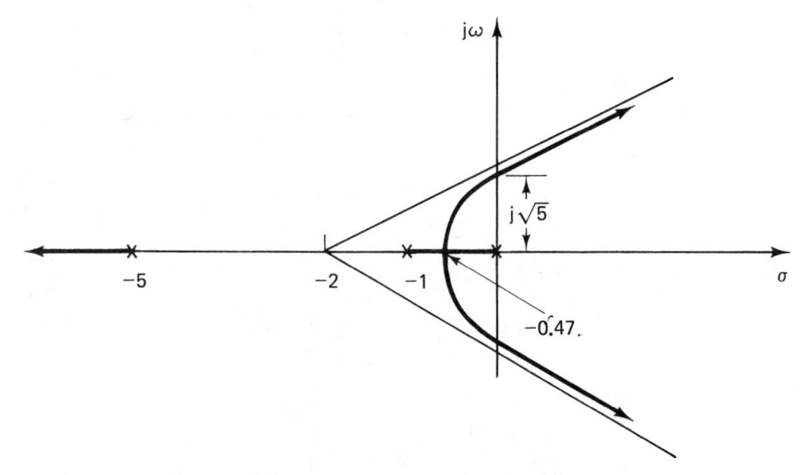


**Figure 7.20.** Root Locus for Problem 7A-1

Since the breakaway point must be between 0 and $-1$; hence we have

$$\sigma_b = -0.47$$

The complete root locus is shown in Figure 7.20.

PROBLEM 7A-2

Sketch the root locus for the system of Problem 7A-1 assuming that parameter $a$ varies from zero to infinity, with

$$K = 4 \qquad \text{and} \qquad b = 3.75$$

SOLUTION

The characteristic equation is

$$K + s(s+a)(s+b) = 0$$

We write it so that parameter $a$ appears as a gain factor. Expanding, we get

$$\left[K + s^2(s+b)\right] + as(s+b) = 0$$

The required form is obtained by dividing through the term in brackets:

$$1 + a\frac{s(s+b)}{s^3 + bs^2 + K} = 0$$

Thus the corresponding open-loop transfer function is

$$\tilde{G}(s) = \frac{as(s+b)}{s^3 + bs^2 + K}$$

Taking

$$K = 4 \qquad b = 3.75$$

we get

$$\tilde{G}(s) = \frac{as(s+3.75)}{s^3 + 3.75s^2 + 4}$$

Thus we have the following zeros:

$$z_1 = 0 \qquad z_2 = -3.75$$

Note that we can factor the denominator as

$$s^3 + 3.75s^2 + 4 = (s+4)(s^2 - 0.25s + 1)$$

The poles are given by

$$p_1 = -4$$

$$p_{2,3} = \frac{0.25 \pm \sqrt{(0.25)^2 - 4}}{2}$$

$$= 0.125 \pm j0.992$$

Since there is one more pole than finite zeros, we have only one locus terminating at infinity. The angle of the asymptote is 180°. The Routh–Hurwitz criterion provides the intersection with the imaginary axis on the basis of the characteristic equation

$$s^3 + bs^2 + K + as(s + b) = 0$$
$$s^3 + (3.75 + a)s^2 + 3.75as + 4 = 0$$

The Routh array is thus

$$\begin{array}{c|cc} s^3 & 1 & 3.75a \\ s^2 & 3.75 + a & 4 \\ s^1 & \dfrac{3.75a(3.75 + a) - 4}{3.75 + a} & \\ s^0 & 4 & \end{array}$$

The intersection with the imaginary axis occurs for

$$3.75a(3.75 + a) - 4 = 0$$

or

$$3.75a^2 + (3.75)^2 a - 4 = 0$$
$$a = \frac{-(3.75)^2 \pm \sqrt{(3.75)^4 + 4(3.75)}}{2(3.75)}$$
$$= 0.0698 \quad \text{or} \quad -3.8198$$

As we are interested in positive values of $(a)$, the crossover occurs at

$$a = 0.0698$$

The points of intersection on the imaginary axis are obtained from

$$(3.75 + 0.0698)s^2 + 4 = 0$$

Thus

$$s = j\omega$$
$$j\omega = \pm j1.023$$

The departure angle from the complex pole $p_2 = 0.125 + j0.992$ is calculated from

$$\theta_D = 180° + \arg(\tilde{G}')$$
$$\tilde{G}' = \left.\frac{s(s + 3.75)}{(s + 4)(s - p_3)}\right|_{p_2}$$
$$= \frac{p_2(p_2 + 3.75)}{(p_2 + 4)(p_2 - p_3)}$$
$$= \frac{(0.125 + j0.992)(3.875 + j0.992)}{(4.125 + j0.992)(2j \times 0.992)}$$
$$\arg(\tilde{G}') = (82.82°) + (14.36°) - (13.52) - (90°)$$
$$= -6.343°$$

Thus

$$\theta_D = 173.65$$

For the break-in point we have the characteristic equation written in the following form:

$$a = -\frac{s^3 + 3.75s^2 + 4}{s(s + 3.75)}$$

Thus

$$a(\sigma) = \frac{-(\sigma^3 + 3.75\sigma^2 + 4)}{\sigma(\sigma + 3.75)}$$

We will have to satisfy

$$f(\sigma) = \frac{\partial a}{\partial \sigma} = 0$$

Thus

$$f(\sigma) = \sigma(\sigma + 3.75)(3\sigma^2 + 7.5\sigma) - (\sigma^3 + 3.75\sigma^2 + 4)(2\sigma + 3.75) = 0$$

This reduces to

$$f(\sigma) = \sigma^4 + 7.5\sigma^3 + 14.0625\sigma^2 - 8\sigma - 15 = 0$$

We now have to solve this fourth-order equation in $\sigma$. We will use Newton–Raphson method. We will assume an initial guess of $\sigma^{(0)} = -1$. The iterations proceed as follows:

$$f(\sigma^{(0)}) = -15.4375$$
$$f'(\sigma) = 4\sigma^3 + 22.5\sigma^2 + 28.125\sigma - 8$$

Thus

$$\Delta\sigma^{(0)} = \frac{-f(\sigma^{(0)})}{f'(\sigma^{(0)})} = \frac{15.4375}{-17.63} = -0.87589$$
$$\sigma^{(1)} = -1.87589$$
$$\Delta\sigma^{(1)} = \frac{-12.36671}{-7.98753} = 1.54825$$
$$\sigma^{(2)} = -0.32764$$
$$\Delta\sigma^{(2)} = \frac{11.12163}{-14.94017} = -0.74441$$
$$\sigma^{(3)} = -1.07205$$
$$\Delta\sigma^{(3)} = \frac{-1.81839}{-17.22077} = 0.10559$$
$$\sigma^{(4)} = -0.96645$$
$$\Delta\sigma^{(4)} = \frac{0.03136}{-17.77657} = -0.00176$$
$$\sigma^{(5)} = -0.96822$$
$$\Delta\sigma^{(5)} = \frac{0.0001}{-38.86174} = 1.670 \times 10^{-7}$$

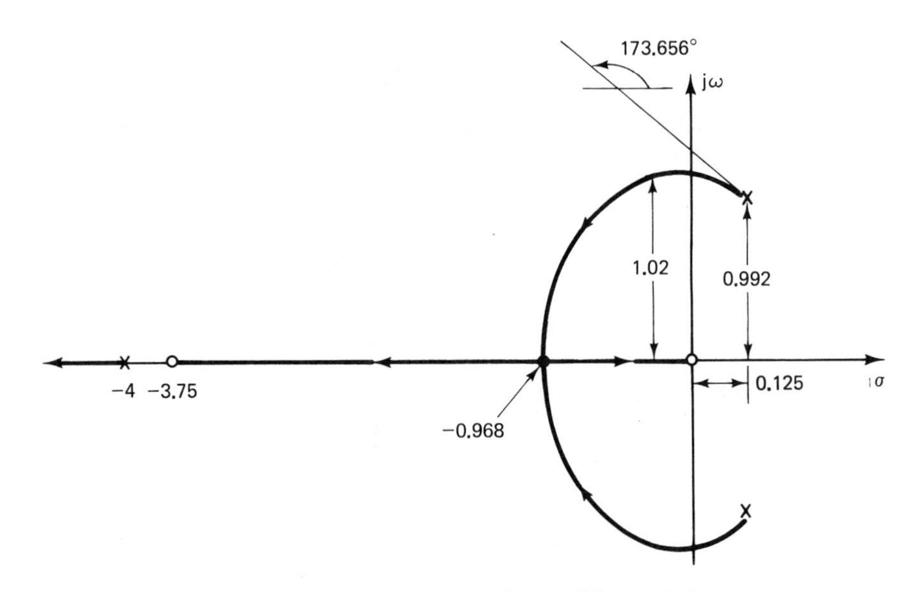


**Figure 7.21.** Root Locus for Problem 7A-2

Thus we conclude that a reasonable approximation is

$$\sigma_{b_1} = -0.96822$$

To find the second root, we factor $f(\sigma)$ to

$$f(\sigma) = (\sigma + 0.96822)(\sigma^3 + 6.5032\sigma^2 + 7.7666\sigma - 15.5192)$$

Again by an iterative method we get a root to the third-order polynomial at

$$\sigma_{b_2} = 1.01045$$

Again the function $f$ is factored to

$$f(\sigma) = (\sigma + 0.96822)(\sigma - 1.01045)(\sigma^2 + 7.51365\sigma + 15.358168)$$

The second-order polynomial has complex roots and since $\sigma_{b_2}$ is in the right half of the real axis where we established the nonexistence of a locus segment, we conclude that the only break-in point is at

$$\sigma_b = -0.96822$$

The complete root locus is shown in Figure 7.21.

PROBLEM 7A-3

Consider the system shown in Figure 7.22. Determine the gain K and the velocity feedback coefficient ($b$) so that the closed-loop poles are $s = -1 \pm j\sqrt{3}$. Use the value of $b$ so chosen to sketch the root-locus plot.

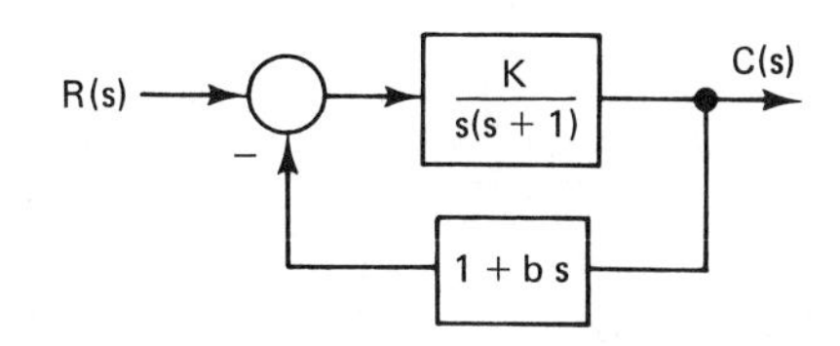


**Figure 7.22.** System for Problem 7A-3

SOLUTION

The closed-loop transfer function is given by

$$\frac{C(s)}{R(s)} = \frac{K}{s(s+1) + K(1+bs)}$$

Thus the characteristic equation is

$$s^2 + (1 + Kb)s + K = 0$$

The closed-loop poles are thus

$$s_{1,2} = \frac{-(1+Kb) \pm \sqrt{(1+Kb)^2 - 4K}}{2}$$

Thus for

$$s_{1,2} = -1 \pm j\sqrt{3}$$

we require that

$$1 + Kb = 2$$

Thus

$$Kb = 1$$

Also,

$$3 = \frac{-\left[(1+Kb)^2 - 4K\right]}{4}$$

$$-12 = 4 - 4K$$

Thus

$$K = 4$$
$$b = 0.25$$

These are the required values.

The characteristic equation with $K$ allowed to vary is

$$s^2 + (1 + 0.25K)s + K = 0$$

or

$$s^2 + s + K(0.25s + 1) = 0$$

This reduces to

$$1 + \frac{K(1 + 0.25s)}{s(s + 1)} = 0$$

As a result, we deal with

$$\tilde{G}(s) = \frac{1 + 0.25s}{s(s + 1)}$$

There is one zero at

$$z_1 = -4$$

Two poles are at

$$p_1 = 0$$
$$p_2 = -1$$

The Routh array is given by

$$\begin{array}{c|cc} s^2 & 1 & K \\ s^1 & 1 + 0.25K & 0 \\ s^0 & K & \end{array}$$

There is no intersection with the imaginary axis as no element of the first column changes sign.

Breakaway points are obtained using the form

$$K(\sigma) = \frac{-\sigma(\sigma + 1)}{1 + 0.25\sigma}$$

Application of the condition

$$\frac{\partial K}{\partial \sigma} = 0$$

results in

$$(1 + 0.25\sigma)(2\sigma + 1) - 0.25\sigma(\sigma + 1) = 0$$

This reduces to

$$0.25\sigma^2 + 2\sigma + 1 = 0$$

The roots are

$$\sigma_b = \frac{-2 \pm \sqrt{4 - 1}}{0.5} = -4 \pm 2\sqrt{3}$$
$$= -0.54 \quad \text{and} \quad -7.46$$

Thus there are two breakaway and break-in points. The root locus is shown in Figure 7.23.

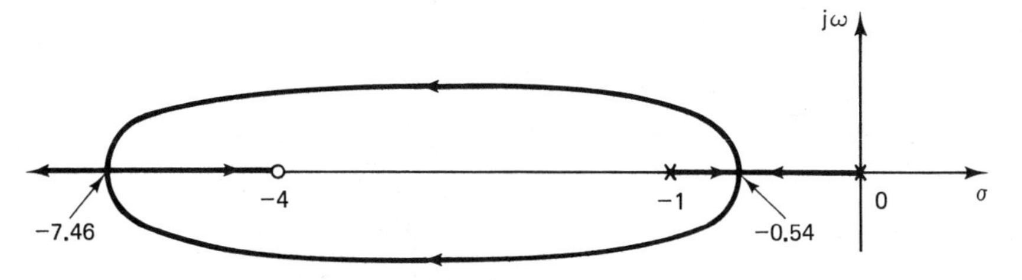


**Figure 7.23.** Root Locus for Problem 7A-3

PROBLEM 7A-4

Sketch the root locus for a system with an open-loop transfer function given by

$$G(s) = \frac{K}{s^2(s+2)}$$

Show that the system is unstable for all positive values of the gain $K$. This system can be stabilized by adding a zero on the negative real axis to modify $G(s)$ to $G_1(s)$ given by

$$G_1(s) = \frac{K(s+a)}{s^2(s+2)}$$

Find the limits on $a$ for stability. Complete the block diagram shown in Figure 7.24 to realize the stabilizing effect.

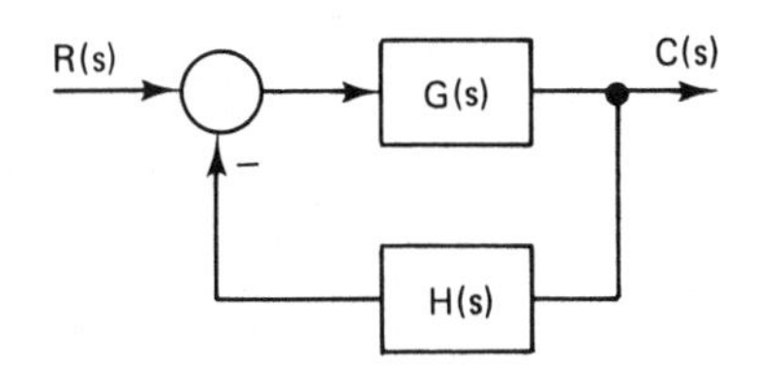


**Figure 7.24.** System for Problem 7A-4

SOLUTION

For the first case there are three poles at

$$p_1 = 0$$
$$p_2 = 0$$
$$p_3 = -2$$

The asymptotes intersect at

$$\sigma_A = \frac{-2}{3}$$

The angles of the asymptotes are

$$\phi_A = 60°, 180°, 300°$$

The characteristic equation of the original system is

$$s^3 + 2s^2 + K = 0$$

The Routh array is given by

$$\begin{array}{c|cc} s^3 & 1 & 0 \\ s^2 & 2 & K \\ s^1 & \frac{-K}{2} & \end{array}$$

For all positive $K$, the system is unstable. The root locus is sketched in Figure 7.25.

With the addition of a zero, we have

$$z_1 = -a$$
$$p_1 = 0$$
$$p_2 = 0$$
$$p_3 = -2$$

Now there are two asymptotes, intersecting at

$$\sigma_A = \frac{-2 + a}{2}$$

The asymptotes' angles are

$$\phi_A = 90° \text{ and } 270°$$

For $a < 2$, the locus remains in the left-hand side of the $s$-plane.

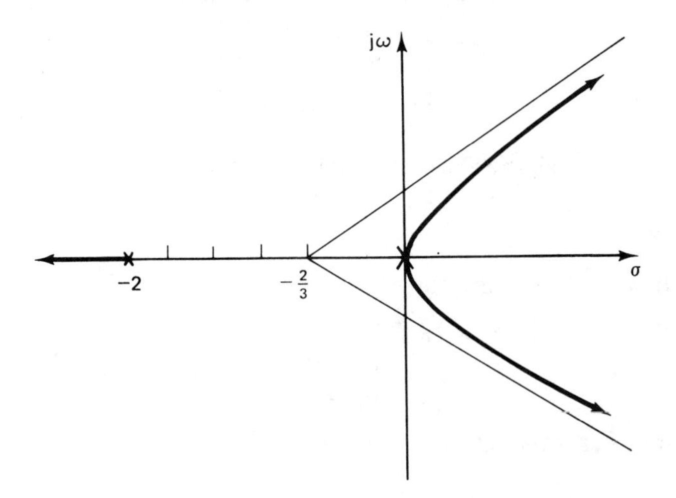


**Figure 7.25.** Root Locus for Problem 7A-4

The characteristic equation of the system is

$$s^3 + 2s^2 + Ks + Ka = 0$$

The Routh array is thus given by

$$\begin{array}{c|cc} s^3 & 1 & K \\ s^2 & 2 & Ka \\ s^1 & \dfrac{2K - Ka}{2} & \end{array}$$

For stability we require that

$$2K - Ka > 0$$

Thus

$$a < 2$$

The root locus for the stabilized system is shown in Figure 7.26.

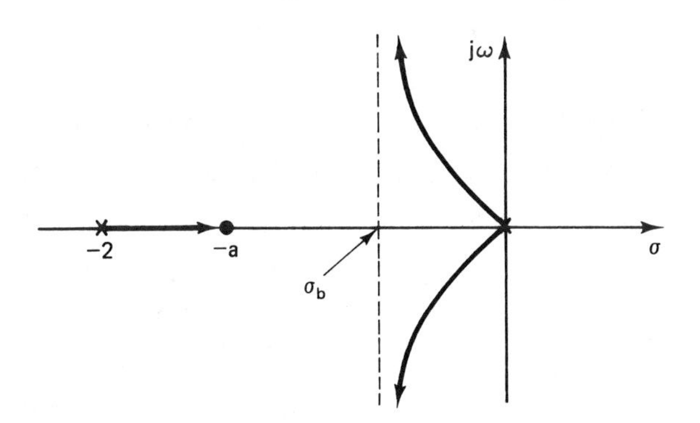

**Figure 7.26.** Root Locus for the Stabilized System of Problem 7A-4

The block diagram shown in Figure 7.24 gives a characteristic equation

$$1 + G(s)H(s) = 0$$

Thus for the two systems to match,

$$G_1(s) = G(s)H(s)$$

By comparison we conclude that

$$H(s) = s + a$$

PROBLEM 7A-5

Sketch the root locus for the system with unity feedback and open-loop (forward) transfer function given by

$$G(s) = \frac{(s + 4)(s + 40)}{s^3(s + 200)(s + 900)}$$

SOLUTION

The poles are given by

$$\begin{aligned} p_1 &= p_2 = p_3 = 0 \\ p_4 &= -200 \\ p_5 &= -900 \end{aligned}$$

The zeros are given by

$$\begin{aligned} z_1 &= -4 \\ z_2 &= -40 \end{aligned}$$

As a result, the point of intersection of the asymptotes is given by

$$\sigma_A = \frac{-1100 + 44}{3} = -352$$

The angles of the asymptotes are

$$\begin{aligned} \phi_A &= (2q+1)(60^\circ) \\ &= 60^\circ, 180^\circ, 300^\circ \end{aligned}$$

The characteristic equation of the closed-loop system is obtained as

$$s^5 + 1100s^4 + 180{,}000s^3 + Ks^2 + 44Ks + 160K = 0$$

The Routh array is now constructed as

$$\begin{array}{c|ccc} s^5 & 1 & 1.8 \times 10^5 & 44K \\ s^4 & 1100 & K & 160K \\ s^3 & c_1 & b_1 & \\ s^2 & c_2 & 160K & \\ s^1 & c_3 & & \\ s^0 & 160K & & \end{array}$$

In the array we have

$$c_1 = \frac{1.98 \times 10^8 - K}{1100}$$

Thus, for stability, we require that $c_1 \geq 0$; thus

$$K \leq 1.98 \times 10^8$$

We also have

$$b_1 = \frac{(1100)(44K) - 160K}{1100} = 43.8545K$$

$$\begin{aligned} c_2 &= \frac{c_1 K - 1100 b_1}{c_1} \\ &= \frac{-K(K - 1.44936 \times 10^8)}{1.98 \times 10^8 - K} \end{aligned}$$

Thus a second requirement for stability is

$$K \leq 1.44936 \times 10^8$$

Finally, we have

$$c_3 = \frac{c_2 b_1 - 160Kc_1}{c_2}$$

This reduces to

$$c_3 = \frac{-43.999(K^2 - 1.45766 \times 10^8 K + 1.296 \times 10^{14})}{1.44936 \times 10^8 - K}$$

The roots of the quadratic are

$$K = \frac{1.45766 \times 10^8 \pm \sqrt{(1.45766 \times 10^8)^2 - 4(1.296 \times 10^{14})}}{2}$$

$$= 1.44871 \times 10^8 \quad \text{and} \quad 8.94588 \times 10^5$$

Thus

$$c_3 = \frac{44(K - 1.44871 \times 10^8)(K - 8.94588 \times 10^5)}{K - 1.44936 \times 10^8}$$

For $c_3 \geq 0$, with $K \leq 1.44936 \times 10^8$, we require that

$$8.94588 \times 10^5 \leq K \leq 1.44871 \times 10^8$$

There are clearly two intersections with the imaginary axis. This is obtained as follows:

$$c_2 s^2 + 160K = 0$$

$$\frac{K(K - 1.44936 \times 10^8)}{K - 1.98 \times 10^8} s^2 + 160K = 0$$

or

$$\frac{K - 1.44936 \times 10^8}{K - 1.98 \times 10^8} s^2 + 160 = 0$$

For $K = 8.94588 \times 10^5$, we have

$$0.73078 s^2 + 160 = 0$$

Thus

$$s = \pm j14.7967$$

For $K = 1.44871 \times 10^8$, we have

$$1.223 \times 10^{-3} s^2 + 160 = 0$$

Thus

$$s = \pm j361.634$$

To determine the breakpoints, we have

$$-K = \frac{\sigma^3(\sigma + 200)(\sigma + 900)}{(\sigma + 4)(\sigma + 40)}$$
$$= \frac{\sigma^5 + 1100\sigma^4 + 1.8 \times 10^5\sigma^3}{\sigma^2 + 44\sigma + 160}$$

The requirement

$$\frac{\partial K}{\partial \sigma} = 0$$

yields

$$(5\sigma^4 + 4400\sigma^3 + 5.4 \times 10^5\sigma^2)(\sigma^2 + 44\sigma + 160)$$
$$-(2\sigma + 44)(\sigma^5 + 1100\sigma^4 + 1.8 \times 10^5\sigma^3) = 0$$

This reduces to

$$3\sigma^2(\sigma^4 + 792\sigma^3 + 1.0866 \times 10^5\sigma^2 + 5.51467 \times 10^6\sigma + 2.88 \times 10^7) = 0$$

or

$$\sigma^2(\sigma + 5.875)(\sigma + 634.27)(\sigma + 76 - j44.3)(\sigma + 76 + j44.3) = 0$$

The roots of the equation above are

$$\sigma_{b_{1,2}} = 0, 0$$
$$\sigma_{b_3} = -5.875$$
$$\sigma_{b_4} = -634.27$$
$$\sigma_{b_{5,6}} = -76 \pm j44.3$$

Only the first two roots are acceptable since $\sigma_{b_3}$, $\sigma_{b_4}$, $\sigma_{b_5}$, and $\sigma_{b_6}$ are not on

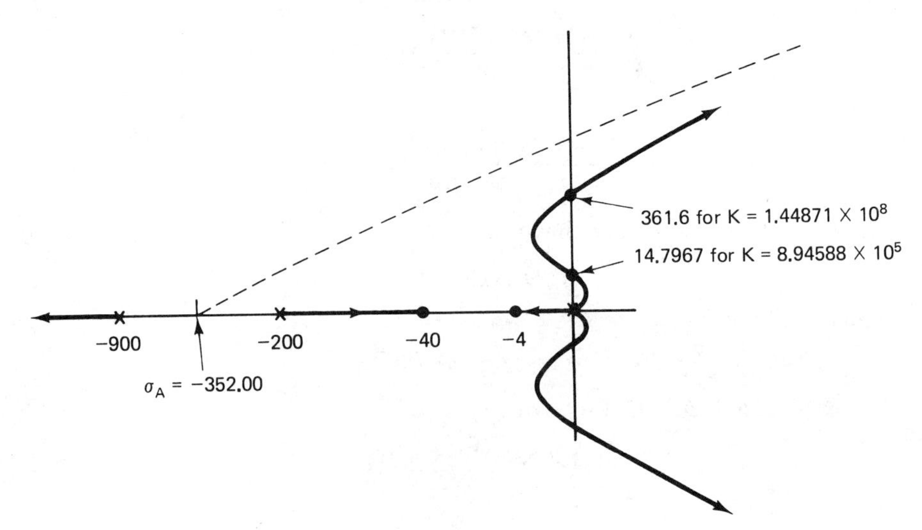


**Figure 7.27.** Root Locus for the System of Problem 7A-5

the locus. Figure 7.27 shows a sketch of the required root locus. The main point of this exercise is that the system is unstable for gains lower than $8.94588 \times 10^5$ and higher than $1.44871 \times 10^8$.

PROBLEM 7A-6

Find the loop transfer function for a unity feedback system with the root locus shown in Figure 7.28. Identify the value of relevant points on the locus.

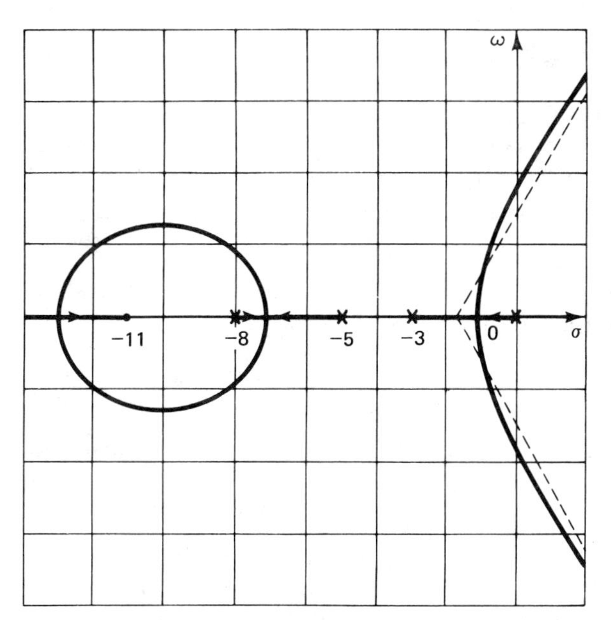


**Figure 7.28.** Root Locus for the System of Problem 7A-6

SOLUTION

By inspection of the locus we obtain

$$p_1 = 0$$
$$p_2 = -3$$
$$p_3 = -5$$
$$p_4 = -8$$
$$z_1 = -11$$

Thus the required transfer function is

$$G(s) = \frac{s + 11}{s(s + 3)(s + 5)(s + 8)}$$

The characteristic equation is

$$s(s + 3)(s + 5)(s + 8) + K(s + 11) = 0$$

This reduces to

$$s^4 + 16s^3 + 79s^2 + (120 + K)s + 11K = 0$$

The Routh array is given by

$$\begin{array}{c|ccc} s^4 & 1 & 79 & 11K \\ s^3 & 16 & 120 + K & \\ s^2 & \dfrac{1144 - K}{16} & 11K & \\ s^1 & c_1 & & \\ s^0 & 11K & & \end{array}$$

Here we have

$$c_1 = \frac{1144 - K}{16}(120 + K) - 16(11K) \Big/ \frac{1144 - K}{16}$$
$$= \frac{(73.59 - K)(1865.59 + K)}{1144 - K}$$

For stability we require that limits on $K$ be satisfied due to the first columns of the Routh array as follows:

$$\begin{array}{c|l} s^2 & K \leq 1144 \\ s^1 & K \leq 73.59 \end{array}$$

Thus the auxiliary polynomial for the $s^2$ row is obtained as

$$\frac{1144 - 73.59}{16}s^2 + 11(73.59) = 0$$
$$s^2 = \frac{-11(73.59)(16)}{1144 - 73.59}$$
$$= -12.10$$

The intersection with the imaginary axis is thus at

$$s = \pm j3.48$$

The breakaway points are obtained from

$$-K = \frac{s^4 + 16s^3 + 79s^2 + 120s}{s + 11}$$

Applying the condition

$$\frac{\partial K}{\partial s} = 0$$

we obtain

$$(s + 11)(4s^3 + 48s^2 + 158s + 120) - (s^4 + 16s^3 + 79s^2 + 120) = 0$$

This reduces to

$$3s^4 + 76s^3 + 607s^2 + 1738s + 1320 = 0$$

This has a solution at

$$s = -1.172$$

As a result, we obtain

$$(s + 1.172)(3s^3 + 72.484s^2 + 522.0494s + 1126.1588) = 0$$

A second root is obtained at

$$s = -12.989$$

Thus our condition is

$$(s + 1.172)(s + 12.989)(3s^2 + 33.5158s + 86.6984) = 0$$

The second-order term gives the factoring

$$3(s + 1.172)(s + 12.99)(s + 4.068)(s + 7.1) = 0$$

As a result, the breakaway points are at

$$\sigma_{b_1} = -1.172 \qquad \sigma_{b_2} = -7.1 \qquad \sigma_{b_3} = -12.99$$

We have thus completely defined the relevant points on the locus.

PROBLEM 7A-7

(a) Consider the feedback control system with unity feedback and a forward transfer function given by

$$G(s) = \frac{e^{-\tau s}}{s^n}$$

where $n$ is an integer and $\tau \leq 1$. Assume that

$$\begin{aligned} s &= r\exp(j\theta) \\ &= \sigma + j\omega \end{aligned}$$

Show that on the root locus for the system we have

$$\tan n\theta = -\tan \omega\tau$$

or

$$n\theta = q\pi - \omega\tau$$

(b) Assume that for the system of part (a), we have $n = 5$. It is required to verify the sketch of the root locus shown in Figure 7.29 using the following procedure:

1. Find the poles of the open-loop transfer function $G(s)$.
2. Find the intersection of the loci with the imaginary axis and the corresponding gains.
3. Find the breakaway point on the real axis and the corresponding gain.
4. Show that for $q = 3$, the maximum angle $\theta$ is given by $\theta_{\max} = 108°$.

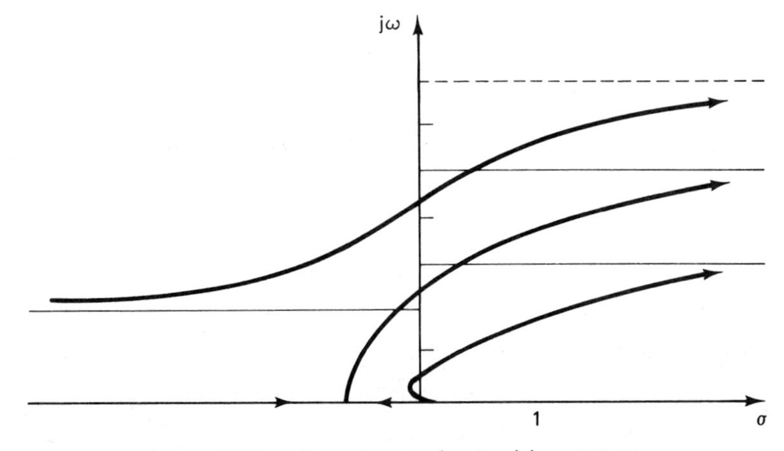


**Figure 7.29.** Root Locus for Problem 7A-7

SOLUTION

(a) The characteristic equation is given by

$$s^n + Ke^{-\tau s} = 0$$

Thus in polar form we have

$$r^n e^{jn\theta} + Ke^{-\tau(\sigma + j\omega)} = 0$$

The real part gives

$$r^n \cos n\theta + Ke^{-\tau\sigma} \cos \omega\tau = 0$$

The imaginary part gives

$$r^n \sin n\theta - Ke^{-\tau\sigma} \sin \omega\tau = 0$$

As a result, we conclude that

$$\tan n\theta = -\tan \omega\tau$$

The solution is

$$n\theta = q\pi - \omega\tau \qquad q = 0, 1, \ldots$$

(b) For $n = 5$, we have

$$G(s) = \frac{e^{-\tau s}}{s^5}$$

1. There are five poles at the origin and one at minus infinity.
2. To find the intersection with the imaginary axis, we have

$$(j\omega)^5 + Ke^{-j\omega\tau} = 0$$

Thus

$$j\omega^5 + K(\cos \omega\tau - j \sin \omega\tau) = 0$$

The real part gives

$$\cos \omega\tau = 0$$

while the imaginary part gives

$$K = \frac{\omega^5}{\sin \omega\tau}$$

For positive $K$, we have the solutions given by

$$\omega\tau = \frac{\pi}{2}, \frac{5\pi}{2}, \ldots$$

$$K = \omega^5$$

3. To find the breakaway points, we write the characteristic equation as

$$-K = s^5 e^{\tau s}$$

The condition

$$-\frac{\partial K}{\partial s} = 0$$

gives

$$s^5 \tau e^{\tau s} + 5s^4 e^{\tau s} = 0$$

or

$$s^4 e^{\tau s}(5 + \tau s) = 0$$

Thus

$$s = \frac{-5}{\tau}$$

As a result,

$$K_{\max} = \frac{5^5}{\tau^5} e^{-5} = \frac{21.0568}{\tau^5}$$

4. Applying the condition of item 1, we have

$$5\theta = 3\pi - \omega\tau$$

Thus

$$\theta = \frac{3\pi - \omega\tau}{5} = 108° - \frac{\omega\tau}{5}$$

The value of $\theta_{\max}$ occurs for $\omega = 0$; thus we obtain

$$\theta_{\max} = 108°$$

## PROBLEM 7A-8

Consider a control system with unity feedback and forward transfer function given by

$$G(s) = \frac{e^{-s}}{s(s+1)}$$

It is required to find the limit of the gain $K$ for stability.

(a) Write the characteristic equation of the system and find the critical value of $K$.

(b) Padé all-pass approximants are commonly used to simulate time delays. A second-order approximant is given by

$$G_{22}(s) \simeq e^{-s} = \frac{s^2 - 6s + 12}{s^2 + 6s + 12}$$

With this we have an approximation to $G(s)$ given by

$$G_b(s) = \frac{G_{22}(s)}{s(s+1)}$$

Use the Routh criterion to find the approximate critical value of $K$.

(c) Repeat part (b) for the Chebyshev approximant

$$G_{22}(s) \simeq \frac{s^2 - 3.67084s + 8.15734}{s^2 + 3.67084s + 8.15734}$$

SOLUTION

(a) The characteristic equation is given by

$$s(s+1) + Ke^{-s} = 0$$

For

$$s = j\omega$$

we obtain

$$j\omega(j\omega + 1) + Ke^{-j\omega} = 0$$

This reduces to

$$-\omega^2 + j\omega + K(\cos\omega - j\sin\omega) = 0$$

The real part yields

$$K\cos\omega = \omega^2$$

The imaginary part yields

$$\omega = K\sin\omega$$

As a result, we have

$$\tan\omega = \frac{1}{\omega}$$

The first solution is obtained as

$$\omega = 0.86 \text{ rad/sec}$$

Consequently, the value of the critical gain is found as

$$K_{\text{crit}} = \frac{\omega}{\sin \omega} = 1.135$$

(b) We have

$$G_b(s) = \frac{s^2 - 6s + 12}{s^2 + 6s + 12} \frac{1}{s(s+1)}$$

Thus the characteristic equation is given by

$$(s^2 + 6s + 12)(s+1)s + K(s^2 - 6s + 12) = 0$$

This reduces to

$$s^4 + 7s^3 + (18 + K)s^2 + 6(2 - K)s + 12K = 0$$

The Routh array is obtained as

$$\begin{array}{c|ccc} s^4 & 1 & 18 + K & 12K \\ s^3 & 7 & 12 - 6K & \\ s^2 & c_2 & 12K & \\ s^1 & c_1 & 0 & \\ s^0 & 12K & & \end{array}$$

The elements $c_2$ and $c_1$ are obtained as

$$c_2 = \frac{7(18 + K) - (12 - 6K)}{7}$$

$$= \frac{114 + 13K}{7}$$

$$c_1 = \left( \frac{114 + 13K}{7}(12 - 6K) - 84K \right) / c_2$$

$$= \frac{-78[K + 15.44][K - 1.136]}{114 + 13K}$$

For

$$c_1 \geq 0 \qquad K \leq 1.136$$

(c) We have

$$G_c(s) = \frac{s^2 - 3.67084s + 8.15734}{s^2 + 3.67084s + 8.15734} \frac{1}{s(s+1)}$$

Thus the characteristic equation is given by

$$(s^2 + s)(8.15734 + 3.67084s + s^2) + K(8.15734 - 3.67084s + s^2) = 0$$

This reduces to

$$s^4 + 4.67s^3 + (11.828 + K)s^2 + (8.157 - 3.67K)s + 8.15734K = 0$$

The Routh array is thus

$$
\begin{array}{c|ccc}
s^4 & 1 & K + 11.828 & 8.157K \\
s^3 & 4.67 & 8.157 - 3.67\text{ K} & \\
s^2 & c_2 & 8.157K & \\
s^1 & c_1 & & \\
s^0 & 8.157K & &
\end{array}
$$

The elements $c_2$ and $c_1$ are obtained as

$$
\begin{aligned}
c_2 &= \frac{4.67(K + 11.828) - (8.157 - 3.67K)}{4.67084} \\
&= 1.7859K + 10.0817 \\
c_2c_1 &= (8.157 - 3.67K)(1.786K + 10.08) - (4.67)(8.157K) \\
&= -6.556(K^2 + 9.235K - 12.545)
\end{aligned}
$$

Thus

$$
\begin{aligned}
K_{\text{crit}} &= \frac{-9.235 \pm \sqrt{(9.23)^2 + (4)(12.5566)}}{2} \\
&= \frac{-9.23486 \pm 11.63879}{2} = 1.2 \quad \text{or} \quad -10.437
\end{aligned}
$$

We conclude that for stability

$$K < 1.2$$

It can be concluded that the limiting value of $K$ obtained in part (a) is the most safe value. In any case a prudent engineer would design the system for values of $K$ less than unity. This conclusion can be drawn from either of the three parts of the problem.

# Problems

PROBLEM 7B-1

Consider the system shown in Figure 7.30 and assume that $a = 2$. Sketch the root locus of the system as $K$ varies from zero to infinity

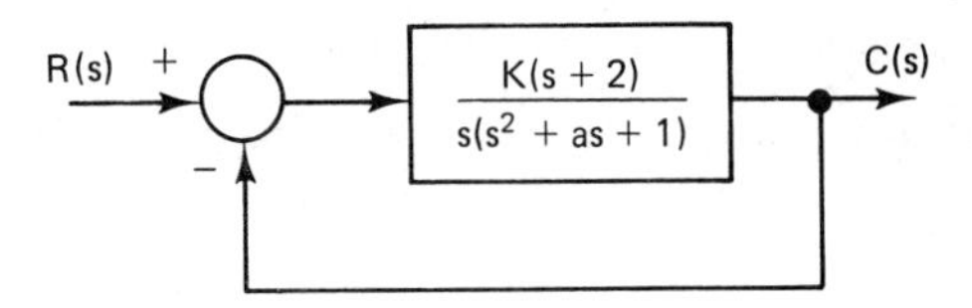


**Figure 7.30.** System for Problems 7B-1 and 7B-2

according to the following:

(a) Find the poles and zeros of the open-loop transfer function.

(b) Find segments on the real axis.

(c) Find the point of intersection of the asymptotes and their angles.

(d) Find the breakaway point.

(e) Find the range of $K$ for stability of the system.

PROBLEM 7B-2

Consider now the same system in Figure 7.30, and assume that $K = 2$. Sketch the root locus of the system as $a$ varies from zero to infinity according to the following:

(a) Convert the characteristic equation to the form

$$1 + a\tilde{G}(s) = 0$$

(b) Find the poles and zeros of $\tilde{G}(s)$.

(c) Find the segments of the root locus on the real axis.

(d) Find the intersection of the loci with the imaginary axis.

(e) Find the departure angle from the complex poles.

(f) Find the range of $a$ for stability of the system.

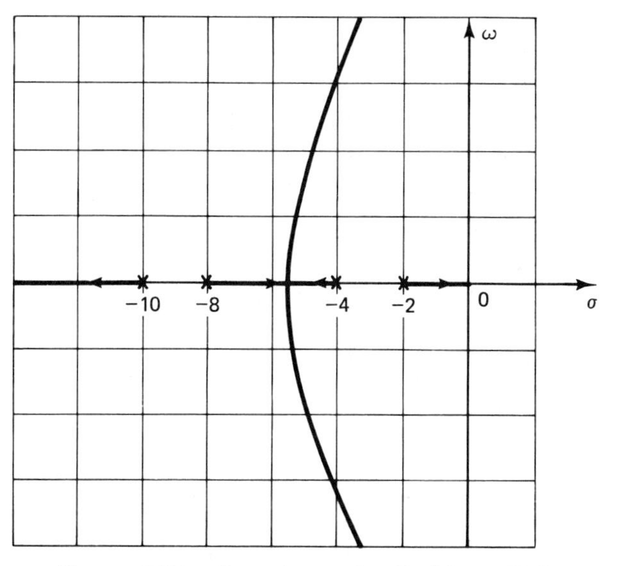


**Figure 7.31.** Root Locus for Problem 7B-3

PROBLEM 7B-3

Find the loop transfer function $\tilde{G}(s)$ for the system whose root locus is shown in Figure 7.31. Find the relevant points on the locus.

PROBLEM 7B-4

Sketch the root locus for the unity feedback system with the following open-loop transfer function:

$$G(s) = \frac{(s+3)(s+30)}{s^2(s+0.01)(s+300)(s+1000)}$$

PROBLEM 7B-5

Sketch the root locus for a unity feedback system with a forward transfer function given by

$$G(s) = \frac{K(s+4)(s+40)}{s(s+0.1)(s+0.2)(s+100)(s+200)}$$

PROBLEM 7B-6

Repeat Problem 7B-5 for the transfer function

$$G(s) = \frac{K(s+1.6)(s+0.45)}{s(1+s/2)(1+s/6)(s+3.9)(s+0.12)}$$

PROBLEM 7B-7

Repeat Problem 7B-5 for the transfer function

$$G(s) = \frac{K(5s+1)(4s+1)}{s(10s+1)(8s+1)(2s+1)(s+1)}$$

PROBLEM 7B-8

Sketch the root locus for a unity feedback control system with an open-loop transfer function given by

$$G(s) = \frac{e^{-\tau s}}{s^3}$$

PROBLEM 7B-9

Repeat Problem 7B-8 for the transfer function

$$G(s) = \frac{e^{-\tau s}}{s^4}$$

PROBLEM 7B-10

Repeat Problem 7B-9 for the transfer function

$$G(s) = \frac{(s + 0.3)e^{-s}}{s^2}$$

PROBLEM 7B-11

Sketch the root locus for the unity feedback system with an open-loop transfer function

$$G(s) = \frac{e^{-0.1s}}{s(s + 2)(s + 3)}$$

PROBLEM 7B-12

Sketch the root locus for the system of Problem 3B-5 with $M$ as the varying parameter.

PROBLEM 7B-13

Sketch the root locus for the system of Problem 3B-6 with $K_2$ as the varying parameter.

PROBLEM 7B-14

Sketch the root locus for the system of Problem 3B-6 with $a$ as the varying parameter.

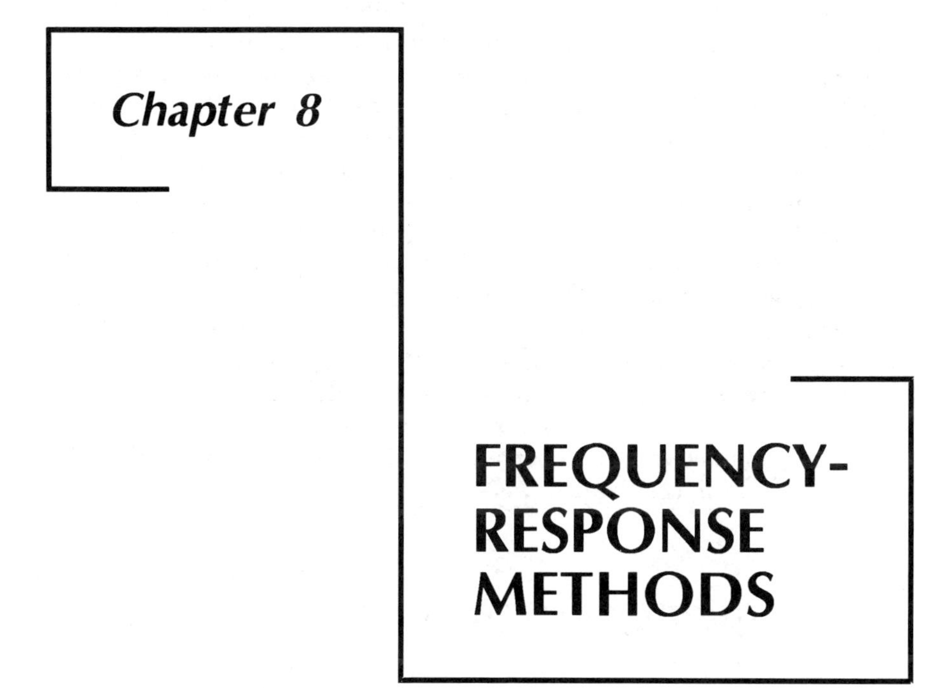

# Chapter 8

# FREQUENCY-RESPONSE METHODS

## 8.1 Introduction

This chapter is devoted to the study of frequency-response analysis methods. As mentioned in Chapter 1, since their inception, the frequency-response techniques due to the work of Nyquist and Bode gained wide acceptance as valuable tools in control engineering practice.

The basic idea of a frequency-response function is simple. Assuming that the driving function is sinusoidal, the transfer function turns out to be a complex function, as there will be a change in amplitude as well as phase angle in the output from that of the input function. The polar plot of a transfer function is treated first in this chapter. This is followed by the logarithmic or Bode plot. It turns out that a powerful stability method is available to infer the stability status of the closed-loop system from the open-loop frequency response. This is the celebrated Nyquist stability criterion. The basis of the criterion is the fundamental work by Cauchy. Our discussion attempts to highlight the connection using simple examples.

There are two principal advantages to frequency-response methods. The first is that we need not have a mathematical model for the open-loop system. It is sufficient to have results of frequency-response tests of the existing facility to construct the complex plots. This is of considerable value in practice. A second advantage is that not only can we assess closed-loop system stability, but we also come up with an answer to the question of relative stability: How stable will this stable system be if the parameters change? Thus arise the concepts of gain and phase margins. The reader should keep in mind that this chapter and the following one should be dealt with as one unit in the thought process.

# 8.2 Frequency Response

Consider a linear system represented by its transfer function $G(s)$. The Laplace transform of the input is $E(s)$ and that of the output is $C(s)$. Thus

$$\frac{C(s)}{E(s)} = G(s)$$

Assume that the input is a sinusoidal function of time given by

$$e(t) = E_m \sin \omega t$$

The output $c(t)$ can be obtained using the inverse Laplace transform

$$c(t) = \mathscr{L}^{-1}\{G(s)E(s)\}$$

The output contains a transient component and a steady-state component. For a stable system the transient component will decay while the steady-state component will dominate.

In this chapter our interest is in the sinusoidal steady-state performance analysis. In this case the output $c_{ss}(t)$ is given by

$$c_{ss}(t) = E_m|G(j\omega)|\sin(\omega t + \phi)$$

The output is therefore a sinusoid with a different magnitude $C_m$ and phase $\phi$.

$$C_m = E_m|G(j\omega)|$$

The sinusoidal transfer function $G(j\omega)$ is obtained by simply substituting $j\omega$ for $s$ in $G(s)$.

$$G(j\omega) = |G(j\omega)| \angle \phi$$

Note that the magnitude $|G(j\omega)|$ and phase $\phi$ are functions of $\omega$. The sinusoidal transfer function itself is a complex-valued function with $\omega$ as a parameter.

The frequency-response characteristics of control systems can be studied using three representations. In the first representation a polar plot is used. Here the magnitude of $G(j\omega)$ and its phase $\phi$ are plotted in the complex plane as the radian frequency $\omega$ is varied. The polar plot is referred to as a Nyquist diagram. In the second representation, referred to as the logarithmic or Bode diagram, two separate plots are employed. The first gives the logarithm of the magnitude of the sinusoidal transfer function versus the frequency, while the second gives the phase angle versus frequency. The third representation utilizes the log magnitude versus phase plot. The following section discusses the first representation.

# 8.3 Polar Plots

The polar plot is the locus of the vectors (phasors) $|G(j\omega)| \angle G(j\omega)$ in the complex plane as $\omega$ is varied from zero to infinity. We use the standard notation of a positive phase angle measured counterclockwise from the positive real axis. Each point of the polar plot of $G(j\omega)$ is the tip of the phasor $G(j\omega)$ for a particular $\omega$.

In some specific cases it is possible to obtain an analytic expression to describe the polar plot, as shown in some of the examples given in this section. This is achieved in some instances by first expressing $G(j\omega)$ in rectangular form

$$G(j\omega) = X(\omega) + jY(\omega)$$

The next step involves eliminating $\omega$ between the two expressions for $X$ and $Y$ to obtain the formula for the locus. In general, this exercise is not practical and a general procedure for sketching the plot as outlined later in this section should be adopted.

EXAMPLE 8.1

Consider the transfer function of an integrating element given by

$$G(s) = \frac{1}{s}$$

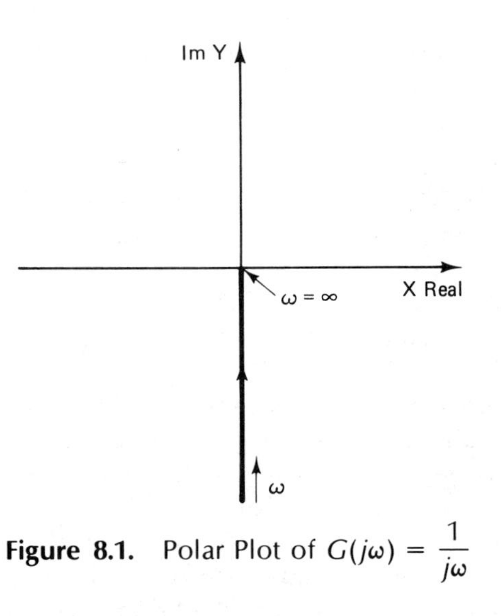


**Figure 8.1.** Polar Plot of $G(j\omega) = \frac{1}{j\omega}$

The sinusoidal transfer function is given by

$$G(j\omega) = \frac{1}{j\omega} = \frac{-j}{\omega}$$

In rectangular coordinates it is clear that

$$X(\omega) = 0$$

$$Y(\omega) = \frac{-1}{\omega}$$

As a result, we conclude that the plot lies on the imaginary axis as shown in Figure 8.1.

$$\lim_{\omega \to 0} Y(\omega) = -\infty$$

$$\lim_{\omega \to \infty} Y(\omega) = 0$$

EXAMPLE 8.2

Consider the transfer function of a first-order system given by

$$G(s) = \frac{1}{1 + \tau s}$$

We can find the polar plot of this function by substituting $s = j\omega$ to obtain

$$G(j\omega) = \frac{1}{1 + j\omega\tau} = \frac{1 - j\omega\tau}{1 + \omega^2\tau^2}$$

The real part of $G(j\omega)$ is given by

$$X = \frac{1}{1 + \omega^2\tau^2}$$

The imaginary part is given by

$$Y = \frac{-\omega\tau}{1 + \omega^2\tau^2}$$

By squaring the two equations above, we can conclude that

$$X^2 + Y^2 = X$$

The equation above can be rearranged into the form

$$\left(X - \tfrac{1}{2}\right)^2 + Y^2 = \left(\tfrac{1}{2}\right)^2$$

This is the equation of a circle with center at (0.5, 0) and radius 0.5, as shown in Figure 8.2.

It is important to associate a frequency orientation with our plot. We start by considering $\omega = 0$, for which we have by direct substitution

$$G(0) = 1$$

Let us take a value for $\omega$ such that $\omega\tau = 1$, we thus have

$$G\left(j\frac{1}{\tau}\right) = \frac{1}{1 + j1} = 0.707\angle{-45°}$$

As $\omega \to \infty$ we have

$$\lim_{\omega \to \infty} G(j\omega) = \lim \frac{1}{1 + j\omega\tau}$$
$$= 0\angle{-90°}$$

The discussion above leads us to conclude that for positive $\omega$ values, the polar plot is a semicircle in the fourth quadrant.

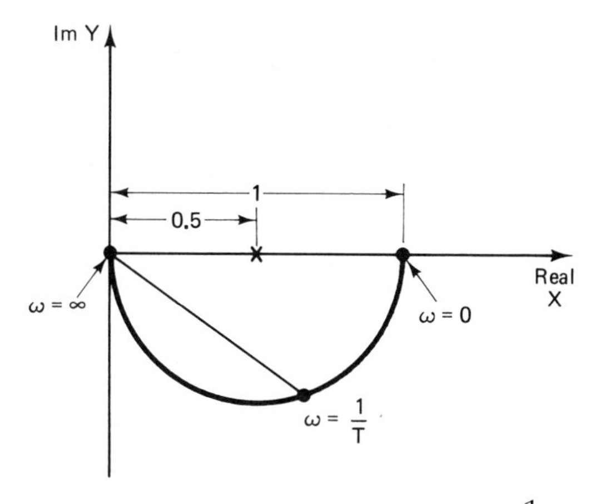


**Figure 8.2.** Polar Plot of $G(j\omega) = \dfrac{1}{1 + j\tau\omega}$

EXAMPLE 8.3

Consider the transfer function given by

$$G(s) = \frac{K}{s(1 + \tau s)}$$

We obtain for $s = j\omega$

$$G(j\omega) = \frac{K}{j\omega(1 + j\omega\tau)}$$

In polar form we have

$$G(j\omega) = |G|\angle\phi$$

where

$$|G| = \frac{1}{\omega\sqrt{1 + \omega^2\tau^2}}$$

$$\phi = -(90° + \tan^{-1}\omega\tau)$$

For $\omega \to 0$ we have

$$\lim_{\omega \to 0} |G| = \infty$$

$$\lim_{\omega \to 0} \phi = -90°$$

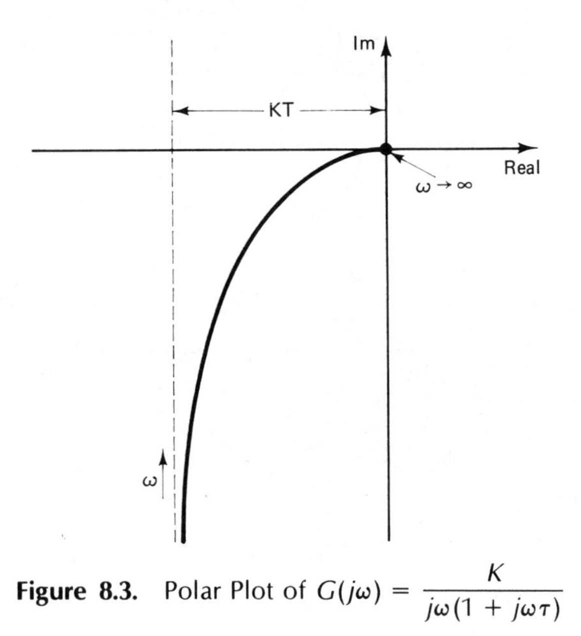


**Figure 8.3.** Polar Plot of $G(j\omega) = \dfrac{K}{j\omega(1 + j\omega\tau)}$

For $\omega \to \infty$ we have

$$\lim_{\omega \to \infty} |G| = 0$$
$$\lim_{\omega \to \infty} \phi = -180°$$

This enables us to sketch the polar plot as shown in Figure 8.3.

If we are interested in pursuing the matter further, we can show that in rectangular coordinates we have

$$X = \frac{-K\tau}{1 + \omega^2\tau^2}$$
$$Y = \frac{-K}{\omega(1 + \omega^2\tau^2)}$$

Note that $\lim_{\omega \to 0} X = -K\tau$; this provides us with the asymptote shown in the diagram. Eliminating $\omega$, we get

$$Y^2(-KT - X) = X^3$$

This is a cissoid of Diocles.

## General Properties of Polar Plots

A general form of the sinusoidal transfer function $G(j\omega)$ is given by

$$G(j\omega) = \frac{K_m N(j\omega)}{(j\omega)^m D(j\omega)}$$

As defined in Chapter 5, $m$ is the system type. The numerator and denominator functions $N(j\omega)$ and $D(j\omega)$ can be assumed to be given in factored form as

$$N(j\omega) = (1 + j\omega\tau_a)(1 + j\omega\tau_b) \cdots (1 + j\omega\tau_M)$$
$$D(j\omega) = (1 + j\omega\tau_1)(1 + j\omega\tau_2) \cdots (1 + j\omega\tau_P)$$

The presence of the $(j\omega)^m$ term in the denominator contributes a constant angle $\phi_c$ to the total angle $\phi$ of $G(j\omega)$ for all frequencies.

$$\phi_c = -\frac{m\pi}{2}$$

The denominator function $D(j\omega)$ contributes an angle $\phi_D$ which is frequency dependent:

$$\phi_D = -(\phi_1 + \phi_2 + \cdots + \phi_P)$$

where

$$\phi_1 = \tan^{-1}\omega\tau_1$$
$$\phi_2 = \tan^{-1}\omega\tau_2$$
$$\vdots$$
$$\phi_P = \tan^{-1}\omega\tau_p$$

The numerator function $N(j\omega)$ contributes an angle $\phi_N$ which is also frequency dependent,

$$\phi_N = \phi_a + \phi_b + \cdots + \phi_M$$

where

$$\begin{aligned} \phi_a &= \tan^{-1} \omega\tau_a \\ \phi_b &= \tan^{-1} \omega\tau_b \\ &\vdots \\ \phi_M &= \tan^{-1} \omega\tau_M \end{aligned}$$

The total angle is then given by

$$\phi = \phi_c + \phi_D + \phi_N$$

The low-frequency portion of the polar plot is obtained by setting $\omega \to 0$. For system types other than zero, the magnitude of $G(j\omega)$ tends to infinity, as can be seen from

$$\lim_{\omega \to 0} \frac{K_m |N(j\omega)|}{\omega^m |D(j\omega)|} = \infty$$

For type 0 systems, $m = 0$ and as a result

$$\lim_{\omega \to 0} \frac{K_0 |N(j\omega)|}{|D(j\omega)|} = K_0$$

The angles $\phi_D$ and $\phi_N$ are zero for $\omega \to 0$, and thus the total angle $\phi$ is equal to the constant angle contribution of the $(j\omega)^m$ term

$$\lim_{\omega \to 0} \phi = \phi_c = -m\frac{\pi}{2}$$

It is thus clear that the zero frequency point on the polar plot for a type 0 system is on the real axis with a magnitude of $K$. Example 8.1 belongs in this category.

For a type 1 system, the magnitude of the low-frequency portion of the polar plot approaches infinity as $\omega$ approaches zero. The angle $\phi$ for this portion is clearly $-90°$. From Example 8.3 it is clear that there is an asymptotic line parallel to the negative imaginery axis but displaced to the left of the origin to which $G(j\omega)$ tends as $\omega$ approaches zero. The true asymptote is obtained by an evaluation of $X(\omega)$ as $\omega$ approaches zero.

The low-frequency portion of the polar plot of a type 2 system has a magnitude that approaches infinity as $\omega$ approaches zero. The angle, however, is $-180°$.

The high-frequency portion of the polar plot can be determined by letting $\omega$ approach infinity in the expression for $G(j\omega)$. If the number of poles exceeds that of the zeros of the transfer function, the magnitude of

$G(j\omega)$ approaches zero as $\omega$ approaches infinity. The angle $\phi$ will have the following components:

$$\lim_{\omega \to \infty} \phi_D = -P\frac{\pi}{2}$$
$$\lim_{\omega \to \infty} \phi_N = +M\frac{\pi}{2}$$

Thus

$$\lim_{\omega \to \infty} \phi = (M - P - m)\frac{\pi}{2}$$

The plot therefore ends at the origin tangent to the axis determined by the condition described above.

The frequencies at the points of intersection of the plot with real and imaginary axes are obtained by solving

$$X(\omega) = 0$$
$$Y(\omega) = 0$$

It is appropriate now to take another example.

EXAMPLE 8.4

Consider the transfer function of the lead–lag network given by

$$G(s) = \frac{(1 + \tau_1 s)(1 + \tau_2 s)}{(1 + \tau_1 s)(1 + \tau_2 s) + \tau_{12} s}$$

As a function of frequency we have

$$G(j\omega) = \frac{(1 - \omega^2\tau_1\tau_2) + j\omega(\tau_1 + \tau_2)}{(1 - \omega^2\tau_1\tau_2) + j\omega(\tau_1 + \tau_2 + \tau_{12})}$$

A few manipulations result in the following expressions for the real part of $G(j\omega)$ denoted by $X$ and the imaginary part denoted by $Y$:

$$X = \frac{a^2 + \omega^2 b(b + \tau_{12})}{a^2 + \omega^2(b + \tau_{12})^2}$$
$$Y = \frac{-\omega a \tau_{12}}{a^2 + \omega^2(b + \tau_{12})^2}$$

where

$$a = 1 - \omega^2\tau_1\tau_2$$
$$b = \tau_1 + \tau_2$$

We can carry out a few algebraic steps to eliminate $\omega$ and obtain the equation of the polar plot. The result is a circle with center at $(X_0, 0)$ on the real axis and radius denoted by $A$. The circle's equation is

$$Y^2 + (X - X_0)^2 = A^2$$

with

$$X_0 = \frac{\tau_{12} + 2(\tau_1 + \tau_2)}{2(\tau_1 + \tau_2 + \tau_{12})}$$

$$A = \frac{\tau_{12}}{2(\tau_1 + \tau_2 + \tau_{12})}$$

It is clear that $X_0 > A$ for positive values of $\tau_1$ and $\tau_2$. Figure 8.4 shows the polar plot of this function.

The plot starts for $\omega = 0$ at the point $(1, 0)$ regardless of the values of the time constants $\tau_1$, $\tau_2$, and $\tau_{12}$. As the frequency is increased, the angle $\phi$ is lagging and decreases to a minimum at $\omega = \omega_m$, as shown on the polar plot. The angle $\phi_m$ is defined by

$$\phi_m = -\sin^{-1}\frac{A}{X_0} = -\sin^{-1}\frac{\tau_{12}}{\tau_{12} + 2(\tau_1 + \tau_2)}$$

The real part of $G(j\omega)$ at $\omega = \omega_m$ is given by

$$X_m = \frac{2(\tau_1 + \tau_2)}{\tau_{12} + 2(\tau_1 + \tau_2)}$$

The value of $\omega_m$ is obtained from the solutions to

$$\tau_1^2\tau_2^2\omega^4 - \omega^2\left[(\tau_1 + \tau_2)(\tau_1 + \tau_2 + \tau_{12}) + 2\tau_1\tau_2\right] + 1 = 0$$

This is a second-order equation in $\omega^2$ which has two positive roots. The smaller root is $\omega_m^2$, the larger corresponds to the leading portion (first quadrant).

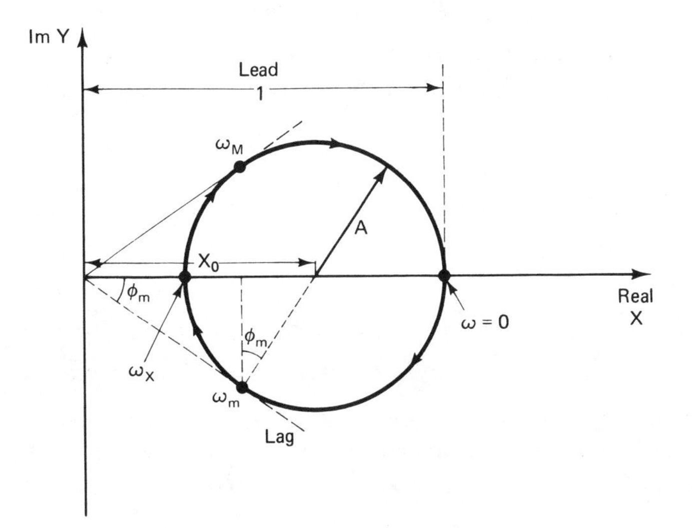


**Figure 8.4.** Polar Plot for a Lead–Lag Transfer Function

As the frequency is increased beyond $\omega_m$ up to the radian frequency $\omega_x$, the angle $\phi$ is still lagging but increases up to a value $\phi = 0$ at $\omega = \omega_x$. Clearly, $\omega_x$ is obtained by setting $Y = 0$, and as a result we conclude that

$$\omega_x = \frac{1}{\sqrt{\tau_1 \tau_2}}$$

With a further increase in frequency beyond $\omega_x$, the phase angle increases and is leading. The maximum value of $\phi$ is attained at $\omega = \omega_M$, which is obtained from the solution to the quartic equation in $\omega$ given above. It should be noted that the plot is symmetric and hence that

$$\phi_M = \sin^{-1} \frac{A}{X_0}$$

EXAMPLE 8.5

Consider the sinusoidal transfer function of a pure time delay $\tau$, which is given by

$$G(j\omega) = Ke^{-j\omega\tau}$$

The magnitude of this function is

$$|G(j\omega)| = K$$

Its phase angle is

$$\phi(\omega) = -\omega\tau \text{ radians}$$

It is thus clear that the polar plot of the pure time-delay transfer function is a circle with radius of unity, centered at the origin. For $\omega = 0$ we have $\phi(\omega) = 0$. The locus is described over the circle and over again as $\omega$ is increased. For example, for $\omega = 2\pi/\tau$, the same point as that for $\omega = 0$ is obtained. The plot is shown in Figure 8.5.

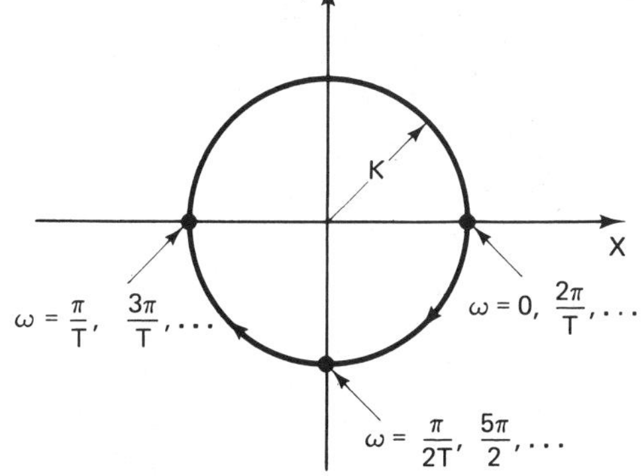

**Figure 8.5.** Polar Plot of a Time-Delay Function

EXAMPLE 8.6

Consider the transfer function of a delayed integrator given by

$$G(s) = \frac{e^{-Ds}}{s}$$

The sinusoidal transfer function is

$$G(j\omega) = \frac{e^{-j\omega D}}{j\omega}$$

Clearly,

$$|G(j\omega)| = \frac{1}{\omega}$$

$$\phi = -\omega D - \frac{\pi}{2}$$

$$X = \frac{-\sin \omega D}{\omega}$$

$$Y = \frac{-\cos \omega D}{\omega}$$

In the limit as $\omega \to 0$, we have

$$\lim_{\omega \to 0} X = -D$$

$$\lim_{\omega \to 0} Y = -\infty$$

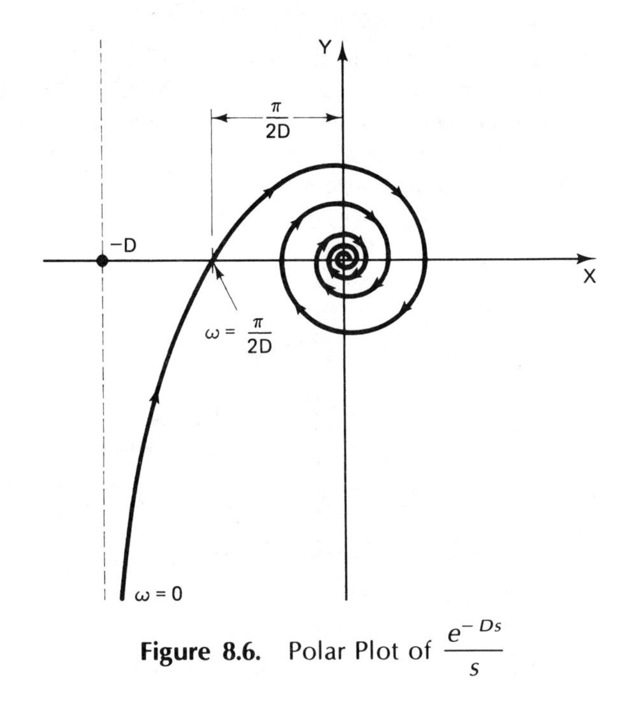


**Figure 8.6.** Polar Plot of $\frac{e^{-Ds}}{s}$

The polar plot will inherit the cyclic effect from the delay $D$, and will spiral into the $\omega \to \infty$ point at the origin, as shown in Figure 8.6. The curious reader may like to find exactly some (not all, of course) of the intersections with the real and imaginary axes. For the first intersection with the real axis, set $Y = 0$ to obtain $\omega D = \pi/2$; thus $X = -1/\omega = -\pi/2D$, and so on.

# 8.4 Bode Logarithmic Plots

The sinusoidal transfer function $G(j\omega)$ is a complex function expressed in polar form as

$$G(j\omega) = |G(j\omega)|e^{j\phi(\omega)}$$

An alternative to the polar plot representation is to use one plot depicting the variation of the magnitude of $G(j\omega)$ with the radian frequency and a second plot depicting the variation of the phase angle $\phi$ with the radian frequency. A slight, but tremendously advantageous, modification to this concept is to use the logarithm of $G(j\omega)$ and that of $\omega$. This is the contribution of H. W. Bode, who used this concept extensively in his pioneering studies of feedback amplifiers. Taking the natural logarithm of $G(j\omega)$, we get

$$\ln G(j\omega) = \ln|G(j\omega)| + j\phi(\omega)$$

In terms of the logarithm to the base 10, we have

$$\log G(j\omega) = \log|G(j\omega)| + j0.434\phi(\omega)$$

Thus the logarithm of the complex function $G(j\omega)$ resolves to the sum of a real part given by the logarithm of the magnitude and an imaginary part given by a scaled version of the phase angle. The real part is normally represented in terms of the logarithmic gain in decibels (dB) denoted by $A_{\text{dB}}$, defined as

$$A_{\text{dB}} = 20\log|G(j\omega)|$$

Note that $A_{\text{dB}}$ is a function of $\omega$. The horizontal axis in Bode plots is $\log \omega$. Use of semilog paper eliminates the need to take logarithms of many numbers. This also expands the low-frequency range.

The beauty of Bode plots stems from the simple fact that multiplication and division operations are transformed to addition and subtraction. It will become clear to us that work with Bode plots is mainly graphical and provides tremendous insight into the effects of terms incorporated in the system transfer function. Let us consider the general form of $G(j\omega)$ given by

$$G(j\omega) = \frac{K_m N(j\omega)}{(j\omega)^m D(j\omega)}$$

The magnitude of the function is

$$|G(j\omega)| = \frac{K_m|N(j\omega)|}{\omega^m|D(j\omega)|}$$

Taking logarithms, we have

$$\log|G(j\omega)| = \log K_m + \log|N(j\omega)| - \left[m \log \omega + \log|D(j\omega)|\right]$$

Thus if we have the four ingredients of the right-hand side, we are able to assemble the $A_{dB}$ for the system by simple additions and subtractions. The magnitudes of numerator and denominator functions are, in turn, products of elementary or basic terms such as

$$|(1 + j\omega\tau)|$$

$$\left|\left[1 + 2\xi\frac{j\omega}{\omega_n} + \left(\frac{j\omega}{\omega_n}\right)^2\right]\right|$$

It is thus clear that the first order of business in the study of Bode plots is to examine the plots of its basic building blocks. This we presently do.

## First Building Block: Constant gain $K_m$

A constant gain $K_m$ is a positive real number, and as a result no phase angle is associated with it. The contribution to the log magnitude characteristic is a constant which may be positive (for $K_m > 1$) or negative (for $K_m < 1$). The gain in decibels is thus

$$A_{dB} = 20 \log_{10} K \text{ dB}$$

This is a horizontal straight line. The constant raises or lowers the log magnitude curve of the complete sinusoidal transfer function by a fixed amount. Figure 8.7 shows the Bode plots for a constant gain.

It is appropriate at this juncture to make some observations about Figure 8.7. The horizontal axis in both characteristics is labeled $\omega$ but in actual fact it is a logarithmic scale. Note that the division between $\omega = 1$ and $\omega = 10$ is equal to that between $\omega = 10$ and $\omega = 100$. Frequency ratios are expressed in terms of decades or octaves. A *decade* is a frequency band from $\omega$ to $10\omega$, where $\omega$ is any radian frequency. An *octave* is a frequency band from $\omega$ to $2\omega$, where again $\omega$ is any radian frequency. The second observation is that $A_{dB}$ and $\phi$ are in a linear scale. Note that the base value for the ordinate of our phase characteristic is $-180°$. This will provide us with a convenient means of drawing certain conclusions from the plots. The significance of this statement will not remain a mystery for long, as it will be explained (or discovered) later on.

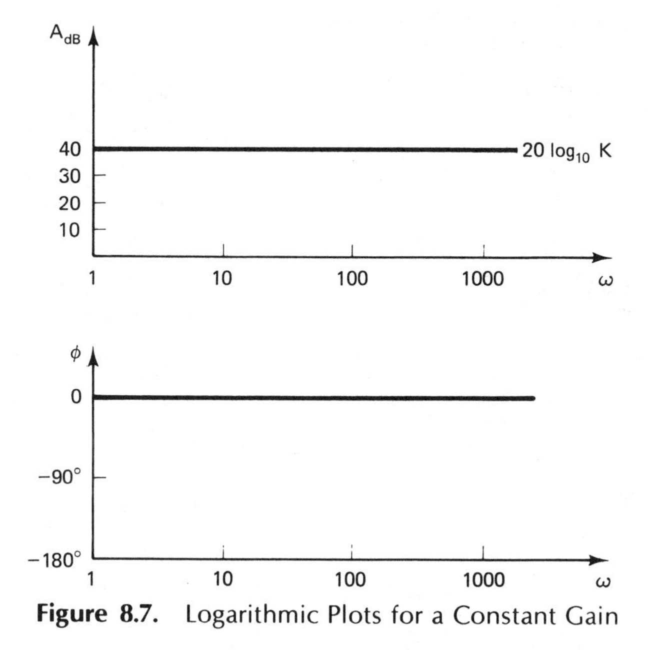


**Figure 8.7.** Logarithmic Plots for a Constant Gain

## Second Building Block: Factors $(j\omega)^m$

The factor $(j\omega)^m$ appearing in the denominator contributes an angle given by

$$\phi = -m\frac{\pi}{2}$$

The log magnitude is

$$\begin{aligned} A_{dB} &= 20\log\frac{1}{|[j\omega]^m|} = 20\log\frac{1}{\omega^m} \\ &= -20m\log\omega \end{aligned}$$

When plotted against log $\omega$, the log magnitude is a straight line with a slope of $-20m$. Let us now look at a practical aspect of the development for which we will benefit by dealing with the simplest case first.

Consider the case with a type 1 system, where $m = 1$. Our log magnitude is then given by

$$A_{dB} = 20\log\frac{1}{|j\omega|} = -20\log\omega$$

The slope of the straight line $A_{dB}$ versus log $\omega$ is $-20$. The question is: How do we go about locating this line in the $A_{dB}$–log $\omega$ plot? It is clear that the line passes through the origin; that is, 0 dB gain is achieved for log $\omega = 0$ or

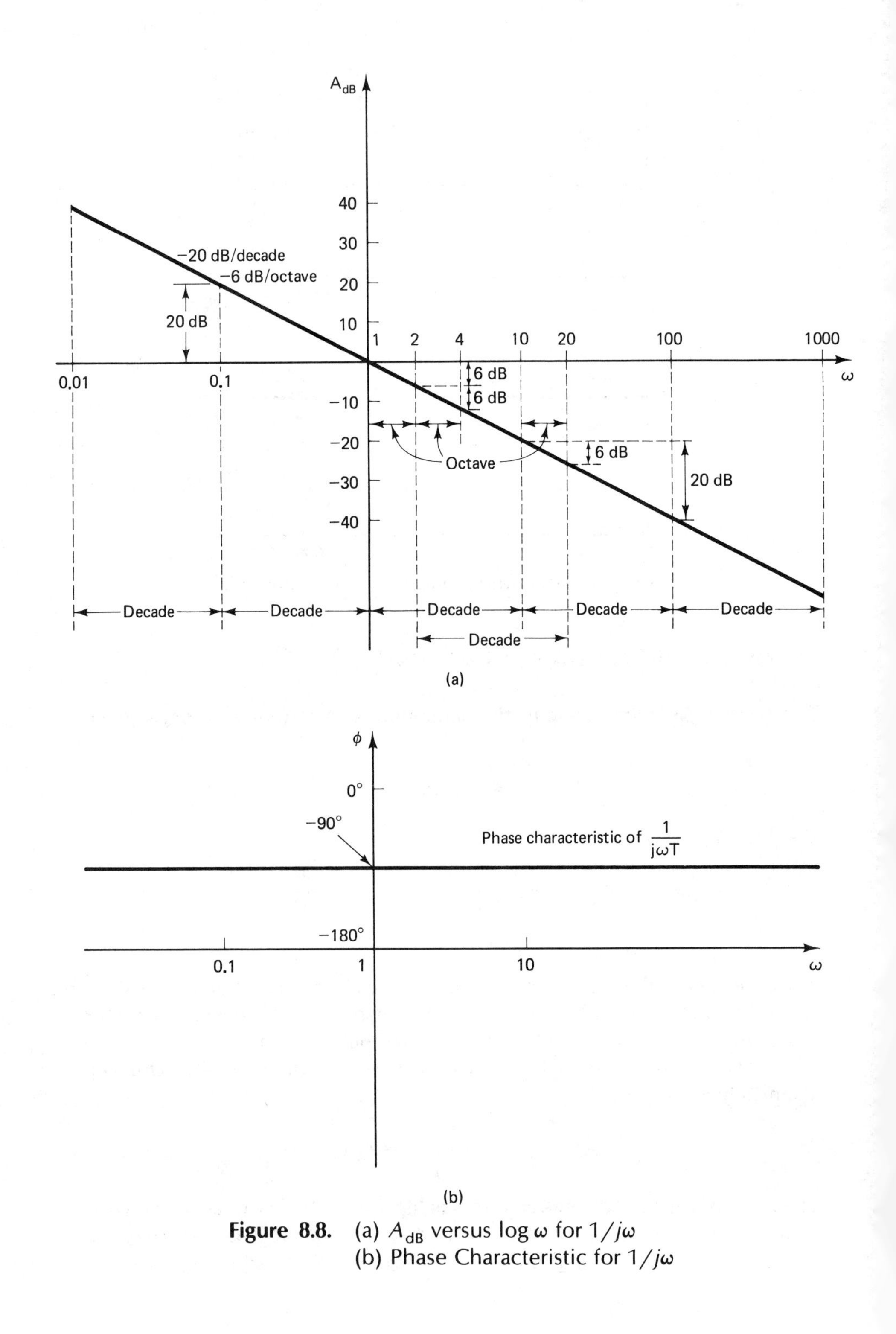


**Figure 8.8.** (a) $A_{dB}$ versus log $\omega$ for $1/j\omega$
(b) Phase Characteristic for $1/j\omega$

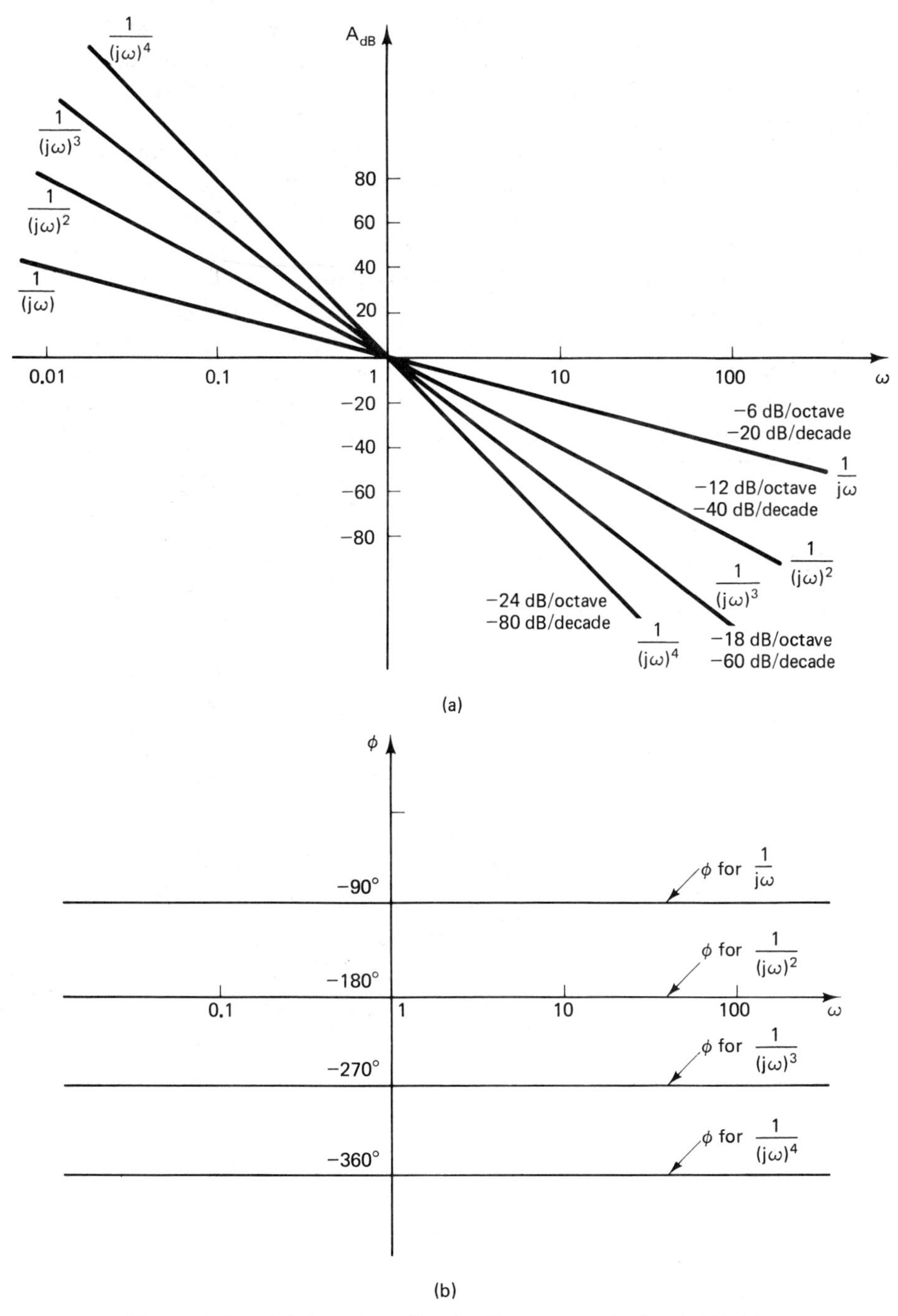


**Figure 8.9.** (a) Log Amplitude Characteristic for $1/(j\omega)^m$
(b) Phase Angle Plot for $1/(j\omega)^m$

at $\omega = 1$. Next we take $\omega = 10$, for which we have $A_{dB} = -20$ dB. The information above is sufficient to draw the straight-line characteristic, which is simply a $-20$-dB/decade slope, as shown in Figure 8.8a. We talked about octaves earlier; at $\omega = 2$ we have $A_{dB} = -20 \log 2 = -6$ and we can equivalently say that the characteristic also has a $-6$-dB/octave slope. For the sake of clarity we illustrate the decade and octave concepts in Figure 8.8a. In Figure 8.8b we show the phase characteristic for this case, which is a constant of value $-90°$ for all frequencies.

A moment's reflection on the state of affairs for $m > 1$ shows that the slope for $m = 2$ is double that for $m = 1$ (that was easy to figure out). Thus we have $-40$ dB/decade or $-12$ dB/octave for $1/(j\omega)^2$, and so on. Our assertions for these cases are shown in Figure 8.9a. Similarly, the angle for $m = 2$ is $-180°$, and so on, as shown in Figure 8.9b. In the case of differentiation, that is, $j\omega$ is in the numerator, we have Figure 8.10, and in this case the slope is positive.

It is tempting to keep going and deal with the next building blocks, but we will probably benefit from an example at this juncture. We will also learn a little graphical trick that can save time later.

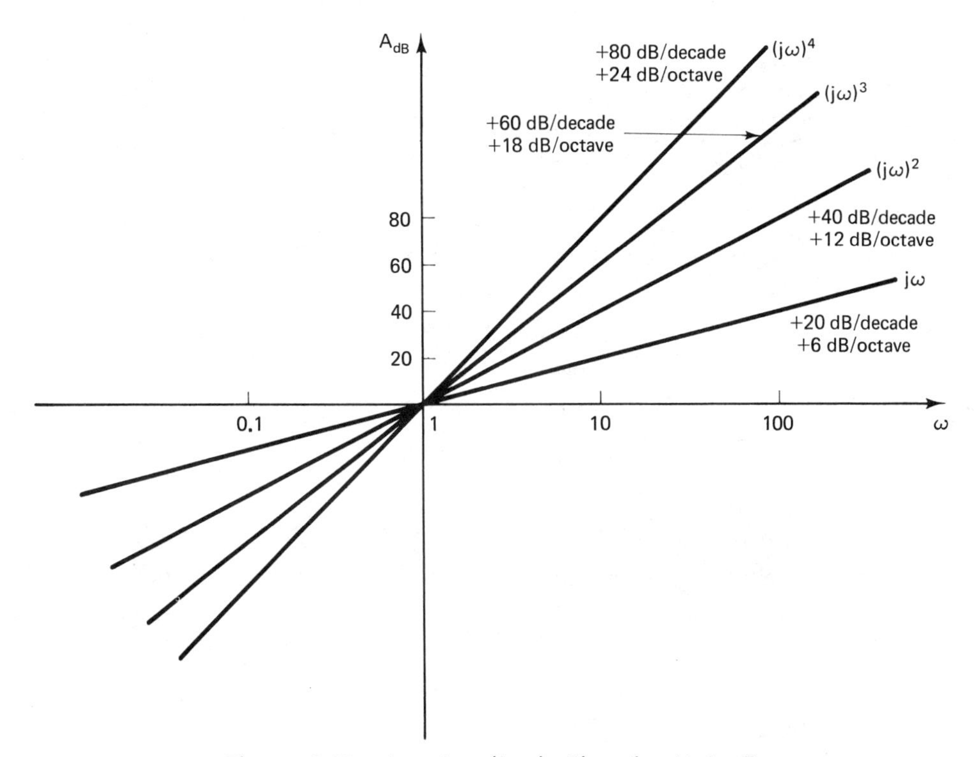


**Figure 8.10.** Log Amplitude Plots for $(j\omega)^{-m}$

EXAMPLE 8.7

Construct the Bode logarithmic plots for the sinusoidal transfer function given by

$$G(j\omega) = \frac{K}{(j\omega)^3}$$

Assume that $K = 10$.

SOLUTION

There are two terms to contend with. The first term is $K = 10$, with contribution

$$A_1 = 20 \log 10 = 20 \text{ dB}$$

The second is the triple integral. Since we have just dealt with this in Figure 8.9, we simply transfer the result to Figure 8.11a, as shown on line $AOB$. The gain $K$ contributes the horizontal line $DE$. All we do now to obtain the required characteristics is to add for each $\omega$ the values of $DE$ and that of $AOB$. This provides us with the required line $A'O'B'$. Note that this

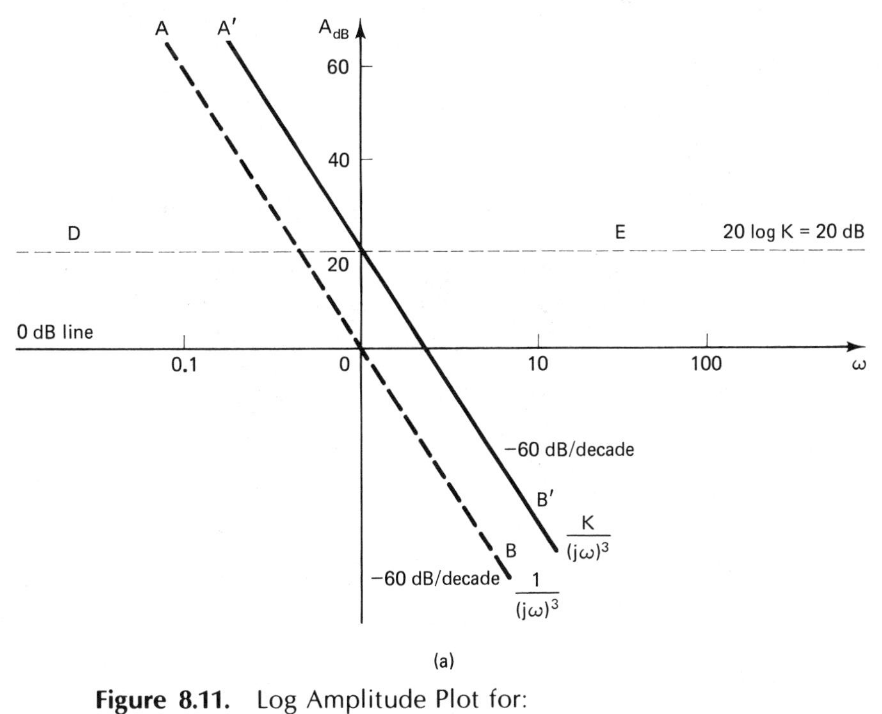


**Figure 8.11.** Log Amplitude Plot for:
(a) Example 8.7
(b) Example 8.7 with 0-dB Line Shift Technique

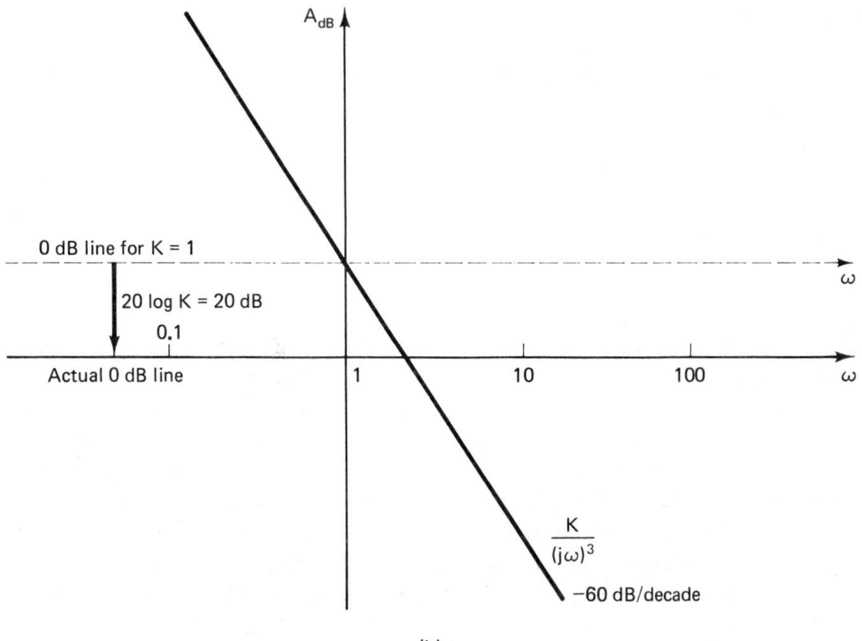


**Figure 8.11** (continued). Log Amplitude Plot For:
(a) Example 8.7
(b) Example 8.7 with 0-dB Line Shift Technique

is parallel to *AOB* and is simply an upward shift of *AOB* by the amount $20 \log K$.

Figure 8.11b is intended to deliver on the promise of a graphical shortcut. The idea here is to save on drawing one line with a slope of $-60$ dB/decade. The method involves drawing the set of axes $A_{\text{dB}}$ and the log $\omega$ axis and identifying the latter as the 0-dB line for $K = 1$. Now take the contribution of the triple integration, which has a slope of $-60$ dB/decade and has a value of 0 dB at $\omega = 1$. The actual 0-dB line is obtained by a downward shift of the 0-dB line for $K = 1$ by an amount equal $20 \log K$, as indicated by the arrow in Figure 8.11b.

We can now return to our remaining building blocks.

## Third Building Block: Factors $(1 + j\omega\tau)$

Consider first the simple phase-lag case in which we have

$$G(j\omega) = \frac{1}{1 + j\omega\tau}$$

The log magnitude is

$$\begin{aligned} A_{\text{dB}} &= 20\log\frac{1}{\sqrt{1+\omega^2\tau^2}} \\ &= -20\log\sqrt{1+\omega^2\tau^2} \end{aligned}$$

It is possible to avoid a large number of computations and arrive at an appropriate but intuitively appealing representation of the characteristic if we observe that for $\omega\tau \gg 1$ (i.e., the low-frequency portion) we have

$$A_{\text{dB}} \simeq -20\log\sqrt{1+0} = 0 \text{ dB}$$

For the high-frequency portion $\omega\tau \gg 1$, we have

$$A_{\text{dB}} \simeq -20\log\sqrt{\omega^2\tau^2} = -20\log\omega\tau$$

Thus the function is a straight line with a negative slope of $-20$ dB/decade and has a zero value for $\omega\tau = 1$. The value of $\omega$ for which this asymptote is zero is called the corner frequency $\omega_c$:

$$\omega_c = \frac{1}{\tau}$$

We can thus assert that our asymptotic representation is 0 dB up to the corner frequency, where a $-20$-dB/decade asymptotic line begins.

The exact curve is shown together with our asymptotes in Figure 8.12a. Note that the exact curve is 3 dB below the asymptotes at the corner frequency (this is simply because $20\log\sqrt{2} = 3$). Our figure assumes that $\tau = 0.2$ and thus $\omega_c = 5$ rad/sec and thus at $\omega_c = 5$, the exact curve is at $-3$ dB. For $\omega = \omega_c/2$ and $\omega = 2\omega_c$ (i.e., an octave below and an octave above $\omega_c$), the exact curve is 1 dB below the asymptotes. Thus for $\omega = \omega_c/2$ we have an exact value of $-1$ dB (actually $-0.97$), while for $\omega = 2\omega_c$, we have an exact value of $-7$ dB ($-6.99$ dB) as compared with an asymptote value of $-6.02$ dB.

The phase curve has the following relevant points:

$$\begin{aligned} &\text{For } \omega \to 0: && \phi \to 0 \\ &\text{For } \omega = \omega_c: && \phi = -45^\circ \\ &\text{For } \omega \to \infty: && \phi = -90^\circ \end{aligned}$$

An exact curve may be constructed using some selected frequencies. A commonly used approximation is to draw a straight line with a slope of $-45^\circ$ per decade through the $(-45^\circ, \omega_c)$ point. This results in $0^\circ$ phase for $\omega = 0.1\omega_c$ and $-90^\circ$ phase for $\omega = 10\omega_c$. The two characteristics are shown in Figure 8.12b.

Figure 8.13 iliustrates the log magnitude characteristics for factors of the form $(1 + j\omega\tau)^n$ in the denominator. Note that all curves have the same

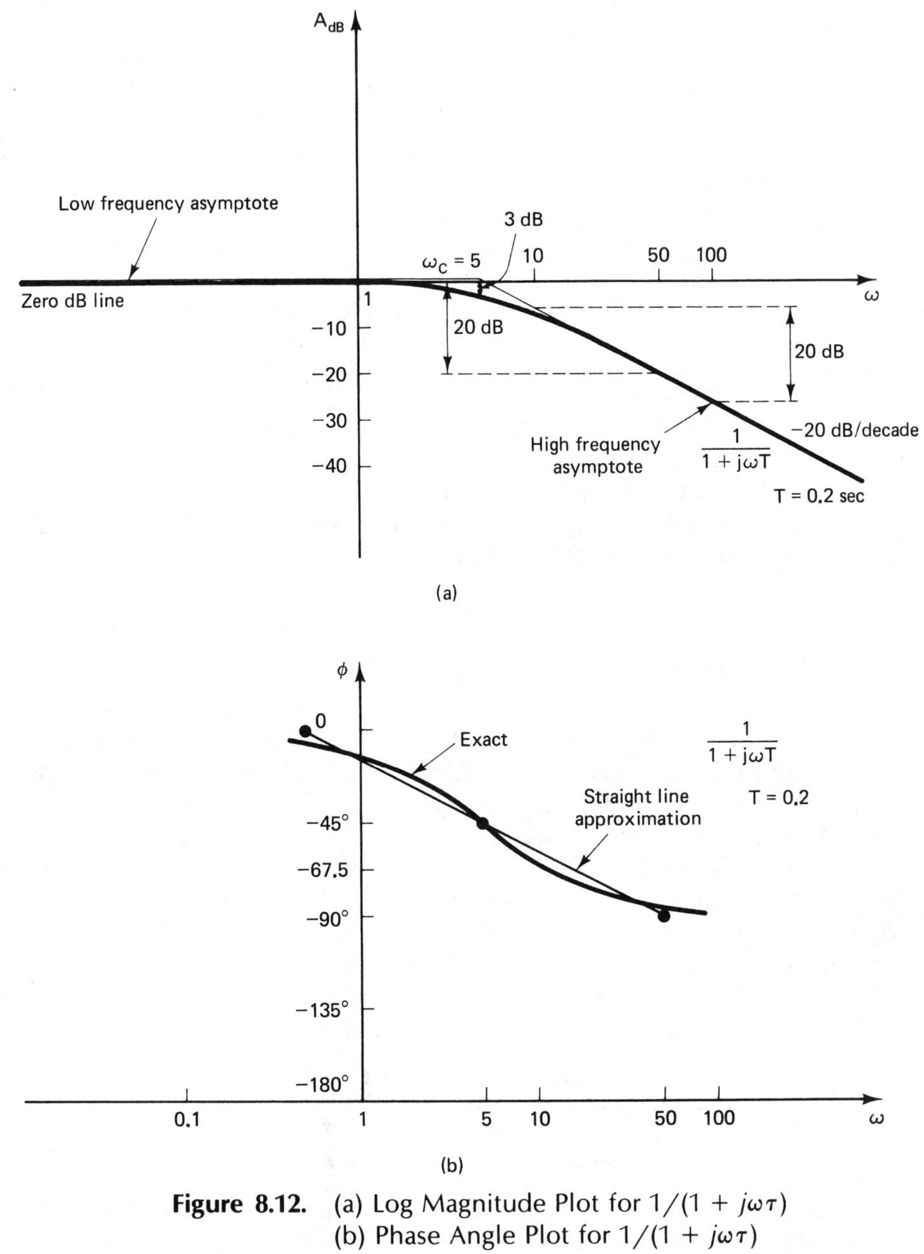

**Figure 8.12.** (a) Log Magnitude Plot for $1/(1 + j\omega\tau)$
(b) Phase Angle Plot for $1/(1 + j\omega\tau)$

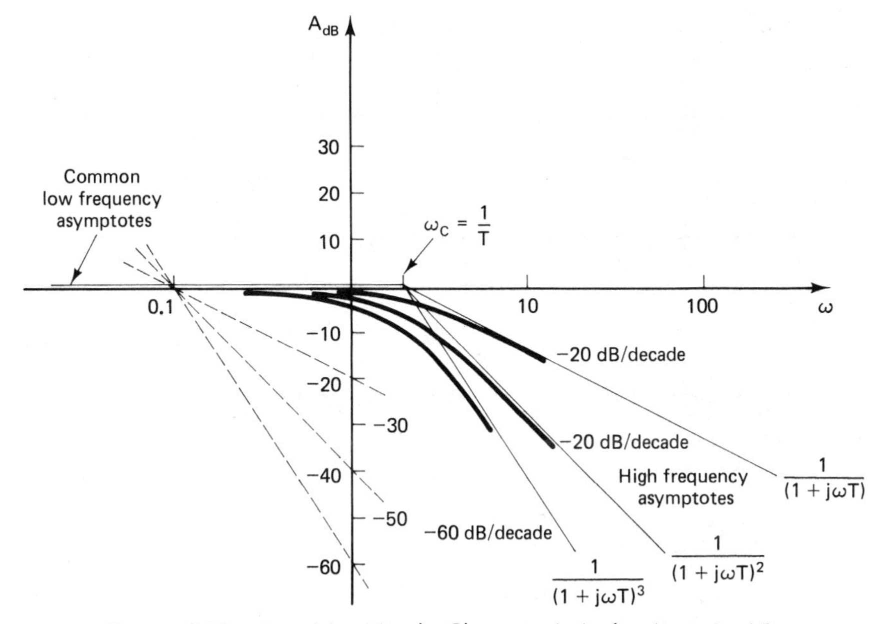


**Figure 8.13.** Log Magnitude Characteristic for $(1 + j\omega\tau)^n$

corner frequency and 0 dB for the low-frequency portion. For the high-frequency portion the slopes are $-20n$ dB/decade. The angle characteristic is not shown but can easily be deduced from the observation that the low-frequency angles are almost the same with $\lim_{\omega\to 0}\phi = 0$. At the corner frequency the angle is $-n(45°)$ and for high frequency the angle tends to $-n(90°)$.

It is clear that for factors of the form $(1 + j\omega\tau)^n$ in the numerator, the foregoing assertions can be repeated but with a sign reversal. Again we take a pair of examples.

EXAMPLE 8.8

Construct the Bode logarithmic plot for the sinusoidal transfer function given by

$$G(j\omega) = \frac{1}{j\omega(1 + j\omega)(1 + j0.2\omega)}$$

SOLUTION

The factors involved in this function have already been discussed. We deal first with the log magnitude characteristic. The first factor is $j\omega$ in the denominator, for which the characteristic is shown in the top plot of Figure 8.14a with a $-20$-dB/decade line passing through the $\omega = 1$ and 0 dB gain

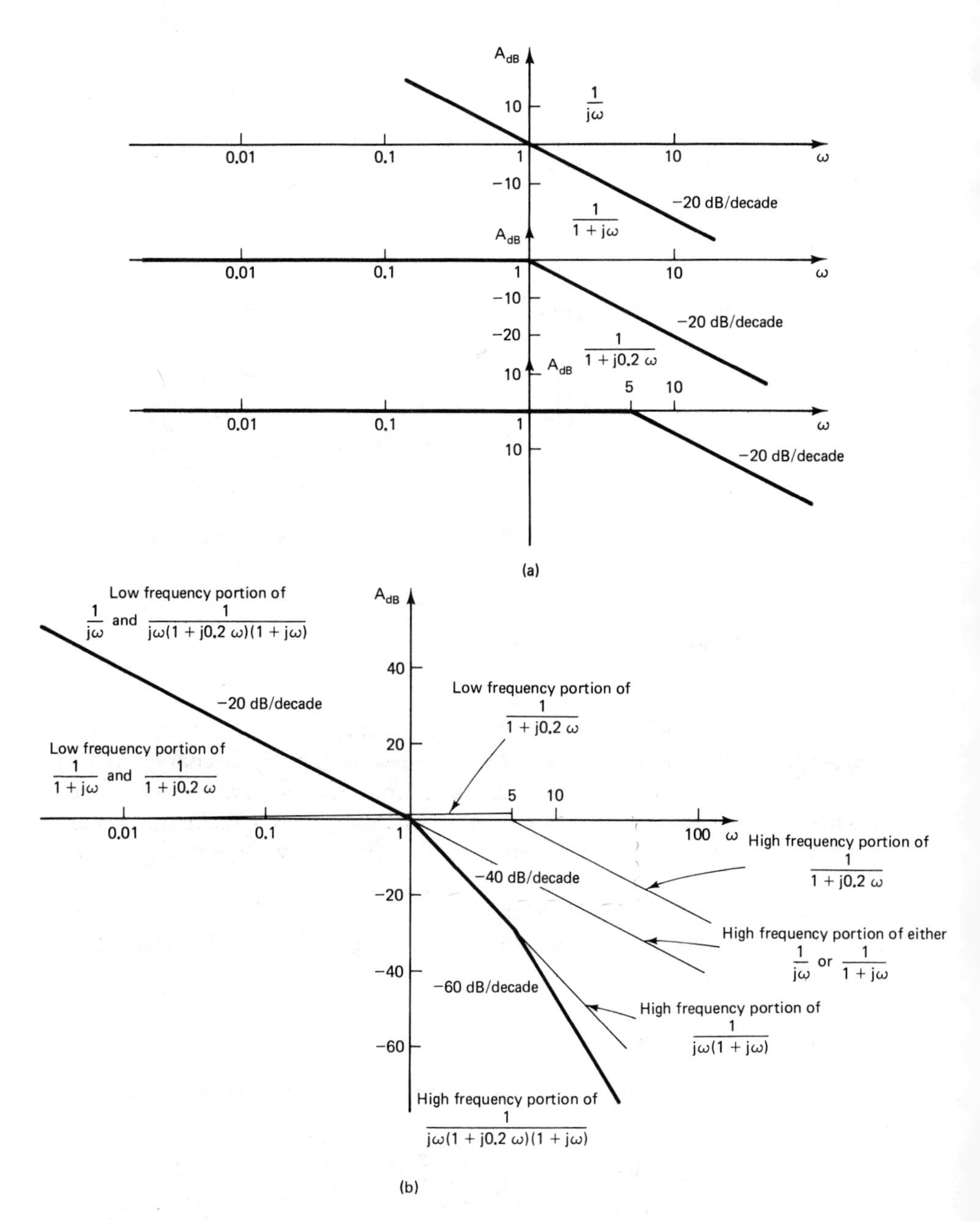


**Figure 8.14.** (a) Ingredients of the Log Magnitude Characteristic of Example 8.8
(b) Log Magnitude Characteristic of $1/[j\omega(1 + j0.2\omega)(1 + j\omega)]$
(c) Complete Log Magnitude Characteristic for Example 8.8

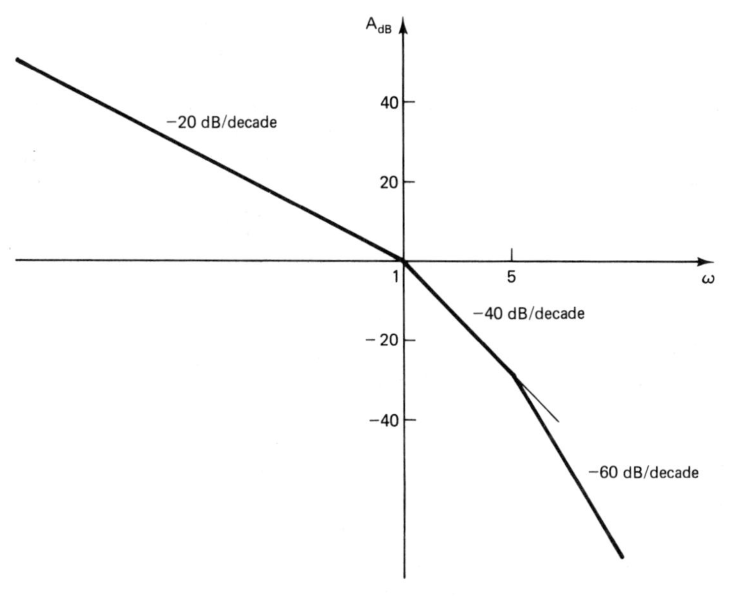


**Figure 8.14** (continued). (a) Ingredients of the Log Magnitude Characteristic of Example 8.8
(b) Log Magnitude Characteristic of $1/[j\omega(1 + j0.2\omega)(1 + j\omega)]$
(c) Complete Log Magnitude Characteristic for Example 8.8

point. The second factor is $1 + j\omega$ in the denominator, with a corner frequency of $\omega = 1$. Therefore, the middle plot of Figure 8.14a shows a gain of 0 dB up to the corner frequency, followed by a straight line with slope $-20$ dB/decade originating at $\omega = 1$ and 0 dB. The third factor's contribution is shown in the bottom portion of Figure 8.14a and corresponds to $1 + j0.2\omega$ in the denominator. The corner frequency here is $\omega = 5$ rad/sec.

Having now obtained the three ingredients, the task at hand is to assemble the plot. This is simply graphical addition. We do this by starting at a low frequency such as $\omega = 0.1$; the only nonzero contribution is that due to $j\omega$ in the denominator. Thus for $\omega = 0.1$, the $A_{dB}$ value is simply $+20$ dB. At $\omega \simeq 0.3$, the $A_{dB}$ value of $j\omega$ in the denominator is approximately 10 dB, and thus for the overall characteristic the $A_{dB}$ value at $\omega \simeq 0.3$ is 10 dB. Continuing on to $\omega = 1$, we find that the overall characteristic is coincident with that of $(j\omega)^{-1}$, as shown in Figure 8.14b. Continuing beyond $\omega = 1$, we find that $1 + j\omega$ in the denominator has a nonzero contribution. Thus the two lines for $1 + j\omega$ and $j\omega$ in the denominator add. The slope between $\omega = 1$ and $\omega = 5$ is thus $-40$ dB/decade. Beyond $\omega = 5$, we find that $1 + j0.2\omega$ in the denominator has a nonzero contribution and should be involved. The result is a line with $-60$ dB/decade.

It is instructive at this point to list the factors, together with their relevant frequencies of action and contributions to slope.

| FACTOR | CORNER FREQUENCY | CONTRIBUTION TO SLOPE (dB/DECADE) | NET SLOPE (dB/DECADE) |
|---|---|---|---|
| $\dfrac{1}{j\omega}$ | $\omega = 0$ | $-20$ | $-20$ |
| $\dfrac{1}{1 + j\omega}$ | $\omega = 1$ | $-20$ | $-40$ |
| $\dfrac{1}{1 + j0.2\omega}$ | $\omega = 5$ | $-20$ | $-60$ |

What we have just developed leads us to a description of a fast procedure for constructing the log magnitude characteristic. What we do is to consider the transfer function and place the corner frequencies in an ascending order. The term $j\omega$ in the denominator will make it necessary that we have a $-20$-dB/decade slope passing through $\omega = 1$ for low frequencies up to the first corner frequency. In the present example this is at $\omega_c = 1$, corresponding to the term $1 + j\omega$ in the denominator. This term now contributes $-20$ dB/decade, which is added to the slope maintained already of $-20$ dB/decade. What we do now is to break from the $-20$ dB/decade at $\omega = 1$ and draw a straight line with $-40$ dB/decade, which should continue until we encounter the next corner frequency at $\omega = 5$, corresponding to the term $1 + j0.2\omega$ in the denominator. This contributes an additional $-20$ dB/decade and thus we have to draw a line at $\omega = 5$, starting on the $-40$ dB/decade line but with the new slope $-60$ dB/decade. Since we have exhausted all terms, the job is done as shown in Figure 8.14c.

The phase plot for this transfer function is considered now. We use the straight-line approximations mentioned earlier to arrive at the plots shown in Figure 8.15a. The top plot is exact and represents the angle of $j\omega$ in the denominator. The middle plot corresponds to $1 + j\omega$ in the denominator with a corner frequency at $\omega = 1$. We assume that for $\omega \leq 0.1$, the phase contribution of this term is zero. For $\omega > 10$, the phase angle is $-90°$. The approximation between $\omega = 0.1$ and $\omega = 10$ is a straight line with a slope of $-45°$/decade. For the third factor of $1 + j0.2\omega$ in the denominator we have the bottom plot. The center frequency is $\omega = 5$ and for $\omega \leq 0.5$, the phase angle contribution of this term is zero. For $\omega \geq 50$ we have an angle of $-90°$. Again the slope of the line between $\omega = 0.5$ and $\omega = 50$ is $-45°$/decade. Since angles add, we invoke the normal graphical addition procedure utilized with the log amplitude case to arrive at the approximate phase plot shown in Figure 8.15b.

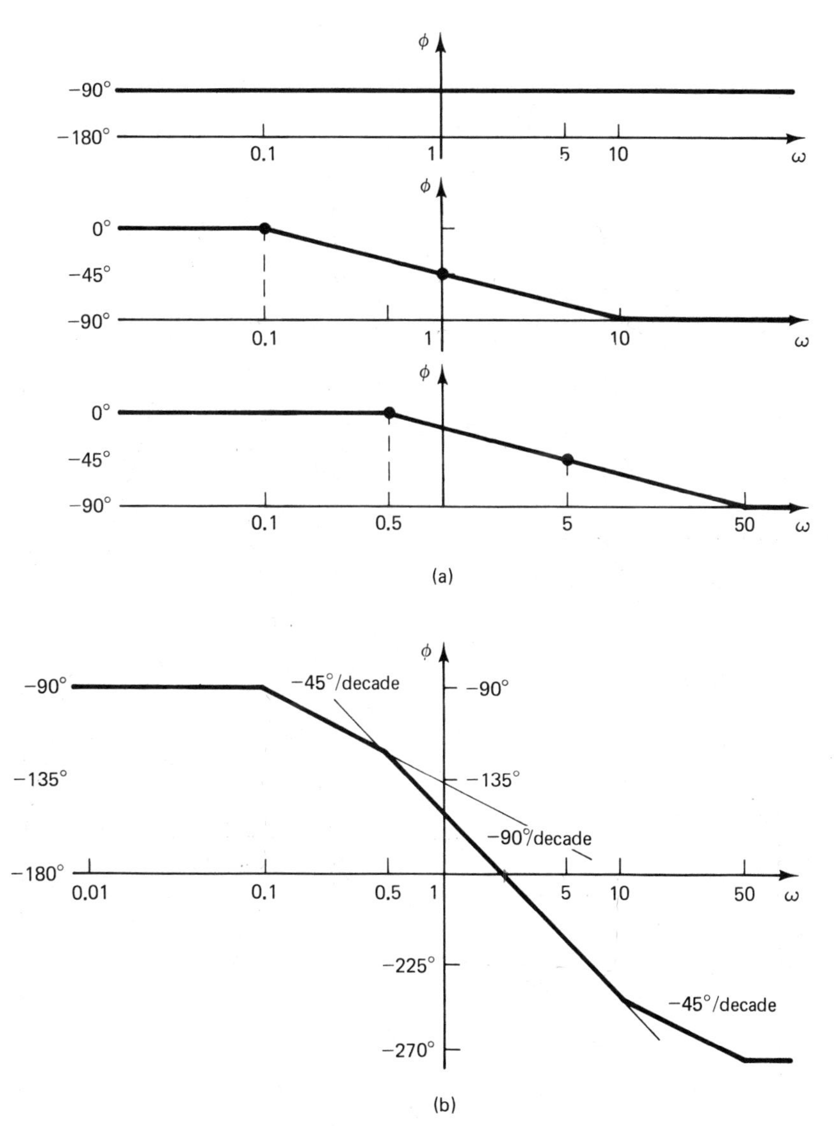


**Figure 8.15.** (a) Phase Plots for the Elements of Example 8.8
(b) Complete Approximate Phase Plots for Example 8.8

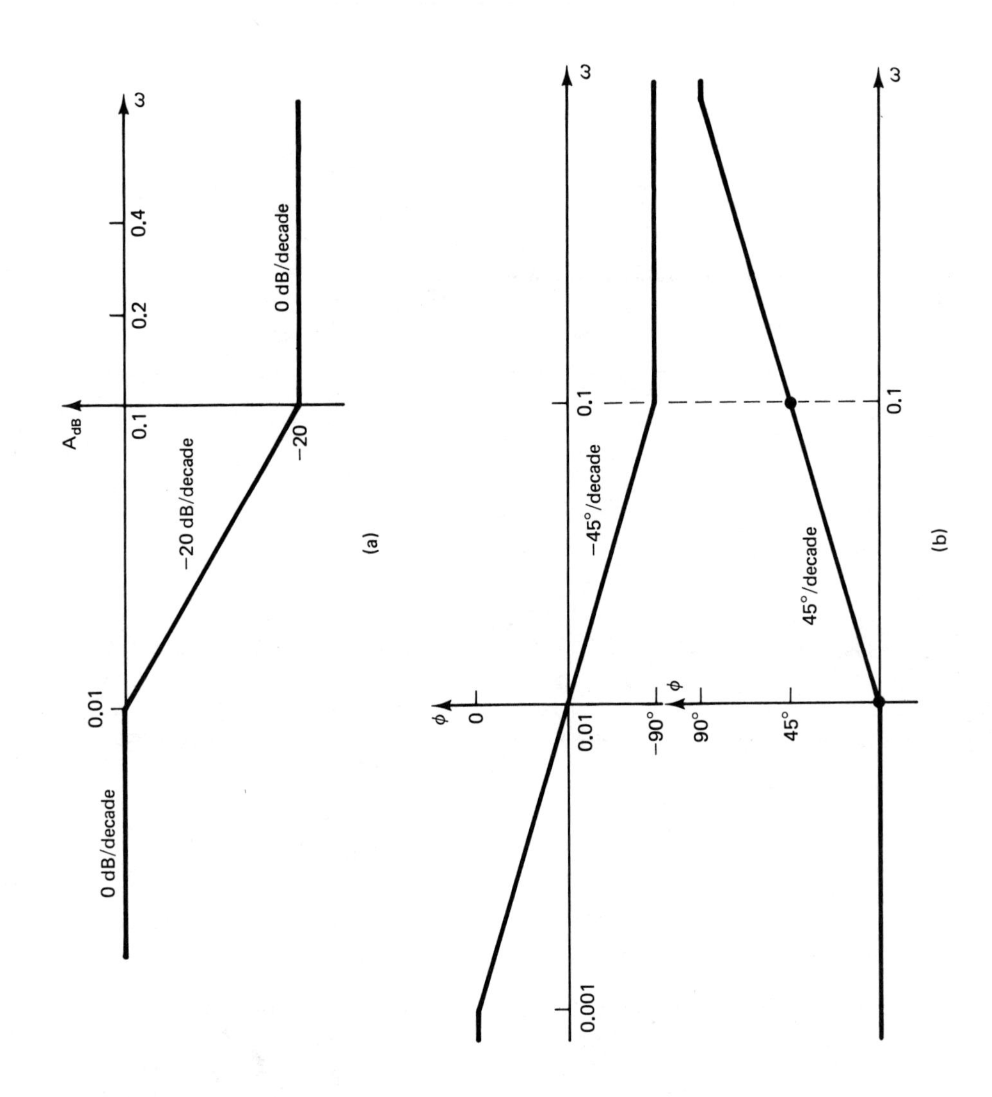

$A_{dB}$
0 dB/decade
0.01
−20 dB/decade
0.1
−20
0.2
0.4
0 dB/decade
ω
(a)
φ
0
0.001
0.01
−90°
−45°/decade
0.1
ω
φ
90°
45°
45°/decade
0.1
ω
(b)

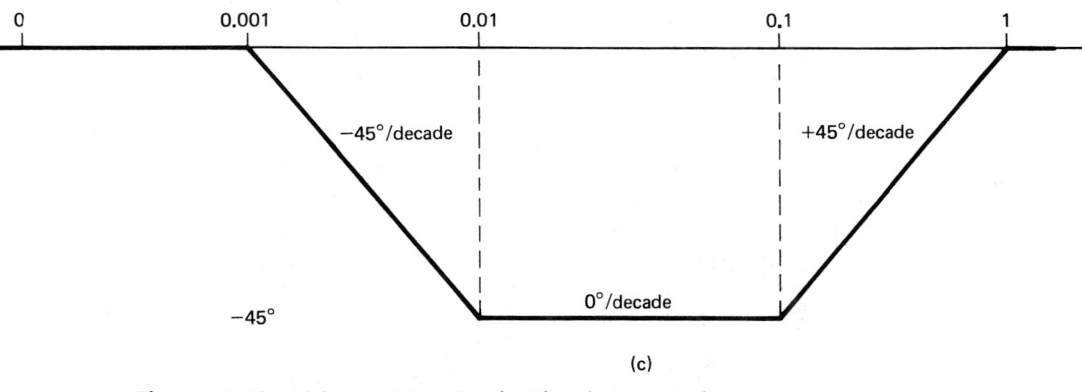


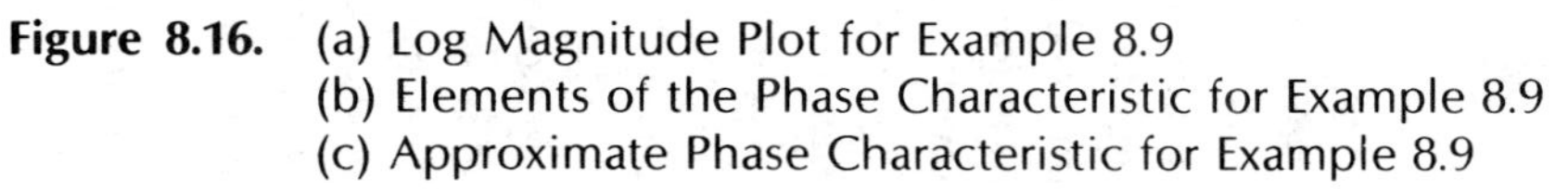

**Figure 8.16.** (a) Log Magnitude Plot for Example 8.9
(b) Elements of the Phase Characteristic for Example 8.9
(c) Approximate Phase Characteristic for Example 8.9

We will now take a look at a less involved example but with a numerator term, a case that we have not dealt with yet by way of example.

EXAMPLE 8.9

Construct the Bode logarithmic plot for the lag network with transfer function

$$G_c(j\omega) = \frac{1 + j10\omega}{1 + j100\omega}$$

SOLUTION

For this particular example we will simply use the fast procedure outlined at the end of Example 8.8. In this case we have two factors in the transfer function. The first is the $(1 + j100\omega)$ term in the denominator, which has a corner frequency of 0.01 rad/sec. Now let us note that Example 8.8 dealt with a type 1 system for which a denominator term of $j\omega$ is present and hence the plot commenced with a $-20$-dB/decade slope for low frequencies. For the transfer function at hand, this is a type 0 system and the log magnitude characteristic has a zero value for frequencies lower than the first corner frequency of 0.01 rad/sec. For frequencies above 0.01 rad/sec, the characteristic has a slope of $-20$ dB/decade. This continues until we encounter the next corner frequency, which is associated with the numerator term at $\omega = 0.1$ rad/sec. The numerator term contributes a $+20$ dB/decade. As a result, beyond $\omega = 0.1$ rad/sec, the log magnitude characteristic has a net slope of 0 dB/decade ($0 = -20 + 20$). The log magnitude characteristic is shown in Figure 8.16a.

The phase plot is obtained as shown first in Figure 8.16b, which shows in the top portion the approximate phase of $1 + j100\omega$ in the denominator. The bottom portion shows that of $1 + j10\omega$ in the numerator. Figure 8.16c shows the aggregate approximate phase characteristic.

The next building block to be discussed involves quadratic forms with complex-conjugate roots. Note that if the roots are real, we should use the procedure used in Example 8.8.

## Fourth Building Block: Quadratic Factors

Consider the transfer function

$$G(j\omega) = \frac{\omega_n^2}{(j\omega)^2 + 2\xi\omega_n(j\omega) + \omega_n^2}$$

or

$$G(j\omega) = \frac{1}{1 + 2\xi(j\omega/\omega_n) + (j\omega/\omega_n)^2}$$

Our interest presently is in this function, which can be written as

$$G(ju) = (1 + j2\xi u - u^2)^{-1}$$

where $u = \omega/\omega_n$. Multiplying and dividing by the conjugate of the denominator, we obtain

$$G(j\omega) = \frac{(1 - u^2) - j2\xi u}{(1 - u^2)^2 + 4\xi^2 u^2}$$

Thus

$$|G(j\omega)| = \frac{1}{\sqrt{(1 - u^2)^2 + 4\xi^2 u^2}}$$

The logarithm of the magnitude in decibels is thus

$$A_{\text{dB}} = 20\log|G(j\omega)| = -10\log\left[(1 - u^2)^2 + 4\xi^2 u^2\right]$$

The angle is given by

$$\phi(\omega) = -\tan^{-1}\frac{2\xi u}{1 - u^2}$$

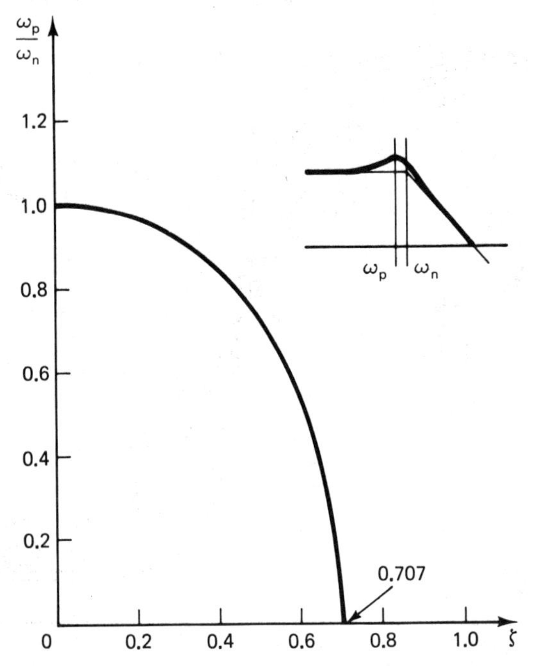


**Figure 8.17.** Frequency Where Peak Amplitude Occurs for the Second-Order System

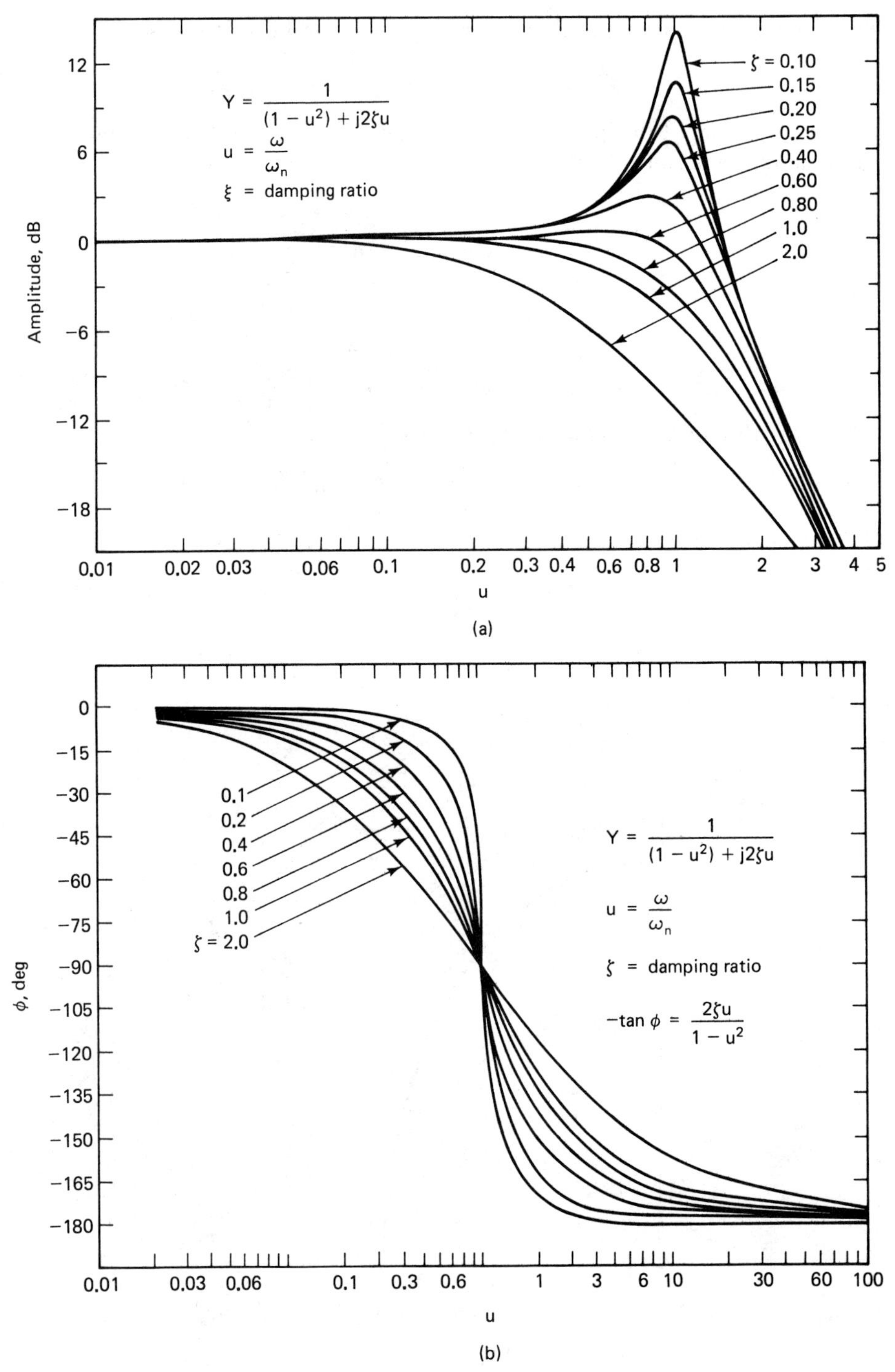


**Figure 8.18.** Logarithmic Characteristics for the Second-Order Factor

The amplitude in decibels for $u \ll 1$ is approximately given by

$$A_{\mathrm{dB}} = -10\log\left[(1-0)^2 + 0\right] = 0$$

and the phase angle approaches zero degrees. When $u \gg 1$, we have

$$A_{\mathrm{dB}} = -10\log u^4 = -40\log u$$

We thus obtain a straight line with a slope of $-40$ dB/decade. The phase angle approaches $-180°$.

We are thus lead to conclude that the asymptotes to the log magnitude plot for the second-order factor are similar to those for the first-order case with the exception of the slope which is $-40$ dB/decade. The corner frequency is at $u = 1$ or $\omega_c = \omega_n$, implying that the corner frequency is coincident with the natural frequency. Note that the asymptotes are independent of the damping ratio $\xi$.

Let us note that the magnitude of the function attains a peak at a frequency $\omega_p$ obtained by setting to zero the derivative of $|G(j\omega)|$ with respect to $\omega$ (or $u$). The result is the condition

$$u^2 = 1 - 2\xi^2$$

Thus the peak frequency is

$$\omega_p = \omega_n\sqrt{1 - 2\xi^2}$$

This is shown in relation to the natural frequency $\omega_n$ in Figure 8.17. The peak value is

$$|G(j\omega)|_p = \frac{1}{2\xi\sqrt{1-\xi^2}}$$

Thus the damping ratio determines the peak value. Errors obviously exist in the straight-line approximation. Figure 8.18a shows the exact log amplitude plot of a quadratic factor for various values of $\xi$. The error is large for small $\xi$. Figure 8.18b shows the angle characteristic for the second-order term exactly.

## Fifth Building Block: Transportation Lag

The transportation lag is characterized by a sinusoidal transfer function given by

$$G(j\omega) = Ke^{-j\omega\tau}$$

The magnitude is always equal to $K$, since

$$|e^{-j\omega\tau}| = 1$$

Therefore, the log magnitude is equal to a constant

$$A_{\mathrm{dB}} = 20\log K$$

The phase angle varies linearly with the frequency, as shown in Figure 8.19. Figure 8.20 is given also for $\omega$ on a logarithmic scale. Note that the angle increases indefinitely.

## System Type and the Bode Plot

There is a relation between system type and the low-frequency portion of the Bode plot. This can be found by examining simple functions that qualify for the system type considered. In the first place let us take a look at a type 0 system represented by

$$G_0(j\omega) = \frac{K_p}{1 + j\omega\tau}$$

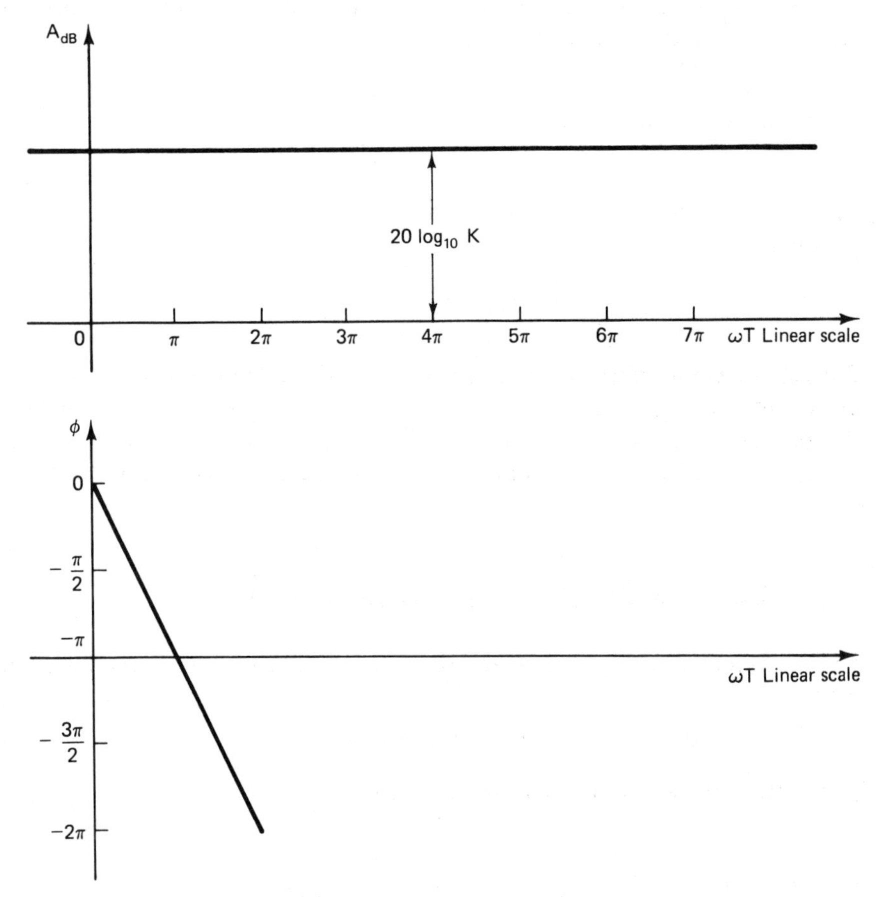


**Figure 8.19.** Log Magnitude Characteristic of Transport Lag, Linear Scale

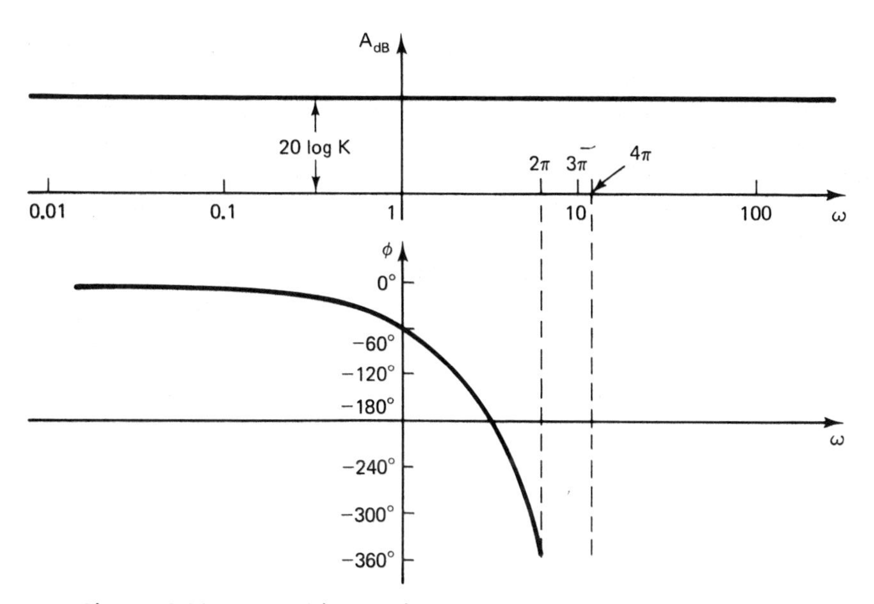


**Figure 8.20.** Logarithmic Characteristic of Transportation Lag

We have seen that below $\omega_c = 1/\tau$, the log amplitude is

$$A_{\text{dB}} = 20 \log K_p$$

The slope of the log amplitude plot is 0 dB/decade below the first corner frequency for any type 0 system. Note that we can obtain the static position error coefficient $K_p$ from the Bode plot as shown in Figure 8.21.

For a type 1 system we can take as an example the sinusoidal transfer function given by

$$G_1(j\omega) = \frac{K_v}{j\omega(1 + j\omega\tau)}$$

The log amplitude characteristic has a slope of $-20$ dB/decade at low frequency, as shown in Figure 8.22. Note that when $\omega = K_v$, the amplitude

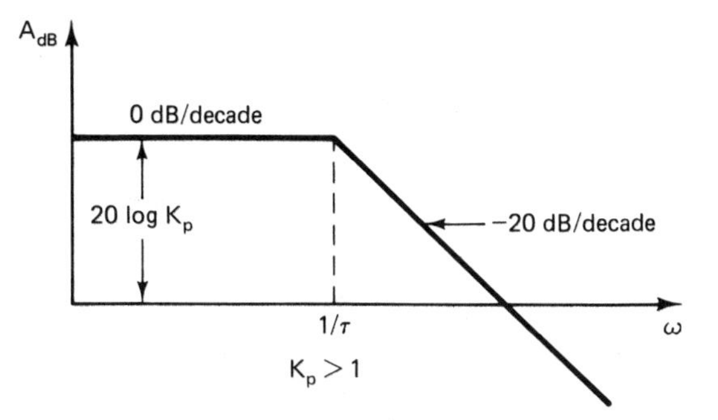


**Figure 8.21.** Log Magnitude Plot for a Type 0 System

in decibels of $K_v/j\omega$ is zero. Thus if the corner frequency $\omega_c = 1/\tau$ is greater than $K_v$, the low-frequency portion of the curve (slope of $-20$ dB/decade) crosses the 0-dB axis at a frequency $\omega_x = K_v$, as shown in Figure 8.22a. On the other hand, if $\omega_c$ is less than $K_v$, the low-frequency portion of the curve, which has a slope of $-20$ dB/decade, may be extended as shown in Figure 8.22b until it crosses the 0-dB axis. Again the value of the frequency where the extension crosses the 0-dB axis is $\omega_x = K_v$.

For a type 2 system let us take the function

$$G(j\omega) = \frac{K_a}{(j\omega)^2(1 + j\omega\tau_a)}$$

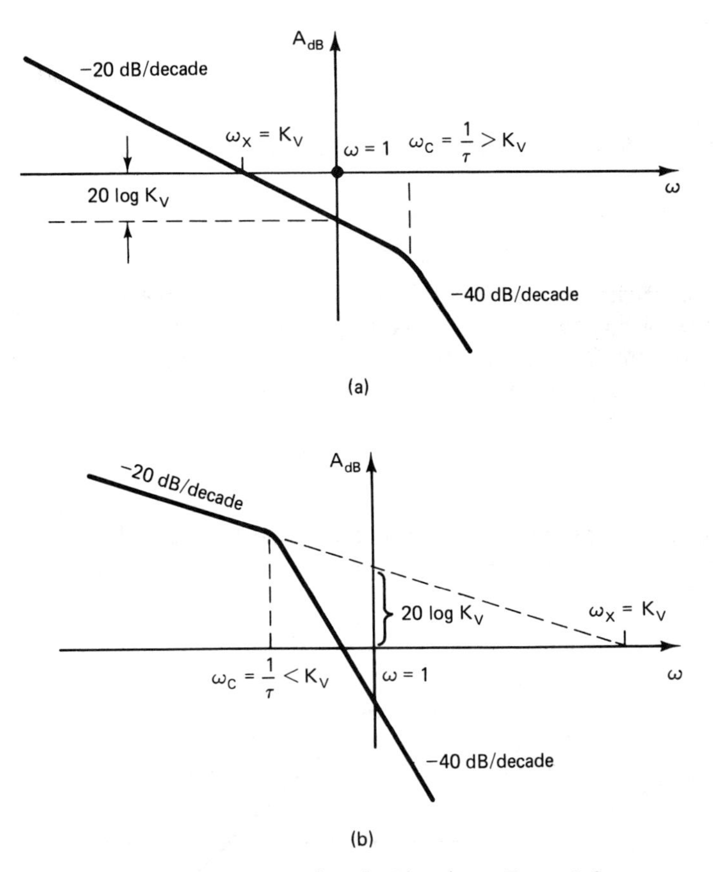


**Figure 8.22.** Log Amplitude Plot for a Type 1 System
(a) For $\omega_c > K_v$
(b) For $\omega_c < K_v$

At low frequency,

$$A_{\text{dB}} = 20 \log \frac{K_a}{\omega^2} = 20 \log K_a - 40 \log \omega$$

Substituting $\omega = 1$, we get

$$A_{\text{dB}}|_{\omega=1} = 20 \log K_a$$

Substituting $A_{\text{dB}} = 0$, we get

$$\omega_y^2 = K_a$$

We can thus conclude that the gain $K_a$ can be obtained as illustrated in Figure 8.23.

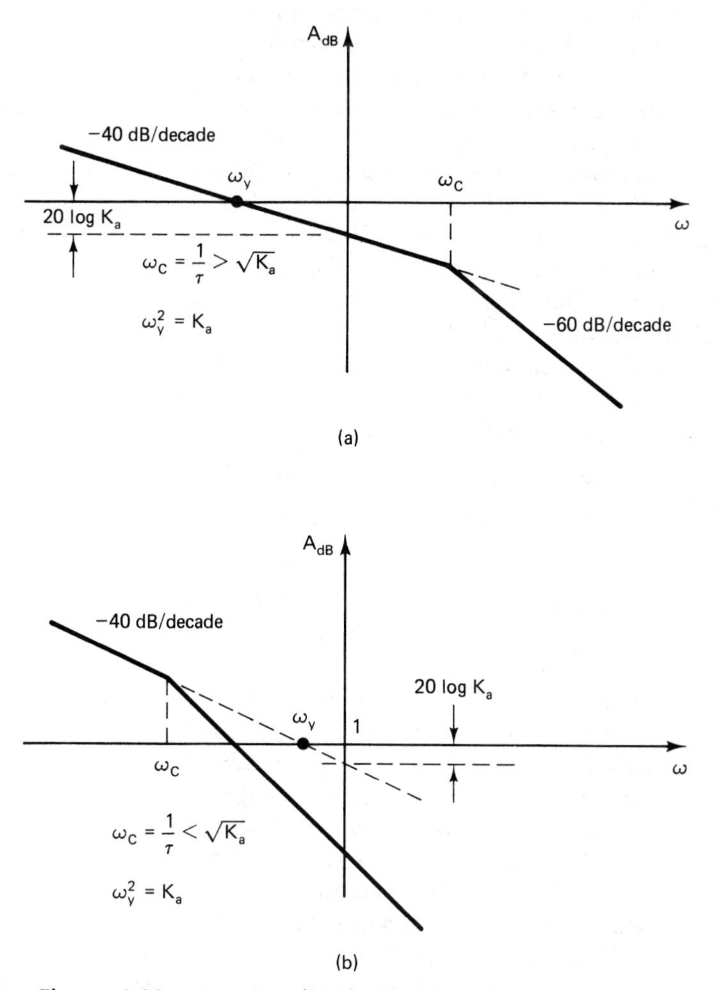


**Figure 8.23.** Log Amplitude Plot for a Type 2 System

## Experimental Evaluation of Transfer Functions

The Bode logarithmic plots can be used effectively to obtain the transfer function of a physical system that can be subjected to a frequency-response test. The procedure involves applying a sinusoidal input at a given frequency and then measuring the steady-state output, which for linear time-invariant systems should be sinusoidal with the same frequency as the input but with different amplitude and phase angle. The ratio of the amplitude of the output to the amplitude of the input is then the amplitude of the transfer function $|G(j\omega)|$ for that particular frequency. The difference between the phase angle of the output and that of the input is the angle of the sinusoidal transfer function $\phi$ for that frequency. The process is repeated for the frequency range of interest and the Bode plot is obtained on a point-by-point basis. This we will refer to as the experimental Bode plot.

With the experimental Bode plot available, we attempt to fit straight lines to the data such that the slopes of the lines are restricted to be positive or negative multiples of (6 dB/octave = 20 dB/decade) or zero. This step is crucial, as control systems specialists should use their experience extensively to arrive at the appropriate combination of lines to approximate the available data. Having obtained the straight-line approximation, the next step is to use the knowledge we have gained of the properties of the log magnitude plot to arrive at a possible transfer function to conform with the available plot.

A number of simple rules can be applied to reconstruct the transfer function from a systematic examination of the plot, starting at a low frequency and going to higher frequencies.

1. The system type can be determined from the slope of the low-frequency portion as indicated by

   Type 0: 0 dB/decade (octave)
   Type 1: $(-20$ dB/decade$) = (-6$ dB/octave$)$
   Type 2: $-40$ dB/decade $= (-12$ dB/octave$)$

2. A change in slope at a certain frequency indicates the presence of a new factor. If the change is by $-20$ dB/decade, we have a first-order term in the denominator of the form $1 + \tau s$, where $\tau = 1/\omega_c$. Naturally, a change by $+20$ dB/decade indicates a corresponding term in the numerator. A change of $\pm 40$ dB/decade indicates a second-order term. The dilemma that may face us is to come up with a decision as to whether this is due to a multiple pole (or zero) of the form $(1 + \tau s)^2$ or $[1 + (2\xi/\omega_n)s + s^2/\omega_n^2]$. This can be resolved by obtaining further data around the relevant corner frequency to observe

the actual shape of the curve. An underdamped system ($\xi < 1$) will exhibit a peak in the proximity of $\omega_n$, as indicated in Figure 8.18a.

3. The transfer function factors are assembled and the gain $K$ can be obtained easily.

Before embarking on our next example, we note that we have not yet examined the phase plot. To illustrate the concepts involved, let us take the following example.

EXAMPLE 8.10

Figure 8.24 shows the log magnitude plot experimentally obtained to determine the transfer function of a physical system. The straight-line approximations are shown together with the actual experimental values. Determine the transfer function using the available information.

SOLUTION

We start by noting that we have a $-6$-dB/octave slope at low frequencies. As a result, we conclude that there is an $s$ factor in the denominator. The

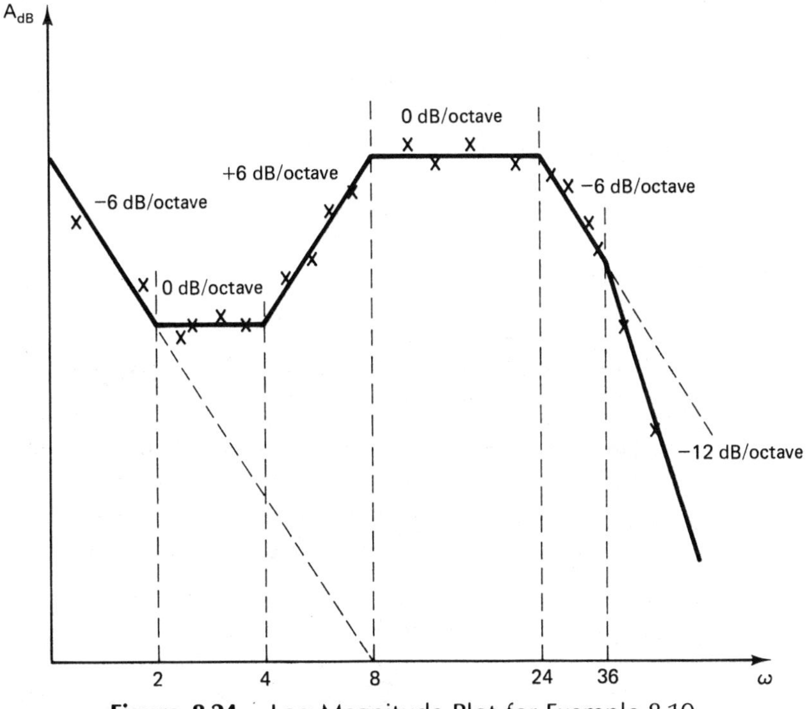


**Figure 8.24.** Log Magnitude Plot for Example 8.10

procedure can best be illustrated by building a tableau listing the frequencies at which a change in slope occurs, the dB/octave change, and the resulting factor in ascending order.

| *FREQUENCY* $\omega_c$ | *NEW SLOPE* | *CHANGE (dB/OCTAVE)* | *FACTOR* | $\tau$ |
|---|---|---|---|---|
| 2 | 0 | +6 | $1 + \tau_1 s$ | 0.5 |
| 4 | +6 | +6 | $1 + \tau_2 s$ | 0.25 |
| 8 | 0 | −6 | $\dfrac{1}{1 + \tau_3 s}$ | 0.125 |
| 24 | −6 | −6 | $\dfrac{1}{1 + \tau_4 s}$ | $\dfrac{1}{24}$ |
| 36 | −12 | −6 | $\dfrac{1}{1 + \tau_5 s}$ | $\dfrac{1}{36}$ |

We can thus conclude that the transfer function is given by

$$G(s) = \frac{K(1 + 0.5s)(1 + 0.25s)}{s(1 + 0.125s)(1 + s/24)(1 + s/36)}$$

One remaining issue is to determine $K$. This we do by observing that at low frequency

$$A_{\mathrm{dB}} = 20 \log \frac{K}{\omega}$$

At $\omega = 2$ we have $A_{\mathrm{dB}} = 12$; thus

$$12 = 20 \log \frac{K}{2}$$

This provides us with

$$K = 8$$

Let us observe here that our derivation of the transfer function assumed implicitly a number of points.

1. We excluded the possibility of terms of the form $(1 - \tau s)$, which should give rise to the same magnitude as $(1 + \tau s)$.
2. We excluded the possibility of the existence of a transportation delay term $e^{-Ds}$. This naturally has a magnitude of 1.

If the system satisfies these two assumptions, it is referred to as a minimum-phase system. To verify whether a system is minimum phase or not, the phase plot obtained by use of the derived function is compared with the actual phase plot. If there is agreement, the system is minimum phase; otherwise, it is not. For a minimum-phase system the log magnitude plot uniquely determines the phase plot. This idea is one of many important contributions of H. Bode.

For non-minimum-phase factors, knowledge of the source as well as the phase plot can be used to resolve the issue. For example, the factor $(1 - j\omega\tau)$ has an angle that varies from zero to $-90°$ as $\omega$ varies from zero to infinity. A delay $e^{-j\omega D}$ is associated with a continually decreasing phase angle. Note that nonminimum-phase factors are associated with poles and zeros in the right half of the $s$-plane.

# 8.5 Log Magnitude Versus Phase Plots

The polar plots and Bode logarithmic plots are extremely useful tools in the analysis and design of control systems. A third tool, which can be thought of as a consequence of both the polar plot and the Bode logarithmic plots, is the log magnitude versus phase plots. In a fashion similar to the polar plot, frequency is shown on the plot as a parameter. The variables displayed are the log magnitude in decibels versus the associated phase angle (linear scale) for each frequency of interest. It is thus clear that if we have an available Bode logarithmic plot for the transfer function considered, the log magnitude versus phase can be obtained easily. To illustrate the concept, we take the following examples.

Consider first the sinusoidal transfer function of a pure integrator

$$G(j\omega) = \frac{1}{j\omega}$$

As we saw earlier in this chapter,

$$A_{\text{dB}} = -20 \log \omega$$

$$\phi = -90°$$

It is thus clear that regardless of the value of $\omega$, the angle is invariant at $-90°$. Figure 8.25b gives us an opportunity to describe the plot. The origin of the plot is taken at $\phi = -180°$ and $A_{\text{dB}} = 0$. The angle is taken to be the horizontal axis and the amplitude in decibels is taken as the vertical axis. A simple graphical procedure to obtain the required plot can be used as shown in the layout of Figure 8.25. The purpose is essentially to eliminate $\omega$. In the present case the log magnitude versus phase angle is a vertical line passing through $\phi = -90°$, and there is no need for an elaborate description of the process.

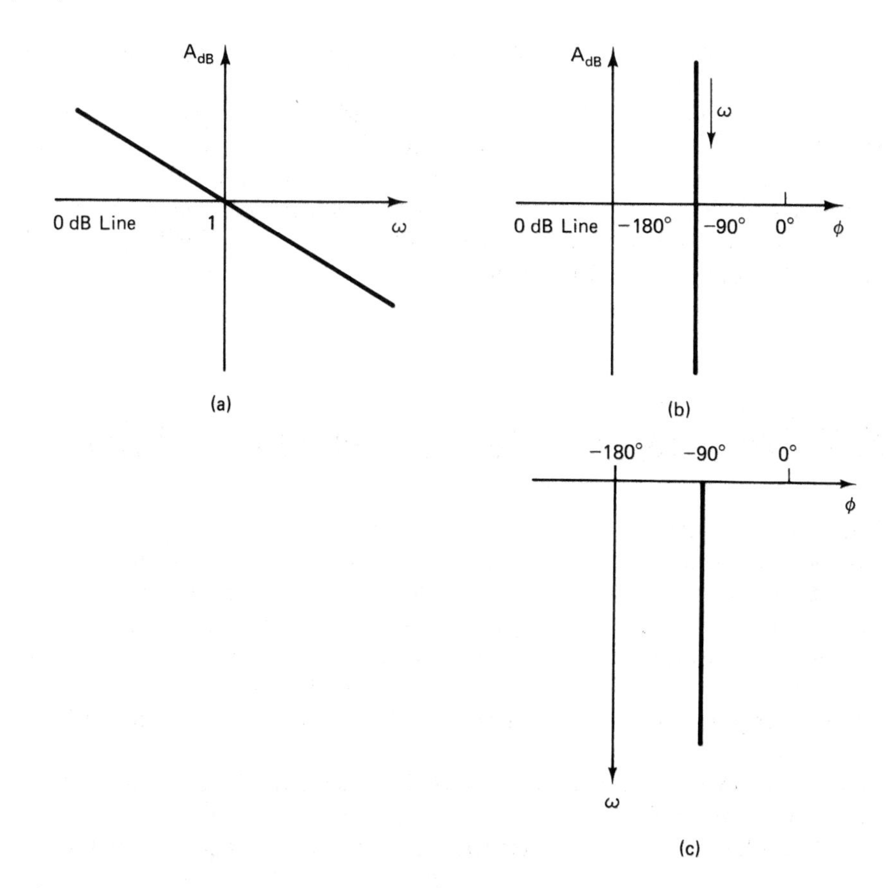


**Figure 8.25.** Construction of Log Magnitude versus Phase Plot (b) from Bode Plots (a) and (c) for $G(j\omega) = \dfrac{1}{j\omega}$
(a) Log Magnitude Plot
(b) Log Magnitude versus Phase Plot
(c) Angle Plot

For our next example take the function

$$G(j\omega) = \frac{1}{1 + j\omega\tau}$$

The asymptotic log magnitude versus frequency plot is shown in Figure 8.26a. The 0-dB line of Figure 8.26b is aligned with that of part (a). The approximate angle characteristic $\phi - \omega$ is illustrated in Figure 8.26c. Note that part (c) is rotated $90°$ clockwise from the normal position to allow for the alignment of the $\phi = -180°$ line in parts (b) and (c). Our task now is to take a given $\omega_1$ and project the corresponding amplitude in decibel $A_1$ from part (a) onto part (b). This is followed by a projection of the corresponding

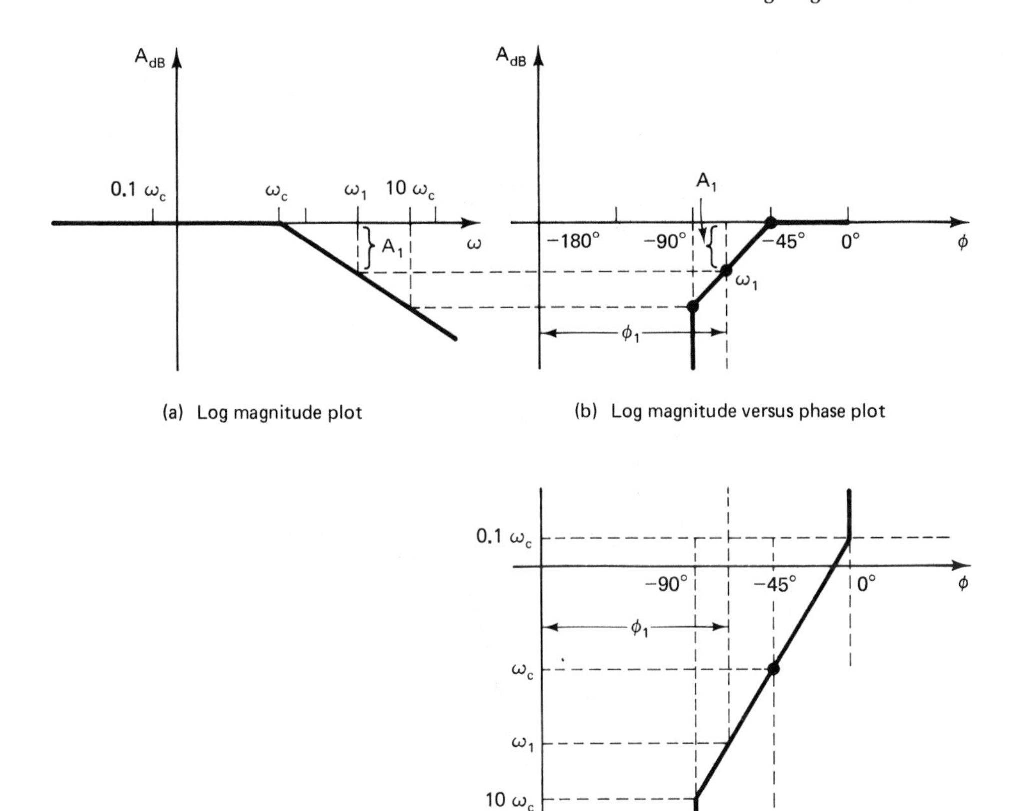


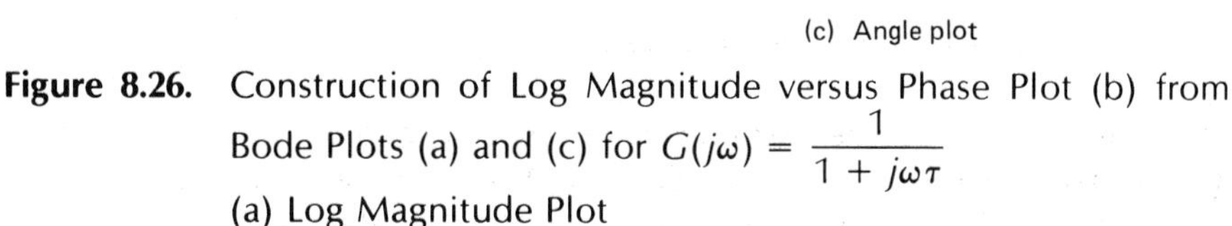


**Figure 8.26.** Construction of Log Magnitude versus Phase Plot (b) from Bode Plots (a) and (c) for $G(j\omega) = \dfrac{1}{1 + j\omega\tau}$
(a) Log Magnitude Plot
(b) Log Magnitude versus Phase Plot
(c) Angle Plot

angle $\phi_1$ from part (c) onto part (b). Note that with an exact Bode plot available we can obtain the desired characteristic point by point using this procedure. Figure 8.27 is designed to compare the approximate result with the exact curve.

Our third example involves the function

$$G(j\omega) = \frac{1}{j\omega(1 + j\omega\tau)}$$

Figure 8.28 shows the construction of the required log magnitude versus phase characteristic. Note that the change in slope at $\omega_c$ in the Bode plot is

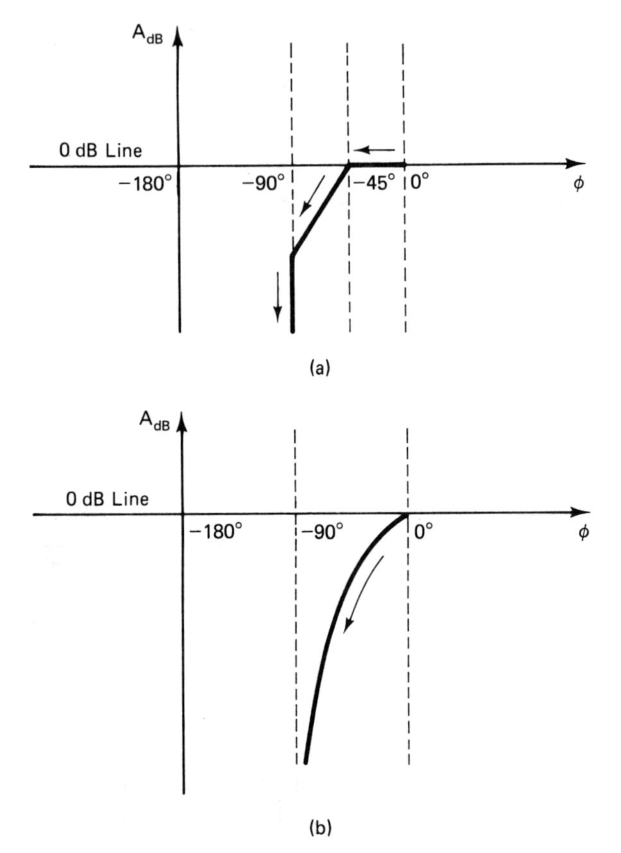


**Figure 8.27.** Comparison of Straight-Line Approximation and Exact Log Magnitude versus Phase Plot for $G(j\omega) = \dfrac{1}{1 + j\omega\tau}$

transformed onto our desired plot. Between $0.1\omega_c$ and $\omega_c$ the $A_{\text{dB}}$–$\phi$ slope is $[(-20 \text{ dB/decade})/(-45°/\text{decade})] = (20/45)$ dB/degree, while that between $\omega_c$ and $10\omega_c$ is double since the slope of the Bode plot doubles.

As a final example, Figure 8.29 depicts the exact log magnitude versus angle for the transfer function of Example 8.8.

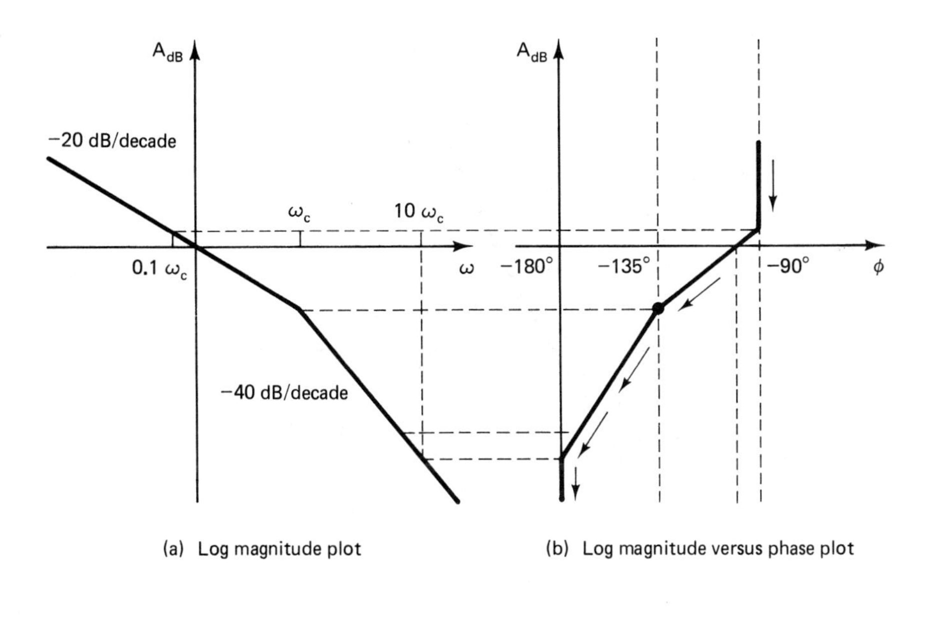


(a) Log magnitude plot

(b) Log magnitude versus phase plot

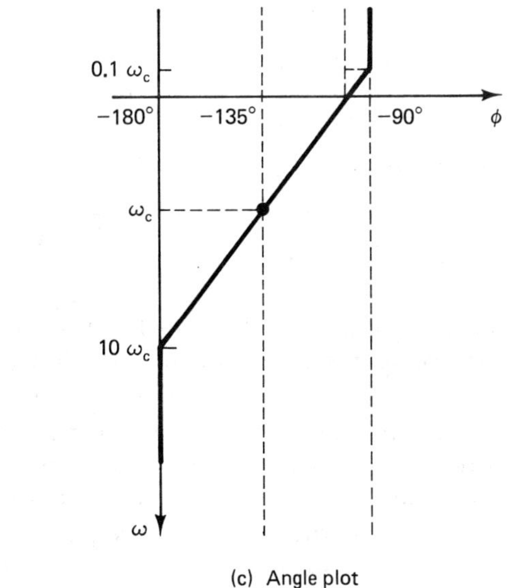


(c) Angle plot

**Figure 8.28.** Construction of Log Magnitude versus Phase Plot (b) from Bode Plots (a) and (c) for $G(j\omega) = \dfrac{1}{j\omega(1 + j\omega\tau)}$

(a) Log Magnitude Plot

(b) Log Magnitude versus Phase Plot

(c) Angle Plot

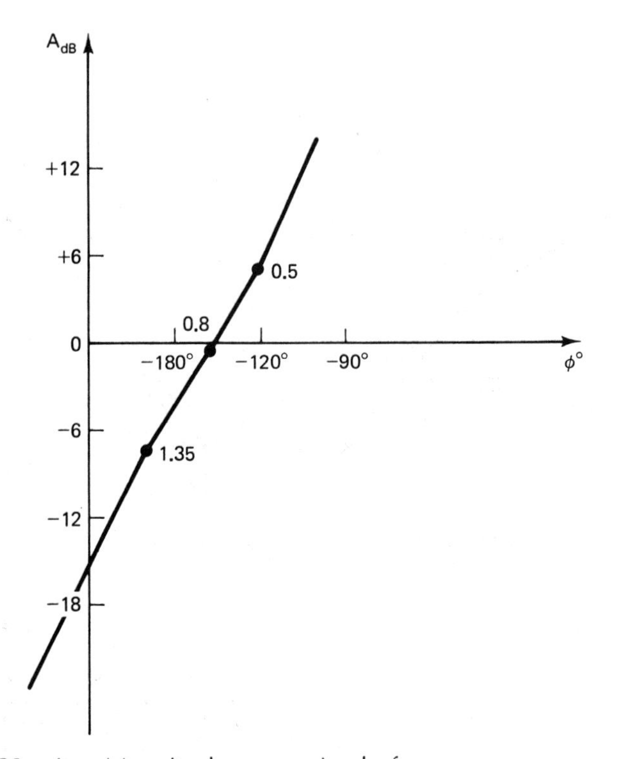


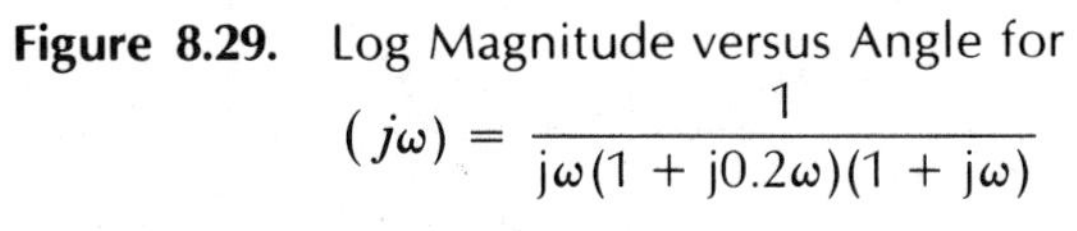

**Figure 8.29.** Log Magnitude versus Angle for

$$( j\omega) = \frac{1}{j\omega(1 + j0.2\omega)(1 + j\omega)}$$

response. We have seen in Chapter 6 that the stability of the feedback control system with forward function $G(s)$ and negative feedback $H(s)$ is determined from the roots of the characteristic equation

$$P(s) = 1 + G(s)H(s) = 0$$

For stability, all roots must lie in the left half of the $s$-plane. The Nyquist criterion for stability relates the loop frequency response $G(j\omega)H(j\omega)$ to the number of zeros and poles of $P(s)$ that lie in the right half of the $s$-plane. This criterion is discussed in the next section and is based on some fundamental results in the theory of complex variables. Our purpose in this section is to use a simple example to illustrate the necessary concepts that lead to the cornerstone of frequency-domain stability analysis known as Cauchy's principle of the argument.

## Contour Mappings: An Example

Consider the loop transfer function of a simple integrator given by

$$G(s)H(s) = \frac{K}{s}$$

The associated characteristic equation is

$$P(s) = 1 + \frac{K}{s} = 0$$

For the complex variable $s$ in the $s$-plane, we assume that rectangular coordinates are used; thus

$$s = \sigma + j\omega$$

As a result,

$$P(s) = \left(1 + \frac{K\sigma}{\sigma^2 + \omega^2}\right) - j\frac{K\omega}{\sigma^2 + \omega^2}$$

We note immediately that $P(s)$ is a complex variable as well. We can express $P(s)$ in rectangular coordinates as

$$P(s) = u + jv$$

Thus we have

$$u = 1 + \frac{K\sigma}{\sigma^2 + \omega^2}$$

$$v = -\frac{K\omega}{\sigma^2 + \omega^2}$$

To be specific, we take $K = 2$ to obtain

$$u = 1 + \frac{2\sigma}{\sigma^2 + \omega^2}$$

$$v = \frac{-2\omega}{\sigma^2 + \omega^2}$$

Consider the point $s = -1 + j$ in the $s$-plane; the associated values of $u$ and $v$ are

$$u = 1 - \frac{2}{2} = 0$$

$$v = \frac{-2}{2} = -1$$

Figure 8.30 illustrates the conclusions of the development so far. Starting with point $A$ in the $s$-plane, the application of $P(s)$ results in point $A'$ in the $P(s)$ plane. Thus $A$ is mapped into $A'$ and the arrow in Figure 8.30 symbolizes this fact.

Let us choose a closed path in the $s$ plane, shown as $ABCDEFGH$ in the $s$-plane of Figure 8.31. We now consider mapping this closed contour into the $P(s)$ plane. We decide on an easy route by simply taking the maps of $A$, $C$, $E$, and $G$. The corresponding points are:

$$s_A = -1 + j1 \qquad u_{A'} + jv_{A'} = 0 - j1$$

$$s_C = +1 + j1 \qquad u_{C'} + jv_{C'} = 2 - j1$$

$$s_E = +1 - j1 \qquad u_{E'} + jv_{E'} = 2 + j1$$

$$s_G = -1 - j1 \qquad u_{G'} + jv_{G'} = 0 + j1$$

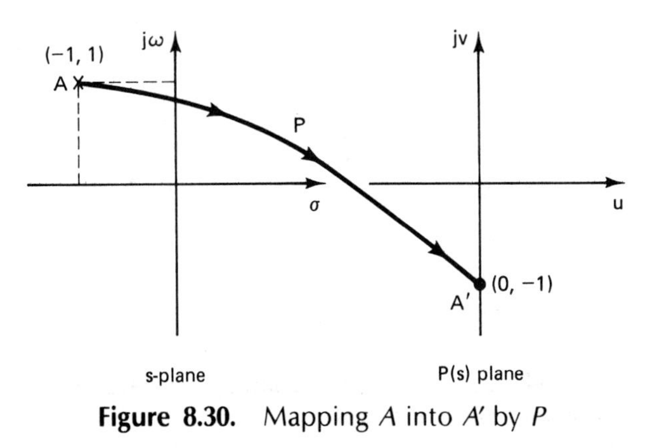


**Figure 8.30.** Mapping $A$ into $A'$ by $P$

The points $A'$, $C'$, $E'$, and $G'$ are not enough to construct the required contour in the $P(s)$ plane. The reason is that with these four points we infer that the contour is a square $A'C'E'G'$. A quick check for point $B$ reveals that

$$s_B = 0 + j1 \qquad u_{B'} + jv_{B'} = 1 - j2$$

Thus $B'$ is not on $A'C'E'G'$. We decide to include $B$, $D$, $F$, and $H$ with the corresponding result shown in Figure 8.31. We are now tempted to connect the points $H'A'B'$ by a straight line, and so on. The result is the diamond $A'B'C'D'E'F'G'H'$. Although this is sufficient for our purposes, it is not accurate enough. Indeed, if the keen reader carries out a quick manipulation, it can be shown that in the $P(s)$-plane the following relations hold.

1. For a variable $\sigma$ and fixed $\omega$, the $P(s)$-plane locus is described by

$$(u - 1)^2 + \left(v + \frac{1}{\omega}\right)^2 = \left(\frac{1}{\omega}\right)^2$$

   Thus the locus is a circle with center at

$$u_c = 1 \qquad v_c = \frac{-1}{\omega}$$

   The radius is

$$r = \frac{1}{\omega}$$

   Thus for the segment $ABC$, with $\omega = 1$, we have a semicircle with center at $1 - j1$ and radius of 1. Similarly, for $EFG$, the center is at $1 + j1$ and radius of 1.

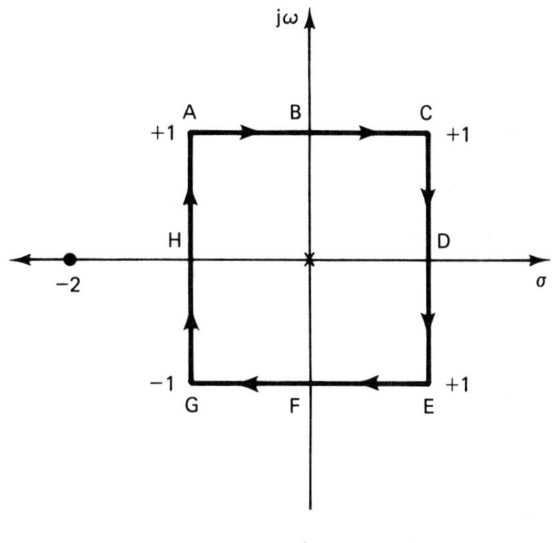

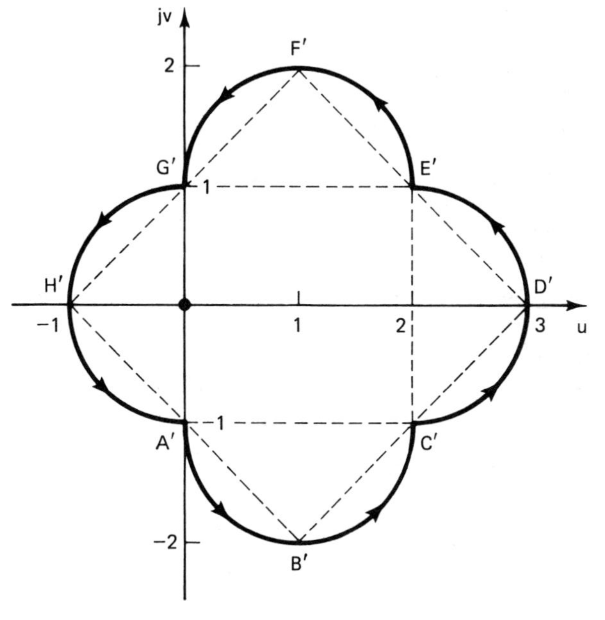


P(s) plane

**Figure 8.31.** Mapping the $s$-Plane Contour into the $P(s)$-Plane with Contour Enclosing the Pole of $P(s)$ in the $s$-Plane

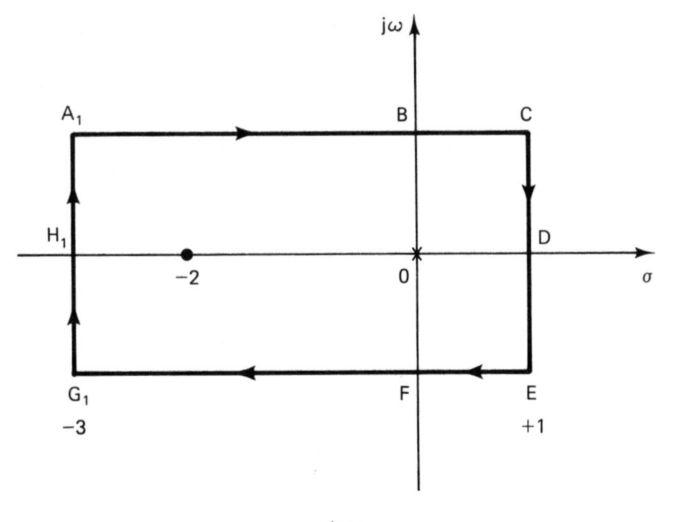


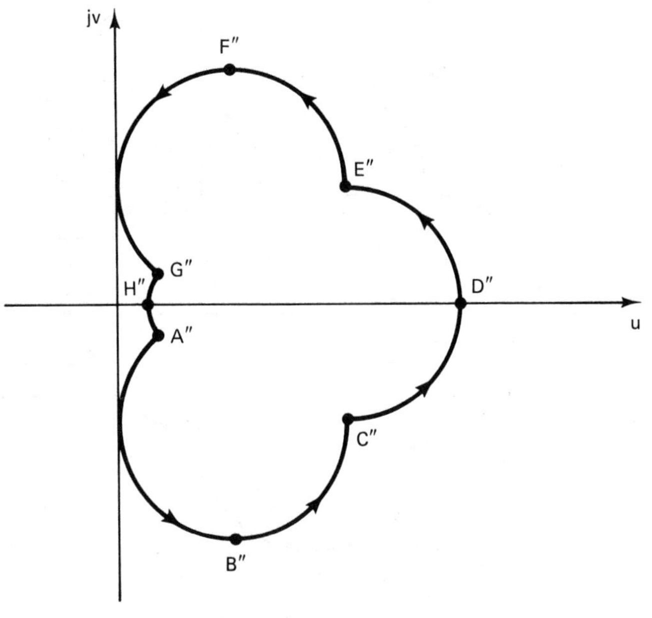


**Figure 8.32.** Mapping the $s$-Plane Contour into the $P(s)$-Plane with the Contour Enclosing Both the Pole and Zero of $P(s)$

2. For a variable $\omega$ and fixed $\sigma$, the $P(s)$-plane locus is described by

$$\left[u - \left(1 + \frac{1}{\sigma}\right)\right]^2 + v^2 = \frac{1}{\sigma^2}$$

Thus we have a circle with center at

$$u_c = 1 + \frac{1}{\sigma} \qquad v_c = 0$$

The radius is

$$r = \frac{1}{\sigma}$$

This applies for $CDE$ and $GHA$.

The conclusion here is that our contour in the $P(s)$-plane comprises the four semicircles shown in Figure 8.31.

A few observations are in order. The function $P(s)$ can be written as

$$P(s) = \frac{K + s}{s}$$

Thus $P(s)$ has a pole at the origin and a zero at $s = -K = -2$. The contour $ABCDEFGH$ in the $s$-plane chosen for Figure 8.31 encloses the pole at the origin and does not enclose the zero as shown. The contour in the $s$-plane rotates clockwise while the mapped contour $A'B'C'D'E'F'G'H'$ is rotating counterclockwise and is observed to enclose the origin of the $P(s)$-plane.

Let us now experiment with the contour in the $s$-plane as to the effect of enclosing the zero as well. In Figure 8.32 we shift $AHG$ to be vertical but at $\sigma = -3$ instead of $\sigma = -1$; we are thus mapping $A_1H_1G_1$. The three other segments are unchanged except for their termination points. The effect now is that the locus for $G''H''A''$ is a circle with a center and radius smaller than the ones for $G'H'A'$. The contour in the $P(s)$-plane no longer encloses the origin!

Our next experiment is to shift $CDE$ so that $\sigma = -1$ instead of $\sigma = +1$. Thus $C_2D_2E_2$ and hence $A_1C_2D_2E_2G_1H_1$ encloses the zero at $\sigma = -2$. Figure 8.33 shows the result of the mapping. Let us note here that the contour in the $P(s)$-plane encloses the origin and rotates clockwise!

We now sum up the conclusions of the foregoing experiments:

1. For a closed contour in the $s$-plane traversing clockwise and enclosing a pole of $P(s)$, the corresponding contour in the $P(s)$ plane encloses the origin and the traversal is in a counterclockwise direction. This is based on Figure 8.31.
2. For a closed contour in the $s$-plane with clockwise traversal and enclosing the pole and zero of $P(s)$, the corresponding

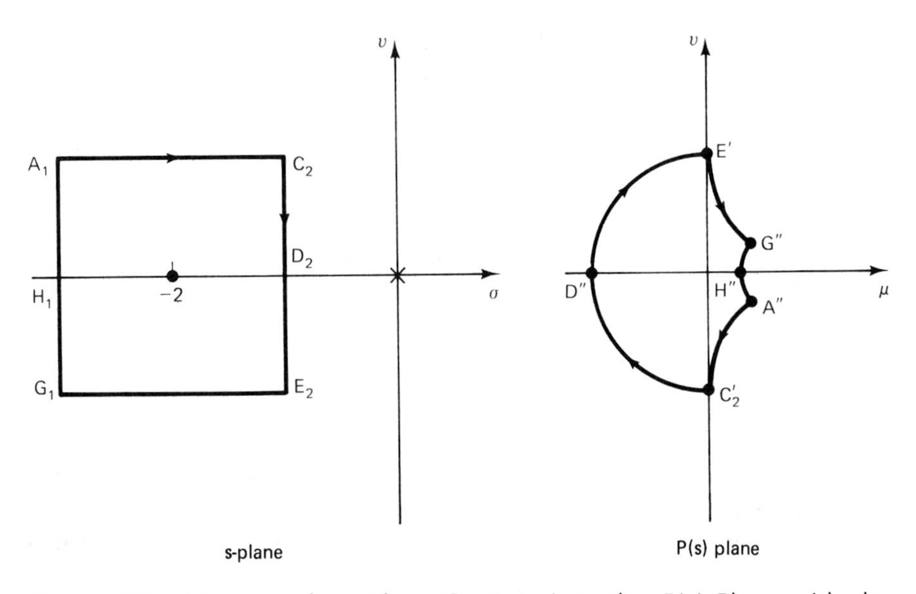


**Figure 8.33.** Mapping the $s$-Plane Contour into the $P(s)$-Plane with the Contour Enclosing the Zero of $P(s)$

contour in the $P(s)$-plane does not enclose the origin and the traversal is in a counterclockwise direction. This is based on Figure 8.32.

3. For a closed contour in the $s$-plane with clockwise traversal and enclosing the zero of $P(s)$, the corresponding contour in the $P(s)$-plane encloses the origin and traversal is clockwise. This is based on Figure 8.33.

## Cauchy's Principle of the Argument

The *principle of the argument* is a theorem due to Cauchy and deals with the relation between the encirclement of the poles and zeros of $P(s)$ in the $s$-domain by a contour, and the encirclement of the origin in the $P(s)$ plane by the mapping of the contour. An engineering-motivated statement of the principle is as follows.

Assume that an $s$-plane contour $\Gamma_s$ and function $P(s)$ are considered. The contour $\Gamma_s$ satisfies the following:

1. $\Gamma_s$ encircles $Z$ zeros and $P$ poles of $P(s)$ in the $s$-plane.
2. $\Gamma_s$ does not pass through any poles or zeros of $P(s)$.
3. The traversal is in the clockwise direction along the contour.

Then the contour $\Gamma_p$ in the $P(s)$-plane satisfies the following:

1. $\Gamma_p$ encircles the origin of the $P(s)$-plane $N$ times, where
$$N = Z - P$$
2. The traversal in the $P(s)$ plane is clockwise $N$ times. Note that a negative $N$ means counterclockwise traversal.

We can relate our findings of the preceding subsection to Cauchy's principle of argument as follows. The function $P(s)$ has a pole at the origin and a zero at $s = -2$.

1. The contour $ABCDEFGH$ in the $s$-plane of Figure 8.31 encircles no zeros and one pole and does not pass through any poles or zeros of $P(s)$ with clockwise traversal. According to Cauchy's principle,
$$N = 0 - 1 = -1$$
Thus in the $P(s)$-plane the contour $A'B' \cdots H'$ encircles the origin once in the counterclockwise direction.
2. The contour $A_1BCDEFG_1H_1$ in Figure 8.32 encircles one zero and one pole of $P(s)$. According to Cauchy's principle,
$$N = 1 - 1 = 0$$
Thus the contour $A''B'' \cdots H''$ in the $P(s)$-plane does not encircle the origin.
3. The contour $A_1C_2D_2E_2G_1H_1$ in Figure 8.33 encircles one zero and no poles of $P(s)$. As a result,
$$N = 1 - 0 = 1$$
Thus the contour $A''C_2'D''E'G''H''$ in the $P(s)$-plane encircles the origin once in the clockwise direction.

It should be noted that Cauchy's principle of the argument is valid for *any* closed contour satisfying the theorem's conditions in the $s$-plane. The discovery that a *specific* closed contour in the $s$-plane leads to conclusions about the stability of the system is due to Nyquist's vision.

# 8.7 Nyquist's Stability Criterion

For a stable system, the roots (zeros) of the characteristic equation $P(s)$ must all lie in the left-hand side of the $s$-plane. The existence of roots of the characteristic equation in the right-hand side of the $s$-plane indicates instability. This fact and Cauchy's principle of the argument were combined by Nyquist to construct a closed contour $Q$ such that all of the right half of

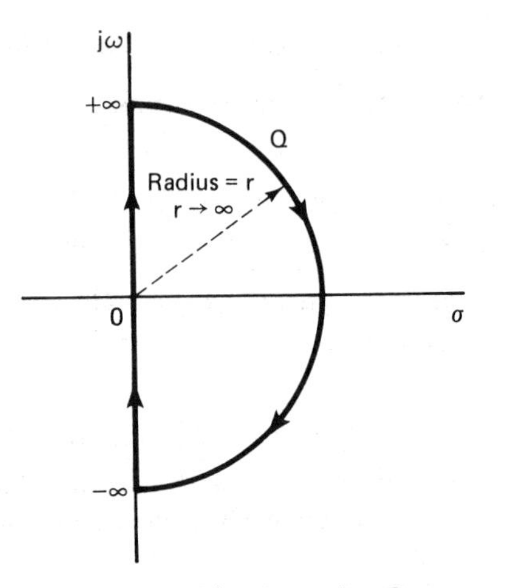


**Figure 8.34.** The Nyquist Contour

the $s$-plane is encircled to detect the presence of poles and zeros of $P(s)$ with positive real parts. The contour $Q$ is referred to as the *Nyquist contour* and is shown in Figure 8.34.

The entire right half of the $s$-plane is included by the closed path $Q$, which is composed of two segments. The first is the whole of the imaginary axis from $-j\infty$ to $+j\infty$ with the exception of poles on the imaginary axis, as discussed later. The second segment is a semicircle of infinite radius that encloses the entire right half of the $s$-plane. The mapping of $Q$ by $P(s)$ is obtained for the first segment by substituting $s = j\omega$ in $P(s)$ and letting $\omega$ vary from $-\infty$ to $+\infty$. For the second segment we substitute $s = re^{j\theta}$ in $P(s)$, the angle $\theta$ is varied from $+\pi/2$ to $-\pi/2$, and the radius $r$ is taken to approach infinity.

With the mapping $P(s)$ of the closed path $Q$ available, we can make the following observations on the basis of the principle of the argument:

1. The total number of zeros $Z_R$ of the function $P(s)$ in the right half of the $s$-plane is equal to the number of clockwise encirclements of the origin of the $P(s)$-plane.
2. The total number of poles $P_R$ of $P(s)$ in the right half of the $s$-plane is equal to the number of counterclockwise encirclements of the origin of the $P(s)$-plane.
3. The total number of poles $P_R$ minus the total number of zeros $Z_R$ of $P(s)$ in the right half of the $s$-plane denoted by $N$ is equal to the net number of encirclements of the origin of $P(s)$:

$$N = P_R - Z_R$$

The number $N$ is positive for counterclockwise encirclement and negative for clockwise encirclement. Note that $Q$ is clockwise in this convention.

Observation 3 leads us to the conclusion that for a stable system the net number $N$ of encirclements of the origin of the $P(s)$-plane must be positive (i.e., counterclockwise) and is equal to the number of poles $P_R$ that lie in the right half of the $s$-plane.

Looked at from the opposite side, we can conclude that if the mapping $P(s)$ encircles the origin in a net clockwise direction, $N$ is negative and there is an excess of zeros $Z_R$ over poles $P_R$ in the right half of the $s$-plane, indicating instability. If $N = 0$, then $Z_R = P_R$ and the system may or may not be stable, according to whether $P_R = 0$ or $P_R > 0$.

In accordance with the principle of the argument's requirements, the Nyquist contour $Q$ must not pass through any poles or zeros of $P(s)$. In terms of a control system with open loop transfer function $G(s)$ and negative feedback with transfer function $H(s)$, we have

$$P(s) = 1 + G(s)H(s) = 0$$

Assume that

$$G(s) = \frac{N_1(s)}{D_1(s)}$$

$$H(s) = \frac{N_2(s)}{D_2(s)}$$

As a result, the characteristic equation is expressed as

$$P(s) = \frac{D_1(s)D_2(s) + N_1(s)N_2(s)}{D_1(s)D_2(s)} = 0$$

It can thus be concluded that the poles of $P(s)$ are the poles of the loop function $G(s)H(s)$.

Since in practical terms, we know the loop function, the following two rules can be utilized.

1. The contour $Q$ should not pass through any poles of the $G(s)H(s)$.
2. The number and location of poles of $P(s)$ is determined from knowledge of the poles of $G(s)H(s)$. Thus $P_R$, the number of poles of $P(s)$ in the right half of the $s$-plane, is the same as the number of poles of $G(s)H(s)$ in the right half of the $s$-plane.

To meet the first rule, the poles of $G(s)H(s)$ on the imaginary axis (first segment of $Q$) should be located and a detour about the poles should be included in the Nyquist contour, as indicated in Figure 8.35 for the loop

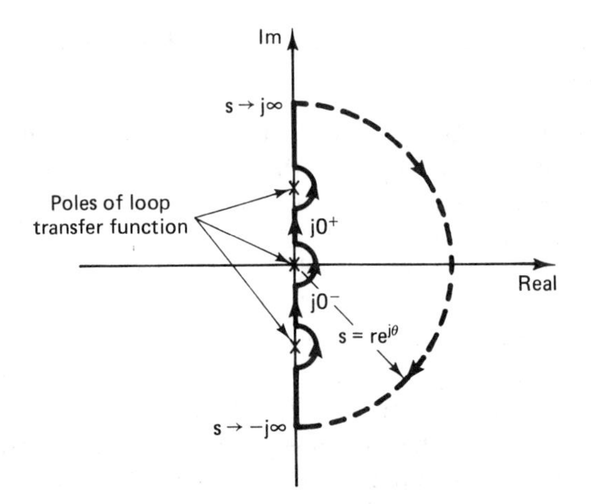


**Figure 8.35.** Modified Nyquist Contour to Avoid Poles on the Imaginary Axis

function

$$G(s)H(s) = \frac{K}{s(s^2 + \omega_n^2)}$$

Note here that we have three poles on the imaginary axis, at

$$p_1 = 0$$
$$p_2 = +j\omega_n$$
$$p_3 = -j\omega_n$$

It seems logical to enquire about the possibility of using the mapping of $G(s)H(s)$ instead of that of $P(s)$ in implementing our conclusions. The answer is affirmative since the origin of the $P(s)$ plane corresponds to the $(-1 + j0)$ point in $G(s)H(s)$. This can be seen from setting $P(s) = 0$ to arrive at

$$1 + G(s)H(s) = 0$$

Thus we should replace the origin of $P(s)$ by

$$G(s)H(s) = -1 + j0$$

in implementing our conclusions. This is advantageous, as normally $G(s)H(s)$ is available together with its polar plot, which as will be seen is part of the mapping considered.

We can now provide statements of the Nyquist stability criterion.

1. *First Version*: $P_R = 0$. A feedback system is stable if and only if the contour $\Gamma_{GH}$ in the $G(s)H(s)$ plane does not encircle the $(-1 + j0)$ point when the number of poles of $G(s)H(s)$ in the right side of the $s$-plane is zero.

2. *Second Version*: $P_R > 0$. A feedback control system is stable if and only if the number of counterclockwise encirclements of the $(-1 + j0)$ point by the contour $\Gamma_{GH}$ is equal to the number of poles of $G(s)H(s)$ with positive real parts.

It is now appropriate to summarize steps involved in the application of Nyquist stability criterion:

1. The loop function $G(s)H(s)$ is examined for the presence of poles in the right side of the plane and hence the number $P_R$ is determined. For $P_R = 0$ we use the first version of the criterion. For $P_R > 0$ we use the second version.
2. The presence of poles of $G(s)H(s)$ on the imaginary axis is evaluated and the Nyquist contour $Q$ is designed accordingly.
3. Construct the mapping of $Q$ by $G(s)H(s)$ and apply the appropriate version of the criterion to arrive at a conclusion as to the stability or instability of the closed-loop feedback system.

After this rather lengthy development, it certainly is appropriate to consider a few examples.

EXAMPLE 8.11

A single-loop control system has the loop transfer function

$$G(s)H(s) = \frac{K}{1 + \tau s}$$

We first observe that there are no poles on the imaginary axis of the $s$-plane. Thus the Nyquist contour of Figure 8.34 is appropriate. For positive $\tau$ there are no poles in the right half of the $s$-plane. We can therefore use version 1 of Nyquist's criterion.

The contour $\Gamma_{GH}$ is obtained in three steps. First, we let $s = +j\omega$, with $\omega$ varying from zero to infinity in $G(s)H(s)$. This provides us with the polar plot of $G(s)H(s)$, which was shown earlier in this chapter to be a semicircle with center at $K/2$ on the real axis of the $GH$ plane. The second segment is

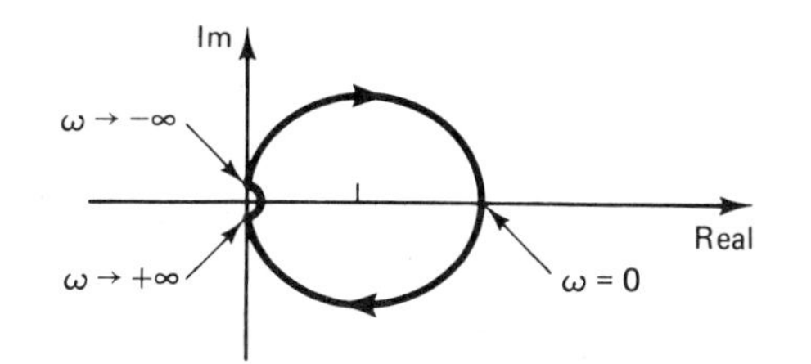


**Figure 8.36.** Nyquist Mapping for Example 8.11

obtained by letting $s = r^{j\theta}$; with $r$ approaching infinity, the angle $\theta$ is varied from $+90°$ to $-90°$. Clearly, the angle $\theta$ of $GH$ will vary from $-90°$ to $+90°$ and the magnitude of $GH$ will approach zero. We finally have to take the negative frequency range $\omega = -\infty$ to $\omega = 0$. This is simply the conjugate of the positive frequency polar plot. The complete contour in the $GH$ plane is shown in Figure 8.36. As there is no encirclement of the $(-1 + j0)$ for any value of $K$, the closed-loop system can be declared stable.

The system of the previous example is of type 0. Other systems of interest with the same type can be studied similarly. Take, for example,

$$G(s)H(s) = \frac{K}{(1 + \tau_1 s)(1 + \tau_2 s)}$$

Figure 8.37 shows the Nyquist mapping of the loop function and we note that this system is stable for all $K$ as well. Let us consider in the next example a higher-order type 0 system.

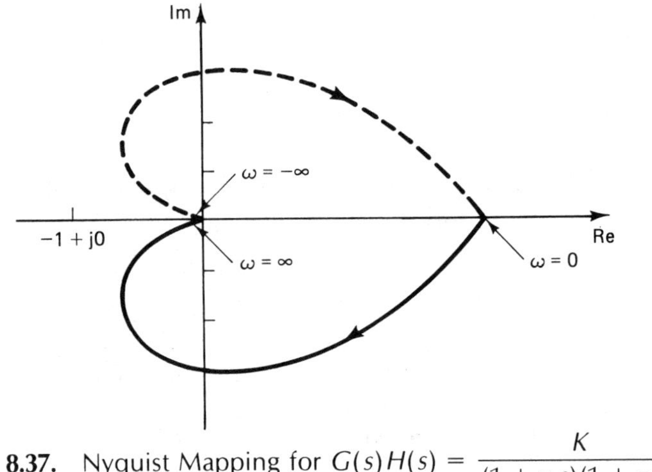


**Figure 8.37.** Nyquist Mapping for $G(s)H(s) = \dfrac{K}{(1 + \tau_1 s)(1 + \tau_2 s)}$

EXAMPLE 8.12

Consider the loop function given by

$$G(s)H(s) = \frac{K}{(1 + \tau_1 s)(1 + \tau_2 s)(1 + \tau_3 s)}$$

Since there are no poles of $G(s)H(s)$ on the imaginary axis and no poles in the right half of the $s$-plane, we will be contending with the polar plot of $G(s)H(s)$ and its conjugate as shown in Figure 8.38. For this system some interesting things happen. In Figure 8.38a we see that the $(-1 + j0)$ point is encircled twice, and according to version 2, the system is unstable. In

Figure 8.38b, the gain $K$ is reduced and the $(-1 + j0)$ point is not circled, indicating a stable system. This cannot be a surprise at this time in the light of the treatment of Chapters 6 and 7.

Let us now switch from a type 0 system to a type 1 system. We will first take the simplest case.

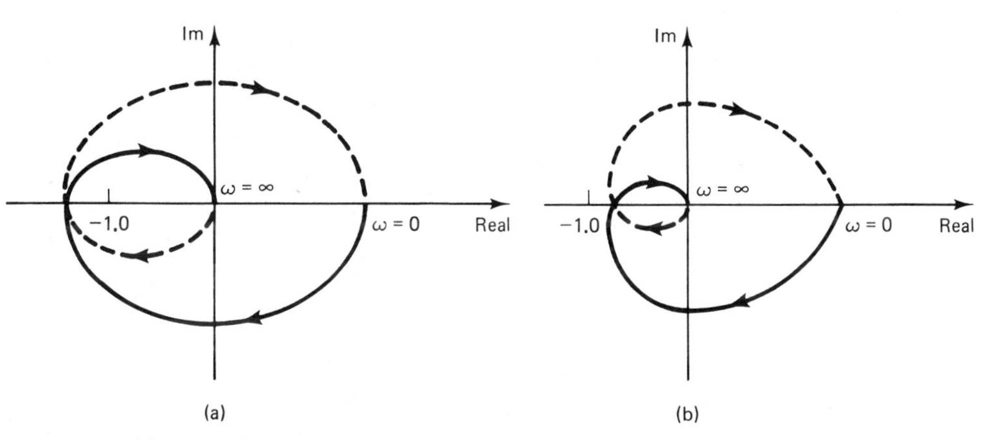


**Figure 8.38.** Nyquist Mapping for System of Example 8.12

EXAMPLE 8.13

Consider the closed-loop system with loop transfer function given by

$$G(s)H(s) = \frac{K}{s(1 + \tau s)}$$

We note here that for this type 1 system there is one pole on the imaginary axis precisely at the origin. A modified Nyquist contour should be used as shown in Figure 8.39a, where a semicircular detour about the origin is indicated.

Around the origin of the $s$-plane, we let

$$s = \varepsilon e^{j\theta}$$

The radius $\varepsilon$ is assumed to approach zero. The term $(1 + \tau s)$ will approach unity and as a result

$$\lim_{\varepsilon \to 0} G(s)H(s) = \lim_{\varepsilon \to 0} \frac{K}{\varepsilon e^{j\theta}} = \lim_{\varepsilon \to 0} \frac{K}{\varepsilon} e^{-j\theta}$$

The magnitude of the mapping will approach infinity. As $\theta$ is varied from $-90°$ to $+90°$, the angle of $GH$ will vary from $+90°$ to $-90°$, passing through $0°$ at $\omega = 0$.

The portion of the Nyquist contour from $\omega = 0$ to $\omega = \infty$ on the imaginary axis is mapped as the real frequency polar plot of $G(s)H(s)$, as

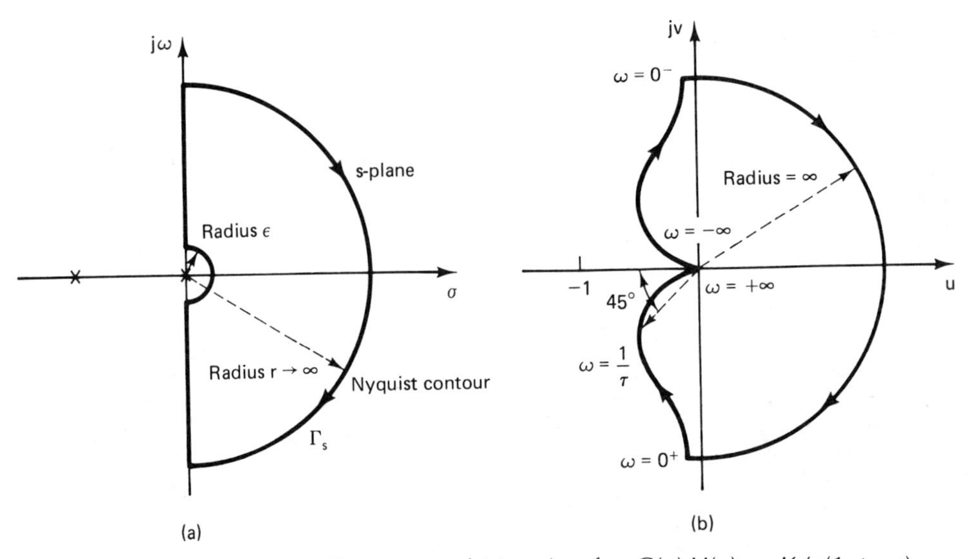


**Figure 8.39.** Nyquist Contour and Mapping for $G(s)H(s) = K/s(1 + \tau s)$

discussed earlier. The portion of the Nyquist contour from $\omega = +\infty$ to $\omega = -\infty$ is represented by

$$s = re^{j\theta}$$

where $r$ approaches infinity. The term $(1 + \tau s)$ will approach $\tau s$, and as a result,

$$\lim_{r \to \infty} G(s)H(s) = \lim_{r \to \infty} \frac{K}{\tau r^2} e^{-j2\theta}$$

Clearly, the magnitude of the mapping will approach zero. The angle of $G(s)H(s)$ will vary from $-180°$ at $\omega = +\infty$ to $+180°$ at $\omega = -\infty$, as $\theta$ is varied from $+90°$ to $-90°$.

The last portion of the Nyquist contour is from $\omega = -\infty$ to $\omega = 0$, that is, on the negative imaginary axis. The mapping is simply the conjugate of the polar plot. This completes the required mapping, as shown in Figure 8.39b.

The number of poles in the right half of the $s$-plane is zero and no encirclement of the $(-1 + j0)$ point occurs. As a result, version 1 of Nyquist stability criterion enables us to conclude that the closed-loop system is stable for all values of $K$.

In the next example we deal with a type 1 system of a higher order. It will become clear to the reader that as shown in Chapters 6 and 8 and Example 8.12, the stability of the system depends on the value of the gain $K$.

EXAMPLE 8.14

Consider the loop function given by

$$G(s)H(s) = \frac{K}{s(1 + \tau_1 s)(1 + \tau_2 s)}$$

We can apply the same procedure as in Example 8.13 to construct the Nyquist mapping as shown in Figure 8.40. In one case the gain $K$ is low enough so that the number of encirclements of the $(-1 + j0)$ is zero and the system is stable. On the other hand, for a higher value of $K$, two encirclements of the $(-1 + j0)$ point occur and the system is unstable.

It is interesting to indicate here how we can go about finding the critical gain $K$ for stability using the Nyquist mapping. The critical point to observe is that the intersection of the polar plot with the real axis is the determining factor. We can write

$$\begin{aligned} G(j\omega)H(j\omega) &= \frac{K}{j\omega(1 + j\omega\tau_1)(1 + j\omega\tau_2)} \\ &= \frac{K\left[j(\omega^2\tau_1\tau_2 - 1) - \omega(\tau_1 + \tau_2)\right]}{\omega\left[(1 - \omega^2\tau_1\tau_2)^2 + \omega^2(\tau_1 + \tau_2)^2\right]} \end{aligned}$$

The intersection with the real axis occurs for

$$\omega^2\tau_1\tau_2 - 1 = 0$$

This simply sets the imaginary part to zero. Thus the frequency at which

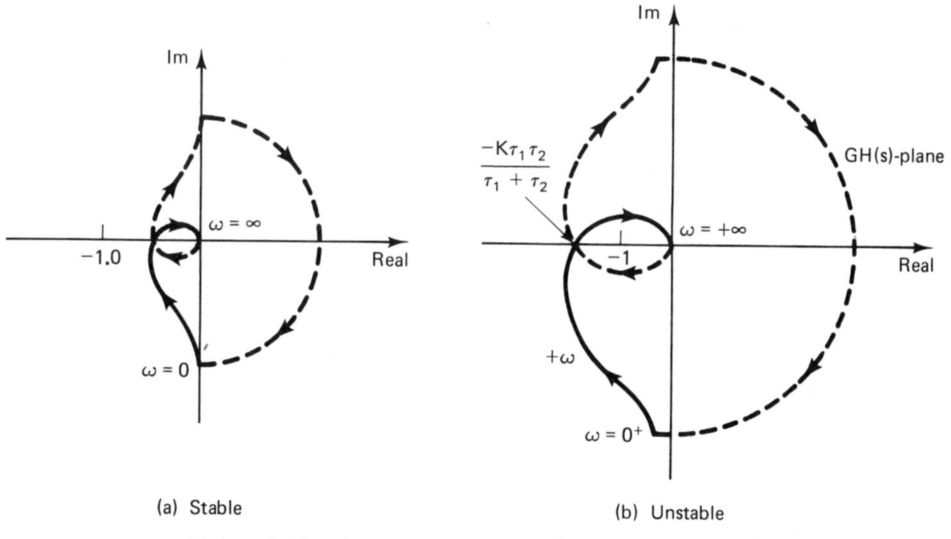


**Figure 8.40.** Nyquist Mapping for Example 8.14

the intersection with the imaginary axis occurs is

$$\omega_x = \frac{1}{\tau_1 \tau_2}$$

The value of the loop function for this frequency can be shown to be given by

$$G(j\omega_x)H(j\omega_x) = \frac{-K\tau_1\tau_2}{\tau_1 + \tau_2}$$

Thus the intersection with the real axis depends on $K$. For stability we require that

$$\frac{-K\tau_1\tau_2}{\tau_1 + \tau_2} > -1$$

or

$$K < \frac{\tau_1 + \tau_2}{\tau_1\tau_2}$$

This conclusion can be verified using the methods of Chapters 6 and 7.

Our next example is a type 2 system with one time constant.

EXAMPLE 8.15

Consider the loop function given by

$$G(s)H(s) = \frac{K}{s^2(1 + \tau s)}$$

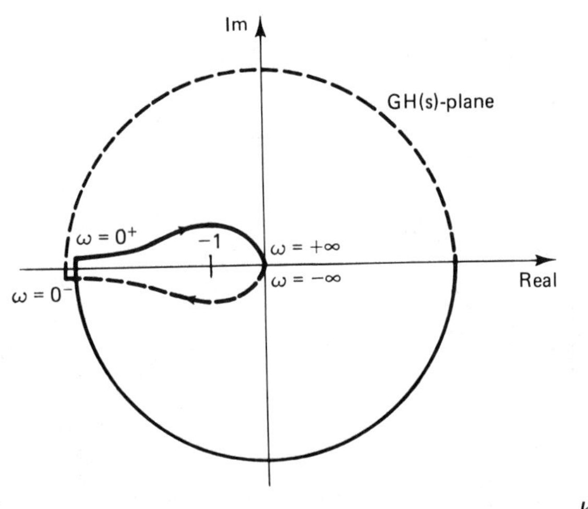


**Figure 8.41.** Nyquist Mapping for $G(s)H(s) = \dfrac{K}{s^2(1 + \tau s)}$

Investigate the stability of the closed-loop system using the Nyquist stability criterion.

The Nyquist mapping of the function is shown in Figure 8.41. The first version of the criterion leads us to conclude that the system is unstable for all values of $K$.

## Conditionally Stable Systems

Consider a closed-loop system with loop transfer function given by

$$G(s)H(s) = \frac{K(1 + \tau_3 s)(1 + \tau_4 s)}{s(1 + \tau_1 s)(1 + \tau_2 s)(1 + \tau_5 s)(1 + \tau_6 s)}$$

Let us assume that

$$\begin{aligned} K &= 900 \\ \tau_1 &= 2.5 \\ \tau_2 &= 1.43 \\ \tau_3 &= 0.4 \\ \tau_4 &= 0.2 \\ \tau_5 &= 0.02 \\ \tau_6 &= 0.0066 \end{aligned}$$

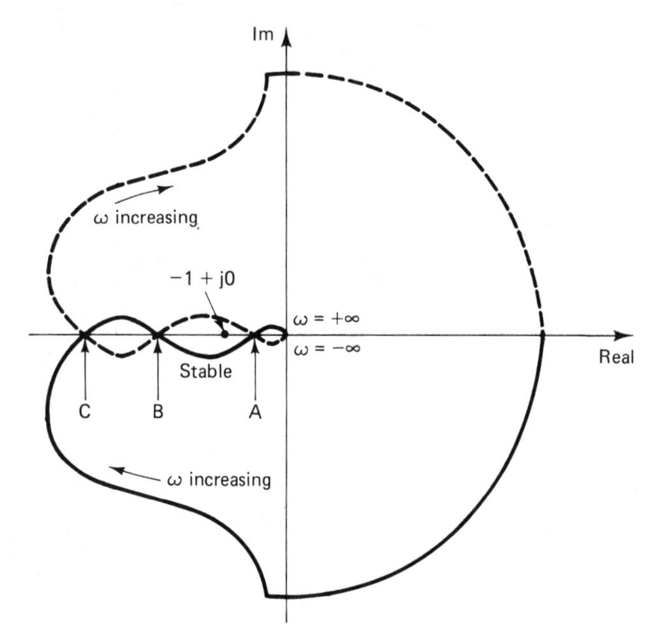


**Figure 8.42.** Nyquist Mapping for a System with

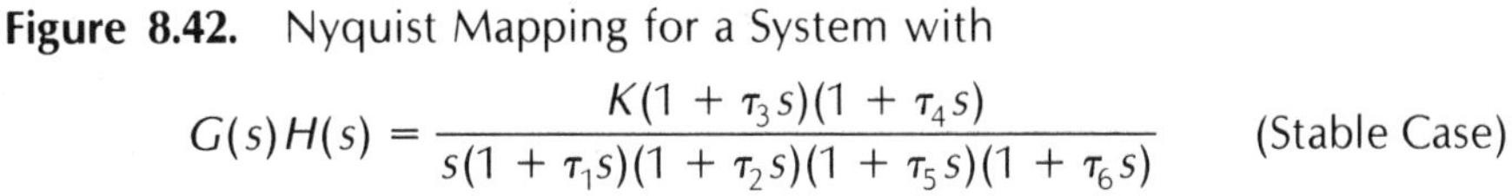

$$G(s)H(s) = \frac{K(1 + \tau_3 s)(1 + \tau_4 s)}{s(1 + \tau_1 s)(1 + \tau_2 s)(1 + \tau_5 s)(1 + \tau_6 s)} \quad \text{(Stable Case)}$$

The Nyquist mapping for this system is shown in Figure 8.42 and the closed-loop system is shown to be stable for the indicated values of gain $K$ and time constants.

From inspection of the Nyquist plot it is clear that encirclement of the $(-1 + j0)$ point can occur by changing the gain $K$ or one of the time constants. Figure 8.43 is used to illustrate the point using the relevant portion of Figure 8.42.

In Figure 8.43a, the polar plot encloses the $(-1 + j0)$ point and hence the closed-loop system is unstable. In Figure 8.43b the $(-1 + j0)$ point is

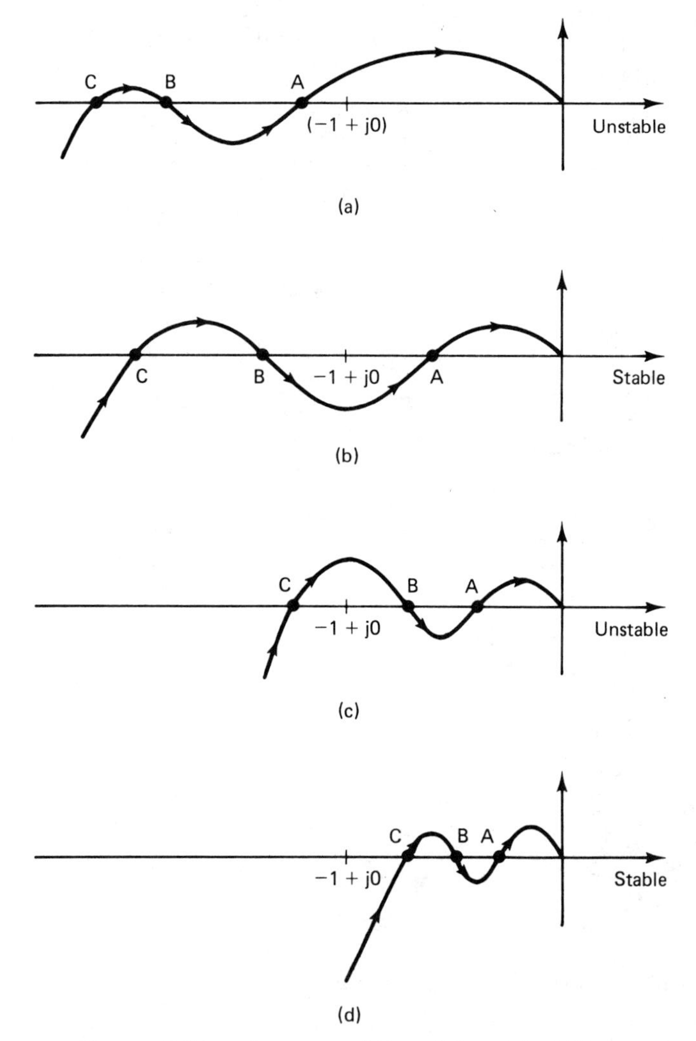


**Figure 8.43.** Concept of Conditional Stability

not enclosed and the system is stable. Yet for another case, as shown in Figure 8.43c, the system is unstable, while in Figure 8.43d we have a stable system. If the gain $K$ is taken as the variable, then $K_a > K_b > K_c > K_d$. It should be noted that instability can also occur for a change in a time constant. Systems that exhibit this characteristic are called conditionally stable systems.

## 8.8 Stability Margins

Before embarking on this topic, let us observe that in order to investigate stability it is sufficient to construct the polar plot, that is, the mapping $G(j\omega)H(j\omega)$ for the frequency range $0 < \omega < \infty$. The Nyquist stability criterion is defined in terms of encirclement of the $(-1 + j0)$ point, which is on the negative real axis of the $GH$-plane. We have seen that a system may or may not be stable, depending on the value of gain $K$ and the time constants. Examples 8.12 and 8.14 deal with systems in this category. We will use the system of Example 8.14 to illustrate the concepts introduced.

Consider the loop function of Example 8.14 given by

$$G(s)H(s) = \frac{K}{s(1 + \tau_1 s)(1 + \tau_2 s)}$$

We have seen that the critical value of the gain $K$ denoted by $K_c$, above which the system is unstable is given by

$$K_c = \frac{\tau_1 + \tau_2}{\tau_1 \tau_2}$$

The polar plot of $G(j\omega)H(j\omega)$ intersects the real axis at

$$\omega_x = \frac{1}{\sqrt{\tau_1 \tau_2}}$$

The magnitude of $G(j\omega)H(j\omega)$ at this frequency is given by

$$a = |G(j\omega_x)H(j\omega_x)| = \frac{K}{K_c}$$

For a given $K$ such that $K < K_c$, the system is stable, as shown in Figure 8.44, as no encirclement of the $(-1 + j0)$ point occurs. A measure of the relative stability of the system is the proximity of the $G(j\omega)H(j\omega)$ locus to the $(-1 + j0)$. Two relative stability margins can be defined if we note that instability may be caused by a change in either the magnitude or phase of the loop function $G(j\omega)H(j\omega)$.

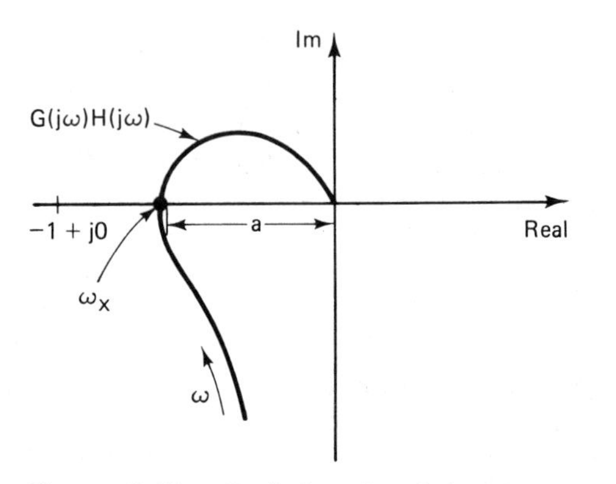


**Figure 8.44.** Defining the Gain Margin

In terms of magnitude, the gain margin (GM) is defined as the inverse of the magnitude of the loop function evaluated at the phase crossover frequency $\omega_x$ for which the phase angle is $-180°$.

$$\mathrm{GM} = \frac{1}{|G(j\omega_x)H(J\omega_x)|} = \frac{1}{a}$$

where $\omega_x$ is obtained from

$$\angle G(j\omega_x)H(j\omega_x) = -180°$$

We frequently express the gain margin in terms of decibels:

$$\mathrm{GM}_{\mathrm{dB}} = 20\log\frac{1}{|G(j\omega_x)H(j\omega_x)|} = -20\log a$$

For the system of Example 8.14 we have a gain margin of

$$\mathrm{GM} = \frac{1}{a} = \frac{K_c}{K}$$

Clearly, a unity gain margin indicates critical stability, while a gain margin of less than 1 indicates instability as $K > K_c$. In terms of decibels, we can conclude that

$$\mathrm{GM}_{\mathrm{dB}} = 20\log\frac{1}{a} = 20\log\frac{K_c}{K}$$

A zero gain margin in decibels indicates critical stability. A positive gain margin in decibels indicates stability, while a negative gain margin in decibels indicates instability. A desirable gain margin in decibels is in the range 4 to 12 dB.

Our second margin of stability is referred to as the phase margin, which is defined as the additional phase lag that can be introduced before the system becomes unstable. For a given $K$, a change in a time constant (or more) results in rotating the point for which the magnitude of $G(j\omega)H(j\omega)$ is unity to coincide with the point $(-1 + j0)$. This concept is best illustrated by reference to Figure 8.45.

An analytical procedure to determine the phase margin starts with finding the frequency $\omega_\phi$ for which the magnitude of $G(j\omega)H(j\omega)$ is unity. This is followed by an evaluation of the corresponding phase angle $\phi$. From this information the phase margin is readily determined.

$$\phi_m = 180° + \phi$$

In order to illustrate the analytical procedure to determine the phase margin, consider the second-order loop function of a two-phase servomotor with inertia

$$G(j\omega)H(j\omega) = \frac{K}{j\omega(1 + j\omega\tau)}$$

The phase angle $\phi$ is given by

$$\phi = -90° - \tan^{-1}\omega\tau$$

The phase margin is obtained at $\omega_\phi$ as

$$\begin{aligned}\phi_m &= 180 + \phi \\ &= 90° - \tan^{-1}\omega_\phi\tau \\ &= \tan^{-1}\frac{1}{\omega_\phi\tau}\end{aligned}$$

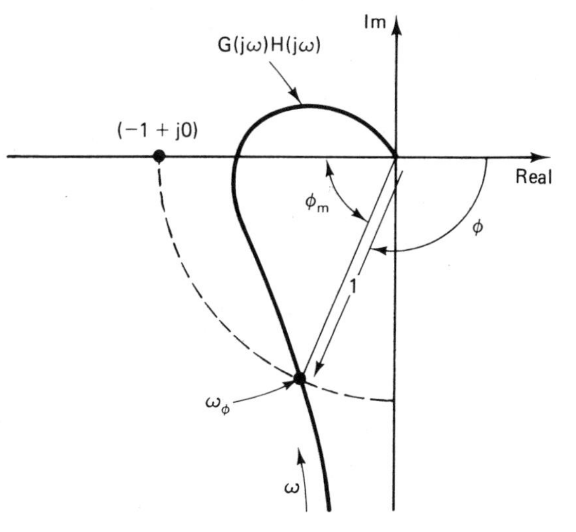


**Figure 8.45.** Defining the Phase Margin

Thus we have

$$\tan \phi_m = \frac{1}{\omega_\phi \tau}$$

The value of $\omega_\phi$ is obtained as the solution to

$$\frac{K^2}{\omega_\phi^2(1 + \omega_\phi^2 \tau^2)} = 1$$

The solution turns out to be

$$\omega_\phi^2 \tau^2 = \frac{1}{2}(\sqrt{1 + 4K^2\tau^2} - 1)$$

As a result, we have a formula for the phase margin for this example:

$$\frac{1}{\tan^2 \phi_m} = \frac{1}{2}(\sqrt{1 + 4K^2\tau^2} - 1)$$

Thus with a given $K$ and a given $\tau$, we can compute the phase margin $\phi_m$.

The development above can be cast in terms of the standard form for a second-order system

$$G(j\omega)H(j\omega) = \frac{\omega_n^2}{j\omega(j\omega + 2\xi\omega_n)}$$

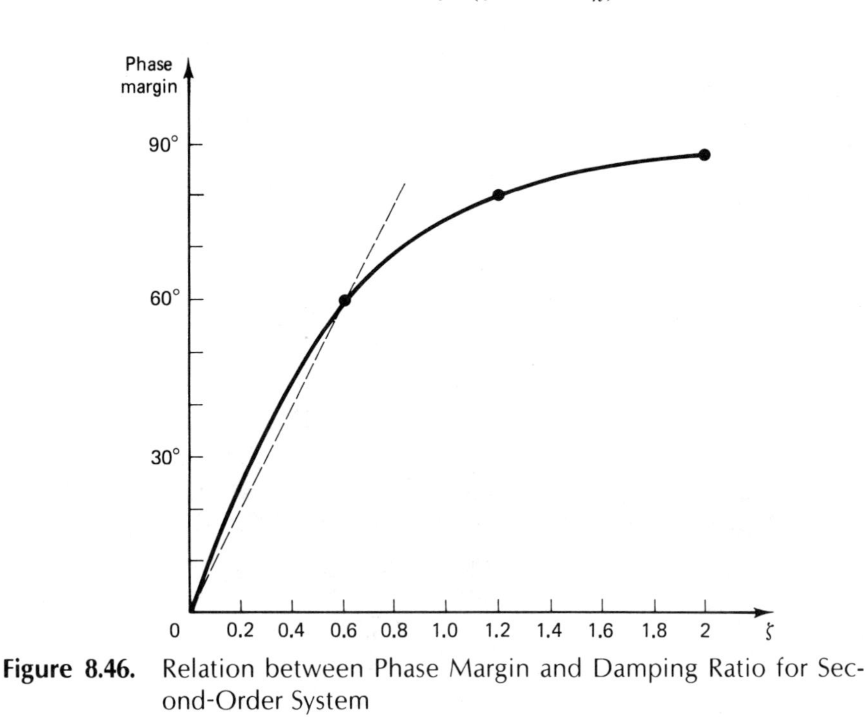


**Figure 8.46.** Relation between Phase Margin and Damping Ratio for Second-Order System

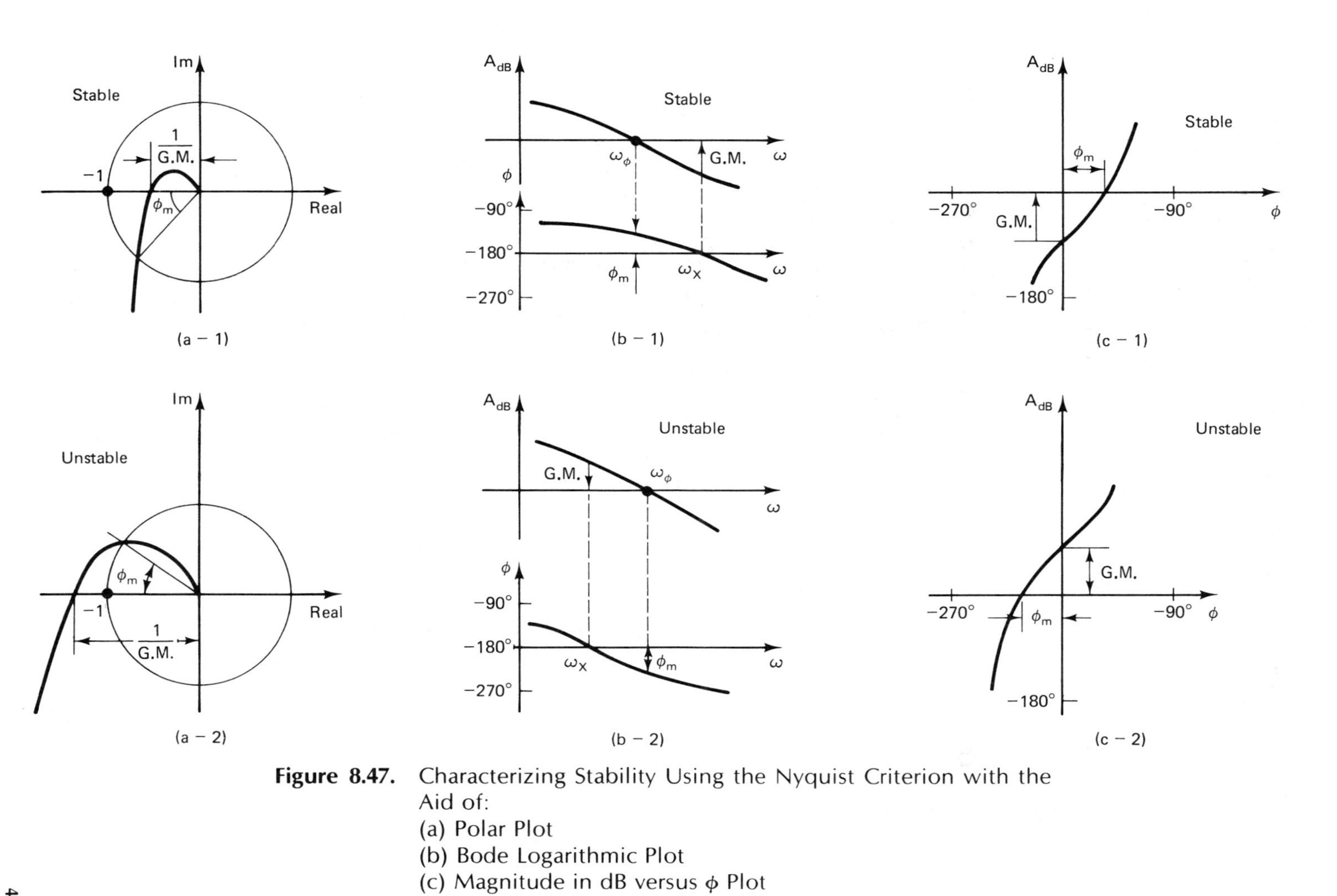


**Figure 8.47.** Characterizing Stability Using the Nyquist Criterion with the Aid of:
(a) Polar Plot
(b) Bode Logarithmic Plot
(c) Magnitude in dB versus $\phi$ Plot

The damping ratio $\xi$ and the undamped natural frequency $\omega_n$ are expressed in terms of $K$ and $\tau$ as

$$\omega_n^2 = \frac{K}{\tau}$$

$$2\xi\omega_n = \frac{1}{\tau}$$

Thus

$$K\tau = \frac{1}{4\xi^2}$$

The relation between phase margin and the damping ratio is thus given by

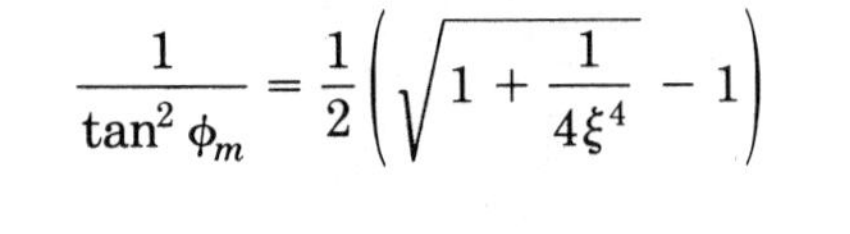

$$\frac{1}{\tan^2 \phi_m} = \frac{1}{2}\left(\sqrt{1 + \frac{1}{4\xi^4}} - 1\right)$$

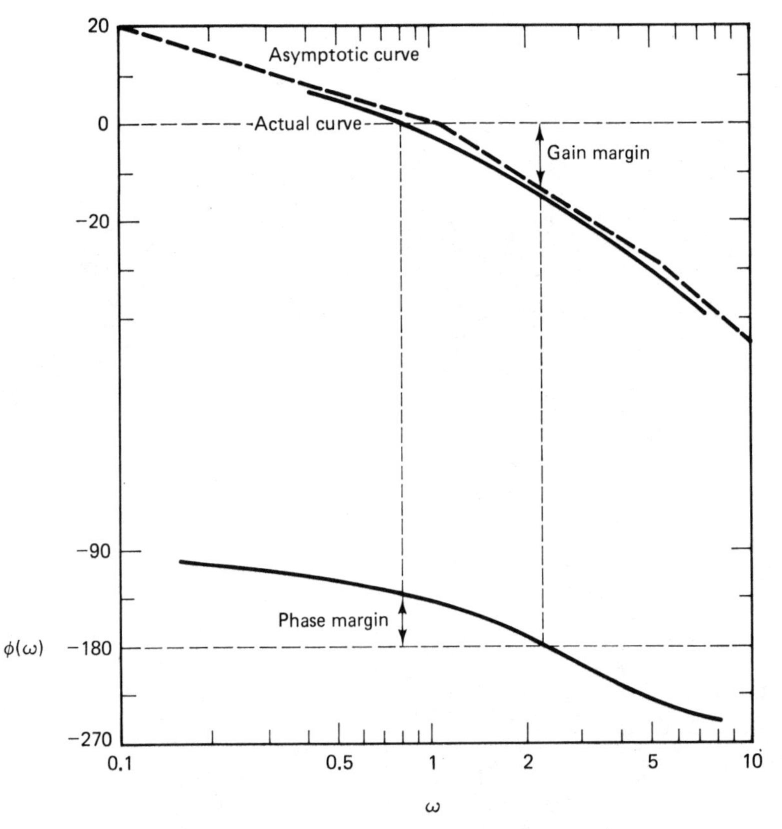


**Figure 8.48.** Bode Plot for Example 8.14 with $\tau_1 = 1$ and $\tau_2 = 0.2$ to Determine the Phase and Gain Margins

Alternatively, we write

$$\tan \phi_m = \frac{2\xi}{\left(\sqrt{4\xi^4 + 1} - 2\xi^2\right)^{1/2}}$$

This relation is depicted in Figure 8.46. Note that for $\xi < 0.6$, we can approximately state that the phase margin in degrees is given by

$$\phi_m = 100\xi$$

The gain and phase margins can be obtained easily from the Bode logarithmic plot as well as the log magnitude versus phase plots. The gain crossover frequency $\omega_\phi$ corresponds to 0 dB gain, and thus the phase margin can readily be obtained as shown in Figure 8.47b. Note that portion (b-1) corresponds to a stable system with positive phase margin, while portion (b-2) corresponds to an unstable system with negative phase margin. The same conclusions can be reached from the $A_{\mathrm{dB}}$–$\phi$ plot as shown in portions (c-1) and (c-2) of the figure. The phase-crossover frequency $\omega_x$ corresponds to a phase of $-180°$, and from the $A_{\mathrm{dB}}$ value we can readily obtain the gain margin.

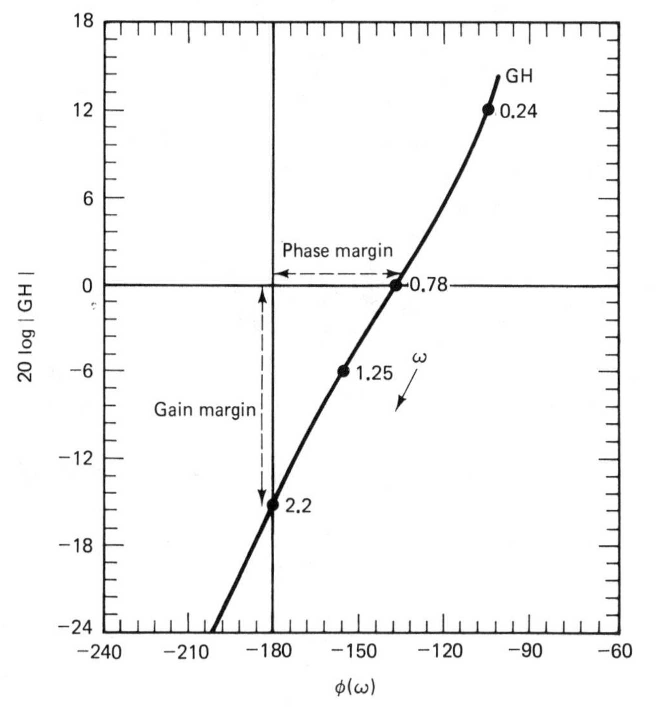


**Figure 8.49.** Log Magnitude–Phase Curve for Example 8.14 with $\tau_1 = 1$ and $\tau_2 = 0.2$ to Find the Phase and Gain Margins

As a specific example, we consider the system of Example 8.14 with $K = 1$, $\tau_1 = 1$, and $\tau_2 = 0.2$. It is clear from inspection of Figures 8.48 and 8.49 that

$$\begin{aligned} \omega_x &= 2.2 \\ \omega_\phi &= 0.78 \\ \text{GM}_{\text{dB}} &= +15 \text{ dB} \\ \phi_m &= 43^\circ \end{aligned}$$

## 8.9 Closed-Loop Frequency Response and the Nichols Chart

In a closed-loop system with unity feedback, the overall transfer function for sinusoidal operation is given by

$$T(j\omega) = \frac{C(j\omega)}{R(j\omega)} = \frac{G(j\omega)}{1 + G(j\omega)}$$

Let us assume that $T(j\omega)$ is expressed in polar form as

$$T(j\omega) = Me^{j\alpha}$$

Clearly, $M$ and $\alpha$ are functions of $\omega$. Assume also that $G(j\omega)$ is expressed in rectangular coordinates as

$$G(j\omega) = X + jY$$

Thus

$$Me^{j\alpha} = \frac{X + jY}{(1 + X) + jY}$$

The complex relation above can be manipulated to provide an expression relating the magnitude $M$, $X$, and $Y$, and another expression relating the phase angle $\alpha$, $X$ and $Y$.

The algebraic manipulations are of secondary importance and we can state the first expression as

$$\left(X + \frac{M^2}{M^2 - 1}\right)^2 + Y^2 = \left(\frac{M}{M - 1}\right)^2$$

The relation above is a circle with center at

$$\begin{aligned} X_{CM} &= \frac{-M^2}{M^2 - 1} \\ Y_{CM} &= 0 \end{aligned}$$

The radius is

$$R_M = \frac{M}{M - 1}$$

The constant $M$ loci in $G(j\omega)$ plane are thus a family of circles, as shown in Figure 8.50a.

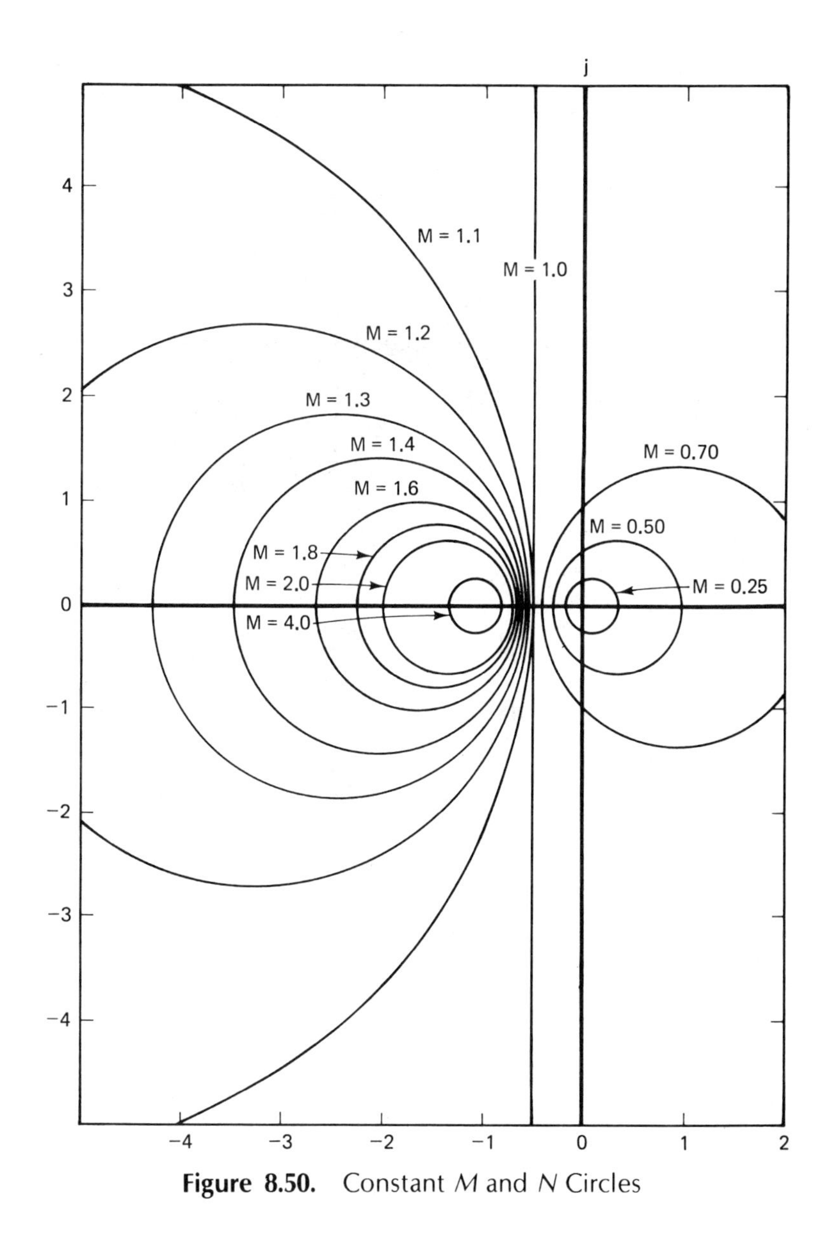


**Figure 8.50.** Constant $M$ and $N$ Circles

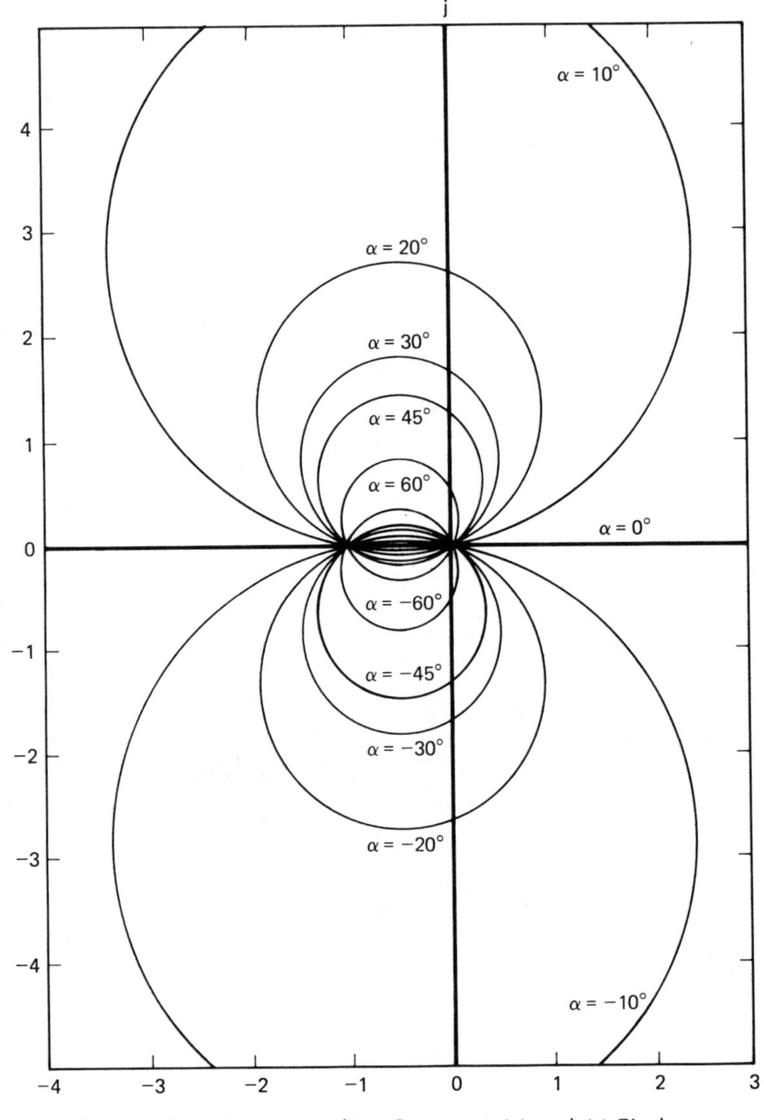


**Figure 8.50** (continued). Constant $M$ and $N$ Circles

The second relation is given by

$$\left(X + \frac{1}{2}\right)^2 + \left(Y - \frac{1}{2N}\right)^2 = \frac{1}{4} + \left(\frac{1}{2N}\right)^2$$

where

$$N = \tan \alpha$$

Again we have the equation of a circle with center at

$$X_{CN} = -\frac{1}{2}$$

$$Y_{CN} = \frac{1}{2N}$$

The radius is

$$R_N = \sqrt{\frac{1}{4} + \left(\frac{1}{2N}\right)^2}$$

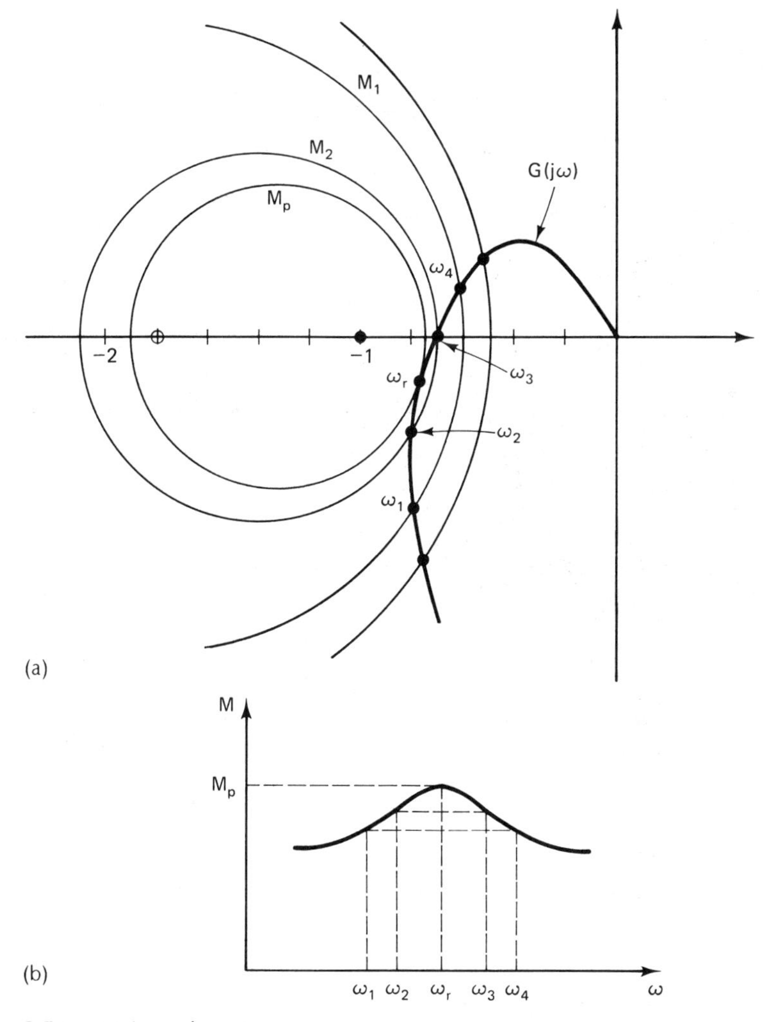

**Figure 8.51.** Using the Constant $M$ Contours with a Polar Plot (a) to Derive the Closed-Loop Frequency Response (b)

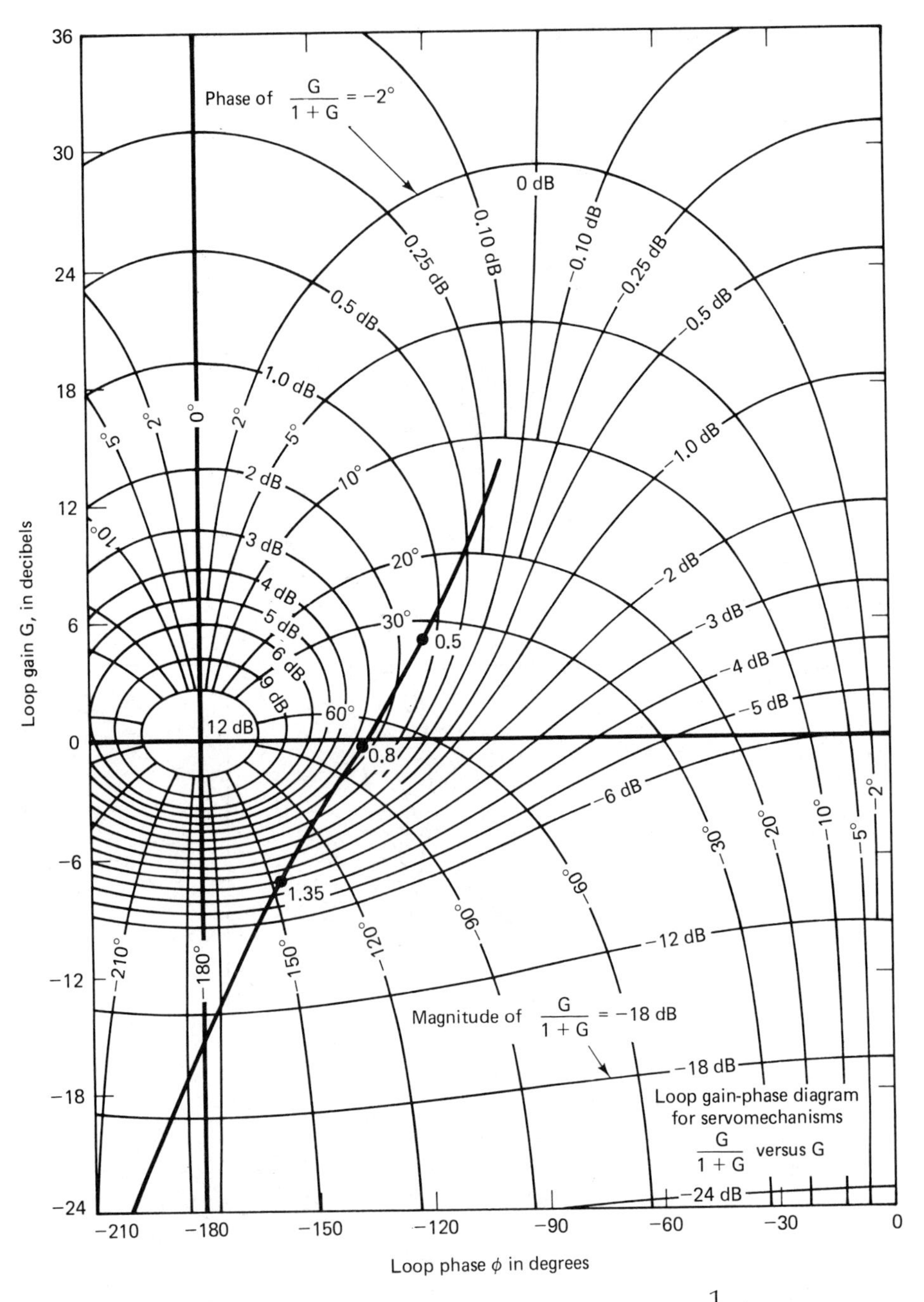


**Figure 8.52.** Nichols Diagram for $G(j\omega) = \dfrac{1}{j\omega(1 + j\omega)(1 + j0.2\omega)}$

The constant $\alpha$ (and hence $N$) loci in the $G(j\omega)$ plane are thus a family of circles that all pass through the origin and the $(-1 + j0)$ point as shown in Figure 8.50b.

The constant $M$ and $N$ contours are useful when overlaid on the polar plot of $G(j\omega)$ as shown in Figure 8.51. The intersection of $G(j\omega)$ plot with a constant $M$ circle such as $M_1$ at frequencies $\omega_1$ and $\omega_4$ indicates the value of the closed-loop gain as $M_1$. Note that the circle tangent to the $G(j\omega)$ plot provides us with the peak magnitude of the closed-loop response denoted by $M_p$ and the corresponding resonant frequency $\omega_r$. The lower portion of Figure 8.51 is derived from the top portion and shows the overall closed-loop frequency response.

An alternative formulation of the closed-loop transfer function in terms of the open-loop function uses the polar form to express $G(j\omega)$. Again, a few manipulations result in

$$M = \left\{ \left[ 1 + \frac{1}{|G(j\omega)|^2} + \frac{2\cos\phi}{|G(j\omega)|} \right]^{1/2} \right\}^{-1}$$

$$\alpha = -\tan^{-1} \frac{\sin\phi}{\cos[\phi + \angle G(j\omega)]}$$

A detailed plot of $G$ in decibels versus $\phi$, with $M$ and $\alpha$ as parameters, is illustrated in Figure 8.52. We have just introduced the Nichols chart, which is a very useful technique for determining both stability and the closed-loop frequency response of a feedback system. The open-loop gain versus phase determines stability, and at the same time the closed-loop frequency response of the system is determined from the contour of constant $M$ (in decibels) and $\alpha$ which are overlaid on the gain–phase plot.

As an example we take the open-loop function of Example 8.14 given by

$$G(j\omega) = \frac{1}{j\omega(1 + j\omega)(1 + 0.2j\omega)}$$

The $G(j\omega)$-locus is plotted on the Nichols chart as shown in Figure 8.52. In a manner similar to that illustrated in Figure 8.51, we can use the Nichols chart of Figure 8.52 to construct the closed-loop frequency response characteristic of the system.

## Some Solved Problems

PROBLEM 8A-1

Sketch the polar plot of the transfer function

$$G(s) = \frac{K(1 + \tau_1 s)^2}{(1 + \tau_2 s)(1 + \tau_3 s)(1 + \tau_4 s)^2}$$

Assume that

$$\tau_2 > \tau_1 > \tau_4 \qquad \tau_3 > \tau_1$$

SOLUTION

Let us substitute $s = j\omega$ to obtain

$$G(j\omega) = \frac{K(1 + j\omega\tau_1)^2}{(1 + j\omega\tau_2)(1 + j\omega\tau_3)(1 + j\omega\tau_4)^2}$$

For $\omega = 0$ we have

$$G(j0) = K$$

For high frequencies we obtain

$$\lim_{\omega \to \infty} G(j\omega) = 0\underline{/-180°}$$

The intersection with the imaginary axis can be obtained by setting the real part of $G(j\omega)$ to zero. We first write $G(j\omega)$ as the quotient

$$G(j\omega) = \frac{N(j\omega)}{D(j\omega)} = \frac{N(j\omega)D^*(j\omega)}{|D(j\omega)|^2}$$

Thus the condition for the intersection with the imaginary axis becomes

$$\text{Re}\{N(j\omega)D^*(j\omega)\} = 0$$

Thus our condition is

$$\text{Re}\{(1 + j\omega\tau_1)^2(1 - j\omega\tau_2)(1 - j\omega\tau_3)(1 - j\omega\tau_4)^2\} = 0$$

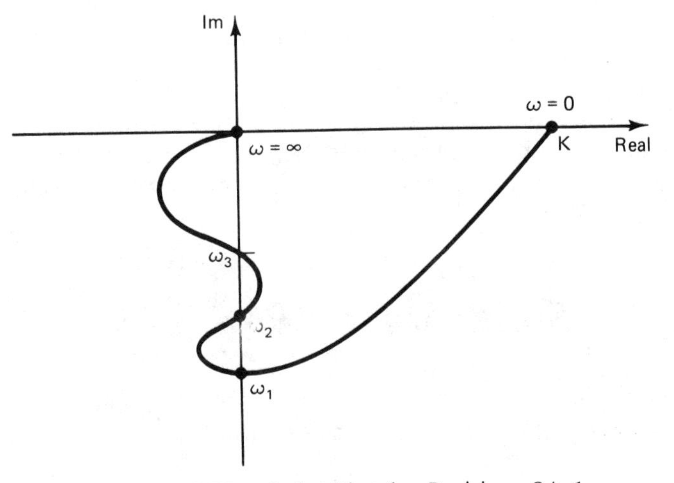


**Figure 8.53.** Polar Plot for Problem 8A-1

A few manipulations of the condition above provide us with a third-order equation in $\omega^2$. As a result, depending on the value of the coefficients, we can obtain intersection with the imaginary axis at three frequencies, $\omega_1$, $\omega_2$, and $\omega_3$, as shown in Figure 8.53.

PROBLEM 8A-2

A typical industrial robot has six joints to provide six degrees of freedom, as shown in Figure 8.54. Each joint is driven hydraulically, pneumatically, or electrically with a feedback control loop. A block diagram of the position-control system for a torque-sensing joint is shown in Figure 8.55. The microcomputer receives input signals consisting of the desired compliant torque $\tau_d$, the anticipated gravity torque $\tau_a$, and the desired angular displacement $\theta_d$. The microcomputer controls the voltage applied to the armature-controlled motor to produce sufficient torque to handle the load. The load torque $\tau_L$, the frictional torque $\tau_f$, and the gravitational

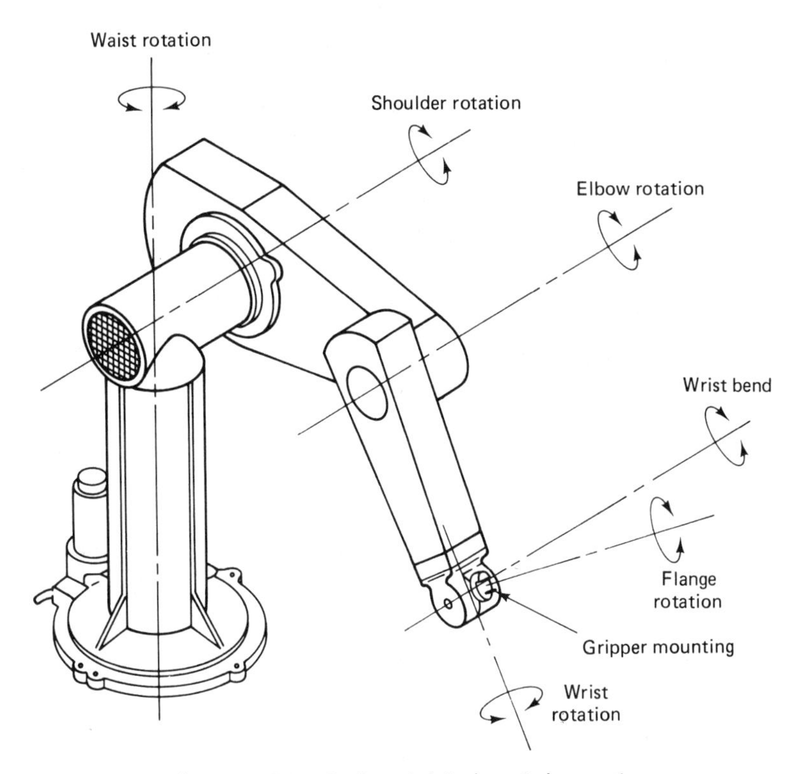


**Figure 8.54.** Industrial Robot Schematic

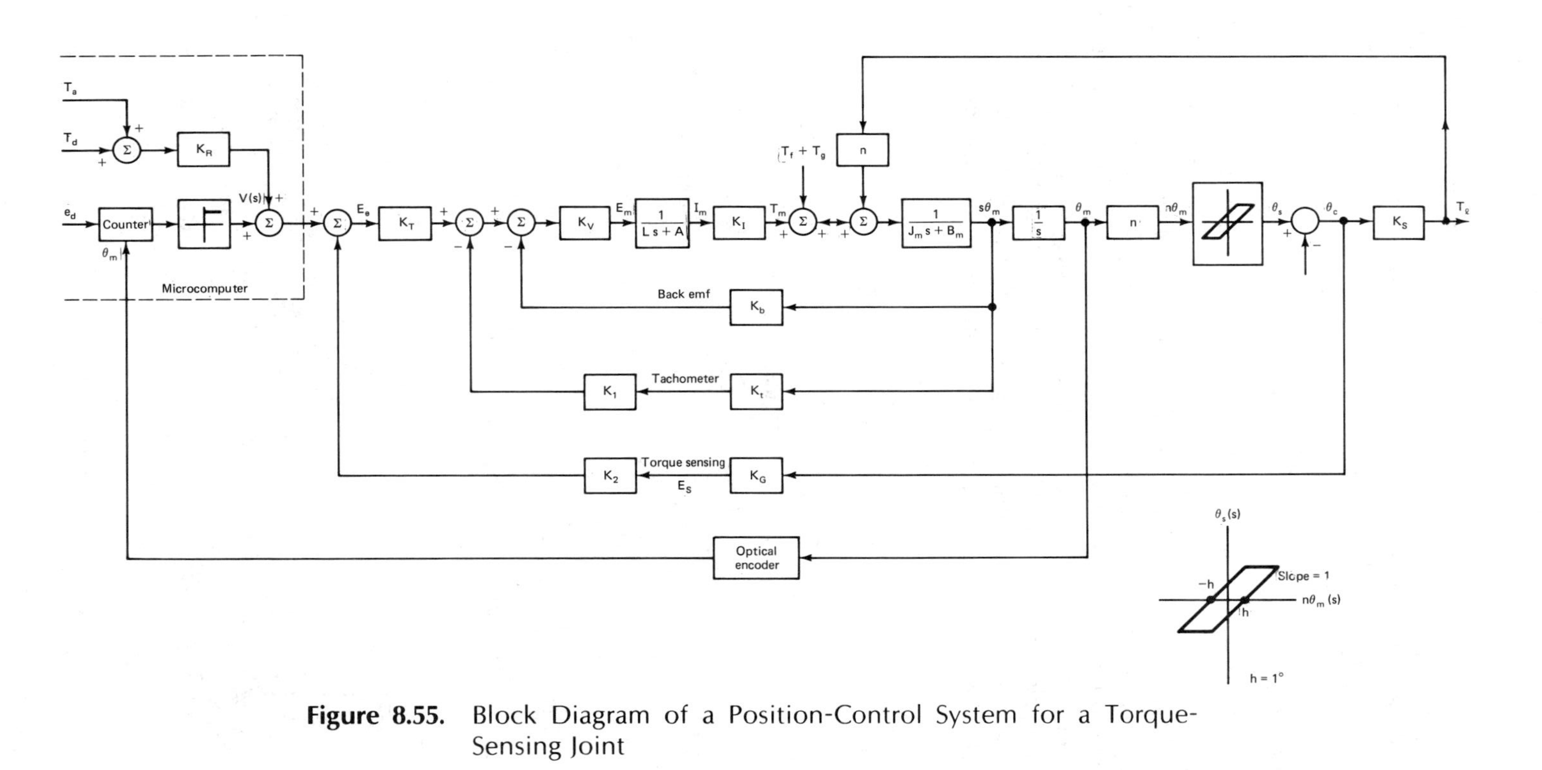


**Figure 8.55.** Block Diagram of a Position-Control System for a Torque-Sensing Joint

torque $\tau_g$ are modeled as external disturbances. Tachometer (velocity) feedback with gain $K_\tau$ is employed as shown. The motor's inductance and resistance are denoted by $L$ and $R$, and its inertia and friction are denoted by $J_m$ and $B_m$, respectively.

An optical encoder is connected to the motor shaft to provide positional feedback. The stiffness of the shaft is denoted by $K_s$. A backlash nonlinearity exists in the gear assembly and is represented as shown in Figure 8.55. Details of the motor, gear train, and load are shown in Figure 8.56.

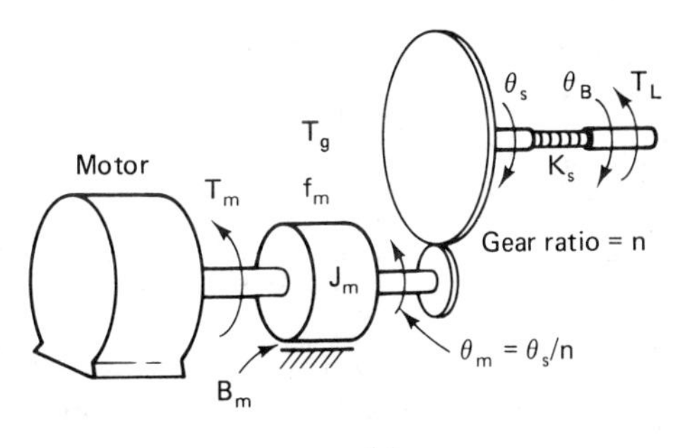


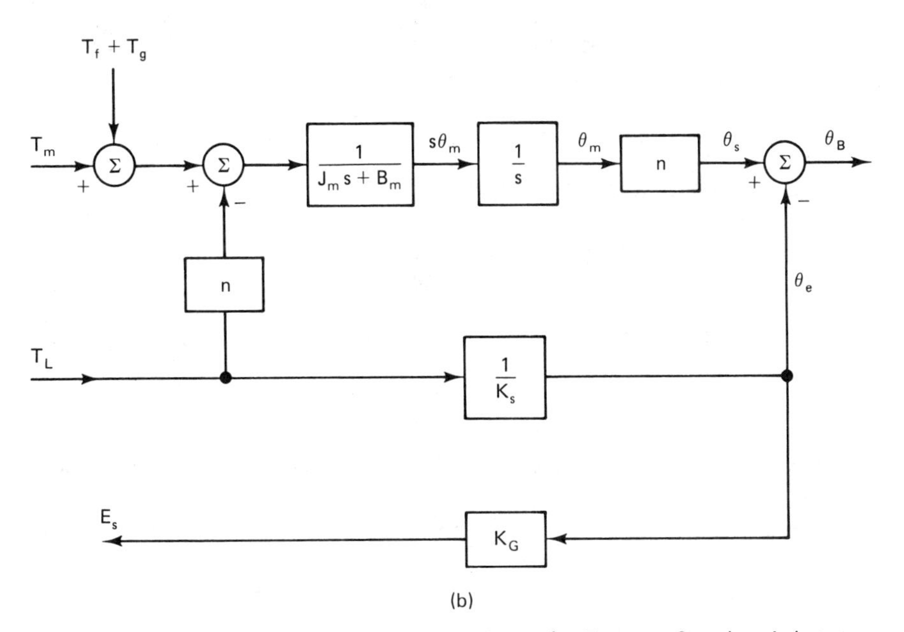


**Figure 8.56.** Schematic Representation of a Torque-Sensing Joint

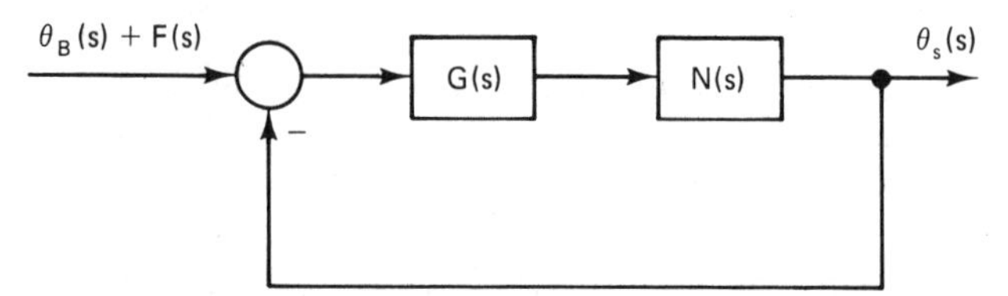


**Figure 8.57.** Equivalent Block Diagram of the Position-Control System of Problem 8A-2

A few manipulations reduce the block diagram of Figure 8.55 to that shown in Figure 8.57. Neglecting the motor armature inductance, we have

$$G_s = \frac{n(nK_s/J_m + K_V K_I K_\tau K_G K_2/RJ_m)}{s\left\{s + \left[\dfrac{B_m}{J_m} + \dfrac{K_V K_I(K_b + K_\tau + K_1)}{RJ_m}\right]\right\}}$$

$$F(s) = \frac{K_V K_I K_\tau[K_R(\tau_a + \tau_d) + V(s)] - (\tau_g + \tau_f)R}{nK_s R + K_V K_I K_\tau K_G K_2}$$

Note that the backlash nonlinearity is included as $N(s)$ in the block diagram. Some typical parameters result in the following transfer function:

$$G(s) = \frac{673 \times 10^4}{s(s + 730)}$$

Sketch the necessary Bode plots for the transfer function $G(s)$. Find the gain and phase margins.

SOLUTION

Our transfer function can be written as

$$G(s) = \frac{9.2192 \times 10^3}{s(1 + 1.37 \times 10^{-3}s)}$$

The gain's contribution is

$$\begin{aligned} 20 \log K &= 20 \log(9.2192 \times 10^3) \\ &= 79.3 \text{ dB} \end{aligned}$$

The low-frequency portion has a slope of $-20$ dB/decade and crosses the $A_{\text{dB}}$ axis at $\omega = 1$ at 79.3 dB. There is a corner frequency at $\omega_c = 730$ rad/sec. The log amplitude asymptotic representation is as shown in Figure 8.58.

The gain crossover frequency can be obtained analytically:

$$\omega_x = 2594.61 \text{ rad/sec}$$

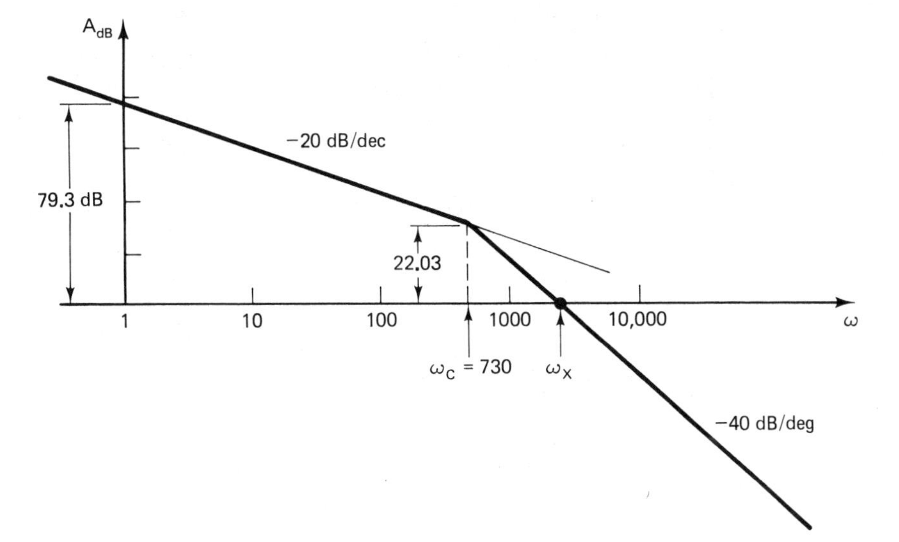


**Figure 8.58.** Log Amplitude Characteristic for Problem 8A-2

The phase angle at $\omega_x$ is given by

$$\begin{aligned}\phi &= -90^\circ - \tan^{-1}(2594.61)(1.37 \times 10^{-3})\\ &= -164.29^\circ\end{aligned}$$

As a result, we find the phase margin as

$$\phi_m = 15.71^\circ$$

The gain margin for this system is infinity, as the phase angle tends to $-180^\circ$ as $\omega$ increases.

PROBLEM 8A-3

The block diagram in Figure 8.59a shows the details of a model for a hydroelectric governor. The variable $Z(s)$ is the gate position and $\Delta\omega(s)$ is the velocity error signal. The hydrogovernor transfer function is denoted $G_1(s)$, as shown in the reduced block diagram shown in Figure 8.59b.

(a) Show that the transfer function $G_1(s)$ is given by

$$G_1(s) = \frac{1 + \tau_r s}{\tau_r \tau_p \tau_g s^3 + \tau_g(\tau_p + \tau_r)s^2 + \left[\tau_g + (\sigma + \delta)\tau_r\right]s + \sigma}$$

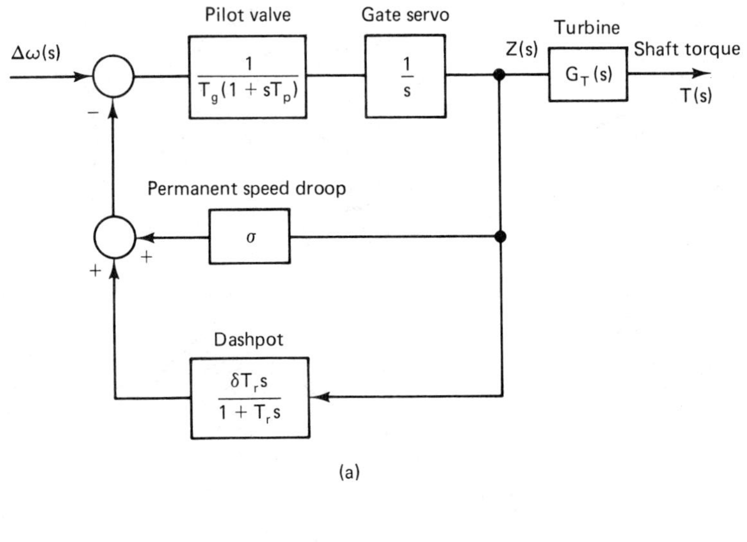


(a)

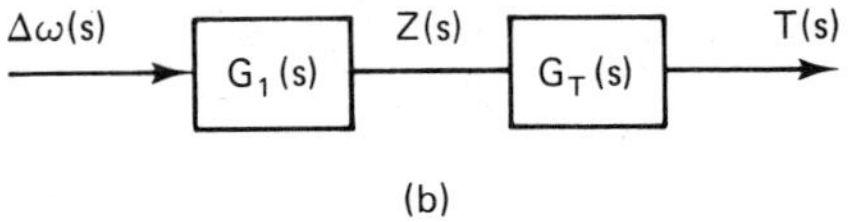


(b)

**Figure 8.59.** Hydroelectric Governor Model for Problem 8A-3

(b) The following typical parameter values are available:

$$\tau_g = 0.2 \qquad \tau_p = 0.1$$
$$\tau_r = 5 \qquad \sigma = 0.04$$
$$\delta = 0.31$$

Sketch the Bode logarithmic plots for $G_1(s)$.

SOLUTION

(a) The transfer function $G_1(s)$ is given by

$$G_1(s) = \frac{Z(s)}{\Delta\omega(s)} = \frac{\dfrac{1}{\tau_g s(1 + \tau_p s)}}{1 + \dfrac{1}{\tau_g s(1 + \tau_p s)}\left(\sigma + \dfrac{\delta\tau_r s}{1 + \tau_r s}\right)}$$

The above reduces to

$$G_1(s) = \frac{1}{\tau_g s(1 + \tau_p s) + \sigma + \dfrac{\delta\tau_r s}{1 + \tau_r s}}$$

This further reduces to

$$G_1(s) = \frac{1 + \tau_r s}{\left[\tau_g s(1 + \tau_p s) + \sigma\right](1 + \tau_r s) + \delta \tau_r s}$$

Expanding, we obtain the desired expression:

$$G_1(s) = \frac{1 + \tau_r s}{\tau_r \tau_g \tau_p s^3 + \tau_g(\tau_p + \tau_r)s^2 + \left[\tau_g + (\sigma + \delta)\tau_r\right]s + \sigma}$$

(b) With the given numerical values, we obtain

$$G_1(s) = \frac{1 + 5s}{0.1s^3 + 1.02s^2 + 1.95s + 0.04}$$
$$= \frac{25(1 + 5s)}{2.5s^3 + 25.5s^2 + 48.75s + 1}$$

We factor the denominator to obtain

$$G_1(s) = \frac{10(1 + 5s)}{(s + 2.518)(s^2 + 7.682s + 0.15672)}$$

This factors further to

$$G_1(s) = \frac{10(1 + 5s)}{(s + 2.518)(s + 7.6615)(s + 0.0205)}$$

In standard form we have

$$G_1(s) = \frac{25(1 + 5s)}{(1 + s/0.0205)(1 + s/2.518)(1 + s/7.6615)}$$

The gain of 25 provides us with

$$20 \log K = 20 \log 25 = 28 \text{ dB}$$

There are four corner frequencies as tabulated below:

| $\omega_c$ | SLOPE (dB/decade) | NET SLOPE (dB/decade) |
|---|---|---|
| 0.0205 | −20 | −20 |
| 0.2 | +20 | 0 |
| 2.518 | −20 | −20 |
| 7.6615 | −20 | −40 |

Figure 8.60 shows the log amplitude and phase angle versus $\omega$ characteristics.

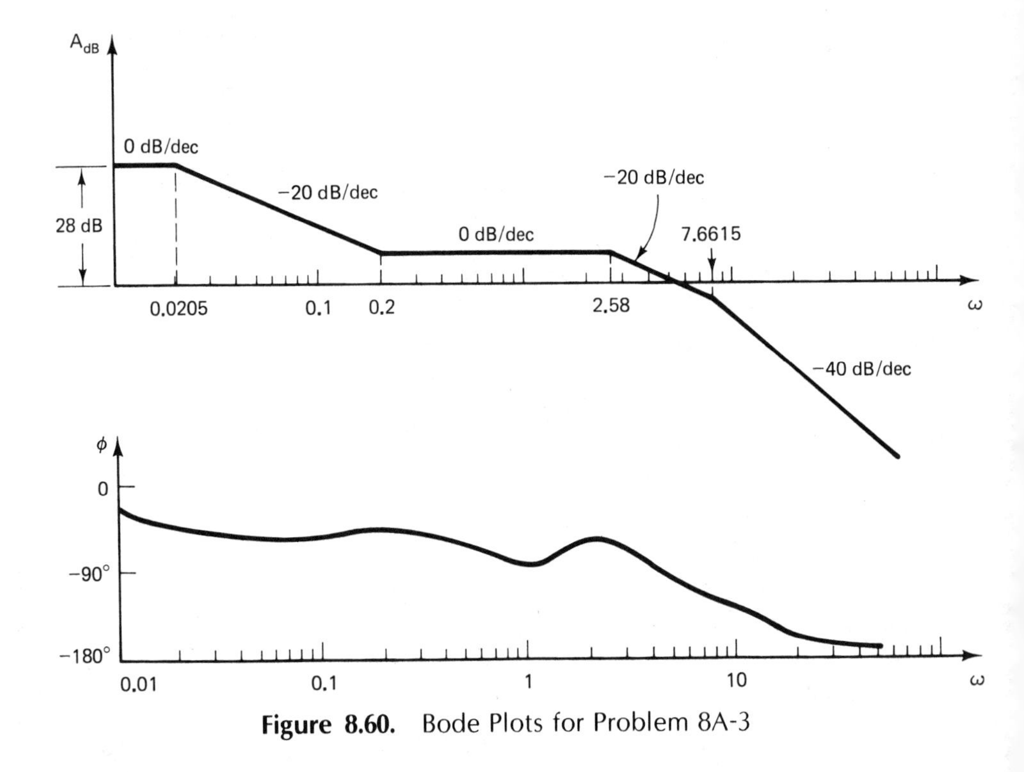


**Figure 8.60.** Bode Plots for Problem 8A-3

PROBLEM 8A-4

The block diagram shown in Figure 8.61 represents a model of a hydroelectric alternator, turbine, and penstock, with the governor being represented by the model of Problem 8A-3. The following parameters are

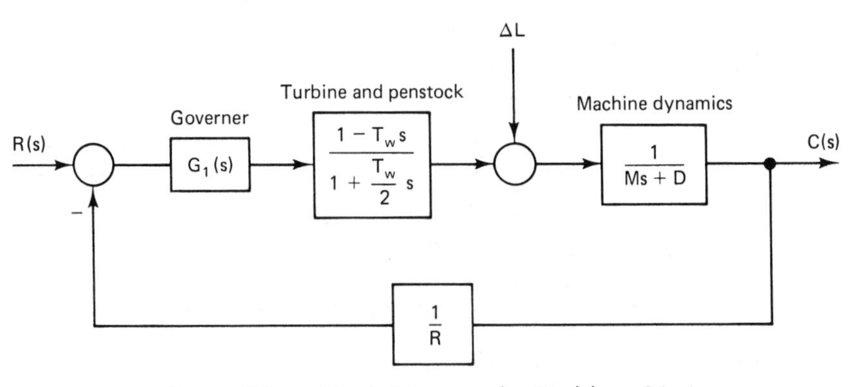


**Figure 8.61.** Block Diagram for Problem 8A-4

available:

$$\tau_w = 1 \qquad R = 0.05$$
$$M = 10 \qquad D = 1.0$$

Sketch the Bode plots for the system and find the gain and phase margins.

SOLUTION

The loop transfer function is obtained as

$$G(s)H(s) = \frac{1 - \tau_w s}{R[1 + (\tau_w/2)s](Ms + D)} G_1(s)$$

Using the available parameters, we get

$$G(s)H(s) = \frac{500(1 - s)(1 + 5s)}{(1 + 0.5s)(1 + 10s)(1 + s/0.0205)(1 + s/2.518)(1 + s/7.6615)}$$

Note that this is a non-minimum-phase system, due to the presence of the $(1 - s)$ term in the numerator.

The gain contribution is obtained as

$$20 \log K = 20 \log 500 = 54 \text{ dB}$$

There are seven corner frequencies, as follows:

| $\omega_c$ | SLOPE (dB/decade) | NET SLOPE (dB/decade) |
|---|---|---|
| 0.0205 | −20 | −20 |
| 0.5 | −20 | −40 |
| 1 | +20 | −20 |
| 2.518 | −20 | −40 |
| 5 | +20 | −20 |
| 7.6615 | −20 | −40 |
| 10 | −20 | −60 |

The log amplitude characteristic is shown in Figure 8.62.

It is probably worthwhile to establish some points analytically on the asymptotic plot. The amplitude at $\omega_c = 0.0205$ is given by 58 dB. The slope between 0.0205 and $\omega_c = 0.1$ is −20 dB/decade. The drop in amplitude is thus

$$A_1 = 20 \log \frac{0.5}{0.0205} = 27.74 \text{ dB}$$

Thus at $\omega_c = 0.5$, the amplitude is

$$A_2 = 58 - 27.74 = 30.26 \text{ dB}$$

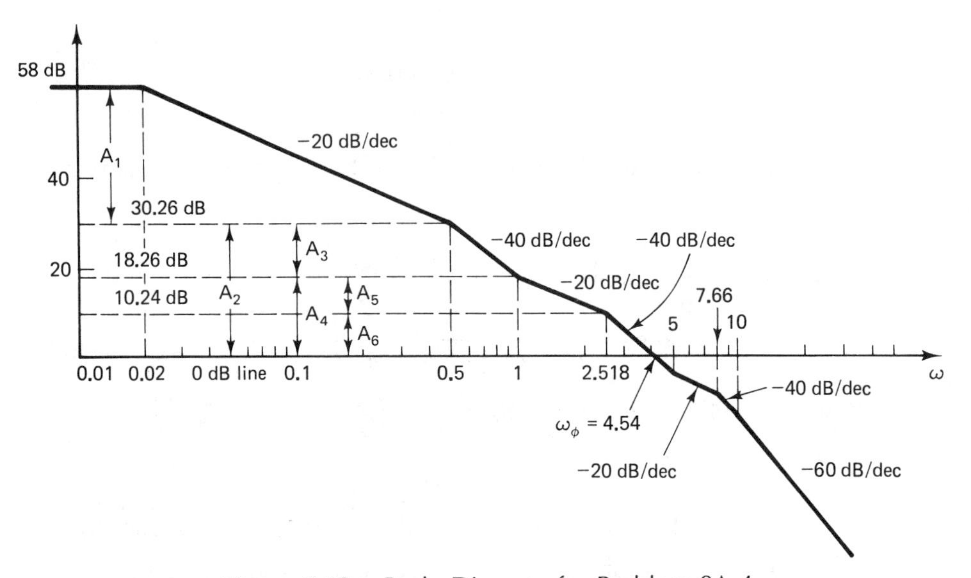


**Figure 8.62.** Bode Diagram for Problem 8A-4

The slope changes to −40 dB/decade at $\omega_c = 0.5$ and is maintained at that level up to $\omega_c = 1$. The drop $A_3$ is thus

$$A_3 = 40 \log \frac{1}{0.5} = 12 \text{ dB}$$

As a result, the decibel level at $\omega_c = 1$ is given by

$$\begin{aligned} A_4 &= A_2 - A_3 \\ &= 30.26 - 12 = 18.26 \text{ dB} \end{aligned}$$

At $\omega_c = 1$, the slope changes to −20 dB/decade up to $\omega_c = 2.518$. The drop $A_5$ is thus given by

$$A_5 = 20 \log \frac{2.518}{1} = 8.02 \text{ dB}$$

Thus the value of $A_6$ is given by

$$\begin{aligned} A_6 &= A_4 - A_5 \\ &= 18.26 - 8.02 \\ &= 10.24 \text{ dB} \end{aligned}$$

Thus we can get a crossover frequency as shown in the figure. We can find the frequency of crossover from the slope of −40 dB/decade as

$$A_6 = 40 \log \frac{\omega_\phi}{2.518}$$

Thus

$$\omega_\phi = 4.54 \text{ rad/sec}$$

The phase angle characteristic is given by

$$\phi = \tan^{-1}(-\omega) + \tan^{-1}5\omega - \left(\tan^{-1}0.5\omega + \tan^{-1}10\omega + \tan^{-1}\frac{\omega}{0.0205} + \tan^{-1}\frac{\omega}{2.518} + \tan^{-1}\frac{\omega}{7.6615}\right)$$

At the crossover frequency the phase angle is

$$\phi(\omega_\phi) = -326.44°$$

The phase margin is thus given by

$$\phi_m = 180 - 326.44 = -146.44°$$

A negative phase margin indicates instability.

The phase angle $\phi$ is approximately $-180°$ at $\omega_x = 0.8$. At that frequency the gain in decibels is computed as 17.36; thus the gain margin is $-17.36$, again indicating instability.

PROBLEM 8A-5

The conclusion of Problem 8A-4 is that with the given parameter values, the hydroelectric unit is unstable. The system can be made stable by changing the loop gain. We thus write the loop transfer function as

$$G(s)H(s) = \frac{K(1-s)(1+5s)}{(1+0.5s)(1+10s)(1+s/0.0205)(1+s/2.518)(1+s/7.6615)}$$

Find the maximum value of $K$ to retain stability using Bode plots.

SOLUTION

For the system to be critically stable, the amplitude in decibels should be zero for $\phi = -180°$. From Problem 8A-4 the angle is $-180°$ for $\omega_x = 0.8$ rad/sec. Figure 8.63 shows a portion of Figure 8.62 without its 0-dB line.

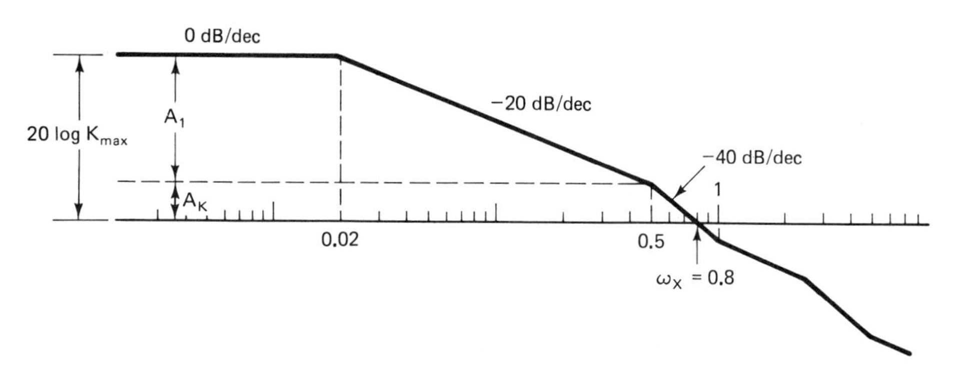


**Figure 8.63.** Construction for Problem 8A-5

The new 0-dB line should intersect the characteristic at $\omega_x = 0.8$ rad/sec, as shown in Figure 8.63. From the geometry of the figure, we have

$$20 \log K_{\max} = A_1 + A_K$$

The value of $A_1$ was obtained as 27.74 dB in Problem 8A-4. The value of $A_K$ is obtained from

$$A_K = 40 \log \frac{0.8}{0.5} = 8.16$$

Thus

$$20 \log K_{\max} = 27.74 + 8.16 = 35.9$$

As a result, the critical value of $K$ is

$$K_{\max} = 62.4$$

PROBLEM 8A-6

The Bode plot of the open-loop transfer function of a control system is shown in Figure 8.64. Determine the system type and its transfer function. If the system is stable, find the phase margin. The amplitude in decibels at $\omega = 1$ is 19.7.

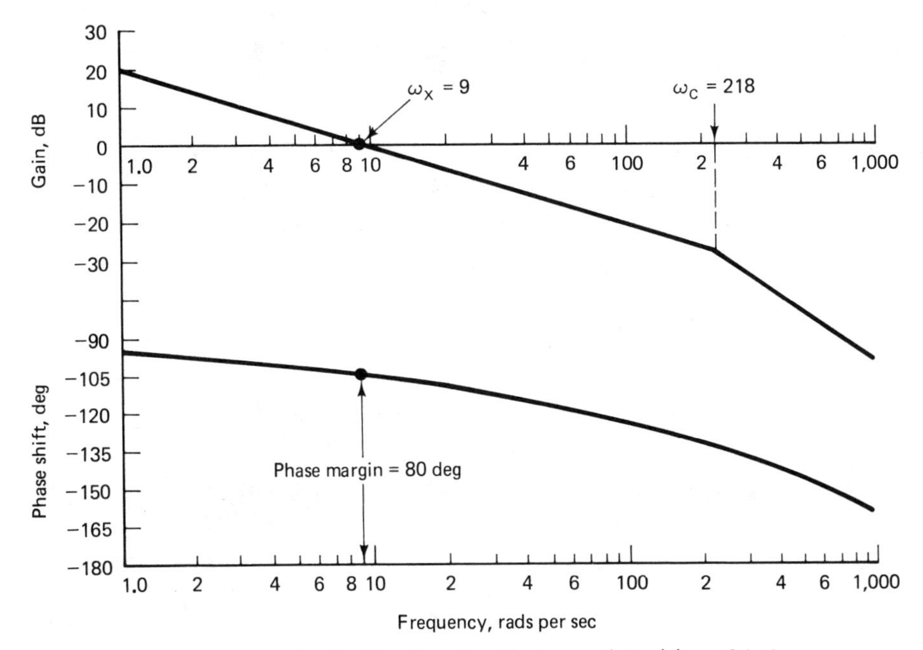


**Figure 8.64.** Bode Plot for the System of Problem 8A-6

SOLUTION

This is a type 1 system with

$$20 \log K_V = 19.7 \text{ dB}$$

Thus

$$K_V = 9.7$$

There is a corner frequency at $\omega_c = 218$. Thus

$$G(s) = \frac{9.7}{s(1 + s/218)}$$

As indicated on the diagram, the phase margin is 80°, indicating a very stable system.

PROBLEM 8A-7

It is sometimes possible to obtain quickly the Bode plot of the closed-loop frequency response function from the corresponding open-loop function. As an example, consider the open-loop transfer function

$$G(s) = \frac{K}{1 + \tau s}$$

Find the Bode plot of the closed-loop function $C(s)/R(s)$ using the Bode plot of $G(s)$. Assume that $K \gg 1$.

SOLUTION

The closed-loop function for unity feedback is obtained as

$$\frac{C(s)}{R(s)} = \frac{K/(K+1)}{1 + [\tau/(K+1)]s}$$

Assuming that $K \gg 1$, then

$$\frac{C(s)}{R(s)} \simeq \frac{1}{1 + (\tau/K)s}$$

The $A_{dB}$ for low frequency is the 0-dB line. The corner frequency is at $\omega_c = K/\tau$.

In Figure 8.65 we show the open-loop amplitude part of the Bode plot. At low frequencies the amplitude in decibels is $20 \log K$. The corresponding closed-loop amplitude in decibels is zero and is shown in heavy line. The open-loop has a corner frequency at $\omega_c = 1/\tau$ with a slope of $-20$ dB/decade. The closed-loop corner frequency is at $K/\tau$, which can be seen to be also on the open-loop characteristics. Beyond $\omega = K/\tau$ the open- and closed-loop characteristics coincide.

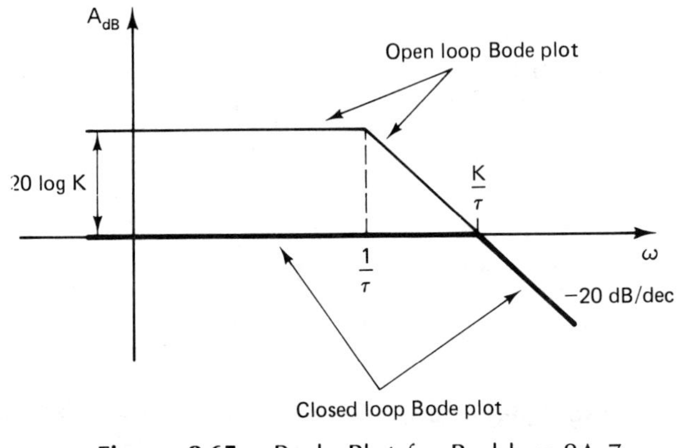


**Figure 8.65.** Bode Plot for Problem 8A-7

# Problems

## PROBLEM 8B-1

Sketch the polar plot of the frequency response for the following transfer functions.

(a) $G_a(s) = \dfrac{1 + \tau_1 s}{1 + \tau_2 s}$

(b) $G_b(s) = \dfrac{1 - \tau_1 s}{1 + \tau_2 s}$

(c) $G_c(s) = \dfrac{-1 + \tau_1 s}{1 + \tau_2 s}$

## PROBLEM 8B-2

Sketch the polar plot of the frequency response for the following transfer functions.

(a) $G(s) = \dfrac{s + 3}{s(s + 2)(s + 4)}$

(b) $G(s) = \dfrac{0.8(1 + 5s)}{s(1 + 0.25s)^2}$

(c) $G(s) = \dfrac{0.333s}{[1 + (s/30)][1 + (s/70)]}$

PROBLEM 8B-3

Sketch the polar plot of the frequency response for the following transfer functions.

(a) $G(s) = \dfrac{Ke^{-\pi s/4}}{s(1 + s)}$

(b) $G(s) = \dfrac{Ke^{-0.5s}}{s(1 + s)(1 + 0.5s)}$

(c) $G(s) = \dfrac{Ke^{-2s}}{s^2 + 2s + 2}$

PROBLEM 8B-4

Sketch the polar plot of the frequency response for the transfer function $G(s)$ for $K = 10$ and $K = 100$.

$$G(s) = \frac{K}{s(1 + 0.25s)(1 + s/16)}$$

PROBLEM 8B-5

Verify that the polar plots shown in Figure 8.66 correspond to the indicated transfer functions.

(a) $G(s) = \dfrac{K}{s^2(1 + \tau_1 s)(1 + \tau_2 s)}$

(b) $G(s) = \dfrac{K}{s(1 + \tau_1 s)(1 + \tau_2 s)(1 + \tau_3 s)}$

(c) $G(s) = \dfrac{1 + 0.5s}{s(1 + 0.25s)}$

(d) $G(s) = \dfrac{1 + 0.25s}{s(1 + 0.5s)}$

PROBLEM 8B-6

Draw the asymptotic Bode plots for the transfer functions of Problem 8B-1. Assume that $\tau_1 = 0.1$ sec and $\tau_2 = 0.01$ sec.

PROBLEM 8B-7

Repeat Problem 8B-6 for the transfer functions of Problem 8B-2.

PROBLEM 8B-8

Repeat Problem 8B-6 for the transfer functions of Problem 8B-3.

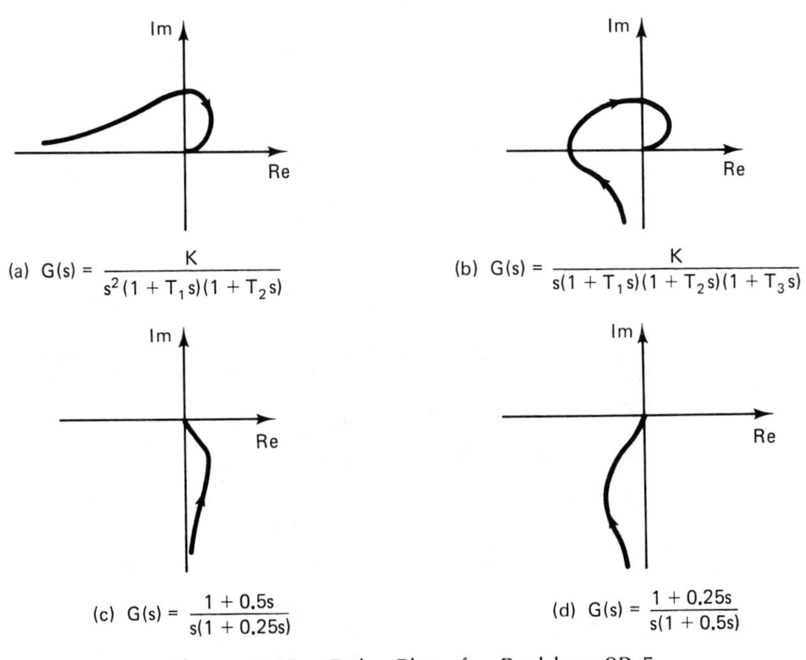


**Figure 8.66.** Polar Plots for Problem 8B-5

PROBLEM 8B-9

Repeat Problem 8B-6 for the transfer function of Problem 8B-4 for $K = 10$.

PROBLEM 8B-10

Sketch the Bode plots for the transfer functions of Problem 8B-5.

PROBLEM 8B-11

Obtain the transfer function for the log amplitude curve shown in Figure 8.67.

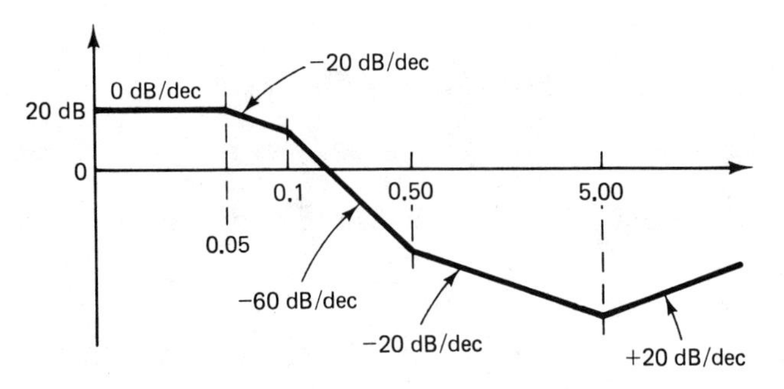


**Figure 8.67.** Log Amplitude Curve for Problem 8B-11

PROBLEM 8B-12

Repeat Problem 8B-11 for the log amplitude curve shown in Figure 8.68.

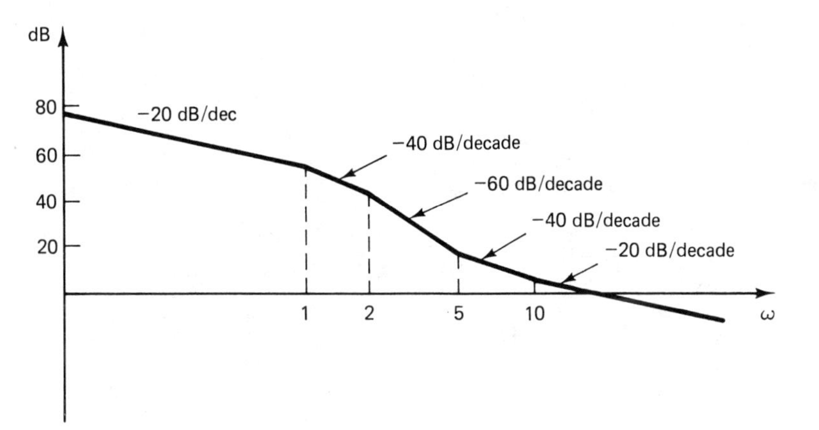


**Figure 8.68.** Log Amplitude Curve for Problem 8B-12

PROBLEM 8B-13

Consider the automatic braking system of Problem 3B-2. With the parameters given, the loop transfer function reduces to

$$H(s)G(s) = \frac{s^2 + s + 10}{s^2}$$

Sketch the polar plot and the Bode plot for the frequency response of the loop transfer function.

PROBLEM 8B-14

Reduce the block diagram of Problem 3B-5 to standard form as shown in Figure 8.69. Sketch the polar and Bode plots of the frequency response of the loop function $H(s)G(s)$.

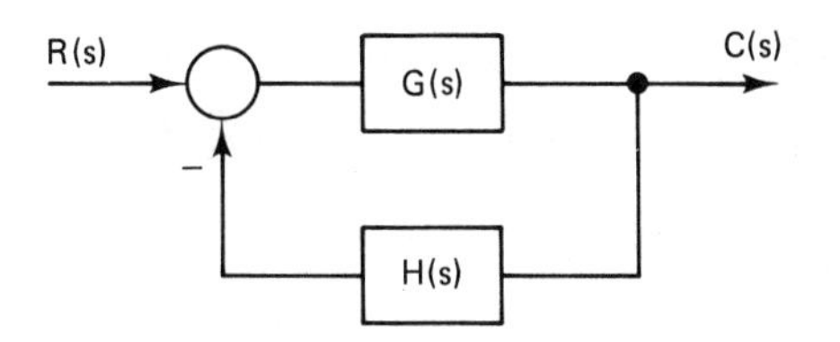


**Figure 8.69.** Block Diagram for Problem 8B-14

PROBLEM 8B-15

Repeat Problem 8B-14 for the system of Problem 3B-6.

PROBLEM 8B-16

A synchronous generator excitation control system is shown in the block diagram of Figure 8.70. The following parameters are available:

$$K_A = 40 \qquad \tau_A = 0.1$$
$$K_E = -0.05 \qquad \tau_E = 0.5$$
$$K_G = 1.0 \qquad \tau_G = 1.0$$
$$K_R = 1 \qquad \tau_R = 0.05$$

Use Bode logarithmic plots to check the stability of the system.

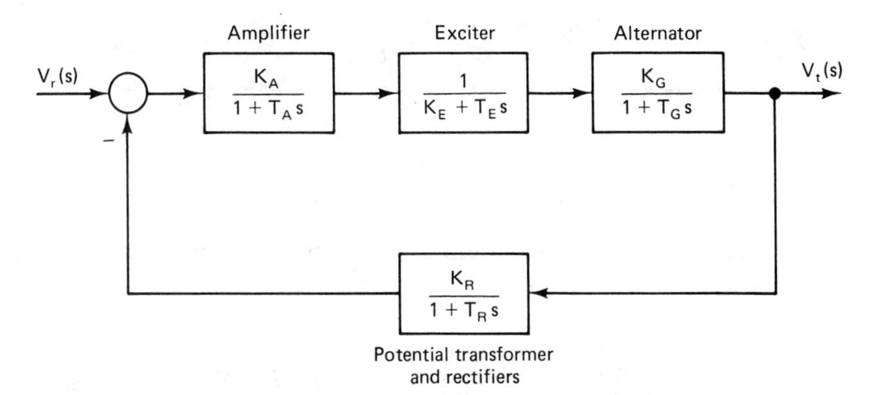


**Figure 8.70.** Excitation Control System for Problem 8B-16

PROBLEM 8B-17

The excitation control system of Problem 8B-16 is provided with a compensating device as shown in Figure 8.71. Assume that

$$K_F = 0.16 \qquad \tau_F = 0.8$$

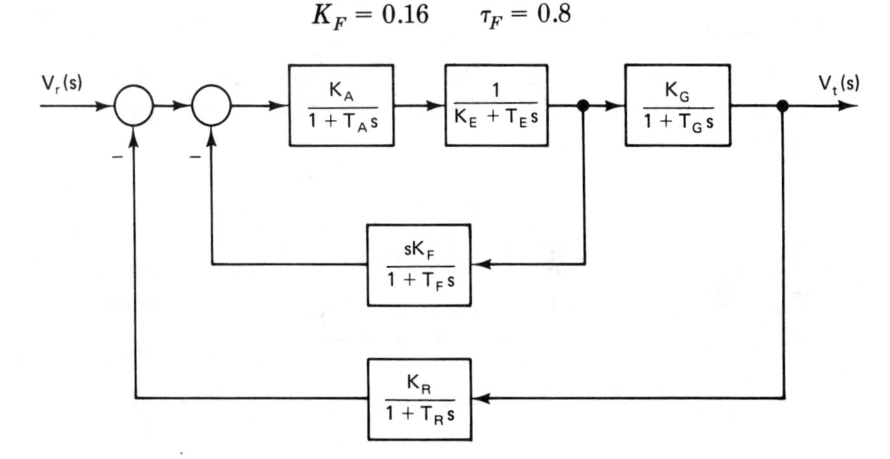


**Figure 8.71.** Compensated System for Problem 8B-17

Use Bode's logarithmic plots to check the stability of the system. Find the gain and phase margins.

PROBLEM 8B-18

The denominator of the transfer function $G_1(s)$ of the hydrogovernor of Problem 8A-3 is a third-order polynomial in $s$. Observing that the pilot valve time constant is $\tau_p = 0.1$, which is much less than the dashpot time constant $\tau_r = 5$, an approximation can be made by setting $\tau_p = 0$. This results in the approximate transfer function

$$\hat{G}_1(s) = \frac{1 + \tau_r s}{\tau_r \tau_g s^2 + \left[\tau_g + \tau_r(\sigma + \delta)\right] s + \sigma}$$

A further approximation results in the transfer function

$$\tilde{G}_1(s) = \frac{1 + \tau_r s}{\sigma(1 + \tau_1 s)(1 + \tau_2 s)}$$

where

$$\tau_1 = \frac{\tau_g + \tau_r(\sigma + \delta)}{\sigma}$$

$$\tau_2 = \frac{\tau_r \tau_g}{\tau_g + \tau_r(\sigma + \delta)}$$

Construct and compare the Bode plots of $\hat{G}_1(s)$ and $\tilde{G}_1(s)$ for the typical parameter values given in Problem 8A-3.

PROBLEM 8B-19

Repeat Problem 8A-4 using the approximate governor representation of Problem 8B-18 denoted by $\tilde{G}_1(s)$.

PROBLEM 8B-20

Sketch the polar and Bode plots of the frequency response for the transfer function

$$G(s) = \frac{K}{s(1 + \tau_1 s)(1 + \tau_2 s)(1 + \tau_3 s)}$$

Show that for this system the crossover frequency $\omega_x$ ($\phi = -180°$) satisfies

$$\omega_x = \left(\tau_1 \tau_2 + \tau_1 \tau_3 + \tau_2 \tau_3\right)^{-1/2}$$

PROBLEM 8B-21

For the log magnitude characteristic shown in Figure 8.72, find the corresponding transfer function. Determine the system type. Find the

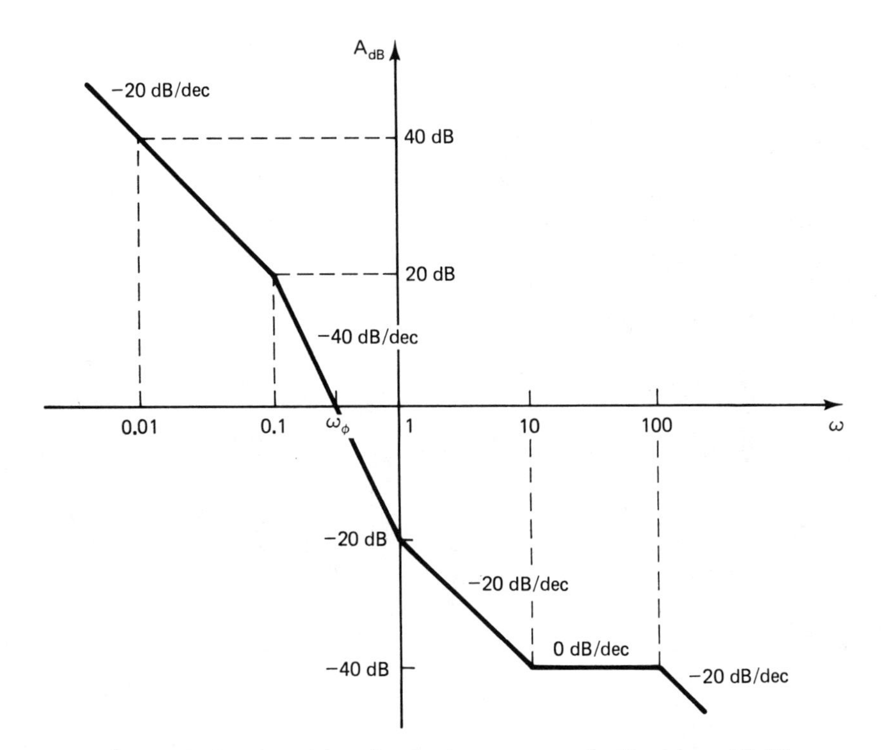


**Figure 8.72.** Log Magnitude Asymptotes for Problem 8B-21

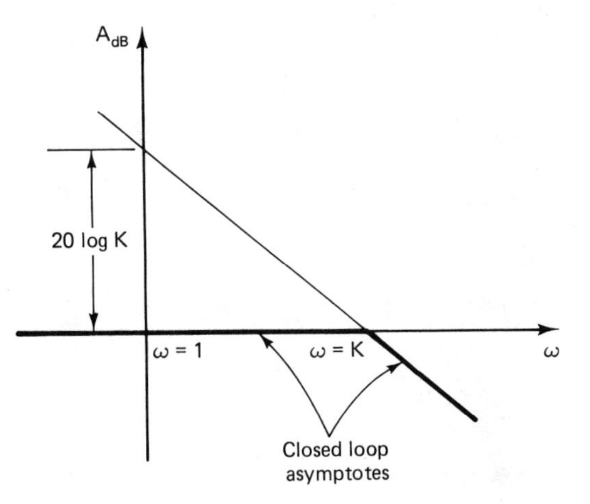


**Figure 8.73.** Bode Plot for Problem 8B-22

frequency $\omega_\phi$. Find the gain and phase margins. Find the velocity error coefficient.

PROBLEM 8B-22

Consider the open-loop transfer function

$$G(s) = \frac{K}{s}$$

Show that the Bode plot of the closed-loop transfer function with unity gain feedback can be obtained from the Bode plot of the open-loop function as shown in Figure 8.73.

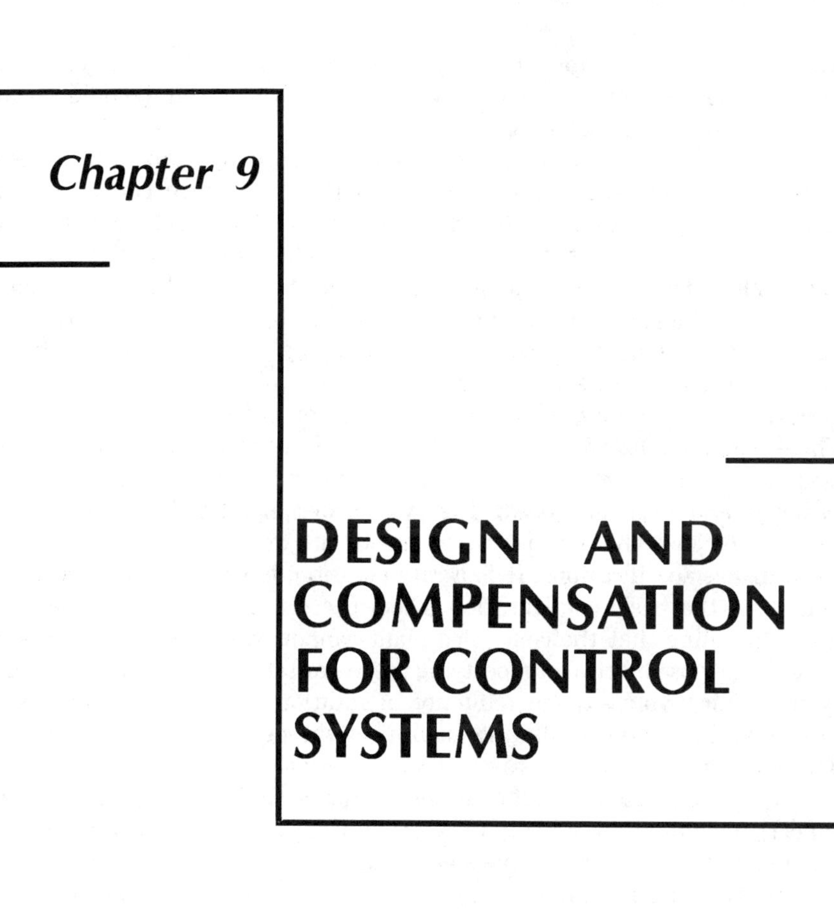

# Chapter 9

# DESIGN AND COMPENSATION FOR CONTROL SYSTEMS

## 9.1 Introduction

In the preceding chapters our study centered on representing and analyzing the performance of feedback control systems. The systems are classified as single input–single output linear time-invariant closed-loop systems. We have considered two approaches to the performance analysis problem. In the first approach, time-domain results were obtained based mainly on the Laplace transform theory in Chapters 4 and 5. The topic of system stability was introduced in Chapter 6 and the root-locus method was discussed in Chapter 7. The second approach involves frequency-domain techniques and was detailed in Chapter 8.

The tools developed in the preceding chapters enable the control systems specialist to evaluate the performance of a given control system configuration such as that shown in Figure 9.1. The block with transfer function $G_p(s)$ denotes the plant or controlled process and is usually unalterable. The first task in performance evaluation is to verify system stability. If the given closed-loop system is unstable, means for its stabilization must be sought. Once we have a stable system, other pertinent performance indicators are considered. A control system is designed to perform a certain task while meeting a set of prespecified requirements. These requirements are generally referred to in control systems specialists' language as *performance specifications*, and relate to accuracy, relative stability, and response speed. The system designer has then to evaluate whether or not the given system meets all or some of the available performance specifications. It is a rare occasion, however, when a designer finds that the basic configuration meets the given performance specifications. Recalling that the controlled plant cannot be modified, the designer looks at the possibilities for modifying the overall structure of the closed-loop system. This involves the introduction of additional components, devices, or subsystem into the overall system to meet the performance specifications. This process is referred to as *system compensation.*

Commonly used time-domain performance specifications include peak time $T_p$, maximum overshoot, and settling time associated with a unit step input. The maximum allowable steady-state error in tracking standard test signals (as detailed in Chapter 5) is also specified. Frequency-domain performance measures commonly used include phase margin, peak of the closed-loop frequency response, resonant frequency, and bandwidth.

There are numerous ways to introduce the compensating function in the closed-loop of Figure 9.1. The first and most popular means of compensation is to introduce the compensating device in the forward loop in cascade with the controlled plant, as shown in Figure 9.2a. The compensating device is represented by the transfer function $G_C(s)$. This is referred to as *cascade* (or *series*) *compensation.* The second means of compensation inserts the compensating device as a feedback element $G_H(s)$ in the system

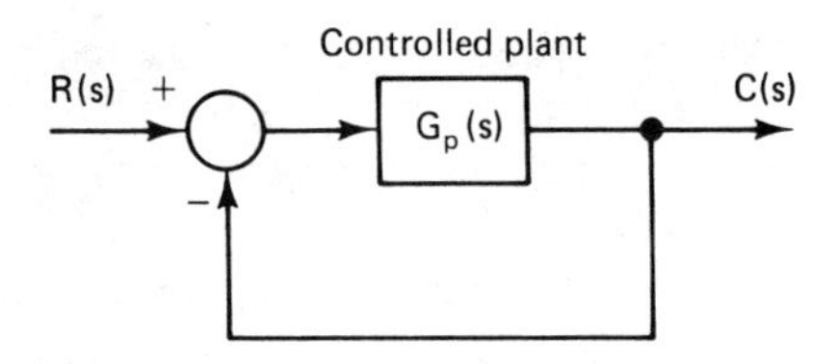


**Figure 9.1.** Basic Unity Feedback Control System

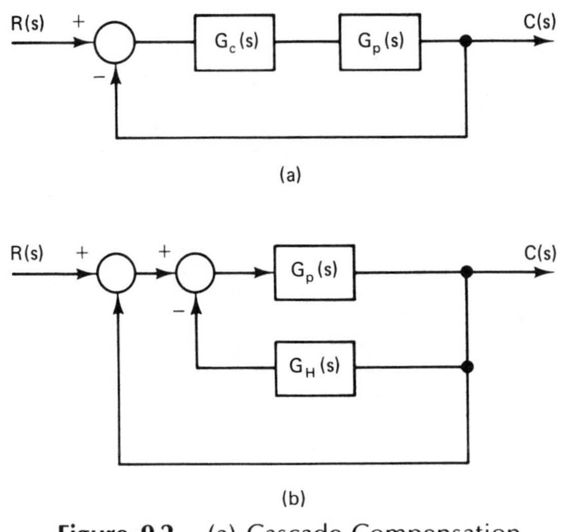


**Figure 9.2.** (a) Cascade Compensation
(b) Feedback Compensation

as shown in Figure 9.2b. This is referred to as *feedback* (or *parallel*) *compensation*. Another form of compensation utilizes a combination of the above in what is referred to as *cascade–feedback compensation*, shown in Figure 9.3.

Sometimes it is useful to consider the insertion of the compensating device at the input side. This is achieved in the input compensation form shown in Figure 9.4a or in the equivalent feedforward compensation form shown in Figure 9.4b. An alternative shown in Figure 9.4c is referred to as the output or load-compensation scheme.

State-feedback compensation, an alternative form of feedback compensation, is based on the state-space representation. Choosing the states to be physically measurable variables, feedback signals derived from the states are utilized for compensation purposes as shown in Figure 9.5.

To choose among the various forms of compensation for a given system, a number of factors should be considered. Among the major

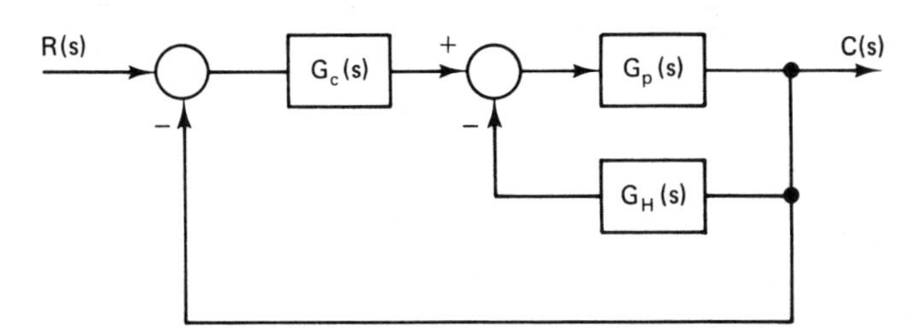


**Figure 9.3.** Cascade – Feedback Compensated Configuration

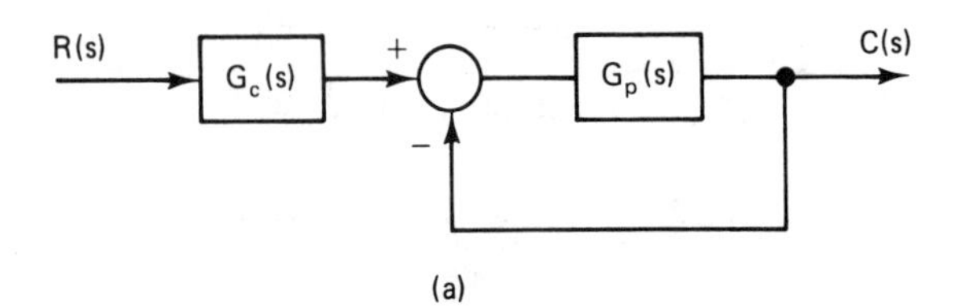


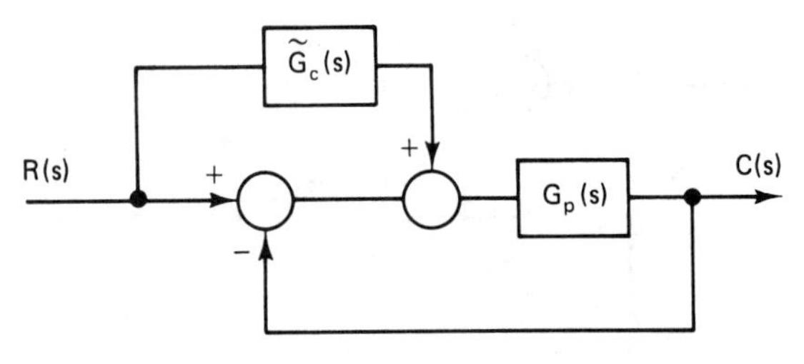


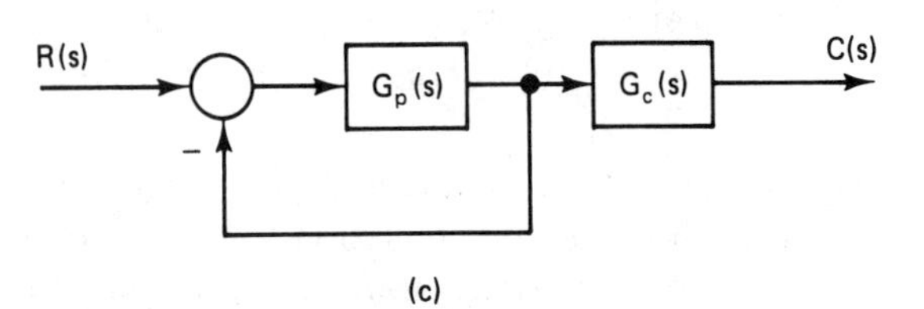


**Figure 9.4.** (a) Input Compensation
(b) Feedforward Compensation
(c) Output or Load Compensation

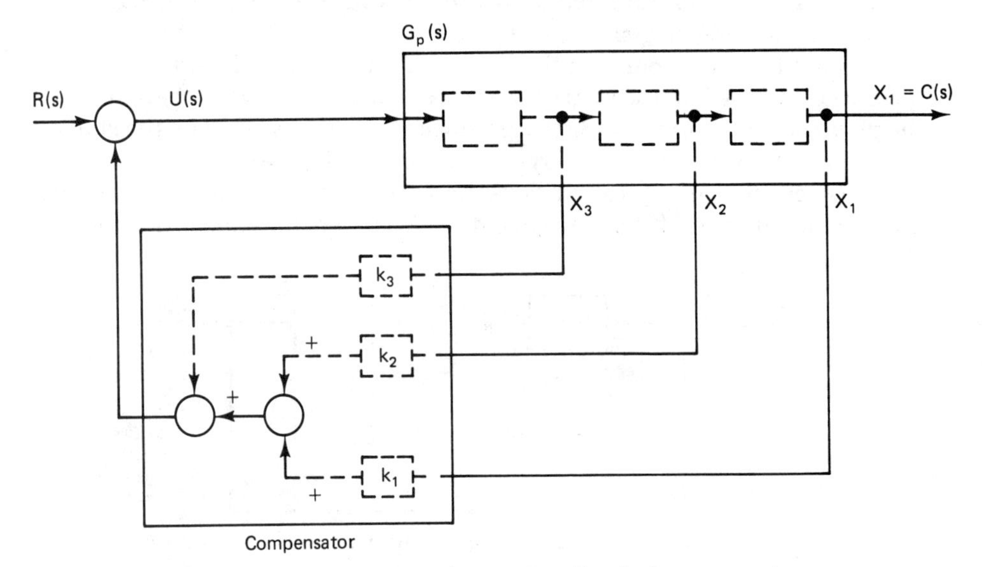

**Figure 9.5.** Concept of State-Feedback Compensation

considerations are the availability of the components required by the design and the designer's experience and preferences. As a general rule, cascade compensation design procedures are more direct and less involved than those for other forms. Many other factors involving the nature (electrical, mechanical, hydraulic, or pneumatic) of the system and its environment will influence the designer's choice. We start our study of design and compensation by considering cascade-compensating elements.

# 9.2 Elementary Cascade-Compensating Networks

A network (or subsystem) intended for cascade compensation may be described by its transfer function $G_C(s)$. A general expression for the transfer function is

$$G_C(s) = \frac{K_C \prod_{i=1}^{M} (s - z_i)}{\prod_{k=1}^{N} (s - p_k)}$$

The gain of the compensator is denoted by $K_C$; $z_i$ and $p_k$ are the zeros and poles associated with the transfer function. The design problem is to determine $M$, the number of zeros; $N$, the number of poles; the values of poles and zeros; and the compensator's gain $K_C$ to meet the performance specifications.

The simplest form of cascade compensation is gain adjustment. In this case we have

$$G_C(s) = K_C$$

The choice of $K_C$ presents no real difficulties using our previously developed analysis tools. However, this form may not be successful in meeting our requirements. We will thus be lead to a higher-order transfer function.

The proportional–derivative (PD) controller of Chapter 3 provides us with the compensator transfer function

$$G_C(s) = K_P + K_D s$$

In this case

$$K_C = K_D$$

$$z_1 = -\frac{K_P}{K_D}$$

The effect of a PD controller is to add a zero at $s = -K_P/K_D$. The

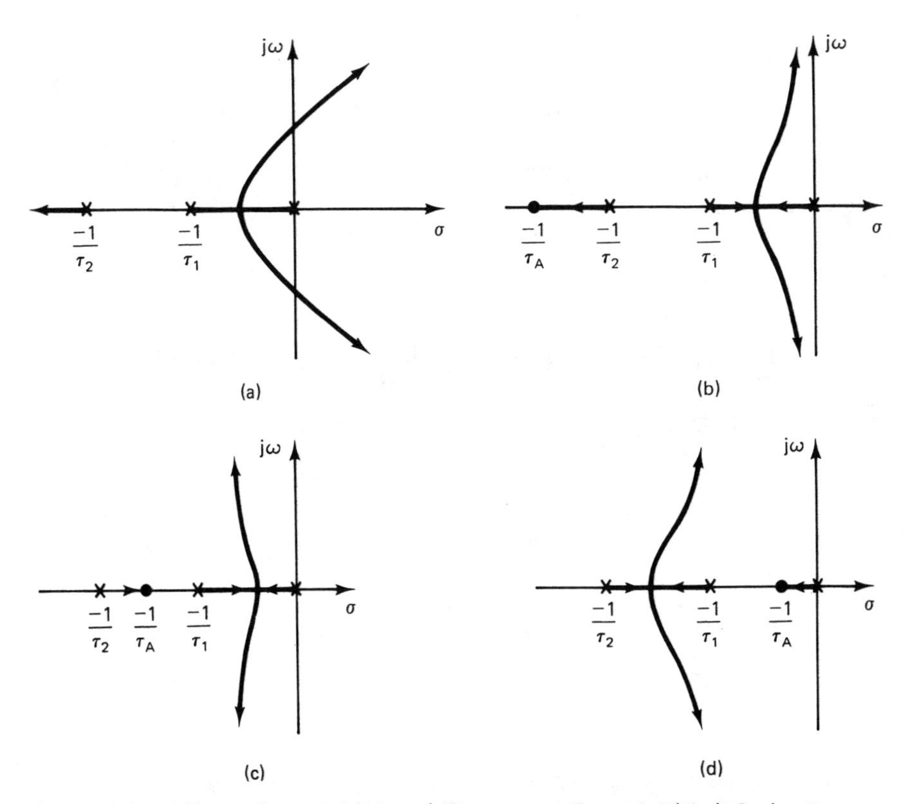


**Figure 9.6.** Effect of an Additional Zero on a Type 1 Third-Order System

addition of a zero to the open-loop transfer function causes a shift of the root locus to the left, as shown in Figure 9.6 for the system described by

$$G_P(s) = \frac{K}{s(1 + \tau_1 s)(1 + \tau_2 s)}$$

In part (a) of the figure we note that the system is unstable for high gain values. Parts (b)–(d) show the root locus for a compensated version such that

$$G_C(s)G_P(s) = \frac{K(1 + \tau_A s)}{s(1 + \tau_1 s)(1 + \tau_2 s)}$$

The value of $\tau_A$ relative to $\tau_1$ and $\tau_2$ is varied as shown, but the conclusion is the same in each case. The phase margin and gain margin are increased due to the addition of a zero to the system, as can be seen from Figure 9.7.

A PD compensator cannot be physically realized using passive circuit elements. It can, however, be realized using operational amplifiers, resistors, and capacitors. A PD compensator acts as a high pass filter, and therefore noise is amplified, which is an undesirable effect.

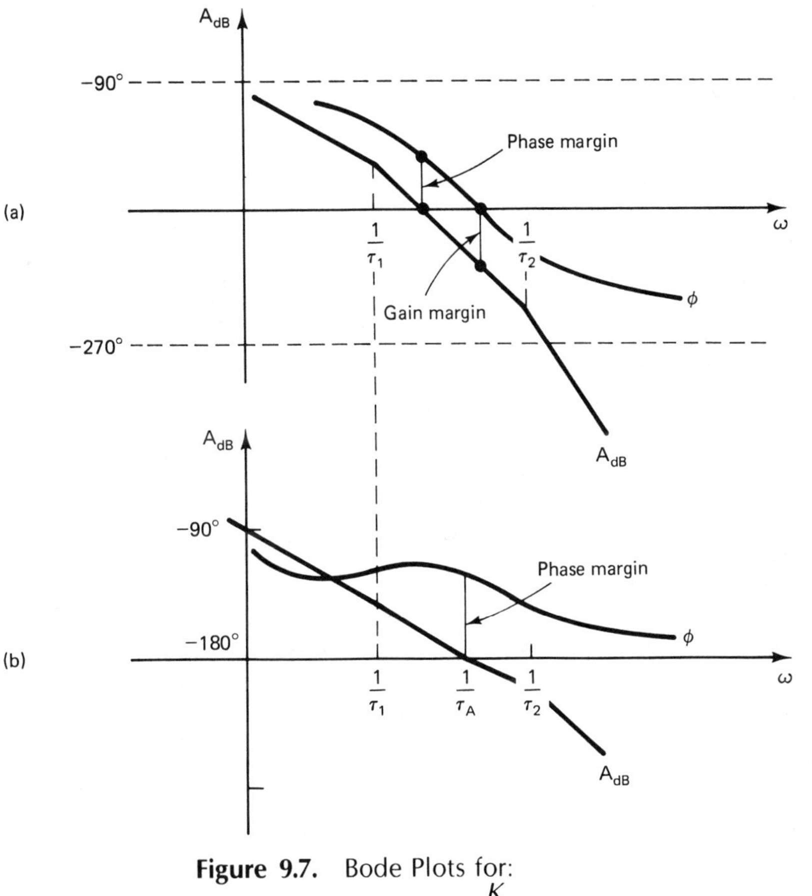


**Figure 9.7.** Bode Plots for:

(a) $\dfrac{K}{s(1 + \tau_1 s)(1 + \tau_2 s)}$

(b) $\dfrac{K(1 + \tau_A s)}{s(1 + \tau_1 s)(1 + \tau_2 s)}$

Cascade compensation can be effected using a proportional–integral (PI) controller with transfer function

$$G_C(s) = K_P + \frac{K_I}{s}$$

This is equivalent to adding a zero at $s = -K_I/K_P$ and a pole at $s = 0$. This increases the system order and type by one. As a result, if the steady-state error of the system to a test signal is constant, the PI controller will result in a zero steady-state error for that same test signal. The increase in system order can result in instability if $K_P$ and $K_I$ are not

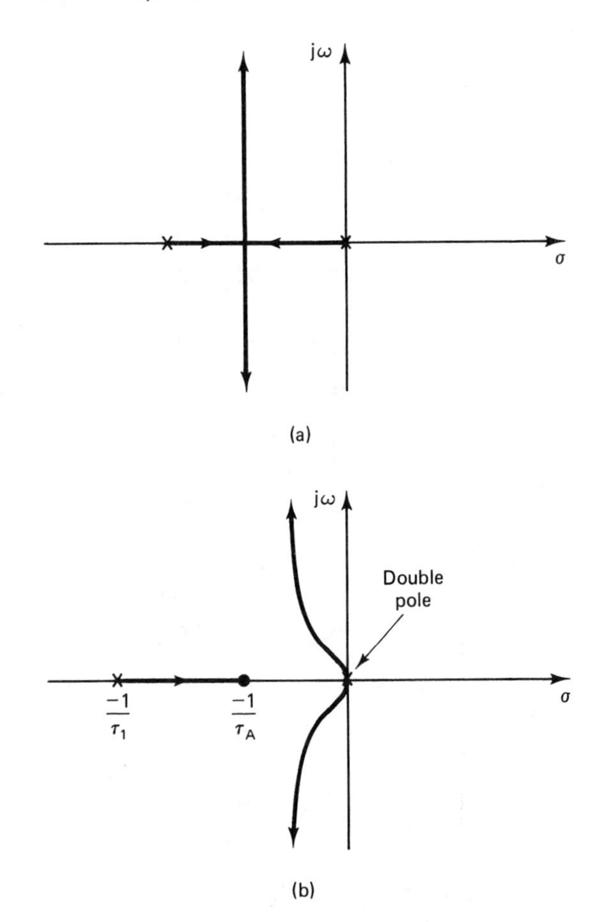


**Figure 9.8.** Root Locus for:
(a) $\dfrac{K}{s(1 + \tau_1 s)}$
(b) $\dfrac{K(1 + \tau_A s)}{s^2(1 + \tau_1 s)}$

chosen properly. Figure 9.8a shows the root locus for

$$G_P(s) = \frac{K}{s(1 + \tau_1 s)}$$

In Figure 9.8b the root locus for the system with a PI controller inserted in cascade is shown.

Figure 9.9 pertains to the open-loop function

$$G_P(s) = \frac{K}{s(1 + \tau_1 s)(1 + \tau_2 s)}$$

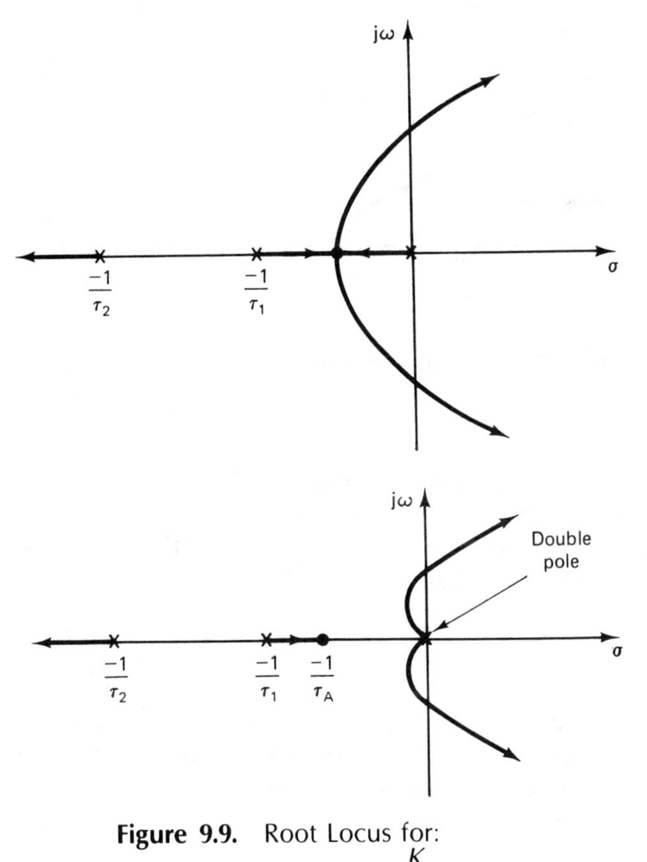


**Figure 9.9.** Root Locus for:
(a) $\dfrac{K}{s(1 + \tau_1 s)(1 + \tau_2 s)}$
(b) $\dfrac{K(1 + \tau_A s)}{s^2(1 + \tau_1 s)(1 + \tau_2 s)}$

whose root locus is shown in portion (a). The effect of the PI controller is shown in Figure 9.9b. In both cases the root locus is pulled to the right.

The proportional-integral controller is essentially a low-pass filter. Thus the gain is high for low frequency and drops to a constant value of $20 \log K_P$ at high frequency. This is illustrated in the Bode plot of Figure 9.10. The converse is true for the proportional–derivative controller; that is, a PDC works as a high-pass filter, as shown in Figure 9.11.

The phase angle of a PI controller is given by

$$\phi = \tan^{-1} \frac{\omega K_P}{K_I} - 90°$$

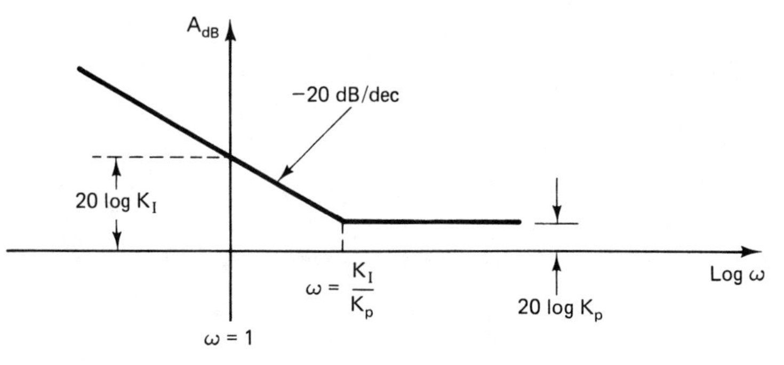


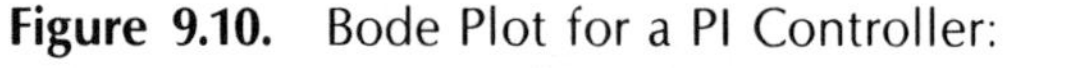

**Figure 9.10.** Bode Plot for a PI Controller:

$$G_C(s) = K_P + \frac{K_I}{s}$$

The phase angle varies from $-90°$ at low frequencies to $0°$; that is, it is always negative. The output always lags the input and a PI controller is sometimes referred to as a *lag controller*. On the other hand, the PD controller has an angle of

$$\phi = \tan^{-1}\frac{\omega K_D}{K_P}$$

The angle varies from $0°$ to $90°$ and is thus positive. The output always leads the input and a PD controller is referred to as a *lead controller*.

It is advantageous to employ compensating networks that consist of passive circuit elements only. The transfer function of the simplest possible network is given by

$$G_C(s) = K_C\frac{s - z_1}{s - p_1}$$

This transfer function has one zero $z_1$ and one pole $p_1$. The associated

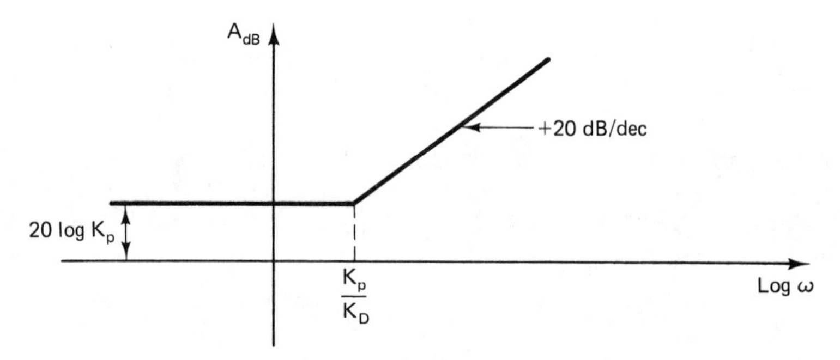


**Figure 9.11.** Bode Plot for a PD Controller:

$$G_C = K_P + K_D s$$

frequency response function is given by

$$G_C(j\omega) = K_C \frac{-z_1 + j\omega}{-p_1 + j\omega}$$

The phase angle is obtained as

$$\phi_C(j\omega) = \tan^{-1}\frac{\omega}{-z_1} - \tan^{-1}\frac{\omega}{-p_1}$$

It is clear that if $z_1 > p_1$, the phase angle is positive and the network is a phase-lead network. If $p_1 > z_1$, the network is a phase-lag network. We start examining these networks now.

# 9.3 Phase-Lead Network

A phase-lead network can be mechanized using two resistors and a capacitor as shown in Figure 9.12. The transfer function of the network can be written as

$$G_C(s) = \frac{V_o(s)}{V_i(s)} = \frac{R_2}{R_2 + \dfrac{1}{(1/R_1) + Cs}}$$

A few algebraic manipulations yield

$$G_C(s) = \frac{R_2(1 + R_1 Cs)}{(R_1 + R_2)\left(1 + \dfrac{R_1 R_2}{R_1 + R_2} Cs\right)}$$

Let us define the ratio $\alpha$ and the time constant $\tau$ by

$$\tau = R_1 C$$

$$\alpha = \frac{R_1 + R_2}{R_2}$$

Note that $\alpha > 1$. As a result, the transfer function of the lead network is

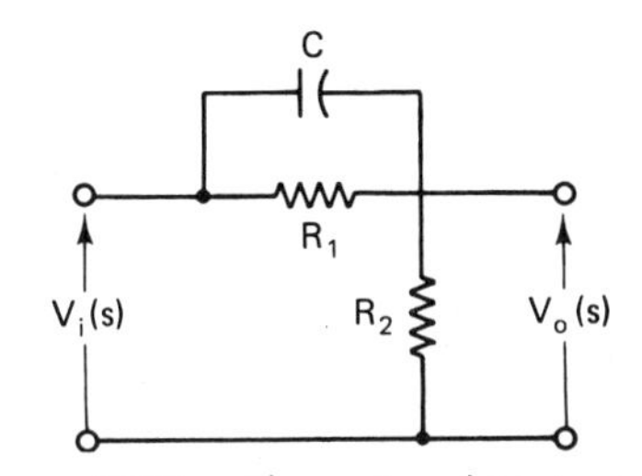


**Figure 9.12.** Phase-Lead Network

written in the standard form:

$$G_C(s) = \frac{1}{\alpha} \frac{1 + \tau s}{1 + \tau s/\alpha}$$

The phase-lead network has one zero at

$$z_1 = \frac{-1}{\tau}$$

and one pole at

$$p_1 = \frac{-\alpha}{\tau}$$

As we have noted that $\alpha$ is greater than 1, the zero is then closer to the origin than the pole on an $s$-plane representation, as shown in Figure 9.13.

The polar plot of the frequency response of the phase-lead network can be obtained from

$$G_C(j\omega) = \frac{1 + j\omega\tau}{\alpha + j\omega\tau}$$

Multiplying and dividing by the complex conjugate of the denominator, we get

$$G_C(j\omega) = \frac{(\alpha + \omega^2\tau^2) + j\omega\tau(\alpha - 1)}{\alpha^2 + \omega^2\tau^2}$$

We denote the real part by $X$ and the imaginary part by $Y$, to obtain

$$X = \frac{\alpha + \omega^2\tau^2}{\alpha^2 + \omega^2\tau^2}$$

$$Y = \frac{\omega\tau(\alpha - 1)}{\alpha^2 + \omega^2\tau^2}$$

To obtain the locus of the polar plot we need to eliminate $\omega\tau$ from the foregoing two relations.

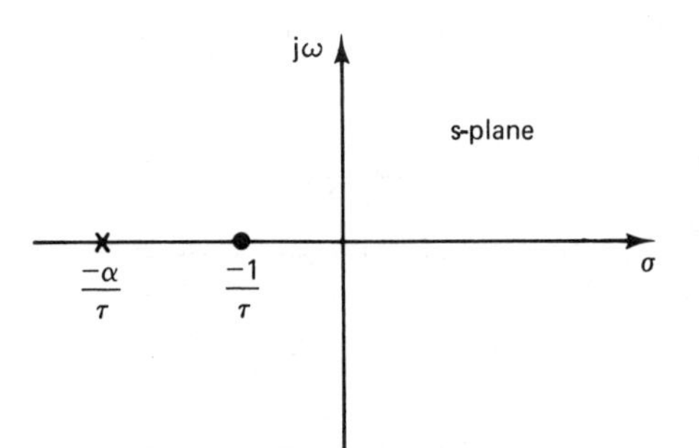


**Figure 9.13.** *s*-Plane Configuration for a Phase-Lead Network

We can show that

$$1 - X = \frac{\alpha(\alpha - 1)}{\alpha^2 + \omega^2\tau^2}.$$

Thus

$$\frac{Y}{1 - X} = \frac{\omega\tau}{\alpha}$$

Substituting the relation above in the equation of the real part, we get after some simplification

$$X = \frac{(1 - X)^2 + \alpha Y^2}{\alpha\left[(1 - X)^2 + Y^2\right]}$$

The above reduces to

$$Y^2 + \left(X - \frac{\alpha + 1}{2\alpha}\right)^2 = \left(\frac{\alpha - 1}{2\alpha}\right)^2$$

As a result we conclude that the polar plot of the phase-lead network is a circle (or part of a circle) with center on the real axis $X_c$ and radius $R_c$

$$X_c = \frac{\alpha + 1}{2\alpha}$$

$$R_c = \frac{\alpha - 1}{2\alpha}$$

Let us observe here that

$$X_c + R_c = 1$$

$$X_c - R_c = \frac{1}{\alpha}$$

The geometry of the polar plot is shown in Figure 9.14.

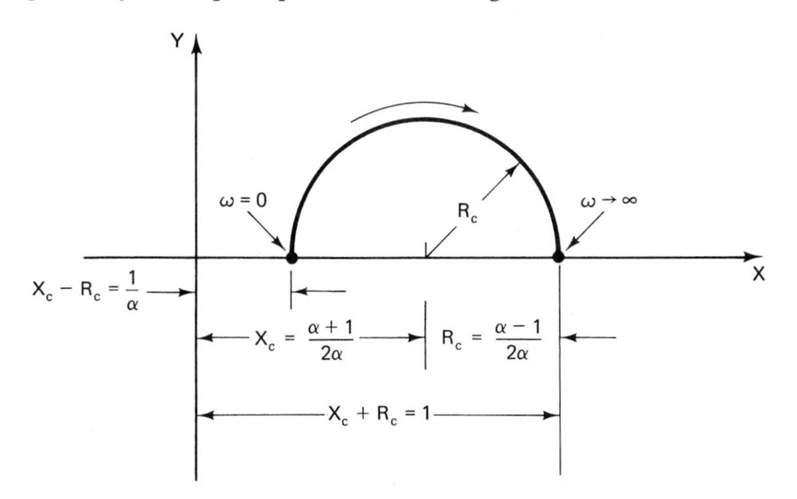


**Figure 9.14.** Geometry of the Polar Plot for a Phase-Lead Network

The polar plot is a semicircle in the first quadrant of the complex plane. At $\omega = 0$ we have

$$G_C(j0) = \frac{1}{\alpha}$$

This point is on the real axis as shown in Figure 9.14. As $\omega$ tends to infinity, we have

$$\lim_{\omega \to \infty} G_c(j\omega) = 1$$

The phase angle at a radian frequency $\omega$ is

$$\phi = \tan^{-1}\omega\tau - \tan^{-1}\frac{\omega\tau}{\alpha}$$

Since $\alpha$ is greater than 1, we conclude that $\phi$ is positive and hence the locus is located in the first quadrant of the complex plane.

The phase angle $\phi$ is zero at $\omega = 0$ and increases as $\omega$ is increased to a maximum angle $\phi_m$, as indicated in the geometry of Figure 9.15. From the figure we can see that the maximum phase angle $\phi_m$ is defined by

$$\sin\phi_m = \frac{R_c}{X_c}$$

Thus

$$\sin\phi_m = \frac{\alpha - 1}{\alpha + 1}$$

The magnitude of the frequency response at the angle of maximum lead is obtained with reference to the geometry of Figure 9.15 as

$$|G(j\omega_m)|^2 = X_c^2 - R_c^2 = \frac{1}{\alpha}$$

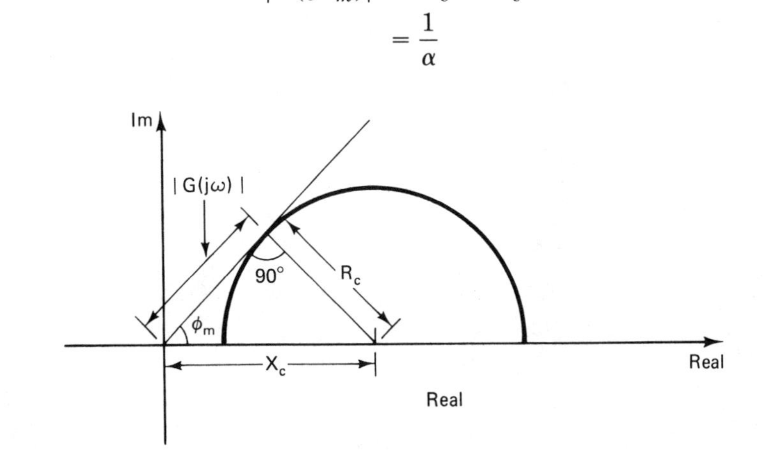


**Figure 9.15.** Defining the Maximum Phase Angle for the Phase-Lead Network

The radian frequency $\omega_m$ at which the maximum lead occurs is obtained from the basic frequency response relation as

$$\frac{1}{\alpha} = \frac{1 + \omega_m^2 \tau^2}{\alpha^2 + \omega_m^2 \tau^2}$$

This reduces to

$$\omega_m = \frac{\sqrt{\alpha}}{\tau}$$

It is probably appropriate at this juncture to comment that since the angle $\phi$ is always positive, the output $V_o(j\omega)$ leads the input voltage $V_i(j\omega)$ and hence the network is appropriately labeled a phase-lead network. In Figure 9.16 we compare the polar plots for three different values of $\alpha$. As $\alpha$ increases so does the angle of maximum phase lead; however, the frequency $\omega_m$ at which the maximum phase lead occurs decreases with an increase in $\alpha$.

The alert reader will by now be anticipating a discussion of the Bode plots of the frequency response of the phase-lead network. This we do now. Our starting point is the transfer function

$$G_C(s) = \frac{1}{\alpha} \frac{1 + \tau s}{1 + \tau s/\alpha}$$

The gain contribution is

$$20 \log K = 20 \log \frac{1}{\alpha} = -20 \log \alpha$$

Since $\alpha$ is greater than 1, the gain contribution is negative up until the first corner frequency occurring at $\omega_{c_1} = 1/\tau$ due to the numerator dynamics. This contributes a slope of $+20$ dB/decade. At the corner frequency

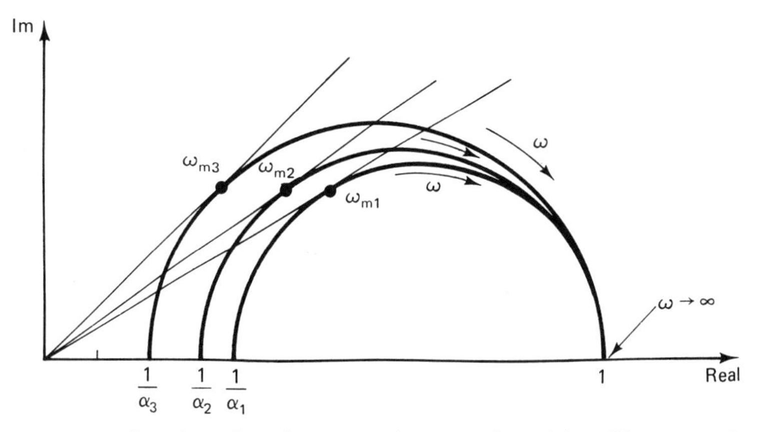


**Figure 9.16.** Polar Plots for Phase-Lead Networks with Different Values of the Parameter $\alpha$

$\omega_{c_2} = \alpha/\tau$, the denominator effect takes place with a $-20$-dB/decade slope, providing us with a net 0-dB/decade slope for the remainder of the frequency range.

The phase angle $\phi$ is given by

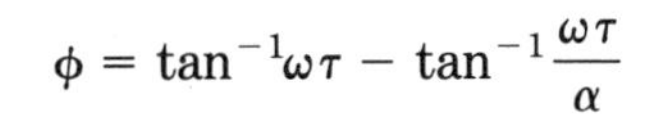

$$\phi = \tan^{-1}\omega\tau - \tan^{-1}\frac{\omega\tau}{\alpha}$$

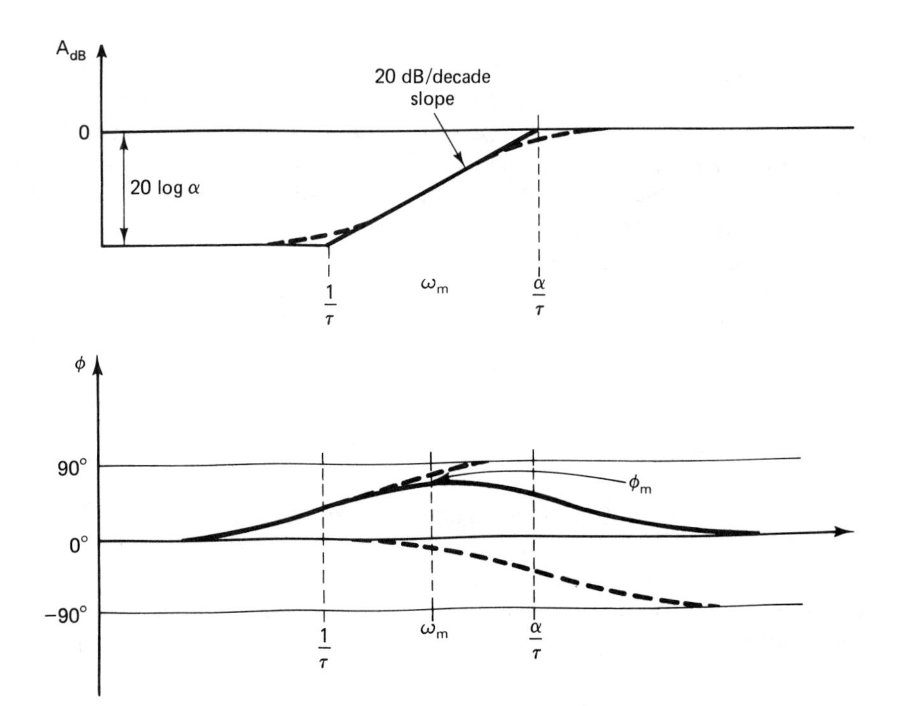


**Figure 9.17.** Bode Plots for a Phase-Lead Network

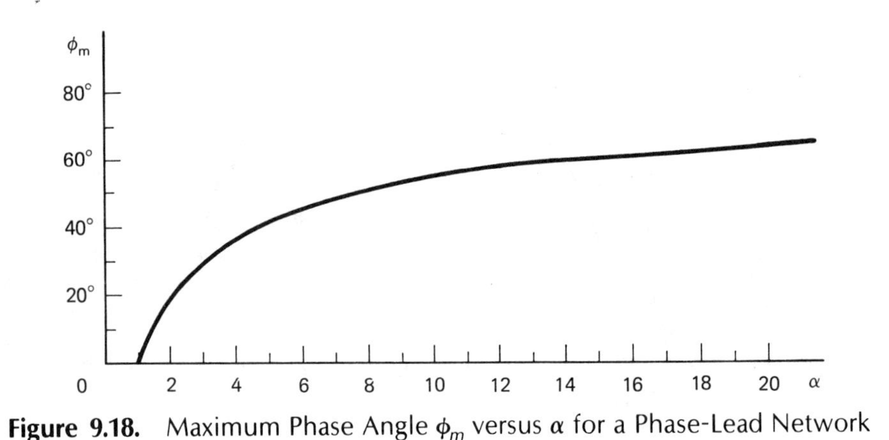


**Figure 9.18.** Maximum Phase Angle $\phi_m$ versus $\alpha$ for a Phase-Lead Network

As we have learned from the polar plot, the angle $\phi$ ranges from zero to $\phi_m$, then decreases to zero again for high frequencies. The complete Bode plot for the phase-lead network is shown in Figure 9.17. The variation of the maximum phase angle for the phase-lead network with the value of $\alpha$ is shown in Figure 9.18.

## Synthesis Problem

Assume that a design specification calls for a required value of gain $|G(j\omega_s)|$ at a given radian frequency $\omega_s$ with a specified phase angle $\phi_s$. The problem is to find $\alpha$ and $\tau$ of the phase-lead network to meet these requirements.

There exist a number of procedures to solve this problem in synthesis. The most straightforward method proceeds by calculating the rectangular components $X_s$ and $Y_s$ using the given information as

$$X_s = |G(j\omega_s)|\cos\phi_s$$

$$Y_s = |G(j\omega_s)|\sin\phi_s$$

We now recall that

$$\frac{Y_s}{1 - X_s} = \frac{\omega_s\tau}{\alpha}$$

Thus

$$\tau = \alpha g_s$$

where

$$g_s = \frac{Y_s}{\omega_s(1 - X_s)}$$

The real component $X_s$ satisfies

$$X_s = \frac{\alpha + \omega_s^2\tau^2}{\alpha^2 + \omega_s^2\tau^2}$$

Substituting for $\tau$ in the equation above and rearranging, we obtain

$$\alpha = \frac{1}{X_s + g_s^2\omega_s^2(X_s - 1)}$$

This resolves the problem. An example will illustrate the procedure.

EXAMPLE 9.1

Find the parameters $\alpha$ and $\tau$ of a phase-lead network such that at $\omega_s = 2.4$ rad/sec, the magnitude is 0.25 and the phase angle is given by $\phi = 41°$.

SOLUTION

From the given specifications we can obtain

$$X_s = 0.25\cos 41° = 0.1887$$
$$Y_s = 0.25\sin 41° = 0.164$$

Thus

$$g_s = \frac{Y_s}{\omega_s(1 - X_s)} = \frac{0.164}{2.4(1 - 0.1887)}$$
$$= 0.0842$$

We can now compute $\alpha$ as

$$\alpha = \frac{1}{X_s + g_s^2\omega_s^2(X_s - 1)}$$
$$= 6.43$$

As a result,

$$\tau = \alpha g_s$$
$$= 0.5416$$

The required transfer function is thus given by

$$G_C(s) = \frac{1 + 0.5416s}{6.43 + 0.5416s}$$

## Normalized Phase-Lead Transfer Function

It is customary in control engineering practice to base the design on a normalized version of the transfer function so that the low-frequency gain is unity. For the phase-lead network this takes the form

$$G_n(j\omega) = \frac{1 + j\omega\tau}{1 + j(\omega\tau/\alpha)}$$

Note that this transfer function is related to the physically based transfer function $G_c(j\omega)$ by

$$G_n(j\omega) = \alpha G_c(j\omega)$$

The polar plot of the normalized phase-lead function is obtained using

$$X_n = \frac{\alpha(\alpha + \omega^2\tau^2)}{\alpha^2 + \omega^2\tau^2}$$
$$Y_n = \frac{\alpha\omega\tau(\alpha - 1)}{\alpha^2 + \omega^2\tau^2}$$

We also have

$$\frac{Y_n}{1 - X_n/\alpha} = \omega\tau$$

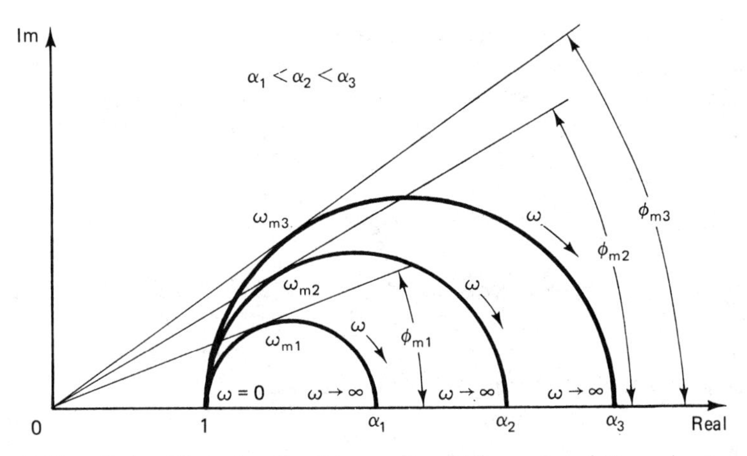


**Figure 9.19.** Polar Plots for the Normalized Phase-Lead Transfer Function for Various Values of $\alpha$

The semicircle equation becomes

$$Y_n^2 + \left(X_n - \frac{\alpha + 1}{2}\right)^2 = \left(\frac{\alpha - 1}{2}\right)^2$$

The center is at

$$X_{c_n} = \frac{\alpha + 1}{2}$$

The radius is

$$R_{c_n} = \frac{\alpha - 1}{2}$$

Figure 9.19 shows the polar plot for the normalized phase-lead function for various values of $\alpha$.

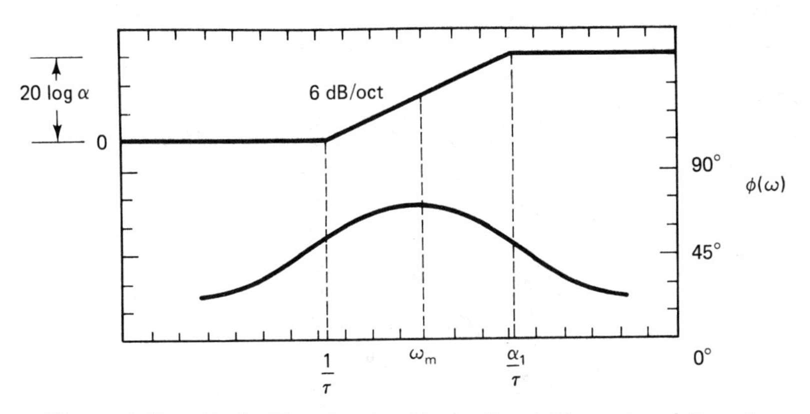


**Figure 9.20.** Bode Plot for the Normalized Phase-Lead Function

The phase angle characteristics are not changed through the normalization process. The Bode plot will have a start at 0 dB for low frequencies and retain the two corner frequencies of $1/\tau$ and $\alpha/\tau$. For high frequencies the gain in decibels is $20 \log \alpha$, as shown in Figure 9.20.

## The Synthesis Problem Again

The synthesis problem requiring finding $\alpha$ and $\tau$ given the magnitude $|G(j\omega_s)|$ and phase angle $\phi_s$ at a specified radian frequency $\omega_s$, discussed earlier, becomes slightly more difficult for the normalized phase-lead function. The key to our earlier proposed procedure is now

$$\frac{Y_{n_s}}{1 - X_{n_s}/\alpha} = \omega_s \tau$$

This is now substituted in

$$X_{n_s} = \frac{\alpha\left(\alpha + \omega_s^2 \tau^2\right)}{\alpha^2 + \omega_s^2 \tau^2}$$

A few simplification steps result in the following quadratic equation in $\alpha$. Note that we drop the subscript $ns$ for convenience.

$$(X - 1)\alpha^2 + \left[2(1 - X)X - Y^2\right]\alpha + X(X^2 + Y^2 - X) = 0$$

An example will illustrate the modified procedure.

EXAMPLE 9.2

Find the parameters $\alpha$ and $\tau$ of the normalized phase-lead transfer function to result in

$$|G(j\omega_s)|^2 = 2.63$$
$$\phi_s = 41.34°$$

with

$$\omega_s = 2.4 \text{ rad/sec}$$

SOLUTION

We calculate

$$X = |G| \cos \phi = 1.2175$$
$$Y = |G| \sin \phi = 1.0714$$

Using these values, our quadratic in $\alpha$ turns out to be

$$0.2175\alpha^2 - 1.6773\alpha + 1.7197 = 0$$

The solution is

$$\alpha = 6.496$$

The corresponding $\tau$ is obtained from

$$\tau = \frac{Y}{\omega_s(1 - X/\alpha)}$$

This turns out to be

$$\tau = 0.55$$

An alternative method available in the literature requires the solution of the following quadratic in $\alpha$:

$$(p^2 - c + 1)\alpha^2 + 2p^2c\alpha + (p^2c^2 + c^2 - c) = 0$$

where

$$p = \tan\phi$$
$$c = |G(j\omega)|^2$$

The value of $\tau$ is obtained from

$$(\omega\tau)^2 = \frac{\alpha^2(1 - c)}{c - \alpha^2}$$

Either of the two methods should provide the same answers.

The following example illustrates the effect of cascade phase-lead compensation on time-domain performance of a simple second-order system.

EXAMPLE 9.3

Consider the simple second-order system shown in Figure 9.21a. The open-loop transfer function is given by

$$G_p(s) = \frac{K_m}{s(1 + \tau_m s)}$$

This transfer function corresponds to many physical systems, such as an armature-controlled dc motor or a two-phase servomotor.

The closed-loop transfer function is given by

$$\frac{C(s)}{R(s)} = \frac{K_m/\tau_m}{s^2 + (1/\tau_m)s + K_m/\tau_m}$$

The standard form for a second-order transfer function in terms of natural frequency $\omega_n$ and damping $\zeta$ is given as

$$\frac{C(s)}{R(s)} = \frac{\omega_n^2}{s^2 + 2\zeta\omega_n s + \omega_n^2}$$

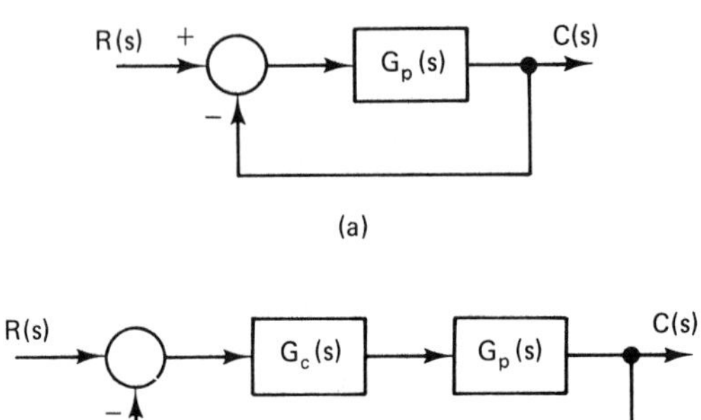


R(s)

$G_c(s)$ $G_p(s)$

C(s)

**Figure 9.21.** Block Diagram for Example 9.3

We thus have

$$\omega_n^2 = \frac{K_m}{\tau_m}$$

$$\zeta = \frac{1}{2\sqrt{K_m \tau_m}}$$

The steady-state error due to a unit ramp input for this type 1 system is given by

$$e_{ss}(t) = \frac{1}{K_m}$$

Note that for a given motor the values of $K_m$ and $\tau_m$ are available and hence we can obtain the undamped natural frequency $\omega_n$, the damping ratio, and the steady-state ramp input error.

Suppose now that a phase-lead compensating network is introduced as shown in Figure 9.21b. The transfer function $G_C(s)$ is given by the normalized form

$$G_C(s) = \frac{1 + \tau s}{1 + (\tau/\alpha)s}$$

It is clear that the compensated open-loop transfer function is given by

$$G_C(s)G_p(s) = \frac{K_m(1 + \tau s)}{s(1 + \tau_m s)[1 + (\tau/\alpha)s]}$$

The velocity error coefficient $K_V$ is given by

$$K_V = \lim_{s \to 0} sG_C(s)G_p(s) = K_m$$

Thus the steady-state error due to a unit ramp input is unchanged and is given by

$$e_{ss}(t) = \frac{1}{K_m}$$

The closed-loop transfer function of the compensated system is given by

$$\frac{C(s)}{R(s)} = \frac{K_m(1 + \tau s)}{s(1 + \tau_m s)[1 + (\tau/\alpha)s] + K_m(1 + \tau s)}$$

The system is now of third order. At frequencies much lower than $\alpha/\tau$ (made possible by a large enough $\alpha$), we can neglect the term $[1 + (\tau/\alpha)s]$. Essentially, we assume that

$$G_C(s) \simeq 1 + \tau s$$

As a result, the closed-loop transfer function is given by

$$\frac{C(s)}{R(s)} = \frac{K_m(1 + \tau s)}{\tau_m s^2 + (1 + K_m\tau)s + K_m}$$

$$= \frac{(K_m/\tau_m)(1 + \tau s)}{s^2 + [(K_m/\tau_m)\tau + 1/\tau_m]s + K_m/\tau_m}$$

In terms of the uncompensated system's natural frequency and damping ratio we have

$$\frac{C(s)}{R(s)} = \frac{\omega_n^2(1 + \tau s)}{s^2 + (\tau\omega_n^2 + 2\zeta\omega_n)s + \omega_n^2}$$

Rearranging, we obtain

$$\frac{C(s)}{R(s)} = \frac{\omega_n^2(1 + \tau s)}{s^2 + 2(\zeta + \tau\omega_n/2)\omega_n s + \omega_n^2}$$

It is clear that the effect of compensation is to increase the damping ratio to $\zeta_{eq}$ given by

$$\zeta_{eq} = \zeta + \frac{\tau\omega_n}{2}$$

The undamped natural frequency remains unchanged.

# 9.4 Bode Plot Phase-Lead Cascade Compensation

The properties of the phase-lead network were studied in the preceding section. We consider now aspects of designing a phase-lead network to achieve prescribed performance specifications for a closed-loop system using cascade compensation. The design problem is to find the time constant $\tau$ and the parameter $\alpha$. The present section deals with Bode plot procedures. The Bode plot is favored over other methods due to its simplicity. The available uncompensated open-loop transfer function is used to establish the Bode plot. The effects of the compensation are obtained by simply adding its amplitude (in decibels) and phase angle to the uncompensated characteristics.

As a general rule, the first step in the design is to adjust the gain of the uncompensated open-loop transfer function to meet steady-state error requirements. Transient response specifications such as settling time and peak overshoot can be translated to phase margin requirements for systems with a second-order transfer function. For higher-order systems we may use these relations to specify phase margin requirements provided that the equivalent second-order system is used. An equivalent second-order system is one that has the dominant poles of the higher-order one, as detailed in Chapter 7.

The phase and gain margins of the uncompensated systems are evaluated from the Bode plot (or analytically if possible) to determine the additional amount of phase-lead required as the compensator's contribution. The amount of phase-lead can then be used to establish the parameter $\alpha$ assuming that the required phase-lead is attained at the maximum phase of the compensating network, $\phi_m$. The proper value of $\tau$ can be obtained by placing the frequency of maximum phase of the compensating network $\omega_m$ at the crossover frequency $\omega_\phi$ of the compensated system. The Bode plot of the compensated system is examined to check that all performance specifications are satisfied. If not, a new value of $\phi_m$ is chosen. Time response specifications should be checked as well for systems of order higher than two. An example will illustrate the ideas.

EXAMPLE 9.4

Design a phase-lead network for cascade compensation of the closed-loop (unity gain) control system with open-loop transfer function

$$G_p(s) = \frac{K}{s^2}$$

The percentage peak overshoot of the system's transient response is to be less than 20%. To satisfy steady-state error requirements, take $K = 10$.

SOLUTION

The uncompensated system has a closed-loop transfer function of

$$\frac{C(s)}{R(s)} = \frac{K}{s^2 + K}$$

This is clearly an oscillatory system. We will assume that the normalized phase-lead transfer function is used for compensation:

$$G_C(s) = \frac{1 + \tau s}{1 + \tau s/\alpha}$$

The specification of the percent overshoot leads us to the specification of the damping ratio $\zeta$. We have seen in Chapter 5 that for a second-order underdamped system,

$$M_p = e^{-(\zeta/\sqrt{1-\zeta^2})\pi}$$

The inverse relation can be written as

$$\zeta = \sqrt{\frac{\nu^2}{1 + \nu^2}}$$

where

$$\nu = \frac{1}{\pi} \ln \frac{1}{M_p}$$

As a result, the requirement that $M_p = 0.2$ results in

$$\zeta = 0.456$$

The phase margin for a second-order system is related to the damping ratio by the following relation derived in Chapter 8:

$$\frac{1}{\tan^2 \phi_m} = \frac{1}{2}\left(\sqrt{1 + \frac{1}{4\zeta^4}} - 1\right)$$

As a result, the corresponding value of $\phi_m$ is given by

$$\phi_m = 48.1477^\circ$$

The phase angle of the uncompensated system is $-180^\circ$ for all frequencies; thus the phase margin will be provided by the compensating network. Assume that the phase margin is obtained for the maximum phase-lead. Thus we use the following relation to obtain $\alpha$:

$$\sin \phi_m = \sin 48.1477^\circ = \frac{\alpha - 1}{\alpha + 1}$$

As a result, we conclude that

$$\alpha = 6.84$$

The open-loop transfer function for the compensated system is given by

$$G_C(s)G_p(s) = \frac{K(1 + \tau s)}{s^2(1 + \tau s/\alpha)}$$

The corresponding frequency response is given by

$$G_C(j\omega)G_p(j\omega) = \frac{K(1 + j\omega\tau)}{-\omega^2(1 + j\omega\tau/\alpha)}$$

The real and imaginary parts of the frequency-response function are given by

$$X(j\omega) = \frac{-K(1 + \omega^2\tau^2/\alpha)}{\omega^2(1 + \omega^2\tau^2/\alpha^2)}$$

$$Y(j\omega) = \frac{-K\omega\tau(1 - 1/\alpha)}{\omega^2(1 + \omega^2\tau^2/\alpha^2)}$$

The phase angle of the compensated system is thus defined by

$$\tan\phi = \frac{(1 - 1/\alpha)\omega\tau}{1 + \omega^2\tau^2/\alpha}$$

We require that at $\omega_\phi$, the phase angle is $\phi_m = 48.1477°$. Thus with $\alpha = 6.48$, we have

$$\tan 48.1477 = \frac{(1 - 1/6.84)\omega_\phi\tau}{1 + \omega_\phi^2\tau^2/6.84}$$

Simplifying, we get a second-order equation in $\omega_\phi\tau$ given by

$$0.1632\omega_\phi^2\tau^2 - 0.8538\omega_\phi\tau + 1.1164 = 0$$

Solving for $\omega_\phi\tau$, we get two solutions,

$$\omega_\phi\tau = 2.6187 \quad \text{or} \quad 2.5737$$

At $\omega_\phi$ the magnitude of $G_C(j\omega)G_p(j\omega)$ is unity; thus

$$1 = \frac{K^2(1 + \omega_\phi^2\tau^2)}{\omega_\phi^4(1 + \omega_\phi^2\tau^2/\alpha^2)}$$

By substituting for $\omega_\phi\tau$ and $K$ in the above, we obtain the value of $\omega_\phi$ from

$$\omega_\phi^4 = \frac{K^2(1 + \omega_\phi^2\tau^2)}{1 + \omega_\phi^2\tau^2/\alpha^2}$$

Numerically, with $K = 10$, we obtain

$$\omega_\phi = 5.1165 \quad \text{or} \quad 5.0836$$

Since we know the product $\omega_\phi\tau$, we conclude that the two possible values of $\tau$ are

$$\tau = \frac{2.6187}{5.1165} = 0.5118$$

or

$$\tau = \frac{2.5737}{5.0836} = 0.5063$$

We can thus conclude that there are two possible designs for the compensating network. The first network provides us with the compensated transfer function

$$G_{C_1}(s)G_p(s) = \frac{10(1 + 0.5118s)}{s^2(1 + 0.0748s)}$$

The asymptotic Bode plot of this transfer function is shown in Figure 9.22. The crossover frequency on the basis of the asymptotic plot is found to be

$$\omega_{\phi_1} = 5.118$$

The phase angle at that frequency is calculated as

$$\phi(j\omega_{\phi_1}) = 48.16 - 180°$$

Thus the phase margin from the Bode plot is given by

$$\text{phase margin} = 48.16°$$

The second network provides us with the compensated transfer function

$$G_{C_2}(s)G_p(s) = \frac{10(1 + 0.5063s)}{s^2(1 + 0.074s)}$$

The crossover frequency from the asymptotic plots is obtained as

$$\omega_{\phi_2} = 5.063$$

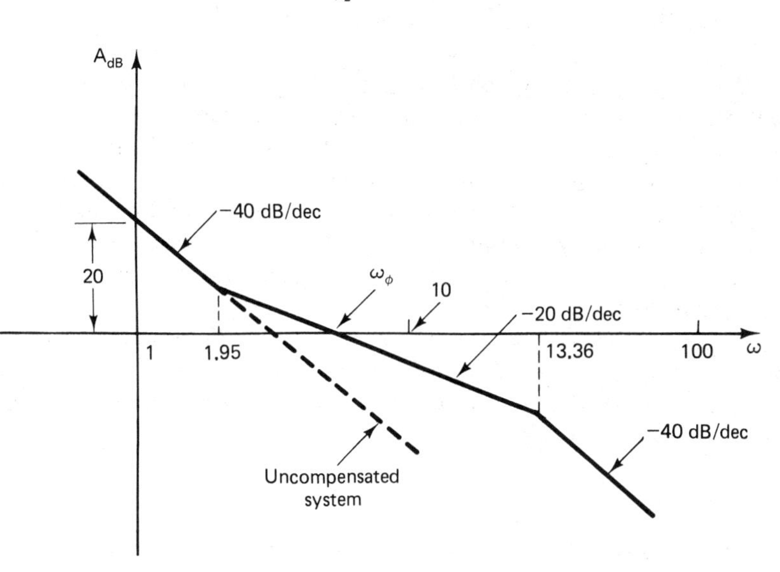


**Figure 9.22.** Asymptotic Bode Plot for $\frac{10(1 + 0.5118s)}{s^2(1 + 0.0748s)}$

The phase margin is given by

$$\text{phase margin} = 48.15°$$

It is clear that for this problem, the two networks are very similar for all engineering purposes.

Let us now consider the compensated system using the first network (the second is so close that the same conclusions would apply). The closed-loop transfer function is given by

$$\frac{C(s)}{R(s)} = \frac{10(1 + 0.5118s)}{s^2(1 + 0.075s) + 10(1 + 0.512s)}$$

This reduces to

$$\frac{C(s)}{R(s)} = \frac{133.69(1 + 0.5118s)}{(s + 5.05)(s^2 + 8.31s + 26.457)}$$

There is a real pole at $s = -5.05$ and two complex-conjugate poles giving rise to $(s^2 + 2\zeta\omega_n s + \omega_n^2)$. The undamped natural frequency $\omega_n$ is

$$\omega_n = 5.14$$

The damping ratio is

$$\zeta = 0.8078$$

The associated peak overshoot is

$$M_p = e^{-(\zeta/\sqrt{1-\zeta^2})\pi}$$
$$= 0.0135$$

This is less than the 20% requirement we started with.

The important point to realize is that we started by assuming that the compensated system can be approximated by a second-order system and required a damping ratio of 0.456 of the approximation system. The required phase margin turned out to be achievable. The actual compensated system (third order with a zero) turned out to have two dominant complex-conjugate poles with a damping ratio of 0.8, which is reasonably higher than what we started with. The requirement on maximum overshoot was met quite reasonably.

The reader should note that our treatment of Example 9.4 was mostly analytical since we are dealing with a simple system. The following example should provide us with an insight to the engineering approach in designing phase-lead compensators.

EXAMPLE 9.5

The open-loop transfer function of a type 1 system is given by

$$G_p(s) = \frac{K}{s(1 + s/20)}$$

It is required to design a cascade lead compensator such that:

1. The steady-state error for sinusoidal inputs with frequency up to 10 rad/sec is less than 1.5%.
2. The phase margin is 40°.

SOLUTION

The sinusoidal steady-state error of the cascade-compensated system is

$$\frac{E(j\omega)}{R(j\omega)} = \frac{1}{1 + G_C(j\omega)G_p(j\omega)}$$
$$\simeq \frac{1}{G_C(j\omega)G_p(j\omega)}$$

The approximation given above is applicable for $G_C(j\omega)G_p(j\omega) \gg 1$. Thus we require that for $\omega \leq 10$, the magnitude of the open-loop transfer function satisfies

$$|G_C(j\omega)G_p(j\omega)| > 66.67$$

Let us note that the first corner frequency of the uncompensated open-loop function is at $\omega = 20$. Assuming that the compensator's frequencies occur for values higher than $\omega = 20$, we can conclude that

$$|G_C(j\omega)G_p(j\omega)| \simeq \frac{K}{\omega} \qquad \text{for } \omega < 10$$

Thus we require that

$$\frac{K}{10} > 66.67$$

As a result, the steady-state error requirement is

$$K > 666.7$$

Let us take $K = 800$ to satisfy this requirement.

We are now ready to consider the phase margin requirements. The uncompensated transfer function is given as

$$G_p(s) = \frac{800}{s(1 + s/20)}$$

The asymptotic Bode plot for the transfer function is shown in Figure 9.23. The gain at $\omega = 1$ is 58 dB. The slope is $-20$ dB/decade until the corner frequency $\omega = 20$ rad/sec. The gain at $\omega = 20$ is 32 dB. At $\omega = 20$, the slope changes to $-40$ dB/decade. The crossover frequency of the uncompensated system occurs for $\omega_\phi = 126.19$ rad/sec. The angle $\phi$ is

$$\phi(j\omega_\phi) = -90° - \tan^{-1}\frac{\omega_\phi}{20}$$
$$= -171°$$

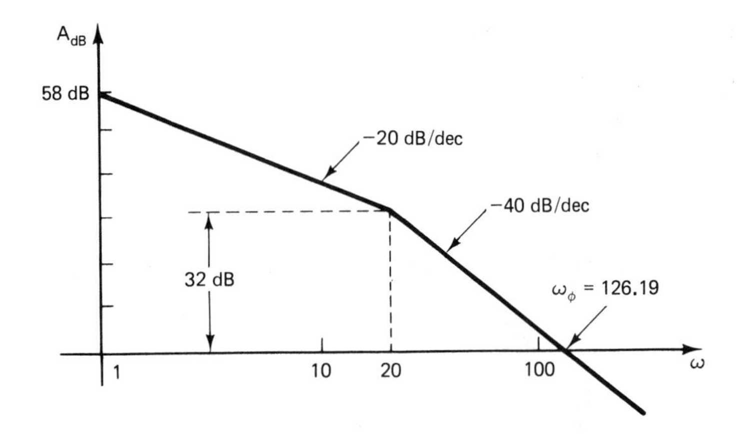


**Figure 9.23.** Bode Plot for the Uncompensated System of Example 9.5

Thus the phase margin of the uncompensated system is 9°, which is not satisfactory.

Our specifications require a phase margin of 40°. We thus need 31° of additional phase lead. To add a margin of safety, we specify an additional phase lead of 35° to be required from the compensator. Assuming that the required phase lead is the maximum for the compensator, we then have

$$\frac{\alpha - 1}{\alpha + 1} = \sin 35^\circ$$

Thus

$$\alpha = 3.69$$

The introduction of the compensator results in a change in slope from −40 dB/decade to −20 dB/decade at $\omega = 1/\tau$. We have not yet determined $\tau$. We will note that the crossover frequency for the compensated system will be higher than 126.19, as can be seen from Figure 9.24. Let us require that the new crossover occur for the frequency of maximum phase lead $\omega_m$, given by

$$\omega_m = \frac{\sqrt{\alpha}}{\tau}$$

From the geometry of Figure 9.24, we have

$$\frac{A_T}{20} = \log \frac{\sqrt{\alpha}/\tau}{1/\tau}$$

Thus the amplitude in decibels at $\omega = 1/\tau$ is required to be

$$\begin{aligned} A_T &= 20 \log \sqrt{\alpha} \\ &= 20 \log \sqrt{3.69} \\ &= 5.67 \text{ dB} \end{aligned}$$

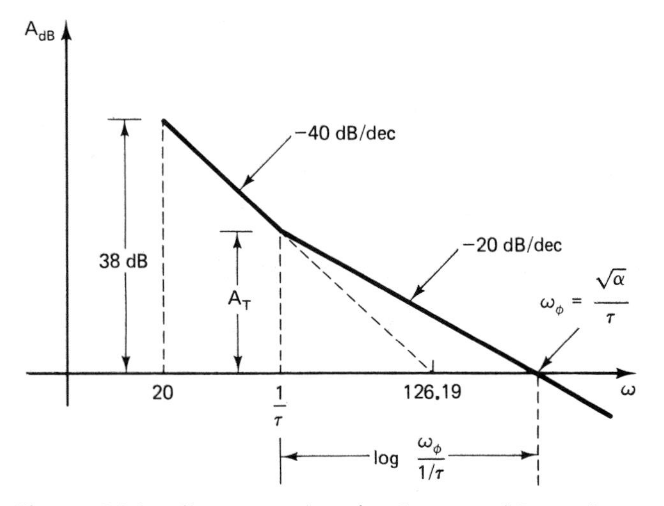


**Figure 9.24.** Compensating the System of Example 9.5

We note that $\omega = 1/\tau$ is also on the $-40$-dB/decade line; thus

$$\frac{38 - A_T}{40} = \log\frac{1/\tau}{20}$$

As a result, we obtain

$$\frac{1}{\tau} = 128.6$$

This completely specifies the compensating network as

$$G_C(s) = \frac{1 + s/128.6}{1 + s/474.564}$$

The open-loop transfer function of the compensated system is now given by

$$G_C(s)G_p(s) = \frac{800(1 + s/128.6)}{s(1 + s/20)(1 + s/474.564)}$$

The crossover frequency is now at

$$\omega_\phi = \frac{\sqrt{\alpha}}{\tau} = 247.05 \text{ rad/sec}$$

The corresponding phase angle is

$$\phi(j\omega_\phi) = -140.37°$$

The new phase margin is

$$\text{phase margin} = 39.63°$$

This is slightly less than the specified value of 40°. The gain at $\omega = 10$ is calculated to be

$$|G(j10)| = 71.756$$

This is satisfactory.

If we are keen on exceeding the phase margin requirement, we should go back and require a higher value of the phase lead: for example, $\phi_m = 40°$; this gives $\alpha = 4.6$. The procedure is continued to obtain $1/\tau = 121.72$. The new compensating network transfer function is given by

$$G_C(s) = \frac{1 + s/121.72}{1 + s/560}$$

The new crossover frequency is at

$$\omega_\phi = 261.0 \text{ rad/sec}$$

The open-loop transfer function has a phase angle $\phi(j\omega_\phi)$ of

$$\phi(j\omega_\phi) = -135.6°$$

As a result, the phase margin is given by

$$\text{phase margin} = 44.39°$$

This value is higher than specified and we should be content with our design. A last check is on the magnitude at $\omega = 10$. This turns out to be

$$|G(j10)| = 71.784$$

Thus the steady-state error requirements are satisfied and our design is complete.

# 9.5 Root-Locus Phase-Lead Cascade Compensation

When transient response specifications are given, we found that to employ the Bode plot for compensation design, we need to translate the specifications to phase margin requirements. The root-locus method is more convenient when time-domain specifications are available. Maximum overshoot, rise time, settling time, damping ratio, and undamped natural frequency requirements can easily be converted to specifications of the desired dominant closed-loop poles.

From the root-locus plot of the uncompensated system one can determine whether or not gain adjustment alone can yield the desired closed-loop poles. The additional angle requirement to place the desired pole on the root locus by satisfying the 180° condition is determined. This angle must be contributed by the phase-lead network. If there are no additional specifications, we will have more than one compensating network to satisfy the angle requirement, as discussed in the next example. The loop gain can be obtained from the magnitude condition. The satisfaction of the performance specification by the compensated system must be verified. If the requirements are not met, the design must be altered and the process is repeated.

EXAMPLE 9.6

Consider the uncompensated system of Example 9.5. Use the root-locus method to design a phase-lead network for cascade compensation such that the dominant poles provide a damping ratio of 0.5 and have an undamped natural frequency $\omega_n = 60$.

SOLUTION

Our specifications provide the location of the desired dominant complex-conjugate poles. An $s$-plane diagram showing the desired pole location and

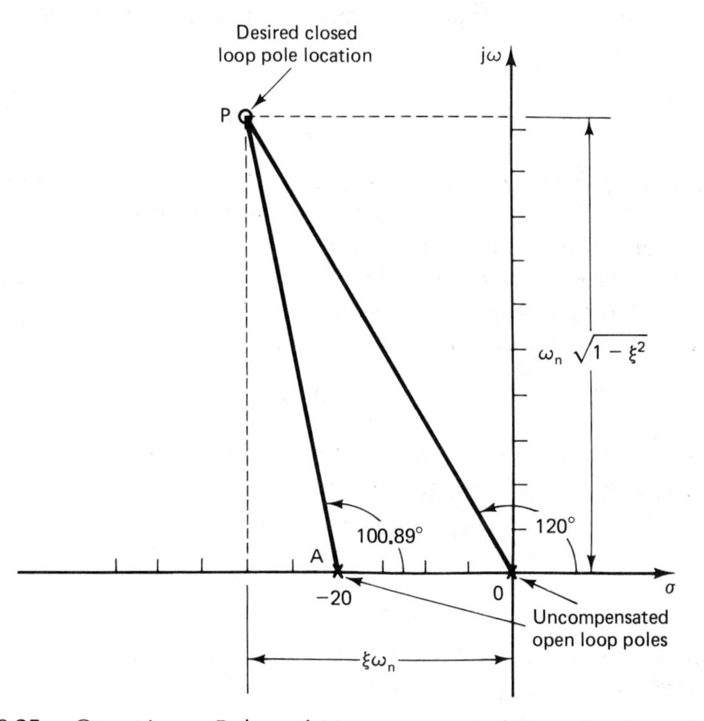


**Figure 9.25.** Open-Loop Poles of Uncompensated Transfer Function and Desired Root Location for Example 9.6

the open-loop poles of the uncompensated transfer function is given in Figure 9.25. In the figure we have

$$\zeta\omega_n = (0.5)(60) = 30$$

$$\omega_n\sqrt{1-\zeta^2} = 60(1-0.25)^{1/2} = 51.96$$

Thus the desired closed-loop pole is at

$$s_1 = -30 + j51.96$$

There are two poles for the open-loop uncompensated system at the origin and at $s = -20$. The total angle contribution of the two poles at $s_1$ is

$$\phi_1 + \phi_2 = -120° - 100.89° = -220.89°$$

This can be obtained graphically (not that accurate, of course) or by using

$$\begin{aligned}\phi_{\text{uc}}(s_1) &= \angle \frac{1}{s_1[1+(s_1/20)]} \\ &= -\tan^{-1}\frac{51.96}{-30} - \tan^{-1}\frac{51.96/20}{1-(30/20)} \\ &= -220.89°\end{aligned}$$

For the point $P$ to be on the root locus, the angle $\phi_t$ should be $-180°$. Thus the lead compensator is required to provide an angle $\phi_c$ such that

$$\phi_{\text{uc}} + \phi_c = -180°$$

As a result,

$$\phi_c = 40.89°$$

The compensating network contributes a zero at $s_z = -1/\tau$ and a pole at $s_p = -\alpha/\tau$. The angle $\phi_c$ is the sum of the angular contributions of the pole and zero of the compensator. Clearly, there is more than one solution to this placement problem, as we do not know $\tau$ and $\alpha$. The simplest procedure is to place the zero immediately under the desired root location. Thus we have for our present example

$$\frac{1}{\tau} = 30$$

By doing this we have the angle contribution of the zero as $90°$. This fixes the angle of the pole at $-\alpha/\tau$, since we have

$$\phi_c = 40.89 = 90 + \phi_p$$

Thus

$$\phi_p = -49.11°$$

We can thus obtain the compensating network's pole location as shown in Figure 9.26. Analytically, we obtain

$$\frac{\alpha}{\tau} - \frac{1}{\tau} = \frac{51.96}{\tan 49.11}$$

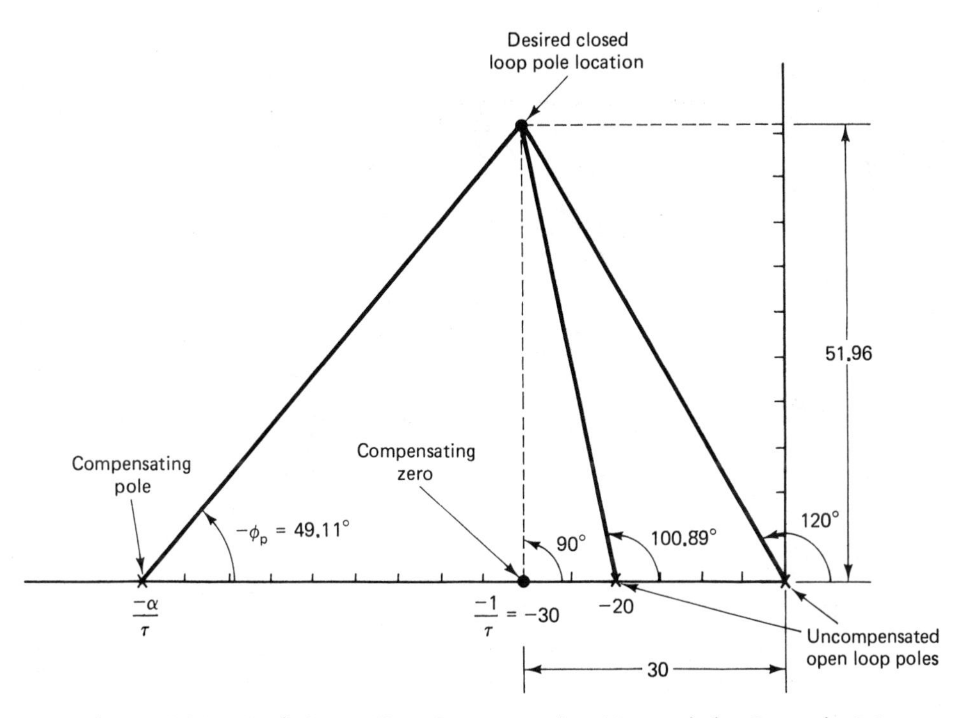


**Figure 9.26.** Defining a First Compensating Network for Example 9.6

Thus we find that

$$\frac{\alpha}{\tau} = 75$$

The value of $\alpha$ is obtained as

$$\alpha = 2.5$$

The compensating network transfer function is thus given by

$$G_c(s) = \frac{1 + s/30}{1 + s/75}$$

There are many other possible ways to locate the compensator's pole and zero. A procedure that yields the largest possible network gain $1/\alpha$ will be outlined with reference to Figure 9.27. First draw a horizontal line passing through the desired root (point $P$) as the line $PF$. Bisect the angle between $PF$ and $PO$ (line from $P$ to origin). Draw the two lines $PC$ and $PD$ making angles $\pm\phi_c/2$ with the bisector $PE$. The intersection of $PC$ and $PD$ with the negative real axis gives the necessary location of the compensator's pole and zero. As a result, $\tau$ and $\alpha$ can be determined. The procedure produces the smallest possible value of the parameter $\alpha$.

Applying the procedure to obtain the minimum possible $\alpha$ to our example is shown in Figure 9.28. From the geometry of the figure we

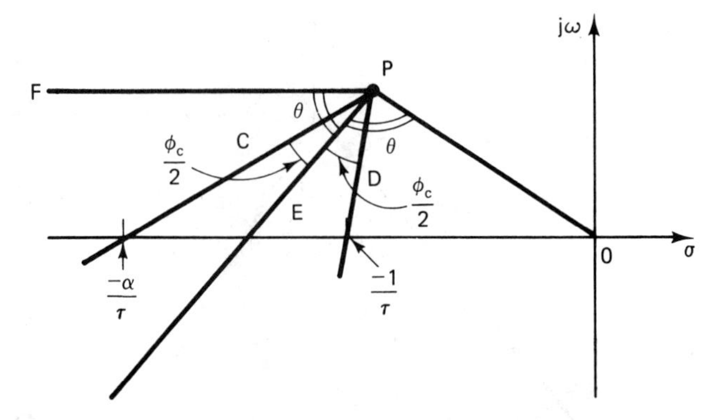


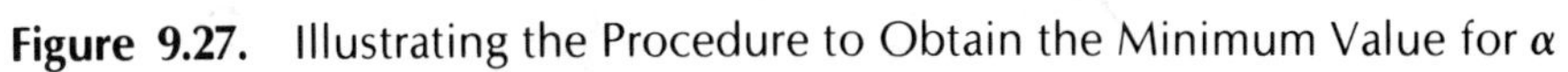

Figure 9.27. Illustrating the Procedure to Obtain the Minimum Value for $\alpha$

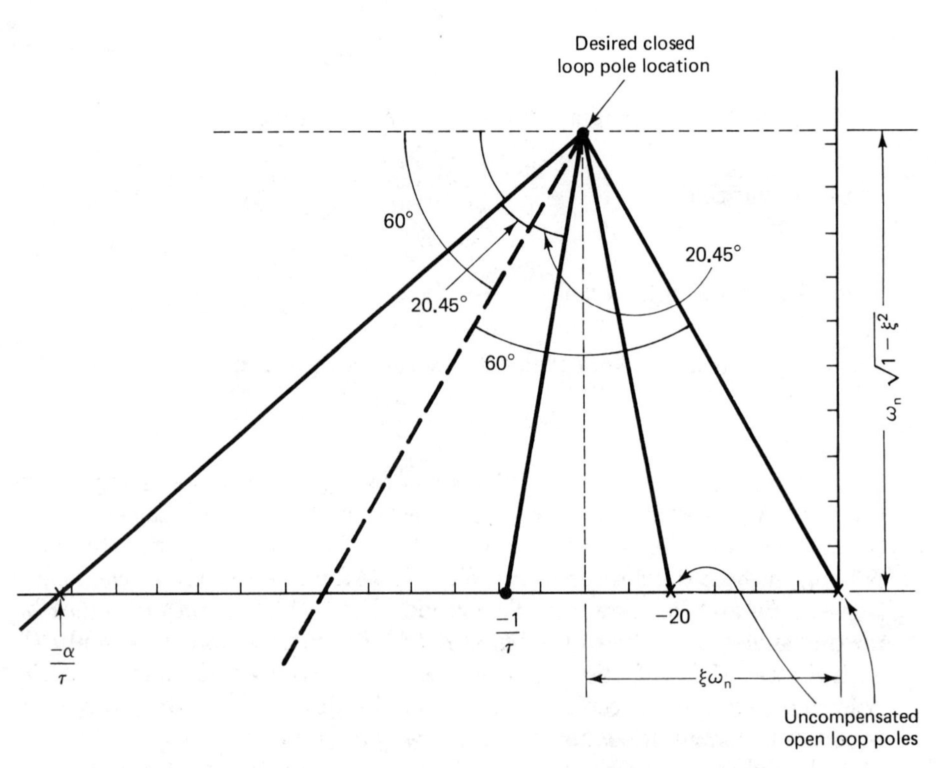


Figure 9.28. Obtaining the Least-$\alpha$ Compensating Network for Example 9.6

conclude that

$$\frac{1}{\tau} = 38.74$$

$$\frac{\alpha}{\tau} = 92.92$$

Thus

$$\alpha = 2.4$$

The compensating network transfer function is thus given by

$$G_c(s) = \frac{1 + s/38.74}{1 + s/92.92}$$

There are some disadvantages to phase-lead compensation. The first is that a higher-frequency gain is increased and thus noise effects are accentuated. The second disadvantage is that there are some systems for which phase lead may not work. Problem 9A-4 illustrates this concept, where we attempt to compensate the system

$$G_p(s) = \frac{80}{s^2(1 + s/20)^2}$$

to achieve a 40° phase margin. The design procedure yields $\alpha \simeq 4000$, which is incredibly high and is impractical to realize. The value of $1/\tau$ is 0.6, and the resulting angle at crossover does not satisfy the phase margin condition.

The basic advantage of lead compensation is that it allows us to achieve transient response requirements without sacrificing the middle-frequency gain.

# 9.6 Phase-Lag Compensation

Phase-lag compensation is intended to provide high gain at low frequencies and thus improve the steady-state error, in contrast to phase-lead compensation, which tends to reduce steady-state accuracy. Lag compensation is used for systems with satisfactory transient response characteristics. The attenuation due to the lag network will shift the gain crossover frequency to a lower-frequency point. This is, of course, the opposite of the result of phase-lead compensation, where the gain crossover frequency is shifted to a higher-frequency point. We discuss in this section the properties of the phase-lag network with transfer function

$$G_c(s) = \frac{1 + \tau s}{1 + \alpha\tau s}$$

with

$$\alpha > 1$$

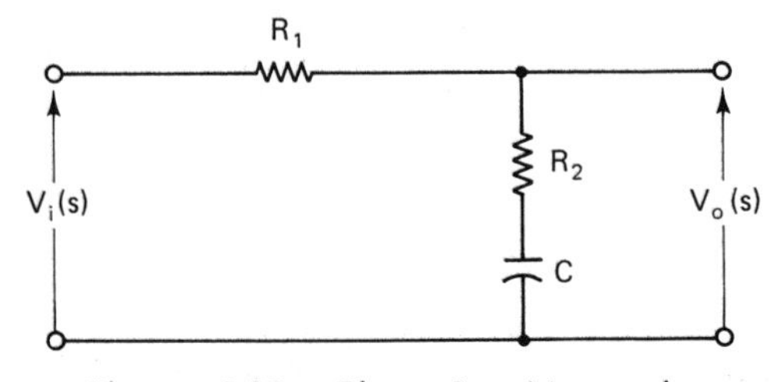


**Figure 9.29.** Phase-Lag Network

## Phase-Lag Network

The phase-lag network can be mechanized using the circuit of Figure 9.29. The transfer function is given by

$$G_c(s) = \frac{V_o(s)}{V_i(s)} = \frac{R_2 + 1/Cs}{R_1 + R_2 + 1/Cs}$$

As a result,

$$G_c(s) = \frac{1 + R_2Cs}{1 + (R_1 + R_2)Cs}$$

We now let

$$\tau = R_2C$$
$$\alpha = \frac{R_1 + R_2}{R_2}$$

Thus

$$G_c(s) = \frac{1 + \tau s}{1 + \alpha\tau s}$$

Again $\alpha > 1$.

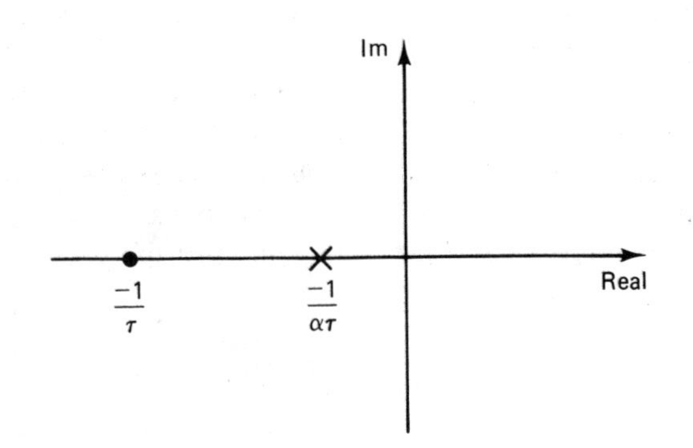

**Figure 9.30.** Pole – Zero Configuration of a Phase-Lag Network

The pole–zero configuration of the phase-lag network is as shown in Figure 9.30. There is a zero at $z = -1/\tau$ and a pole at $p = -1/\alpha\tau$. The pole is closer to the origin of the $s$-plane than the zero.

The polar plot of the frequency response is derived using

$$X = \frac{1 + \alpha\tau^2\omega^2}{1 + \alpha^2\tau^2\omega^2}$$

$$Y = \frac{\omega\tau(1 - \alpha)}{1 + \alpha^2\tau^2\omega^2}$$

We can show that

$$\frac{Y}{1 - \alpha X} = \omega\tau$$

Thus eliminating $\omega\tau$, we obtain

$$\left(X - \frac{\alpha + 1}{2\alpha}\right)^2 + Y^2 = \left(\frac{\alpha - 1}{2\alpha}\right)^2$$

This is the equation of a circle with center on the real axis at

$$X_c = \frac{\alpha + 1}{2\alpha}$$

The radius is given by

$$R_c = \frac{\alpha - 1}{2\alpha}$$

The polar plot of the phase-lag function is shown in Figure 9.31. It is similar to that of the phase-lead polar plot with the exception that the locus is a semicircle in the fourth quadrant.

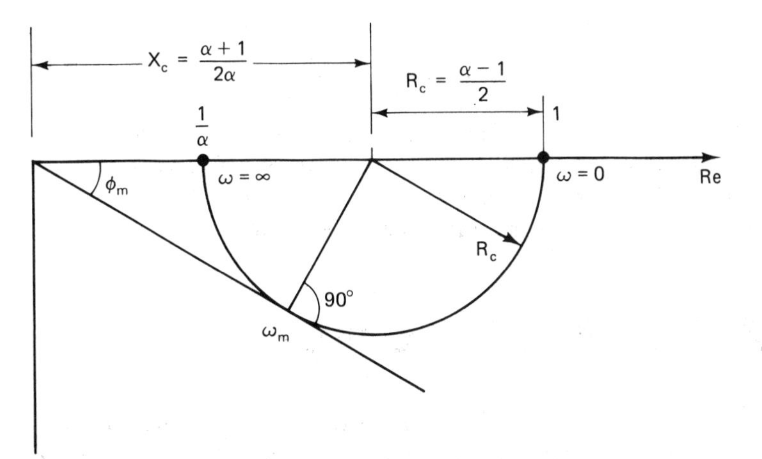


**Figure 9.31.** Polar Plot for the Phase-Lag Network

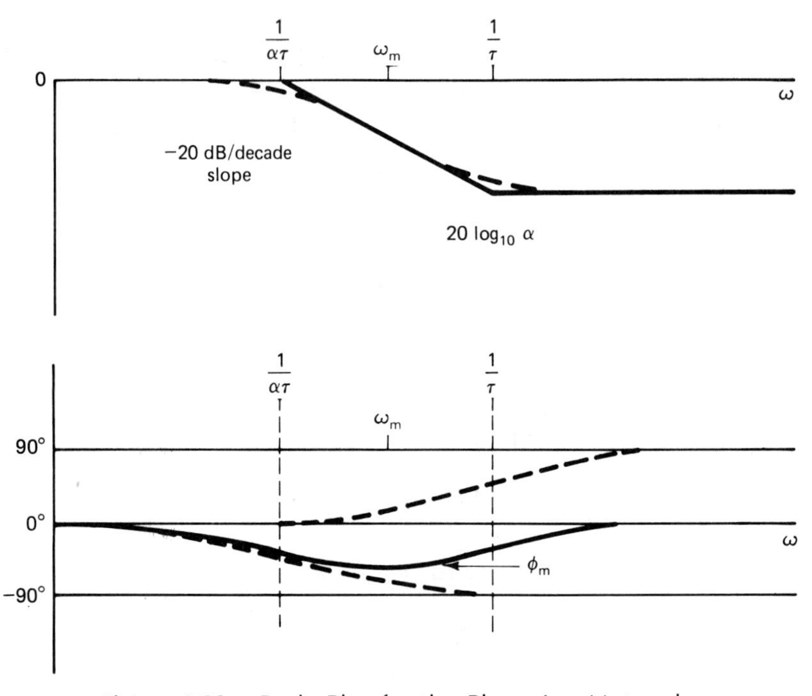


**Figure 9.32.** Bode Plot for the Phase-Lag Network

The phase angle is negative for positive $\omega$, indicating that the output lags the input voltage. The maximum lag can be shown to occur for

$$\omega_m = \frac{1}{\tau\sqrt{\alpha}}$$

The maximum phase lag is

$$\phi_m = \sin^{-1}\frac{1-\alpha}{1+\alpha}$$

The Bode plot for the phase-lag network is shown in Figure 9.32. There are two corner frequencies. At the first, $\omega_c = 1/\alpha\tau$, the slope changes from 0 dB/decade to $-20$ dB/decade. At $\omega_c = 1/\tau$, the slope changes back to 0 dB/decade as the numerator term takes effect.

# 9.7 Bode Plot Lag Compensation

Bode plots are most suitable when performance specifications in terms of frequency-response characteristics such as the phase margin are available. A general procedure for compensation attempts to satisfy steady-state error requirements by gain adjustment first. The Bode plot is then examined to establish the phase margin for the system with only gain adjustment applied. If the phase margin requirements are not satisfied, the frequency $\omega_\phi$ at which the angle (180 − required phase margin) is found. The purpose of the design is to place the corner frequencies $1/\alpha\tau$ and $1/\tau$ in such a fashion that the compensated open-loop transfer function has a gain crossover frequency in the proximity of $\omega_{\phi_0}$. An example will illustrate the procedure.

EXAMPLE 9.7

Consider the open-loop transfer function given by

$$G_p(s) = \frac{10}{s(1 + s/12)(1 + s/20)}$$

It is required to design a compensator so that

1. The velocity error constant $K_v = 50$.
2. The phase margin is at least $45°$.

SOLUTION

To obtain $K_v = 50$, we must have an additional gain so that

$$G_p(s) = \frac{50}{s(1 + s/12)(1 + s/20)}$$

The Bode plot for the open-loop transfer function thus obtained is shown in Figure 9.33. From the geometry of the figure, the amplitude in decibels at $\omega = 1$ is

$$A_0 = 20 \log 50 = 34 \text{ dB}$$

At $\omega = 12$, the amplitude is

$$A_1 = A_0 - 20 \log 12 = 12.42 \text{ dB}$$

The slope is −40 dB/decade between $\omega = 12$ and $\omega = 20$. The amplitude at $\omega = 20$ is

$$A_2 = A_1 - 40 \log \frac{20}{12} = 3.542 \text{ dB}$$

The slope beyond $\omega = 20$ is −60 dB/decade. There will be a crossover at $\omega_\phi$ which is obtained from

$$A_2 = 60 \log \frac{\omega_\phi}{20} = 3.542$$

Thus

$$\omega_\phi = 22.912 \text{ rad/sec}$$

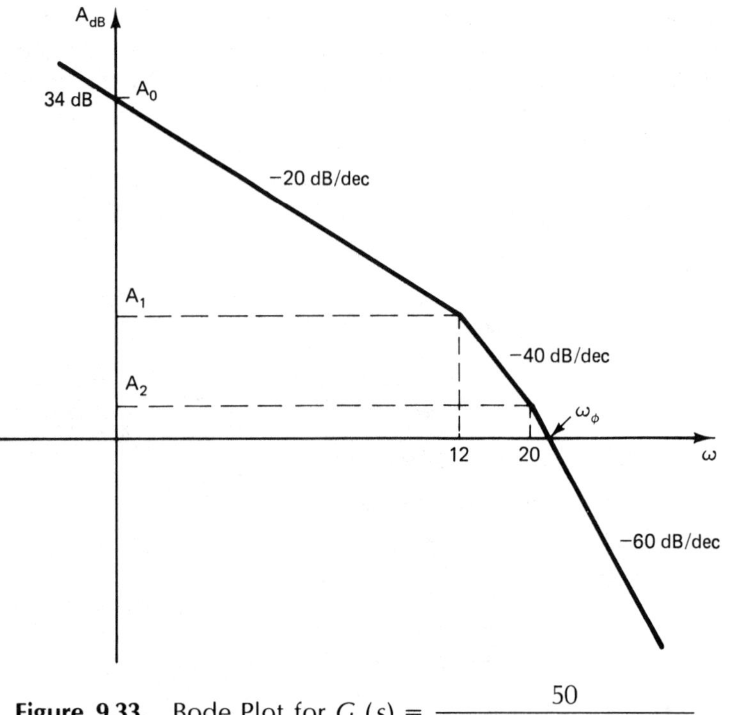


**Figure 9.33.** Bode Plot for $G_p(s) = \dfrac{50}{s(1 + s/12)(1 + s/20)}$

The phase angle at the gain crossover frequency is

$$\begin{aligned}\phi &= -90° - \tan^{-1}\frac{22.912}{12} - \tan^{-1}\frac{22.91}{20}\\ &= -201.237°\end{aligned}$$

The phase margin is $-21.23°$ and this system is unstable as given.

To attain a phase margin of at least 45°, the open-loop transfer function has to be modified, the object being to obtain the gain crossover frequency at a value corresponding to a phase angle larger than $-135°$. The open-loop frequency response has a phase angle given by

$$\phi(j\omega) = -90° - \tan^{-1}\frac{\omega}{12} - \tan^{-1}\frac{\omega}{20}$$

We list the phase angle for a range of frequencies:

$$\begin{aligned}\omega &= 6 \qquad \phi(j6) = -133.26\\ \omega &= 5 \qquad \phi(j5) = -126.66\\ \omega &= 4 \qquad \phi(j4) = -119.74\end{aligned}$$

Thus if the gain crossover is at $\omega = 6$, the phase margin is 46.74°, which is

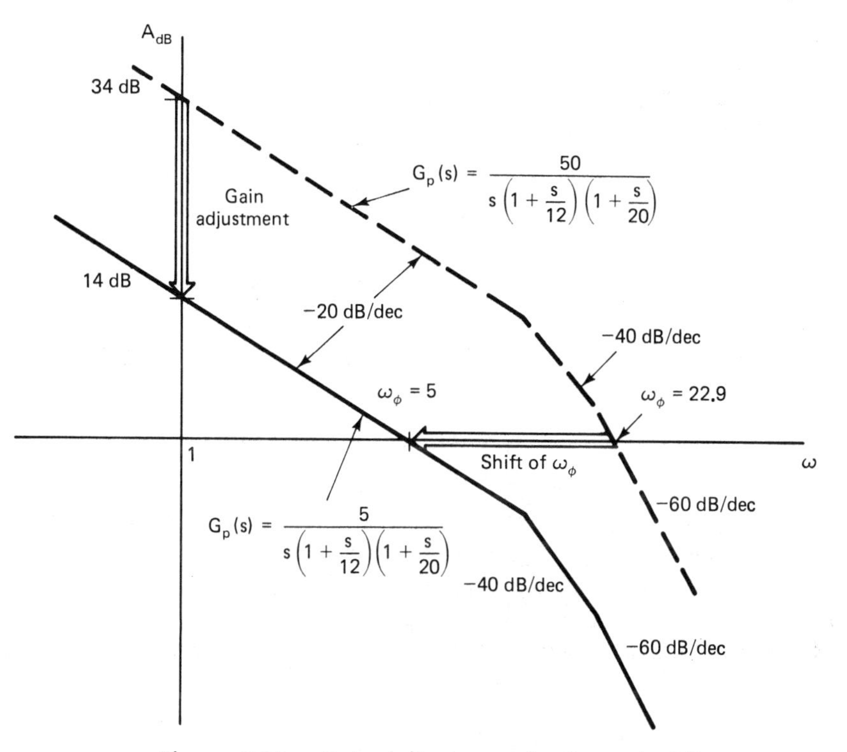


**Figure 9.34.** Gain Adjustment for Example 9.7

good from our specifications point of view. A more generous design takes $\omega = 5$ for gain crossover and provides some good-looking numbers.

We can achieve the crossover at $\omega_\phi = 5$ by a simple gain adjustment as shown in Figure 9.34. For the gain crossover to be at $\omega_\phi = 5$, we have

$$20 \log K_0 = 20 \log \frac{\omega_\phi}{1} = 20 \log 5$$

Thus if the gain $K_0 = 5$, we meet the phase margin requirements. The $K_v$ requirement, however, is not met. We would like to retain $K$ at its value of 50 to meet the steady-state error requirement and would like to move the gain crossover frequency to the proximity of 5 rad/sec. This can be achieved using phase-lag compensation.

The phase-lag transfer function is given by

$$G_c(s) = \frac{1 + \tau s}{1 + \alpha \tau s}$$

The object of the design now is to find $\tau$ and $\alpha$. A lower bound on $\alpha$ is found as 10. This is the ratio $K/K_0$. The crossover should occur in the proximity of 5 rad/sec. Figure 9.35 shows a Bode construction assuming that $1/\tau < 1$,

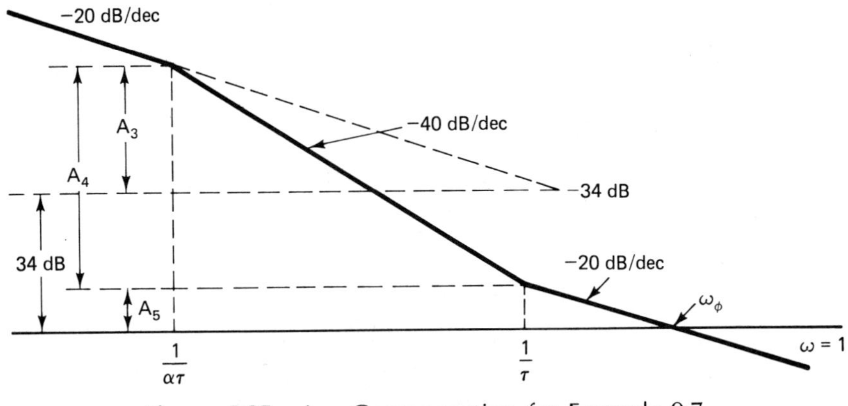


**Figure 9.35.** Lag Compensation for Example 9.7

$\alpha \geq 10$. The denominator term $s$ contributes a $-20$-dB/decade slope up to the corner frequency $\omega = 1/\alpha\tau$, where the slope changes to $-40$ dB/decade. At $\omega = 1/\tau$ the compensating zero comes to play and the slope is $-20$ dB/decade. The gain crossover $\omega_\phi$ is assumed to occur on the $-20$-dB/decade slope as $\omega_\phi$ is assumed less than 12. We can write with reference to the plot

$$A_3 = 20 \log \frac{1}{1/\alpha\tau} = 20 \log \alpha + 20 \log \tau$$

$$A_4 = 40 \log \frac{1/\tau}{1/\alpha\tau} = 40 \log \alpha$$

$$A_5 = 20 \log \frac{\omega_\phi}{1/\tau} = 20 \log \omega_\phi + 20 \log \tau$$

Now we use

$$34 + A_3 = A_4 + A_5$$

or

$$34 + 20 \log \alpha + 20 \log \tau = 40 \log \alpha + 20 \log \omega_\phi + 20 \log \tau$$

The above reduces to

$$34 = 20 \log \alpha + 20 \log \omega_\phi$$

As a result,

$$\alpha\omega_\phi = 50$$

The phase angle of the compensated open-loop transfer function at crossover is given by

$$\phi(j\omega_\phi) = -90° - \tan^{-1}\frac{\omega_\phi}{12} - \tan^{-1}\frac{\omega_\phi}{20} + \tan^{-1}\omega_\phi\tau - \tan^{-1}\alpha\omega_\phi\tau$$

We will require a phase angle of $-125°$ at crossover, corresponding to $55°$ phase margin. As a result,

$$-125° = -90° - \tan^{-1}\frac{\omega_\phi}{12} - \tan^{-1}\frac{\omega_\phi}{20} + \tan^{-1}\omega_\phi\tau - \tan^{-1}50\tau$$

This equation and the condition $\alpha\omega_\phi = 50$ give two relations in the unknowns $\alpha$, $\tau$, and $\omega_\phi$. Note that by construction we assure that $K_v = 50$. We need a third relation to solve for our unknowns. As there are no more specifications, we can arbitrarily fix one parameter and obtain the two others. We thus have a number of methods. We illustrate two methods to resolve the problem.

*Method 1*: *Fix* $\alpha$. We know that $\alpha$ should be greater than 10. Let us assume that $\alpha = 20$, thus $\omega_\phi = 2.5$. As a result,

$$-125° = -90° - \tan^{-1}\frac{2.5}{12} - \tan^{-1}\frac{2.5}{20} + \tan^{-1}2.5\tau - \tan^{-1}50\tau$$

This relation reduces to

$$16.107 = \tan^{-1}50\tau - \tan^{-1}2.5\tau$$

The solution for $\tau$ is obtained by trial and error as

$$\tau = 1.31$$

As a result, our compensating transfer function is given by

$$G_c(s) = \frac{1 + 1.31s}{2 + 26.2s}$$

The open-loop compensated transfer function is given by

$$G_c(s) = \frac{50(1 + 1.31s)}{s(1 + s/12)(1 + s/20)(1 + 26.2s)}$$

The crossover is at $\omega_\phi = 2.5$, and the corresponding phase angle is $-124.998° \simeq -125°$, as required.

If we are not very keen on a trial-and-error solution for $\tau$, we use the arctan approximation to solve

$$16.107° = \tan^{-1}50\tau - \tan^{-1}2.5\tau$$

since $\tau$ is assumed $> 1$, we write

$$\left(\frac{16.107}{180}\right)\pi = \left(\frac{\pi}{2} - \tan^{-1}\frac{1}{50\tau}\right) - \left(\frac{\pi}{2} - \tan^{-1}\frac{1}{2.5\tau}\right)$$

Thus we have approximately

$$\frac{16.107\pi}{180} \simeq \frac{1}{2.5\tau} - \frac{1}{50\tau}$$

Thus

$$\tau \simeq 1.35$$

The compensated transfer function is

$$G_c(s)G_p(s) = \frac{50(1 + 1.35s)}{s(1 + s/12)(1 + s/20)(1 + 27s)}$$

The phase angle at $\omega = 2.5$ is found to be $-124.55°$ for this approximation.

Let us take a smaller $\alpha$, say $\alpha = 12.5$. Thus $\omega_\phi = 4$; the phase angle condition is thus

$$5.26 = \tan^{-1} 50\tau - \tan^{-1} 4\tau$$

The solution for $\tau$ is

$$\tau = 2.5$$

The compensated transfer function is given by

$$G_c(s)G_p(s) = \frac{50(1 + 2.5s)}{s(1 + s/12)(1 + s/20)(1 + 31.25s)}$$

At $\omega_\phi = 4$ we have

$$\phi = -125°$$

This is dead on target.

*Method 2: Fix* $\omega_\phi \tau$. The ratio of the gain crossover frequency $\omega_\phi$ to the corner frequency $1/\tau$ can be specified according to judgment. A good practice is to take

$$\omega_\phi \tau = 10$$

The two additional relations are

$$\alpha\omega_\phi = 50$$

$$35° = \tan^{-1}\frac{\omega_\phi}{12} + \tan^{-1}\frac{\omega_\phi}{20} - \tan^{-1}\omega_\phi\tau + \tan^{-1} 50\tau$$

This reduces to

$$119.29° = \tan^{-1}\frac{10}{12\tau} + \tan^{-1}\frac{10}{20\tau} + \tan^{-1} 50\tau$$

The solution by trial and error is found to be

$$\tau = 2.5$$
$$\alpha = 12.5$$
$$\omega_\phi = 4$$

This is the same as the second result of method 1.

# 9.8 Root-Locus Lag-Compensator Design

A system with satisfactory transient response but poor steady-state performance requires an increase in open-loop gain without altering its transient response characteristics. This is achieved by cascading a lag compensator with pole and zero placed close together near the origin. The result of this on the root locus is negligible if the angle contribution of the lag network is less than a small angle, such as $5°$. The closed-loop poles of the compensated system will be slightly shifted from the corresponding poles of the uncompensated system.

The lag compensator has a transfer function of

$$
\begin{aligned}
G_c(s) &= \frac{1 + \tau s}{1 + \alpha \tau s} \\
&= \frac{1}{\alpha} \frac{s + 1/\tau}{s + 1/\alpha\tau}
\end{aligned}
$$

The parameter $\alpha$ provides the overall gain reduction required for transient response improvement without sacrificing the steady-state error. A first step in the design is to select the value of $\alpha$ so that the required steady-state error specifications are met. The following example illustrates the ideas.

EXAMPLE 9.8

Consider the system of Example 9.7 with uncompensated transfer function given by

$$
G_p(s) = \frac{240K}{s(s + 12)(s + 20)}
$$

Design a phase-lag compensator so that the following criteria are met:

1. The velocity error constant $K_v = 50$.
2. The damping ratio of the dominant closed-loop poles is $\zeta = 0.5$.

SOLUTION

The root locus of the uncompensated system is shown in Figure 9.36. The characteristic equation of the system is

$$
s^3 + 32s^2 + 240s + 240K = 0
$$

Application of the Routh array results in

$$
\begin{aligned}
K_{\text{crit}} &= 32 \\
\omega_{\text{crit}} &= 15.49
\end{aligned}
$$

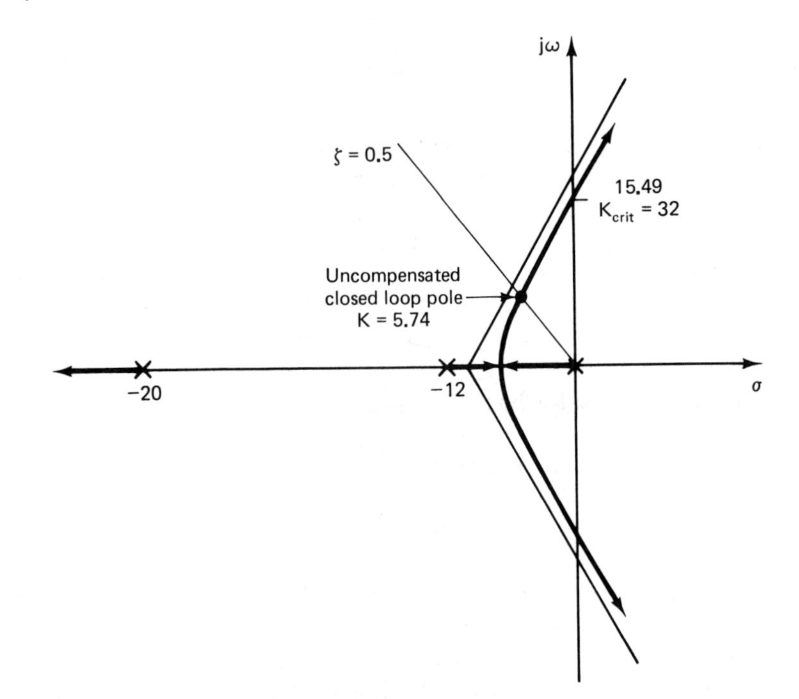


**Figure 9.36.** Root Locus for the Uncompensated System of Example 9.8

The velocity error constant requirement of $K = 50$ results in an unstable system, as was shown in Example 9.7.

For the dominant closed-loop poles to have a damping ratio $\zeta = 0.5$, the root is given by

$$s_1 = \omega_n e^{j120}$$

Substituting in the characteristic equation, we have

$$\omega_n^3 e^{j360} + 32\omega_n^2 e^{j240} + 240\omega_n e^{j120} + 240K_0 = 0$$

The imaginary part of the above yields

$$-32\omega_n^2 \sin 60 + 240\omega_n \sin 60 = 0$$

Thus

$$\omega_n = 7.5$$

The real part gives us

$$240K_0 - 120\omega_n - 16\omega_n^2 + \omega_n^3 = 0$$

As a result,

$$K_0 = 5.74$$

We can thus conclude that for $\zeta = 0.5$, the dominant closed-loop poles are given by

$$s_1 = 7.5e^{\pm j120}$$

The required gain of the uncompensated system is $K_0 = 5.74$.

To obtain a $K_v = 50$, we set

$$\alpha = \frac{K_v}{K_0} = \frac{50}{5.74} = 8.71$$

As a result, the compensated transfer function for the open loop is

$$G_c(s)G_p(s) = \frac{240K(s + 1/\tau)}{\alpha s(s + 12)(s + 20)(s + 1/\alpha\tau)}$$

With $K = K_v = 50$, we have

$$G_c(s)G_p(s) = \frac{240K_0(s + 1/\tau)}{s(s + 12)(s + 20)(s + 1/\alpha\tau)}$$

Our next task is to place the pole of the lag network $(-1/\alpha\tau)$ and the zero at $(-1/\tau)$. We are thus required to choose $\tau$ since we know $\alpha$. The angle condition at a closed-loop pole for the compensated system is

$$\underline{/G_c(s)G_p(s)}\Big|_{s=s_1} = \underline{/s_1 + \frac{1}{\tau}} - \underline{/s_1 + \frac{1}{\alpha\tau}} - \underline{/s_1} - \underline{/s_1 + 12} - \underline{/s_1 + 20}$$

Since $s_1 = 7.5\underline{/120°}$ is on the root locus of $G_p(s)$,

$$\underline{/s_1 + \frac{1}{\tau}} - \underline{/s_1 + \frac{1}{\alpha\tau}} = 0$$

Any value of $\tau$ other than zero will violate the condition above. If we choose $1/\tau$ small enough, we can make this error small. Let us choose

$$\frac{1}{\tau} = 0.1$$

Thus

$$\frac{1}{\alpha\tau} = 0.0115$$

We have

$$s_1 + \frac{1}{\tau} = 7.45\underline{/119.334°}$$

$$s_1 + \frac{1}{\alpha\tau} = 7.494\underline{/119.924°}$$

The difference in angle is 0.59°. With this choice, $s_1$ will not be exactly on the root locus of the compensated system for $K_0 = 5.74$.

The compensated transfer function is now given by

$$G_c(s)G_p(s) = \frac{240K_0(s + 0.1)}{s(s + 12)(s + 20)(s + 0.0115)}$$

with $\alpha = 8.71$. The characteristic equation is given by

$$s^4 + 32.0115s^3 + 240.3673s^2 + (2.755 + 240K_0)s + 24K_0 = 0$$

Choosing $K_0 = 5.74$, we have

$$s^4 + 32.0115s^3 + 240.367s^2 + 1380.36s + 137.76 = 0$$

The roots of the polynomial are obtained as

$$s_{\frac{1}{2}} = -3.7108 \pm j6.4504$$

$$= 7.4416 \angle 119.911°$$

$$s_3 = -0.1016$$

$$s_4 = -24.4884$$

Thus the dominant complex poles have

$$\omega_n = 7.4416$$

The damping ratio is

$$\zeta = \cos(180 - 119.911)$$
$$= 0.4987$$

Note that the damping ratio is slightly less than specified. The value of $K_v$, however, is 50.

If we sacrifice the value of $K_v$ by taking $K_0 = 5.7$, the value of $\zeta$ can be found to be

$$\zeta = 0.5016$$

The corresponding value of $K_v$ is

$$K_v = 49.5652$$

It is thus seen that a trade-off has to be made. For all engineering purposes we can claim that our design is successful. The compensated transfer function is given by

$$G_c(s)G_p(s) = \frac{1377.6(s + 0.1)}{s(s + 12)(s + 20)(s + 0.0115)}$$

The damping ratio is

$$\zeta = 0.4987 \simeq 0.5$$

The velocity error constant is

$$K_v = 50$$

which is the required design.

# 9.9 Lag-Lead Compensation

Lag compensation results in a large increase in gain, which improves the steady-state error characteristics but results in slower response. On the other hand, lead compensation results in a small increase in gain and a large increase in the undamped natural frequency and a faster response. If improvements in both transient and steady-state performance are required, both lead and lag compensation should be employed. There is a passive network that combines both the lag and lead aspects, and this is considered next.

## Lag-Lead Network

Consider the network shown in Figure 9.37. The transfer function is obtained from

$$G_c(s) = \frac{V_o(s)}{V_i(s)} = \frac{R_2 + 1/C_2 s}{\dfrac{1}{(1/R_1) + C_1 s} + R_2 + \dfrac{1}{C_2 s}}$$

A few simplification steps result in

$$G_c(s) = \frac{(1 + R_1C_1s)(1 + R_2C_2s)}{R_1C_1R_2C_2s^2 + (R_1C_1 + R_2C_2 + R_1C_2)s + 1}$$

We may cast the transfer function into one of two standard forms. In the

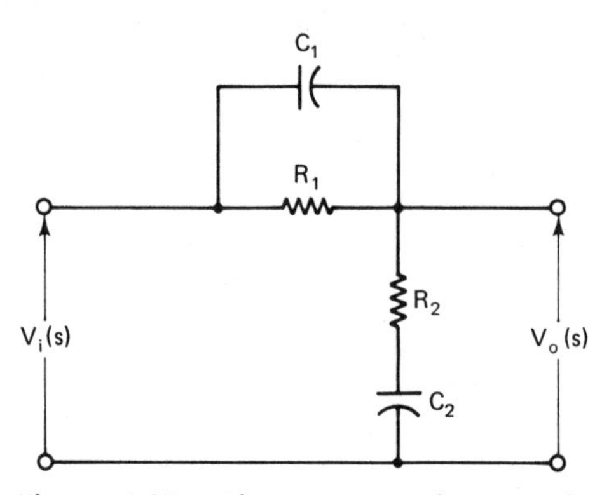


**Figure 9.37.** Phase Lag-Lead Network

first we let

$$\tau_1 = R_1C_1$$
$$\tau_2 = R_2C_2$$
$$\tau_{12} = R_1C_2$$

Thus

$$G_c(s) = \frac{(1 + \tau_1 s)(1 + \tau_2 s)}{(1 + \tau_1 s)(1 + \tau_2 s) + \tau_{12} s}$$

This form is used in Example 8.4 to establish the polar plot for the frequency response of the transfer function.

An alternative form of the lag–lead transfer function which is suited to our purposes is given by

$$G_c(s) = \underbrace{\frac{1 + \tau_1 s}{1 + \alpha\tau_1 s}}_{\text{lag}} \underbrace{\frac{1 + \tau_2 s}{1 + (\tau_2/\alpha)s}}_{\text{lead}}$$

In the above we have

$$\alpha\tau_1 + \frac{\tau_2}{\alpha} = \tau_1 + \tau_2 + \tau_{12}$$

It is clear that the lag portion has a zero at $-1/\tau_1$ and a pole at $-1/\alpha\tau_1$; the lead portion has a zero at $-1/\tau_2$ and a pole at $-\alpha/\tau_2$. The lag portion's pole and zero are placed very close to the origin, with the value of $\alpha$ selected in the neighborhood of 10, as we found out earlier. The zero of the lead network is chosen to coincide and cancel a pole of the given uncompensated system. Figure 9.38 shows the placement of poles and zeros of the network.

The polar plot of the lag–lead transfer function is shown in Figure 9.39. The network acts as a lag network for $0 < \omega < \omega_1$ and as a lead network for $\omega_1 < \omega < \infty$. The frequency $\omega_1$ at which the phase angle is zero

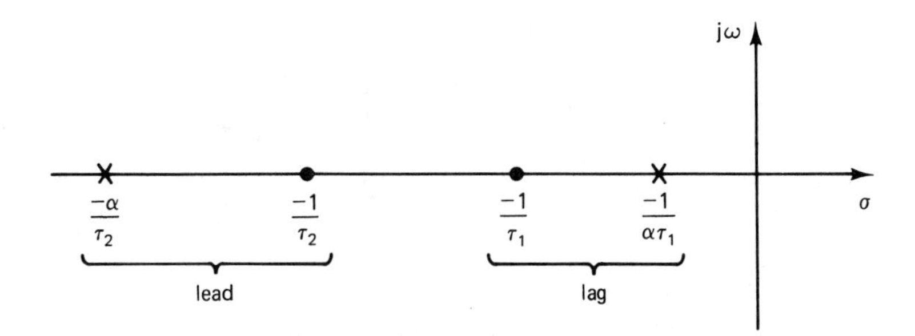


**Figure 9.38.** Pole–Zero Configuration of a Lag-Lead Network

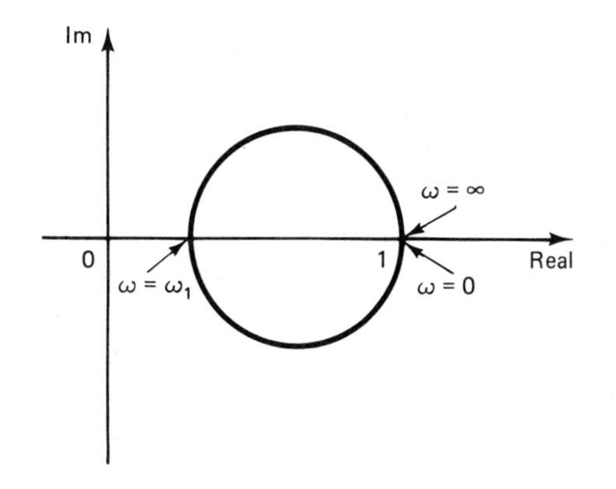


**Figure 9.39.** Polar Plot of a Lag-Lead Network

is given by

$$\omega_1 = \frac{1}{\sqrt{\tau_1 \tau_2}}$$

The Bode plots of the lag–lead transfer function are shown in Figure 9.40 assuming that $\alpha = 10$ and $\tau_2 = 10\tau_1$. The figure shows zero amplitude in decibels at low and high frequencies, as the parameter $\alpha$ does not figure in the magnitude expressions at high and low frequencies.

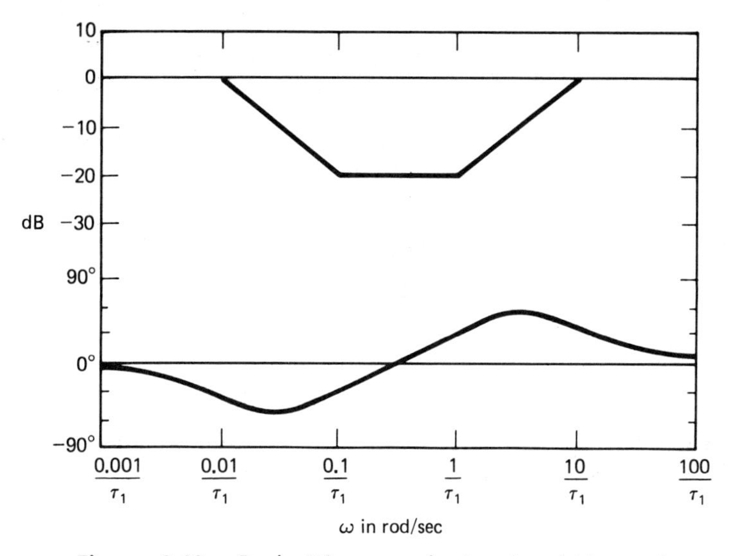


**Figure 9.40.** Bode Diagram of a Lag-Lead Network

# 9.10 Root-Locus Lag-Lead Compensation

The design of a lag–lead compensator to achieve improvements in the steady-state as well as the transient response of a system can be done quite easily in the $s$-plane. Good design practice calls for placing the lag portion close to the origin with a value of $\alpha$ close to 10. The lead portion is designed such that the zero cancels out a pole from the uncompensated system. The following example illustrates the procedure and the benefits of this type of compensation.

EXAMPLE 9.9

Consider the uncompensated system of Example 9.8 with transfer function given by

$$G_p(s) = \frac{K}{s(s+12)(s+20)}$$

The value of the velocity error constant is

$$K_v = \frac{K}{240}$$

The uncompensated system is unstable for $K_v = 50$.

Let us introduce lag–lead compensation with $\alpha = 10$. Choose

$$\frac{1}{\tau_1} = 0.05$$

Thus

$$\frac{1}{\alpha\tau_1} = 0.005$$

The value of $\tau_2$ is chosen to cancel the pole at $s = -12$; thus

$$\frac{1}{\tau_2} = 12$$

$$\frac{\alpha}{\tau_2} = 120$$

The compensator is thus chosen with the transfer function of the compensated system obtained as

$$G_c(s)G_p(s) = \frac{K(s+0.05)}{s(s+20)(s+0.005)(s+120)}$$

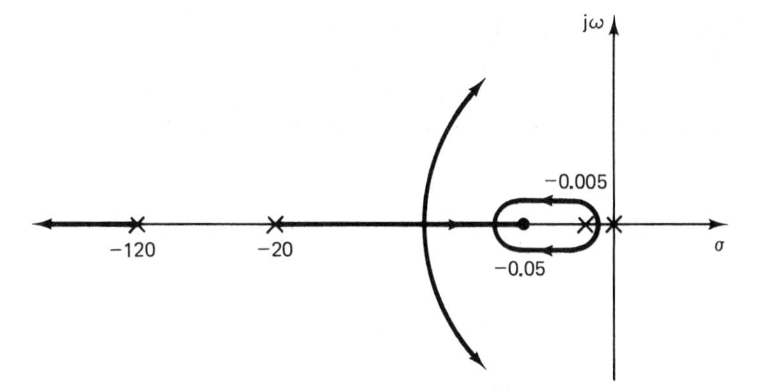


**Figure 9.41.** Root Locus for the Compensated System of Example 9.9

The value of $K_v$ is still

$$K_v = \frac{K}{240}$$

The characteristic equation of the compensated system is

$$s^4 + 140.005s^3 + 2400.7s^2 + (12 + K)s + 0.05K = 0$$

This defines the root locus, which is sketched as shown in Figure 9.41.

If we specify $K_v = 50$, the characteristic equation is

$$s^4 + 140.005s^3 + 2400.7s^2 + 12012s + 600 = 0$$

Solving for the roots, we have the dominant complex poles

$$s_{1,2} = -9.4863 \pm j2.8808 = \omega_n \angle 180° - \cos^{-1}\zeta$$

The other two roots are at $-0.0505$ and $-120.9819$. The dominant roots have

$$\omega_n = 9.914$$
$$\zeta = 0.9569$$

Thus using the lag–lead compensator we can achieve $K_v = 50$ with a much better damping than by using lag compensation alone.

We can specify a much better $K_v$; for example, $K_v = 100$ will result in a damping ratio of $\zeta = 0.6438$. This is inferior to the $\zeta$ corresponding to $K_v = 50$, but is still better than that attained using lag compensation with $K_v = 50$.

# 9.11 Bode Plot Lag-Lead Compensation

Cascade lag–lead compensator design in the frequency domain is accomplished using the Bode plots employing a combination of the techniques covered for lead compensation and lag compensation. The value of $\alpha$ in both lag and lead portions should be the same. The lag portion adds attenuation near and above the gain crossover frequency to improve the steady-state performance. The lead portion improves the phase margin and hence the transient performance. An example will illustrate the procedure.

EXAMPLE 9.10

Consider the open-loop transfer function of Examples 9.7 through 9.9 given by

$$G_p(s) = \frac{K}{s(s+12)(s+20)}$$

Design a lag–lead compensator such that

$$K_v = 50$$
$$\text{phase margin} = 50^\circ$$
$$\text{gain margin} \geq 10 \text{ dB}$$

SOLUTION

The Bode diagram of the uncompensated system was given in Figure 9.33 in connection with Example 9.7. We found that the phase margin of the uncompensated system is $-21.33°$ and the closed-loop system is unstable. The open-loop frequency response has a phase angle of $-180°$ at $\omega_\phi$, obtained from

$$-180^\circ = -90^\circ - \tan^{-1}\frac{\omega}{12} - \tan^{-1}\frac{\omega}{20}$$

Thus

$$90^\circ = \tan^{-1}\frac{\omega}{12} + \tan^{-1}\frac{\omega}{20}$$

Recall that

$$\tan(\theta_1 + \theta_2) = \frac{\tan\theta_1 + \tan\theta_2}{1 - \tan\theta_1 \tan\theta_2}$$

For $\theta_1 + \theta_2 = 90^\circ$ we thus have

$$\tan\theta_1 \tan\theta_2 = 1$$

As a result,

$$\omega_\phi = \sqrt{240} = 15.5 \text{ rad/sec}$$

We choose the new gain crossover frequency to be at 15.5 rad/sec to use a phase-lead angle of $50°$. We choose the zero of the phase-lag portion corresponding to $\omega = 1/\tau_1$ to be 1 decade below the new gain crossover frequency. Thus

$$\frac{1}{\tau_1} = 1.5$$

We choose $\alpha = 10$. Thus the pole of the lag portion is at $\omega = 0.15$. The transfer function of the lag portion is thus

$$\frac{s + 1.5}{s + 0.15}$$

To achieve a gain crossover at $\omega = 15.5$ we first find the gain in decibels at that frequency for the uncompensated system to be 7.97 dB. Thus we need $-7.97$ dB at $\omega = 15.5$ rad/sec to be on the lag–lead network amplitude plot. We draw a 20-dB/decade line passing through the point $-7.97$ dB, 15.5 rad/sec. This intersects the 0-dB line at the corner frequency of the pole of the lead portion. From the construction of Figure 9.42 we see that

$$7.97 = 20 \log \frac{\omega_c}{15.5}$$

Thus

$$\omega_c = 38.8$$

The lead portion will thus have a transfer function of

$$\frac{s + 3.88}{s + 38.8}$$

Our compensated open-loop transfer function is thus given by

$$G_c(s)G_p(s) = \frac{K(s + 1.5)(s + 3.88)}{s(s + 12)(s + 20)(s + 0.15)(s + 38.8)}$$

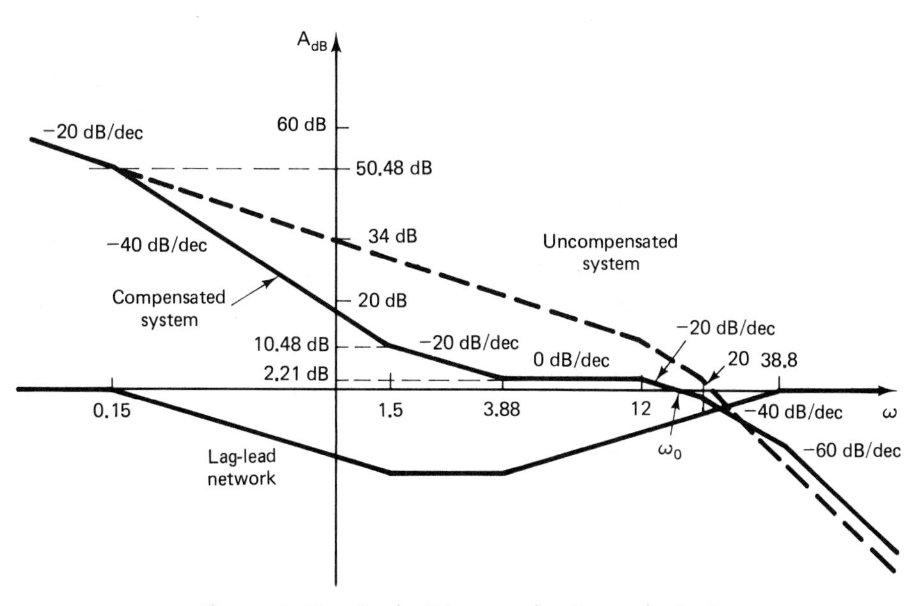


**Figure 9.42.** Bode Diagram for Example 9.10

We have to verify that our gain margin requirements are satisfied. The phase angle is given by

$$\phi = \tan^{-1}\frac{\omega}{1.5} + \tan^{-1}\frac{\omega}{3.88} - \left(90^\circ + \tan^{-1}\frac{\omega}{12} + \tan^{-1}\frac{\omega}{20} + \tan^{-1}\frac{\omega}{0.15} + \tan^{-1}\frac{\omega}{38.8}\right)$$

Calculating the phase angle at a number of frequencies, we find that at

$$\omega = 33.95 \qquad \phi(j33.95) = -180.01$$

The amplitude is obtained as

$$|G_c(j\omega)G_p(j\omega)|_{\omega=33.95} = 0.17$$

As a result, the actual amplitude in decibels is $-15.64$ dB, which meets our specifications. The design is complete.

# 9.12 Minor-Loop Feedback Compensation

Our discussion so far has centered on cascade compensation, as it represents a simple, yet powerful means for achieving required performance specifications. There are situations when placing the compensator in a minor feedback loop is more advantageous, due to the nature of the system, as shown in Figure 9.43. The block $G_c(s)$ represents the discriminator and cascade compensation (if any). The process transfer function $G_p(s)$ is represented as the product of $G_{p_1}(s)$ and $G_{p_2}(s)$ as feedback through $G_H(s)$ takes place after $G_{p_1}(s)$. The transfer function $G_H(s)$ is the minor-loop compensator.

As a specific example, the control system with tachometer feedback is shown in Figure 9.44. The tachometer senses the velocity of the motor $\omega(s)$ and a feedback signal is generated through the tachometer gain $K_T$. Comparing Figures 9.43 and 9.44, we can see that $G_{p_2}(s)$ in the former represents the integration $1/s$ to obtain the angular position $C(s)$ from the velocity $\omega(s)$.

The example system of Figure 9.44 serves to illustrate the effect of tachometer feedback, which is a special case of minor-loop feedback compensation. It is relatively easy to show that the closed-loop transfer function is given by

$$\frac{C(s)}{R(s)} = \frac{\omega_n^2}{s^2 + 2\xi_{\text{eq}}\omega_n s + \omega_n^2}$$

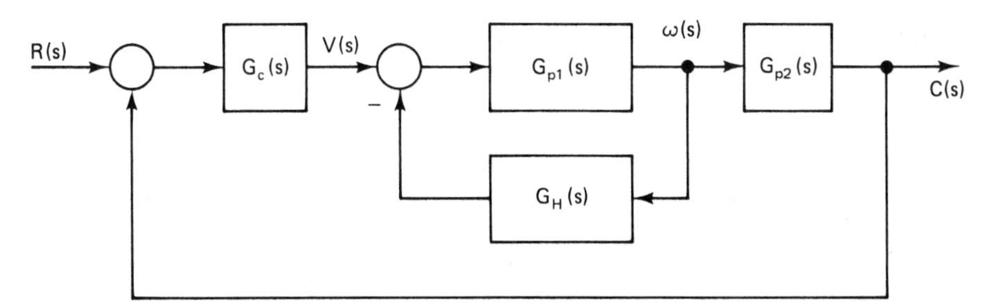


**Figure 9.43.** Block Diagram of a Closed-Loop System with a Minor-Loop Compensator

The equivalent damping ratio $\xi_{eq}$ is given by

$$\xi_{eq} = \xi + \frac{K_T \omega_n}{2}$$

It is thus seen that damping is increased by the amount $K_T\omega_n/2$.

The previous discussion shows a simple problem in minor-loop feedback compensation. It is clear that the design procedure involves simply finding $K_T$ for a specified damping ratio. It is our intention now to discuss the design of $G_H(s)$ in Figure 9.43 using the Bode approach. Although the task may seem to be quite involved, it turns out that a little bit of organization and practical assumptions simplify the problem considerably.

To start, let us designate the minor loop by the subscript $m$ and the major loop by $M$. The minor loop transfer function is defined by

$$G_m(s) = G_{p_1}(s)G_H(s)$$

Without minor loop compensation, $G_H(s) = 0$, and the uncompensated major loop transfer function is given by

$$G_{\text{MU}}(s) = G_c(s)G_{p_1}(s)G_{p_2}(s)$$

The subscript U denotes an uncompensated system. The compensated

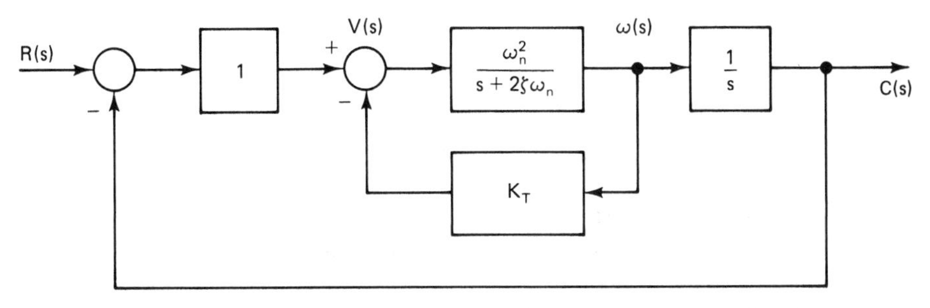


**Figure 9.44.** Control System with Tachometer Feedback

major-loop transfer function is defined as $G_{MC}(s)$ and is clearly given by

$$G_{MC}(s) = \frac{G_c(s)G_{p_1}(s)G_{p_2}(s)}{1 + G_{p_1}(s)G_H(s)}$$

In the above, the subscript C denotes a compensated system. In terms of $G_{MU}(s)$ and $G_m(s)$, we have

$$G_{MC}(s) = \frac{G_{MU}(s)}{1 + G_m(s)}$$

As a result, the compensated transfer function is obtained as the outcome of the division of uncompensated transfer function by $1 + G_m(s)$.

Since it is our intention to use Bode diagrams, it helps to recall the cornerstone of the method employed to arrive at the amplitude asymptotes. In Chapter 8 we saw that the relation between a function of $\omega$ (such as $j\tau\omega$) and unity was used to establish the plots. In a similar manner, we take $G_m(j\omega)$ in relation to 1. We can thus approximate the compensated system by

$$G_{MC}(j\omega) \simeq G_{MU}(j\omega) \qquad \text{for } |G_m(j\omega)| \ll 1$$

$$G_{MC}(j\omega) \simeq \frac{G_{MU}(j\omega)}{G_m(j\omega)} \qquad \text{for } |G_m(j\omega)| \gg 1$$

It is thus clear that the Bode diagrams of the compensated system can be obtained easily from the corresponding diagrams for the uncompensated system and the minor-loop function $G_m(s)$. An example will illustrate the concepts.

EXAMPLE 9.11

Consider the system of Examples 9.7 through 9.10. This time we assume that minor-loop feedback compensation is employed. The representation of Figure 9.43 is adopted. Thus

$$G_c(s) = K_1$$

$$G_{p_1}(s) = \frac{K_2}{(1 + s/12)(1 + s/20)}$$

$$G_{p_2}(s) = \frac{1}{s}$$

Take $K_1 = 50$ and $K_2 = 1$.

Assume that minor-loop feedback compensation is used with

$$G_H(s) = \frac{s}{1 + s/10}$$

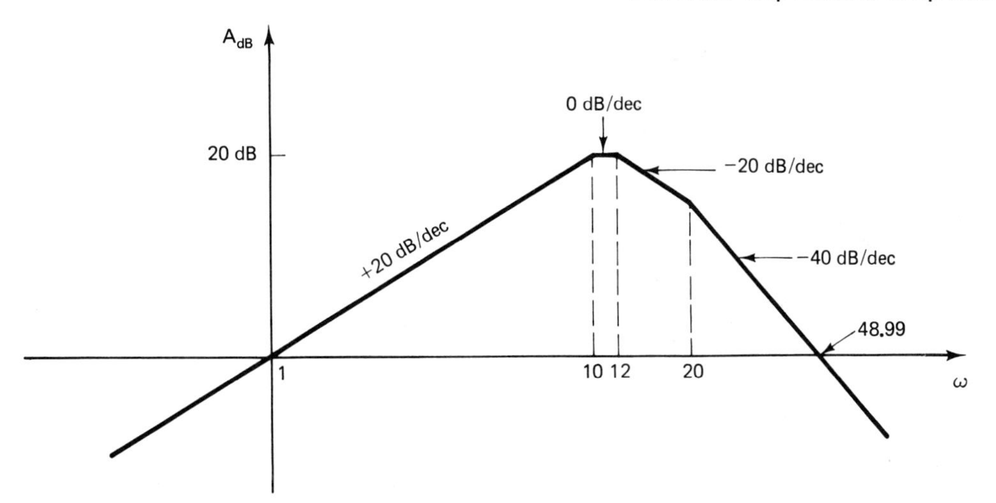


**Figure 9.45.** Bode Plot for $G_m(s)$ of Example 9.11

The minor-loop transfer function is thus given by

$$G_m(s) = G_{p_1}(s)G_H(s)$$

$$= \frac{s}{(1 + s/10)(1 + s/12)(1 + s/20)}$$

The Bode plot of $G_m(s)$ is shown in Figure 9.45. The uncompensated major-loop transfer function is given by

$$G_{\mathrm{MU}}(s) = G_c(s)G_{p_1}(s)G_{p_2}(s)$$

$$= \frac{50}{s(1 + s/12)(1 + s/20)}$$

The Bode plot of $G_{\mathrm{MU}}(s)$ is shown in Figure 9.46.

Let us obtain the exact expression for the major-loop compensated transfer function using

$$G_{\mathrm{MC}}(s) = \frac{G_{\mathrm{MU}}(s)}{1 + G_m(s)}$$

$$= \frac{50(1 + s/10)}{s[(1 + s/10)(1 + s/12)(1 + s/20) + s]}$$

This expression reduces to

$$G_{\mathrm{MC}}(s) = \frac{50 \times 2400(1 + s/10)}{s(s^3 + 42s^2 + 2960s + 2400)}$$

We note that we need to factor the third-order polynomial of the denomina-

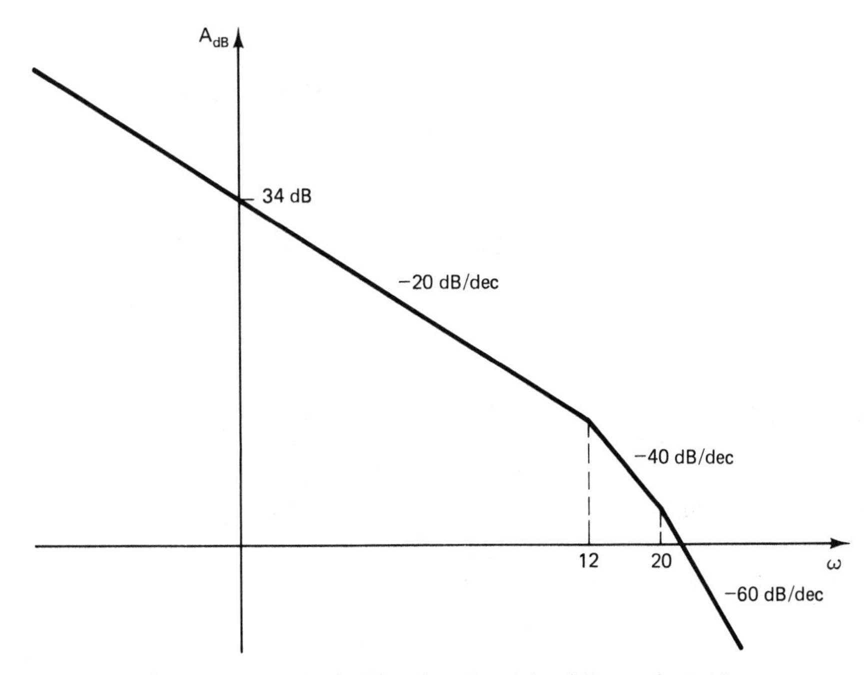


**Figure 9.46.** Bode Plot for $G_{MU}(s)$ of Example 9.11

tor. The result is

$$G_{MC}(s) = \frac{50(1 + s/10)}{s(1 + s/0.8202)\left[1 + 0.0141s + (s/54.09)^2\right]}$$

The Bode plot of $G_{MC}(s)$ is shown in Figure 9.47. From the construction we can show that the gain crossover frequency is given by $\omega_\phi = 6.4114$. The phase angle of $G_{MC}(j\omega)$ at $\omega_\phi$ is $-145.274°$, and thus a phase margin of close to 35° is achieved by minor-loop feedback compensation.

It should be noted that to obtain the Bode plot for $G_{MC}$ as outlined above, we needed to factorize a third-order polynomial. For this reason we use the practical approximation given by

$$\begin{aligned} G_{MC}(j\omega) &= \frac{G_{MU}(j\omega)}{G_m(j\omega)} \qquad |G_m(j\omega)| \gg 1 \\ &= G_{MU}(j\omega) \qquad |G_m(j\omega)| \ll 1 \end{aligned}$$

The key, of course, is in identifying the region where $|G_m(j\omega)| \gg 1$. The Bode plot of $G_m(j\omega)$ provides this information. From Figure 9.45 the amplitude in decibels is greater than zero for the interval $\omega = 1$ to $\omega = 48.99$. Thus to obtain the approximate Bode plot of $G_{MC}(j\omega)$, we subtract the plot of $G_m(j\omega)$ from that of $G_{MU}(j\omega)$, as shown in Figure 9.48.

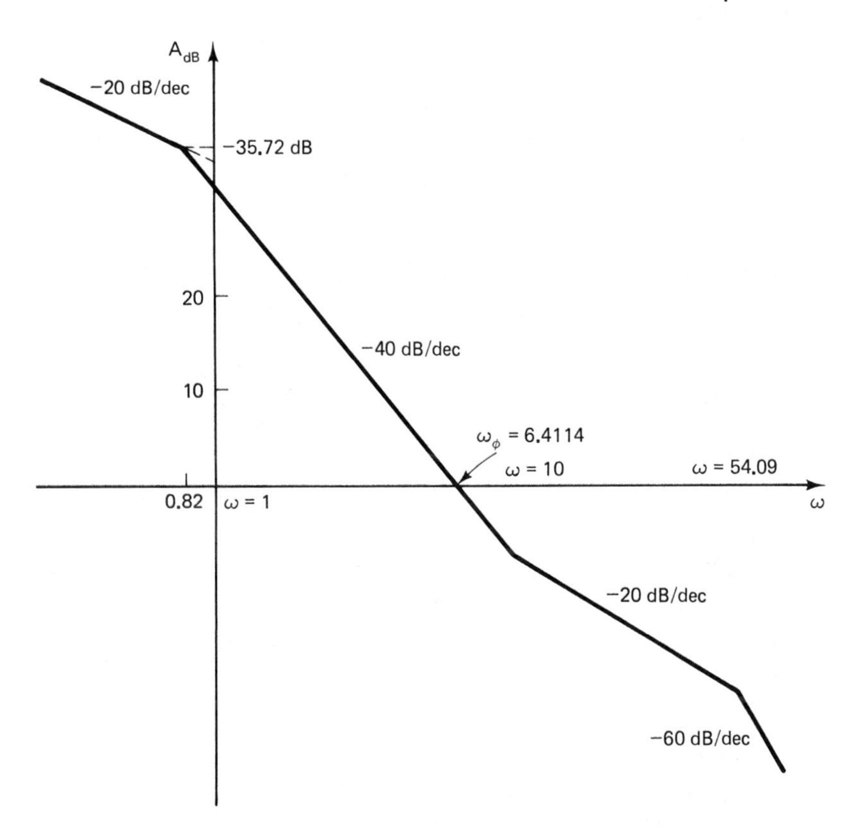


**Figure 9.47.** Bode Plot for Precise $G_{MC}(s)$ of Example 9.11

Figure 9.48 shows the Bode plot of $G_{MU}(s)$ and that of the negative of amplitude of $G_m(s)$ in the interval $\omega = 1$ to 48.99. The approximate $G_{MC}(s)$ Bode plot is obtained by simple addition as shown. The resulting $G_{MC}(s)$ in the interval $\omega = 1$ to 48.99 is

$$G_{MC}(s) = \frac{50(1 + s/10)}{s^2}$$

The approximate gain crossover frequency is found to be $\omega_\phi = 7.07$. The approximate phase angle is

$$\phi = -180° + \tan^{-1}\frac{7.07}{10}$$
$$= -144.72°$$

The phase margin is thus 35.28°, which is close to the exact value. Figure 9.49 shows both approximate and precise Bode plots of $G_{MC}(s)$. Note that the approximation is slightly shifted to the right from the precise diagram. This gives more conservative results.

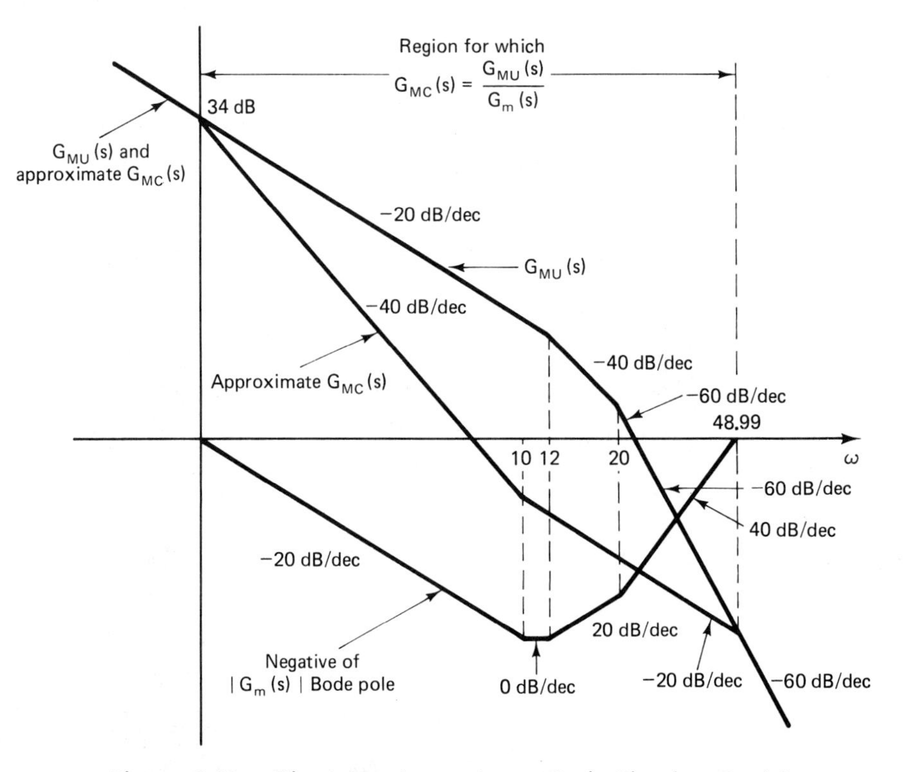


**Figure 9.48.** Obtaining Approximate Bode Plot for $G_{MC}(s)$

In Example 9.11 we discussed the mechanics of approximate Bode plot analysis of systems with minor-loop feedback compensation. One way of approaching the design of such a compensator is to assume that a compensated open-loop transfer function obtained by cascade design is to be realized using minor-loop feedback compensation. The task then is to find $G_H(s)$ from knowledge of $G_{MU}(s)$ and $G_{MC}(s)$. This situation is discussed in Problem 9A-6. A more direct approach is, of course, to start with a compensator $G_H(s)$ with unknown parameters and set out to determine the values that achieve the given performance specifications. The power of our approximation technique will become evident from the following example.

EXAMPLE 9.12

Consider the system of Example 9.7 with uncompensated open-loop transfer function given by

$$G_{MU}(s) = \frac{50}{s(1 + s/12)(1 + s/20)}$$

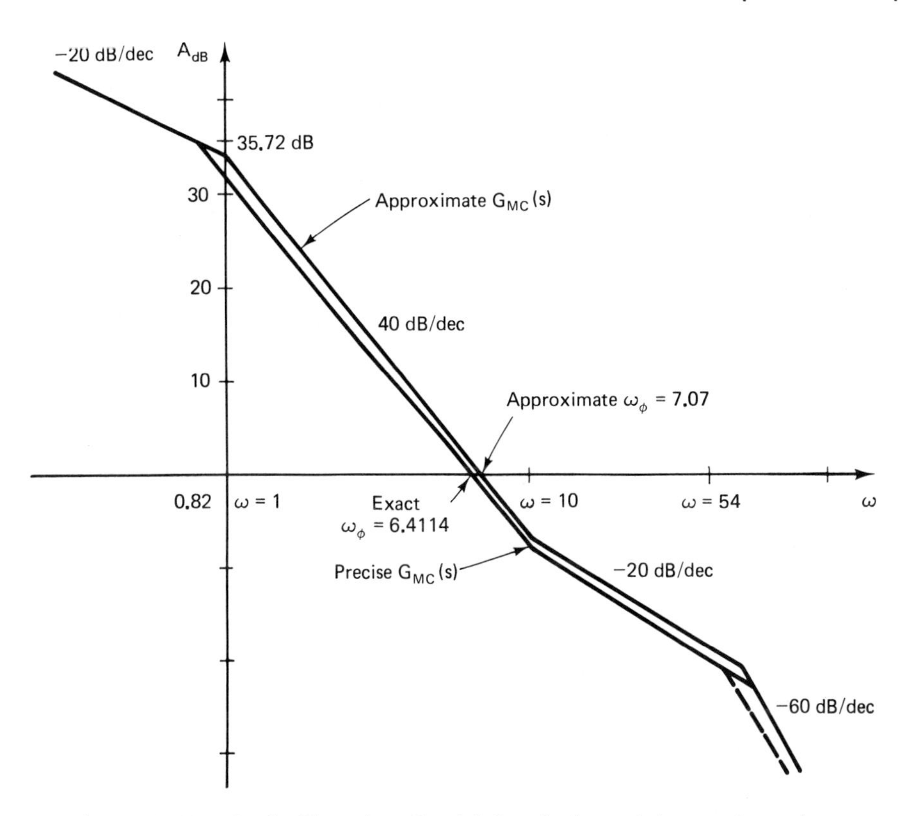


**Figure 9.49.** Bode Plots for $G_{MC}(s)$ Precisely and Approximately

Let us assume that minor-loop feedback is employed with the compensator transfer function given by

$$G_H(s) = \frac{K_t s^2}{1 + \tau s}$$

The tachometric gain $K_t$ and the time constant $\tau$ are then the design parameters.

The minor-loop transfer function $G_m(s)$ is thus given by

$$G_m(s) = G_{MU}(s)G_H(s) = \frac{50K_t s}{(1 + s/12)(1 + s/20)(1 + \tau s)}$$

The Bode plot of $G_m(j\omega)$ is sketched in Figure 9.50 assuming that $1/\tau < 1$ and $K_t > 0$. It is clear from the construction that $|G_m(j\omega)| > 1$, for $\omega_1 < \omega$

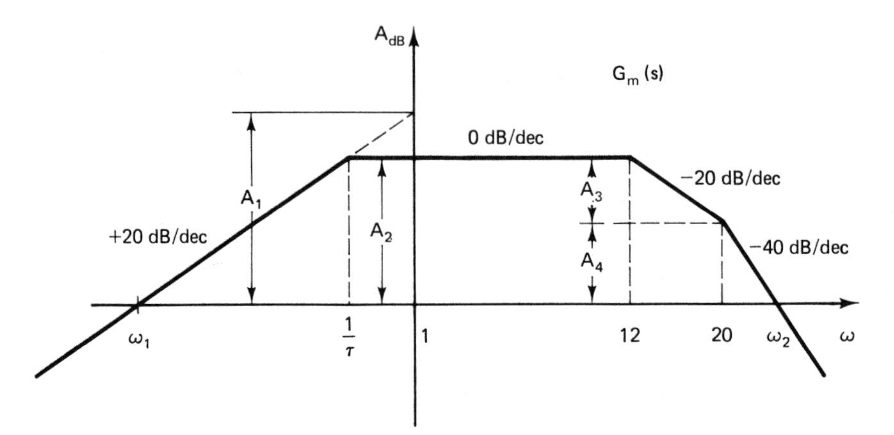


**Figure 9.50.** Bode Plot for $G_m(s)$ of Example 9.12

$< \omega_2$, where $\omega_1$ and $\omega_2$ satisfy the following relations derived from Figure 9.50:

$$A_1 = 20 \log \frac{1}{\omega_1}$$
$$= 20 \log 50K_t$$

Thus

$$\omega_1 = \frac{1}{50K_t}$$

We also have

$$A_2 = 20 \log \frac{1}{\omega_1 \tau}$$

Also,

$$A_2 = A_3 + A_4$$

with

$$A_3 = 20 \log \frac{20}{12}$$
$$A_4 = 40 \log \frac{\omega_2}{20}$$

Thus

$$20 \log \frac{1}{\omega_1 \tau} = 20 \log \frac{20}{12} + 40 \log \frac{\omega_2}{20}$$
$$\frac{1}{\omega_1 \tau} = \frac{20}{12} \left( \frac{\omega_2}{20} \right)^2$$

This can be written as

$$\omega_2^2 = \frac{12{,}000K_t}{\tau}$$

The major-loop transfer function with compensation in place is given approximately by

$$G_{\text{MC}}(s) = \frac{G_{\text{MU}}(s)}{G_m(s)} = \frac{1}{G_H(s)}$$

for $|G_m(j\omega)| \gg 1$. Thus we have

$$\tilde{G}_{\text{MC}}(s) = \frac{1+\tau s}{K_t s^2}$$

The Bode plot for $\tilde{G}_{\text{MC}}(s)$ is shown in Figure 9.51. We assume that $1/\tau < 1$. Part (a) of the figure applies for $K_t > 1$; it can be seen that the gain crossover in this case is at $\omega_\phi < 1/\tau$. Part (b) of the figure applies for $K_t$ values of less than or equal to 1. In this case, $\omega_\phi > 1$. Note that the phase angle of $\tilde{G}_{\text{MC}}(s)$ is given by

$$\tilde{\phi}_{\text{MC}}(j\omega) = -180° + \tan^{-1}\omega\tau$$

Thus the phase margin is given by

$$\text{phase margin} = \tan^{-1}\omega_\phi\tau$$

We require a large value of $\omega_\phi\tau$ for a reasonable phase margin. We thus opt for $0 < K_t \leq 1$, where we have

$$20\log\omega_\phi\tau = 40\log\tau - 20\log K_t$$

Thus

$$\omega_\phi = \frac{\tau}{K_t}$$

Let us summarize our findings so far. The minor-loop transfer function satisfies $|G_m(j\omega)| < 1$ between the limits

$$\omega_1 = \frac{1}{50K_t}$$

$$\omega_2^2 = \frac{12{,}000K_t}{\tau}$$

The gain crossover frequency $\omega_\phi$ is

$$\omega_\phi = \frac{\tau}{K_t}$$

The phase margin is

$$\text{phase margin} = \tan^{-1}\omega_\phi\tau$$

Note that this applies to the approximate system.

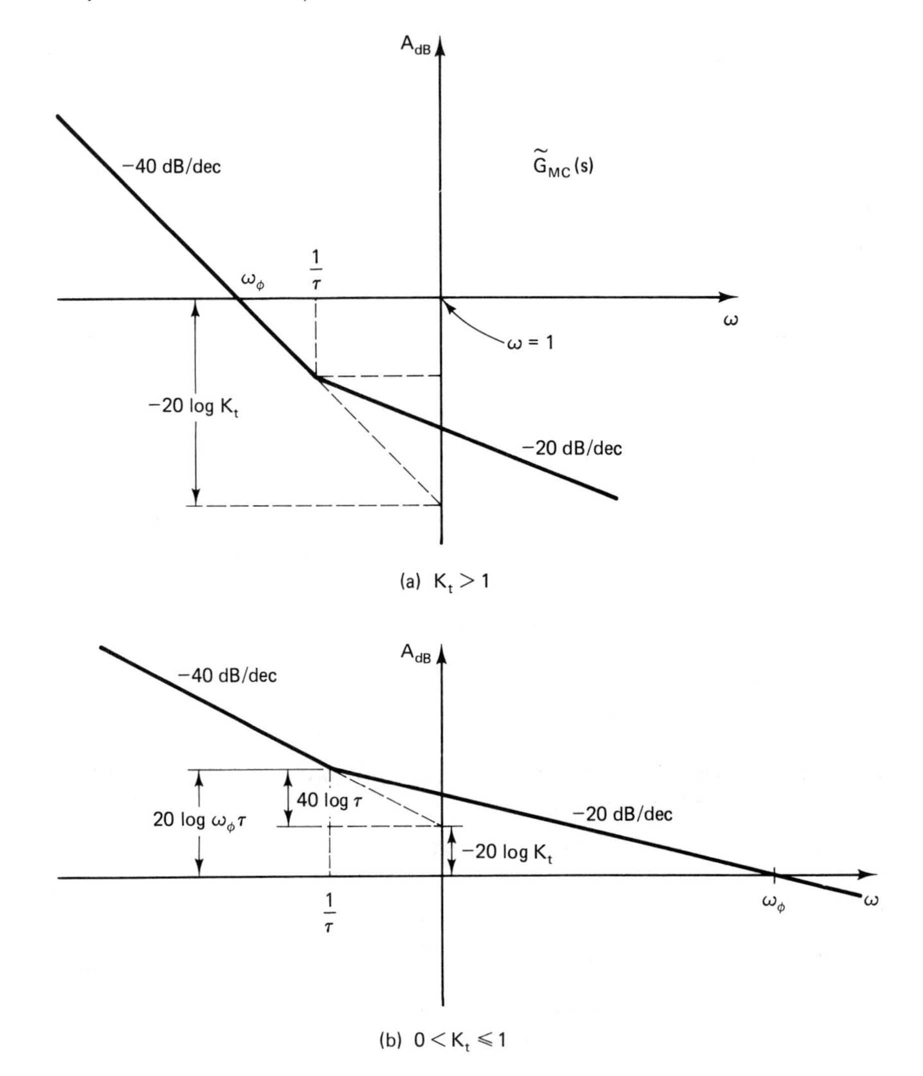


**Figure 9.51.** Bode Plot for $\tilde{G}_{MC}(s)$

Let us assume that

$$K_t = 1$$
$$\tau = 1.25$$

As a result,

$$\omega_1 = 0.02 \text{ rad/sec}$$
$$\omega_2 = 97.98 \text{ rad/sec}$$
$$\omega_\phi = 1.25 \text{ rad/sec}$$
$$\text{phase margin} = 57.38°$$

Note that $\omega_1 < \omega_\phi < \omega_2$ and the phase margin is more than satisfactory. If our choice of $K_t$ and $\tau$ does not provide satisfactory results a trial-and-error procedure should be adopted.

The minor-loop feedback compensator chosen is thus given by

$$G_H(s) = \frac{s^2}{1 + 1.25s}$$

The resulting $G_m(s)$ is

$$G_m(s) = \frac{50s}{(1 + s/12)(1 + s/20)(1 + 1.25s)}$$

We may be concerned about the stability of the minor loop. Its characteristic equation is

$$\left(1 + \frac{s}{12}\right)\left(1 + \frac{s}{20}\right)(1 + 1.25s) + 50s = 0$$

A few manipulations reduce the above to

$$s^3 + 32.8s^2 + 9865.6s + 192 = 0$$

This has roots at

$$s_1 = -0.0195$$
$$s_{2,3} = -16.39 \pm j97.96$$

This indicates stability.

The accurate open-loop compensated transfer function is

$$G_{MC}(s) = \frac{G_{MU}(s)}{1 + G_m(s)}$$

This turns out to be

$$G_{MC}(s) = \frac{50(192)(1 + 1.25s)}{s^3 + 9648.23s^2 + 266.02s + 192}$$

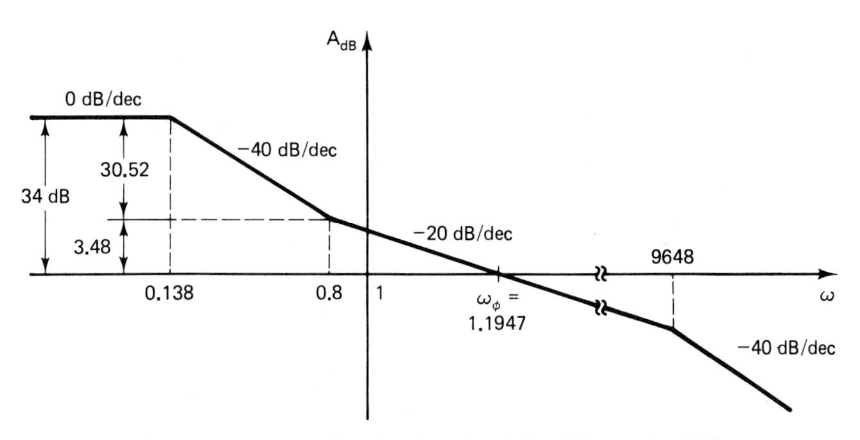


**Figure 9.52.** Bode Plot for $G_{MC}(s)$ of Example 9.12

A factorization of the denominator yields

$$G_{MC}(s) = \frac{(192)(50)(1 + 1.25s)}{(s + 9648.2)\left[s^2 + 0.0276s + (0.1381)^2\right]}$$

The Bode plot of the actual $G_{MC}(s)$ is shown in Figure 9.52. From the diagram we get

$$\omega_{\phi} = 1.1947$$

This is slightly lower than the approximate value of 1.25. The phase margin is found to be equal to 54.85°, which is slightly lower than the approximate value, but is still acceptable. Our design is complete.

Note that Problem 9A-6 carries out details of a design without using approximations.

# *Some Solved Problems*

PROBLEM 9A-1

Many variations of the phase-lead network exist. One such variation is shown in Figure 9.53. Show that the transfer function of this network is given by the standard phase-lead form

$$G_c(s) = G_c(0)\frac{1 + \tau s}{1 + \tau s/\alpha}$$

where

$$\alpha = \frac{R_1 + R_2 + R_3}{R_1 + R_3 + [R_2 R/(R + R_2)]}$$

$$\tau = (R + R_2)C$$

$$G_c(0) = \frac{R_3}{R_1 + R_2 + R_3}$$

Sketch the log amplitude characteristic for this network.

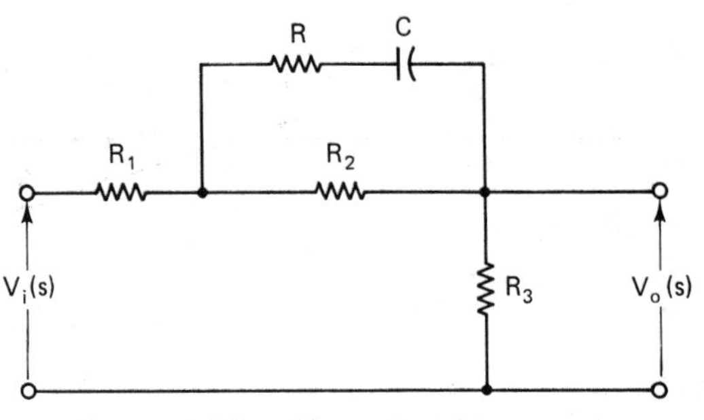


**Figure 9.53.** Phase-Lead Network

SOLUTION

The transfer function is written as

$$G_c(s) = \frac{R_3}{Z_i(s)}$$

The impedance $Z_i(s)$ is given by

$$\begin{aligned} Z_i(s) &= R_1 + R_3 + \frac{1}{\dfrac{1}{R_2} + \dfrac{1}{R + 1/Cs}} \\ &= R_1 + R_3 + \frac{R_2(1 + RCs)}{1 + (R + R_2)Cs} \end{aligned}$$

Thus

$$\begin{aligned} G_c(s) &= \frac{R_3[1 + (R + R_2)Cs]}{(R_1 + R_2 + R_3) + [(R_1 + R_3)(R + R_2) + RR_2]Cs} \\ &= \frac{R_3}{R_1 + R_2 + R_3} \frac{1 + (R + R_2)Cs}{1 + (R + R_2)\dfrac{R_1 + R_3 + [R_2R/(R + R_2)]}{R_1 + R_2 + R_3}Cs} \end{aligned}$$

From the problem statement we can conclude that

$$G_c(s) = G_c(0)\frac{1 + \tau s}{1 + \tau s/\alpha}$$

where

$$\begin{aligned} G_c(0) &= \frac{R_3}{R_1 + R_2 + R_3} \\ \tau &= (R + R_2)C \\ \alpha &= \frac{R_1 + R_2 + R_3}{R_1 + R_3 + R_2R/(R + R_2)} \end{aligned}$$

Note that $\alpha$ is greater than 1.

The Bode diagram will start at low frequencies with a constant value given by

$$A_{\text{dB}} = 20 \log G_c(0)$$

This is negative, as $G_c(0) < 1$. At the corner frequency $\omega_c = 1/\tau$, the characteristic has a positive 20 dB/decade and continues on with this slope up to $\omega_c = \alpha/\tau$, where the slope changes to 0 dB/decade. The asymptote of amplitude in decibels beyond $\alpha/\tau$ is negative and is given by

$$\begin{aligned} 20 \log G_c(\infty) &= 20 \log \alpha G_c(0) \\ &= 20 \log \frac{R_3}{R_1 + R_3 + R_2R/(R + R_2)} \end{aligned}$$

The Bode diagram is shown in Figure 9.54.

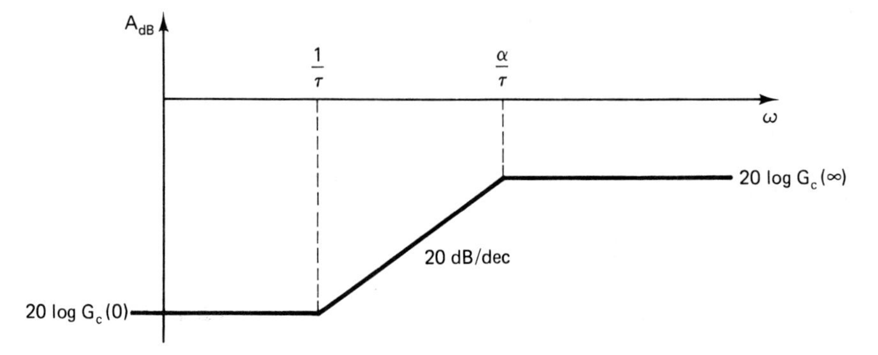

**Figure 9.54.** Bode Diagram for a Phase-Lead Network for Problem 9A-1

PROBLEM 9A-2

The network shown in Figure 9.55 is a variant of the lag–lead network. Show that the transfer function of this network is given by

$$G_C(s) = \frac{(1+\tau_1 s)(1+\tau_2 s)}{\alpha\tau_1\tau_2 s^2 + (\tau_1\beta + \tau_2\delta)s + 1/G_0}$$

where

$$\tau_1 = \frac{R_2 R_4}{R_2 + R_4} C_2$$

$$\tau_2 = (R_1 + R_3)C_1$$

$$\alpha = 1 + \frac{R_1 R_3}{R_2(R_1 + R_3)}$$

$$\beta = 1 + \frac{R_1}{R_2}$$

$$\delta = 1 + \frac{R_1 R_3}{(R_2 + R_4)(R_1 + R_3)}$$

$$G_0 = \frac{1}{1 + R_1/(R_2 + R_4)}$$

Sketch the Bode plot of the frequency response of the transfer function assuming that $\tau_1 > \tau_2$.

SOLUTION

The impedance on the output side is given by

$$Z_o(s) = R_2 + \frac{R_4}{1 + R_4 C_2 s}$$

$$= \frac{R_2 + R_4 + R_2 R_4 C_2 s}{1 + R_4 C_2 s}$$

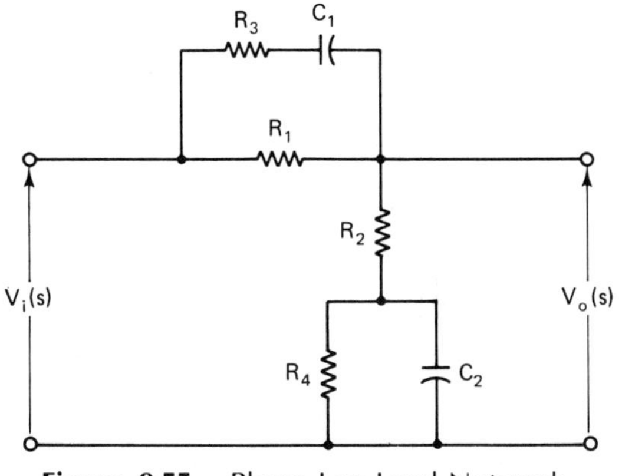


**Figure 9.55.** Phase Lag-Lead Network

Using the given definition of $\tau_1$, we get

$$Z_o(s) = \frac{(R_2 + R_4)(1 + \tau_1 s)}{1 + R_4 C_2 s}$$

The impedance on the input side is given by

$$Z_i(s) = Z_o(s) + \frac{R_1(R_3 + 1/C_1 s)}{R_1 + R_3 + 1/C_1 s}$$

This reduces to

$$Z_i(s) = Z_o(s) + \frac{R_1(1 + R_3 C_1 s)}{1 + (R_1 + R_3)C_1 s}$$

Using the definition of $\tau_2$, we get

$$Z_i(s) = Z_o(s) + \frac{R_1(1 + R_3 C_1 s)}{1 + \tau_2 s}$$

Substituting for $Z_o(s)$, we obtain

$$Z_i(s) = \frac{(R_2 + R_4)(1 + \tau_1 s)}{1 + R_4 C_2 s} + \frac{R_1(1 + R_3 C_1 s)}{1 + \tau_2 s}$$

The transfer function can now be obtained as

$$G_C(s) = \frac{Z_o(s)}{Z_i(s)}$$

$$= \frac{(1 + \tau_1 s)(1 + \tau_2 s)}{(1 + \tau_1 s)(1 + \tau_2 s) + [R_1(1 + R_3 C_1 s)(1 + R_4 C_2 s)/(R_2 + R_4)]}$$

The numerator of the expression above agrees with that in the problem statement. The denominator is denoted by $d(s)$ and is manipulated as follows:

$$d(s) = \left(\tau_1\tau_2 + \frac{R_1R_3C_1R_4C_2}{R_2 + R_4}\right)s^2 + \left(\tau_1 + \tau_2 + \frac{R_1R_3C_1}{R_2 + R_4} + \frac{R_1R_4C_2}{R_2 + R_4}\right)s + \left(1 + \frac{R_1}{R_2 + R_4}\right)$$

A slight rearrangement results in

$$d(s) = \left[\tau_1\tau_2 + \frac{R_1R_3}{R_2(R_1 + R_3)}\frac{R_2R_4C_2}{R_2 + R_4}(R_1 + R_3)C_1\right]s^2 + \left[\tau_1 + \tau_2 + \frac{R_1R_3}{R_2 + R_4}\frac{(R_1 + R_3)C_1}{R_1 + R_3} + \frac{R_1}{R_2}\frac{R_2R_4}{R_2 + R_4}C_2\right] \times s + 1 + \frac{R_1}{R_2 + R_4}$$

This reduces using the given definitions to

$$d(s) = \alpha\tau_1\tau_2 s^2 + (\tau_1\beta + \tau_2\delta)s + \frac{1}{G_0}$$

Thus the denominator is now in the stated form.

We will now consider the log amplitude plot. First, the gain in decibels at low frequencies is given by

$$20 \log G_C(0) = G_0$$

The value of $G_0$ is less than 1. Thus the amplitude in decibels at low frequencies is negative. At high frequencies, we have

$$G_\infty = \lim_{s \to \infty} G_C(s) = \frac{1}{\alpha}$$

As $\alpha$ is greater than 1, the amplitude in decibels is negative for high frequencies.

There are four corner frequencies, corresponding to the numerator and denominator polynomials. The denominator has to be factored as

$$d(s) = \alpha\tau_1\tau_2(s - r_1)(s - r_2)$$

The roots $r_1$ and $r_2$ are given by

$$r_{1,2} = \frac{-(\tau_1\beta + \tau_2\delta) \pm \sqrt{(\tau_1\beta + \tau_2\delta)^2 - (4\alpha\tau_1\tau_2/G_0)}}{2\alpha\tau_1\tau_2}$$

The roots are both real negative valued. The corresponding corner frequencies are at $-r_1$ and $-r_2$. The numerator is already in factored form. The amplitude in decibels is shown in Figure 9.56.

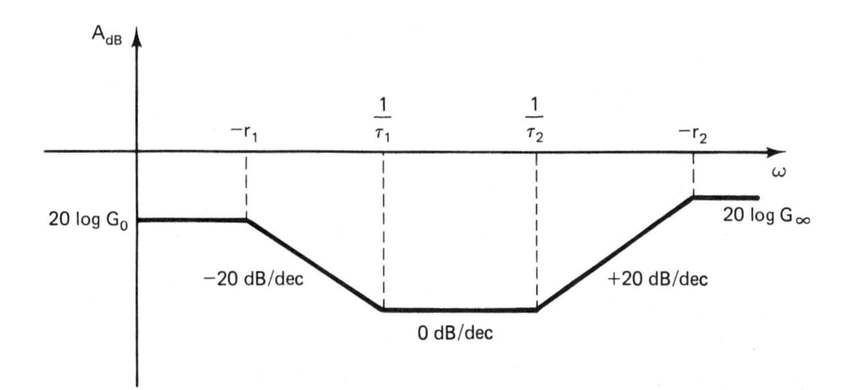


**Figure 9.56.** Bode Diagram for a Lag-Lead Network for Problem 9A-2

PROBLEM 9A-3

The network shown in Figure 9.57 is a 40-dB/decade lead network. Obtain the transfer function and sketch the corresponding Bode plot.

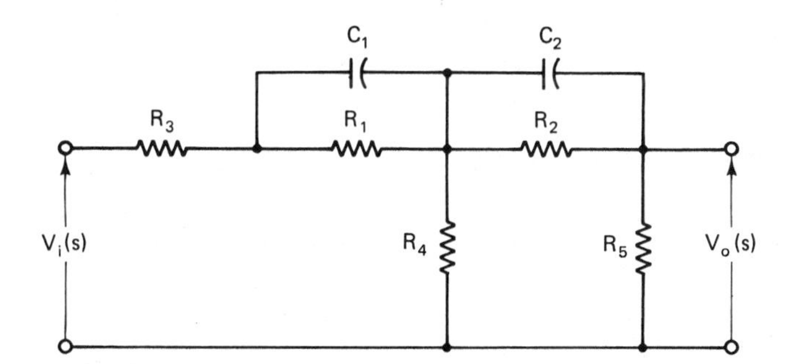


**Figure 9.57.** 40-dB/Decade Lead Network for Problem 9A-3

SOLUTION

A little bit of network reduction helps in here. Figure 9.58 shows the reduced network with

$$Z_1(s) = \frac{R_1}{1 + R_1C_1s} = \frac{R_1}{1 + \tau_1 s}$$
$$Z_2(s) = \frac{R_2}{1 + R_2C_2s} = \frac{R_2}{1 + \tau_2 s}$$

Here we define

$$\tau_1 = R_1C_1 \qquad \text{and} \qquad \tau_2 = R_2C_2$$

We can use any of the techniques explored in Chapter 2 to obtain the transfer function. A reader in an analytic mood will agree that the following

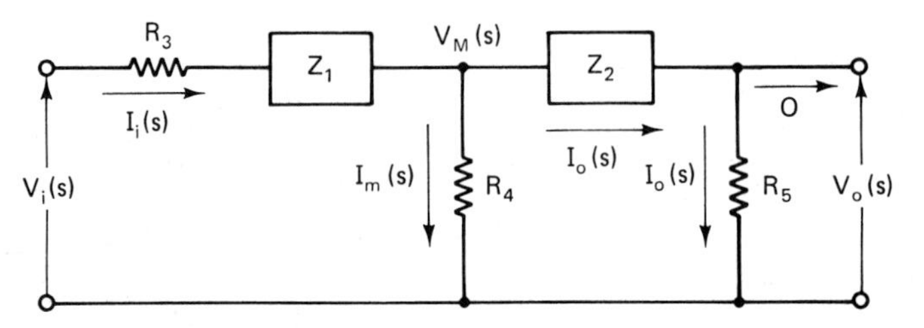


**Figure 9.58.** Reduced Form for Network of Figure 9.57

treatment does the job. We will work backward from $V_o(s)$.

$$V_o(s) = R_5 I_o(s)$$

$$I_o(s) = \frac{V_M(s)}{Z_2(s) + R_5}$$

$$V_M(s) = R_4 I_M(s)$$

$$I_M(s) = I_i(s) - I_o(s)$$

$$I_i(s) = \frac{V_i(s) - V_M(s)}{R_3 + Z_1(s)}$$

Combining the equations above, we obtain

$$G_C(s) = \frac{V_o(s)}{V_i(s)} = \frac{R_5 R_4/(Z_2 + R_5)(R_3 + Z_1)}{1 + \dfrac{R_4}{Z_2 + R_5} + \dfrac{R_4}{R_3 + Z_1}}$$

Thus

$$G_C(s) = \frac{R_5 R_4}{(Z_1 + R_3)(Z_2 + R_5) + R_4[(Z_1 + R_3) + (Z_2 + R_5)]}$$

Substituting for $Z_1$ and $Z_2$, we obtain

$$G_C(s) = \frac{R_4 R_5}{\left(\dfrac{R_1}{1 + \tau_1 s} + R_3\right)\left(\dfrac{R_2}{1 + \tau_2 s} + R_5\right) + R_4\left(R_3 + R_5 + \dfrac{R_1}{1 + \tau_1 s} + \dfrac{R_2}{1 + \tau_2 s}\right)}$$

This reduces to

$$G_C(s) = \frac{(1 + \tau_1 s)(1 + \tau_2 s)}{\alpha \tau_1 \tau_2 s^2 + (\beta \tau_1 + \delta \tau_2)s + 1/G_0}$$

where

$$\alpha = 1 + \frac{(R_4 + R_5)R_3}{R_4R_5}$$

$$\beta = 1 + \frac{R_2}{R_4} + \frac{(R_4 + R_5 + R_2)R_3}{R_5R_4}$$

$$\delta = 1 + \frac{(R_4 + R_5)(R_1 + R_3)}{R_5R_4}$$

$$\frac{1}{G_0} = 1 + \frac{R_2}{R_5} + \frac{(R_5 + R_4 + R_2)(R_1 + R_3)}{R_5R_4}$$

The Bode plot for this network is shown in Figure 9.59.

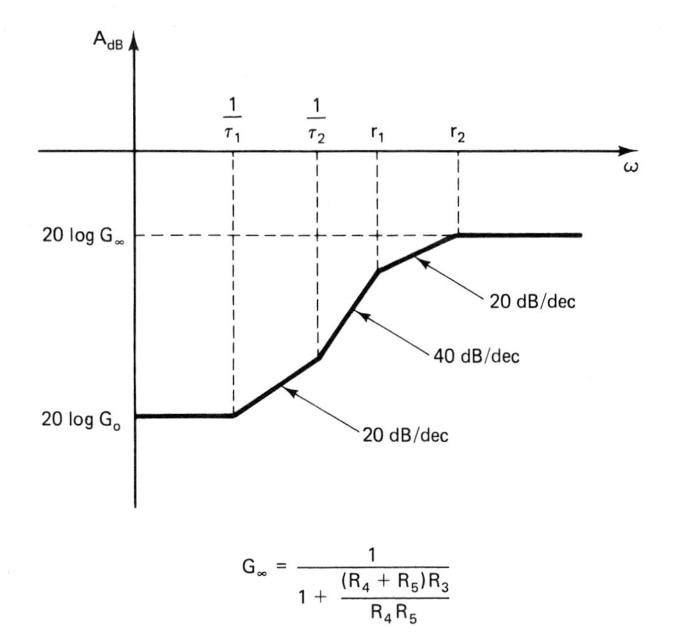


$$G_\infty = \frac{1}{1 + \frac{(R_4 + R_5)R_3}{R_4R_5}}$$

**Figure 9.59.** Bode Plot for a 40-dB/Decade Lead Network

PROBLEM 9A-4

Consider the open-loop transfer function given by

$$G_p(s) = \frac{80}{s^2(1 + s/20)^2}$$

Investigate the possibility of using a phase-lead network to achieve a phase margin of 40°.

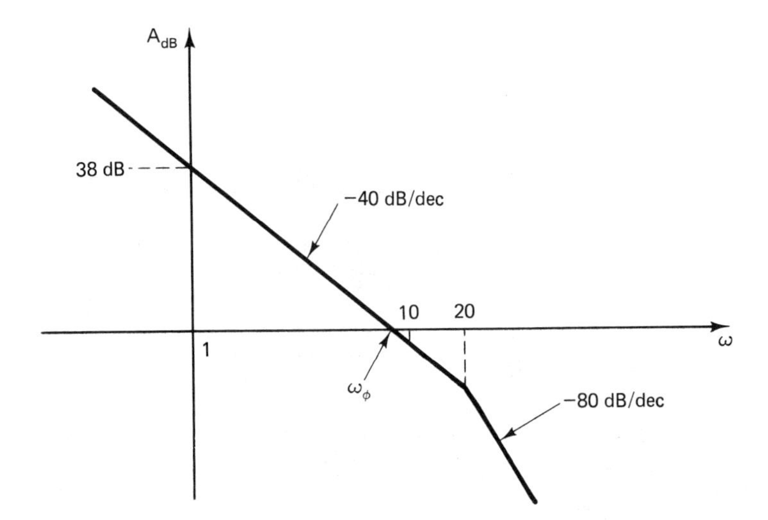


**Figure 9.60.** Bode Plot for Uncompensated System for Problem 9A-4

SOLUTION

We will use the Bode approach. The Bode plot of the uncompensated system is shown in Figure 9.60. The crossover frequency $\omega_\phi$ is given by the relation

$$\frac{20 \log 80}{\log \omega_\phi} = 40$$

Thus

$$\omega_\phi = 8.94$$

The phase angle at $\omega_\phi$ is obtained as

$$\begin{aligned}\phi(j8.94) &= -180° - 2\tan^{-1}\frac{8.94}{20} \\ &= -228.2°\end{aligned}$$

For a phase margin of 40°, the lead network's angle contribution is

$$\phi = 88.2°$$

Thus

$$\alpha = \frac{1 + \sin 88.2}{1 - \sin 88.2} = 4052.2$$

Clearly, this value of $\alpha$ is prohibitively high. Note that since the maximum phase lead obtained from a lead network is 90°, it is impossible to obtain a phase margin of 45°, for example.

In any event we will proceed with the design procedure, taking $\alpha = 4052.2$. The new crossover will be at $\omega_\phi = \sqrt{\alpha}/\tau$. To obtain $\tau$, we have

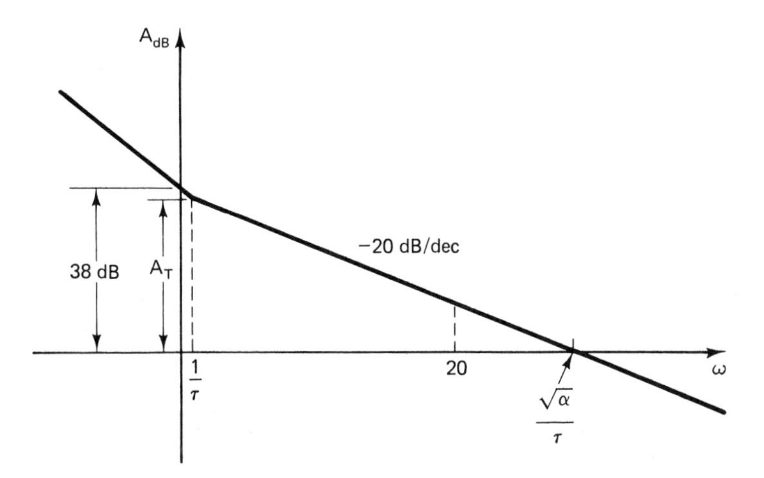


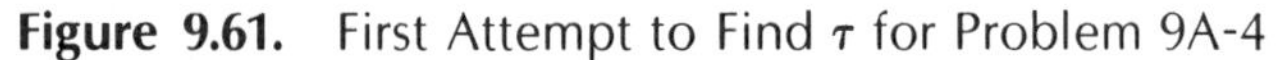
**Figure 9.61.** First Attempt to Find $\tau$ for Problem 9A-4

to assume first that $\omega_\phi < 20$; thus from Figure 9.61,

$$\begin{aligned} A_T &= 20 \log \sqrt{\alpha} \\ &= 20 \log 63.66 \\ &= 36 \text{ dB} \end{aligned}$$

We have

$$\frac{38 - A_T}{40} = \log \frac{1}{\tau}$$

Thus

$$\frac{1}{\tau} = 1.12$$

$$\frac{\alpha}{\tau} = 4526.6$$

The new crossover is at

$$\omega_\phi = \frac{\sqrt{\alpha}}{\tau} = 71.3$$

Since $\omega_\phi > 20$, our calculation is not valid.

We will now assume that $\omega_\phi > 20$, and attempt to find $\tau$. An unscaled construction is shown in Figure 9.62. We can write

$$\begin{aligned} A_1 &= 40 \log \frac{1}{\tau} = -40 \log \tau \\ A_2 &= 20 \log \frac{20}{1/\tau} = 20 \log \tau + 20 \log 20 \\ A_3 &= 60 \log \frac{\omega_\phi}{20} \\ &= 60 \log \frac{\sqrt{\alpha}}{20\tau} = -60 \log \tau + 60 \log \frac{\sqrt{\alpha}}{20} \end{aligned}$$

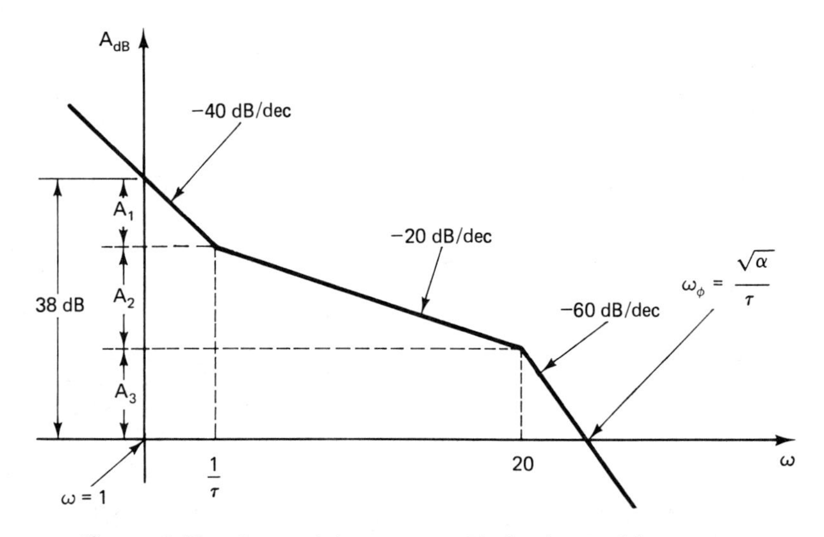

**Figure 9.62.** Second Attempt to Find $\tau$ for Problem 9A-4

Thus

$$\begin{aligned} 38 &= A_1 + A_2 + A_3 \\ &= -80 \log \tau + 56.2 \end{aligned}$$

As a result,

$$80 \log \tau = 18.19$$

This gives us

$$\tau = 1.69$$

or

$$\frac{1}{\tau} = 0.59$$

At this juncture we may throw in the towel, since our construction assumed that $1/\tau > 1$ and what we obtained does not satisfy this condition. A little bit of perseverence helps in here. Let us assume that $1/\tau$ is less than 1. From the unscaled construction of Figure 9.63 we see that

$$\begin{aligned} A_1 &= 20 \log 20\tau = 20 \log \tau + 20 \log 20 \\ A_3 &= 60 \log \frac{\sqrt{\alpha}}{20\tau} = -60 \log \tau + 60 \log \frac{\sqrt{\alpha}}{20} \\ -A_2 &= -40 \log \tau \end{aligned}$$

Thus

$$\begin{aligned} 38 &= A_1 + A_3 - A_2 \\ &= -80 \log \tau + 56.2 \end{aligned}$$

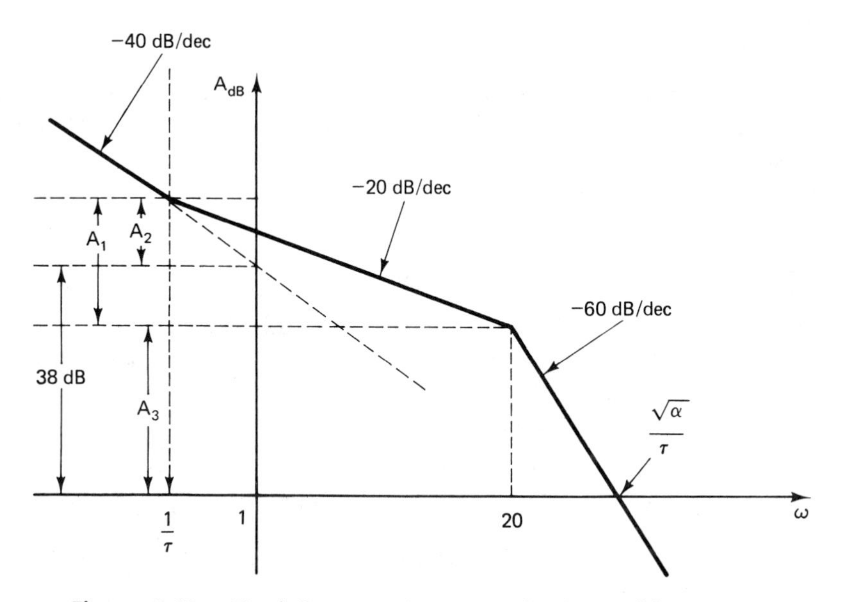


**Figure 9.63.** Final Construction to Find $\tau$ for Problem 9A-4

As a result,

$$\tau = 1.69$$
$$\frac{1}{\tau} = 0.59$$

This is the same result as obtained before and we conclude that the values of $\alpha$ and $\tau$ are given by

$$\alpha = 4052.2$$
$$\frac{1}{\tau} = 0.59$$

The crossover is at

$$\omega_\phi = \frac{\sqrt{\alpha}}{\tau} = 37.56 \text{ rad/sec}$$

The phase angle of the compensated system (is it really compensated?) at the crossover is

$$\begin{aligned}\phi(j\omega_\phi) &= -180^\circ - 2\tan^{-1}\frac{37.56}{20} + \tan^{-1}\frac{37.56}{0.59} - \tan^{-1}\frac{37.56}{(4052.2)(0.59)}\\ &= -215.73^\circ\end{aligned}$$

This clearly does not provide us with the required phase margin and it is impossible to compensate the plant using a phase-lead network.

PROBLEM 9A-5

In solving Example 9.7, we illustrated two methods to design the lag compensation network by providing a third relation to the following two equations in $\alpha$, $\omega_\phi$, and $\tau$:

$$\alpha\omega_\phi = 50$$

$$35° = \tan^{-1}\frac{\omega_\phi}{12} + \tan^{-1}\frac{\omega_\phi}{20} + \tan^{-1}50\tau - \tan^{-1}\omega_\phi\tau$$

A third method is based on minimizing the middle-frequency attenuation by maximizing the corner frequency $1/\alpha\tau$. Determine the lag-compensating network using this method.

SOLUTION

We use the second expression to write approximately

$$\frac{\pi(35)}{180} = \frac{\omega_\phi}{12} + \frac{\omega_\phi}{20} + \left(\frac{\pi}{2} - \frac{1}{50\tau}\right) - \left(\frac{\pi}{2} - \frac{1}{\omega_\phi\tau}\right)$$

Simplifying by use of $\alpha\omega_\phi = 50$, we obtain

$$\frac{1}{\alpha\tau} = \frac{50(0.61\alpha - 6.67)}{\alpha^2(\alpha - 1)}$$

To maximize $1/\alpha\tau$ we take the derivative with respect to $\alpha$ and set the result to zero. Upon simplification we get

$$\alpha^2 - 16.87\alpha + 10.91 = 0$$

There are two solutions:

$$\alpha = 16.2 \quad \text{or} \quad 0.67$$

The second solution is meaningless, as $\alpha$ should be greater than 10. Thus we take

$$\omega_\phi = 3.09$$
$$\tau = 1.532$$
$$\alpha\tau = 24.813$$

The transfer function of the lag-compensating network is thus given by

$$G_c(s) = \frac{1 + 1.532s}{1 + 24.813s}$$

PROBLEM 9A-6

Consider the plant of Example 9.10 with uncompensated transfer function given by

$$G_{MU}(s) = \frac{50}{s(1 + s/12)(1 + s/20)}$$

The compensated transfer function obtained using cascade lag–lead compensation is

$$G_{\mathrm{MC}}(s) = \frac{50(1 + s/1.5)(1 + s/3.88)}{s(1 + s/12)(1 + s/20)(1 + s/0.15)(1 + s/38.8)}$$

Assume that minor-loop feedback compensation is used to compensate the system. Find the required transfer function $G_H(s)$.

SOLUTION

We have

$$G_{\mathrm{MC}}(s) = \frac{G_{\mathrm{MU}}(s)}{1 + G_m(s)}$$

Thus

$$G_m(s) = \frac{G_{MU}(s)}{G_{\mathrm{MC}}(s)} - 1$$

Substituting for $G_{\mathrm{MU}}(s)$ and $G_{\mathrm{MC}}(s)$, we obtain

$$G_m(s) = \frac{(1 + s/0.15)(1 + s/38.8)}{(1 + s/1.5)(1 + s/3.88)} - 1$$

This reduces to

$$G_m(s) = \frac{5.768s}{(1 + s/1.5)(1 + s/3.88)}$$

Recall that

$$G_m(s) = G_{p_1}(s)G_H(s)$$

Assuming that $G_{p_1}(s) = G_{\mathrm{MU}}(s)$, we get

$$G_H(s) = \frac{5.768s}{(1 + s/1.5)(1 + s/3.88)} \, \frac{s(1 + s/12)(1 + s/20)}{50}$$

Thus

$$G_H(s) = \frac{0.1154s^2(1 + s/12)(1 + s/20)}{(1 + s/1.5)(1 + s/3.88)}$$

A number of points have to be considered. The first is the stability of the minor loop. Here we have the characteristic equation given by

$$1 + G_m(s) = 0$$

This turns out to be

$$s^2 + 38.95s + 5.82 = 0$$

The characteristic equation has two roots at $-0.15$ and $-38.8$, which indicates stability. The second point concerns the cancellation of the terms $(1 + s/12)$ and $(1 + s/20)$ between the plant and the compensator. Care must be taken to ensure that a change in plant characteristics does not lead to instability. The third point concerns the ability to realize the $G_H(s)$ transfer function by physical devices. In the present case we can realize $G_H(s)$ in many ways.

# Problems

PROBLEM 9B-1

Two special cases of the phase-lead network treated in Problem 9A-1 are shown in Figure 9.64. Find the expression of $\alpha$, $\tau$, and $G_c(0)$ in the standard transfer function for a phase-lead network. Sketch the Bode plots for the two networks.

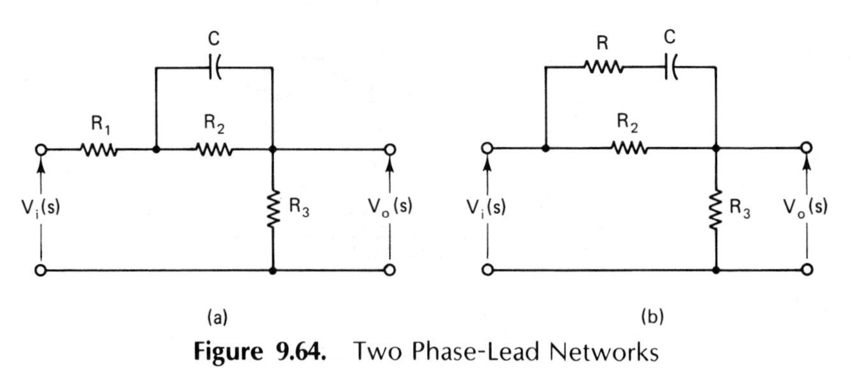


**Figure 9.64.** Two Phase-Lead Networks

PROBLEM 9B-2

The network shown in Figure 9.65 is a phase-lag network. Show that the transfer function is given by the standard phase-lag form

$$G_c(s) = G_c(0)\frac{1 + \tau s}{1 + \alpha\tau s}$$

with

$$G_c(0) = \frac{1}{1 + R_1/(R + R_2) + R_1/R_3}$$

$$\tau = \frac{R_2 R}{R_2 + R}C$$

$$\alpha = \frac{1 + R_1/R_2 + R_1/R_3}{1 + R_1/(R + R_2) + R_1/R_3}$$

Sketch the Bode plot for this network.

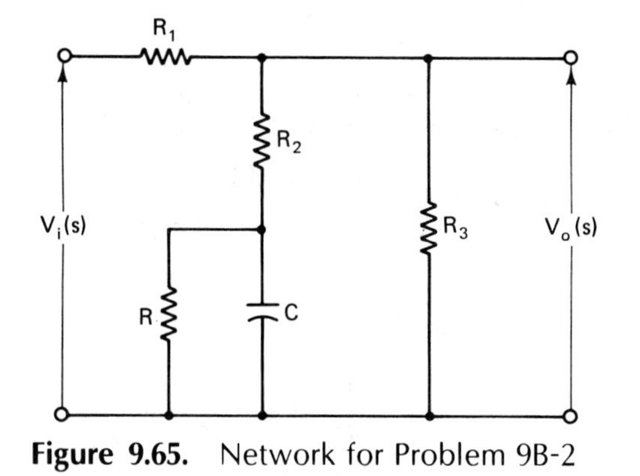


**Figure 9.65.** Network for Problem 9B-2

PROBLEM 9B-3

A lag–lead network is shown in Figure 9.66. Show that the transfer function is given by

$$G_c(s) = \frac{(1 + \tau_1 s)(1 + \tau_2 s)}{\tau_1 \tau_2 s^2 + (\tau_1 \beta + \tau_2)s + 1/G_0}$$

where

$$\tau_1 = \frac{R_2 R_4}{R_2 + R_4} C_2$$

$$\tau_2 = R_1 C_1$$

$$\beta = 1 + \frac{R_1}{R_2}$$

$$G_0 = \frac{1}{1 + R_1/(R_2 + R_4)}$$

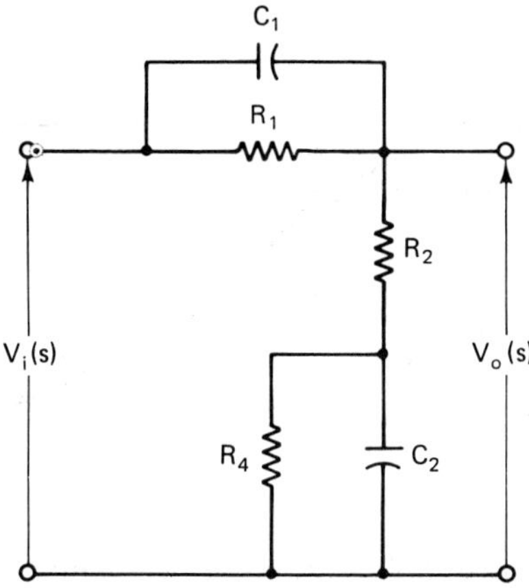

**Figure 9.66.** Network for Problem 9B-3

PROBLEM 9B-4

The network shown in Figure 9.67a is a lag network with a 40-dB/decade slope. Show that the transfer function is given by

$$G_c = \frac{(1 + \tau_1 s)(1 + \tau_2 s)}{\alpha \tau_1 \tau_2 s^2 + (\beta \tau_1 + \delta \tau_2)s + 1}$$

where

$$\tau_1 = R_1 C_1$$
$$\tau_2 = R_2 C_2$$
$$\alpha = \left(1 + \frac{R_3}{R_2}\right)\left(1 + \frac{R_4}{R_1}\right) + \frac{R_3}{R_1}$$
$$\beta = 1 + \frac{R_3}{R_1} + \frac{R_4}{R_1}$$
$$\delta = 1 + \frac{R_3}{R_2}$$

Verify that the Bode plot for this network is as shown in Figure 9.67b.

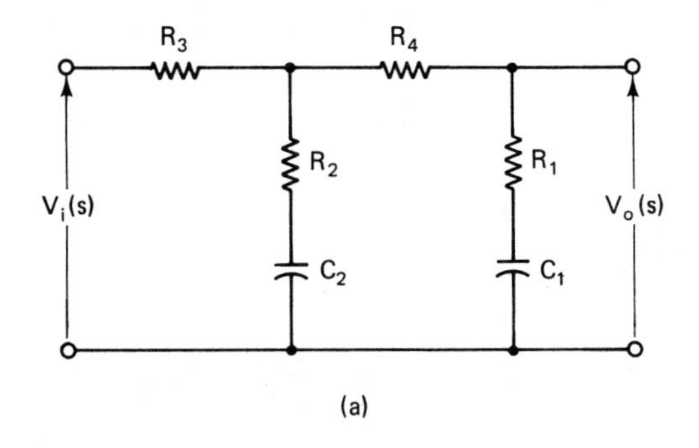


(a)

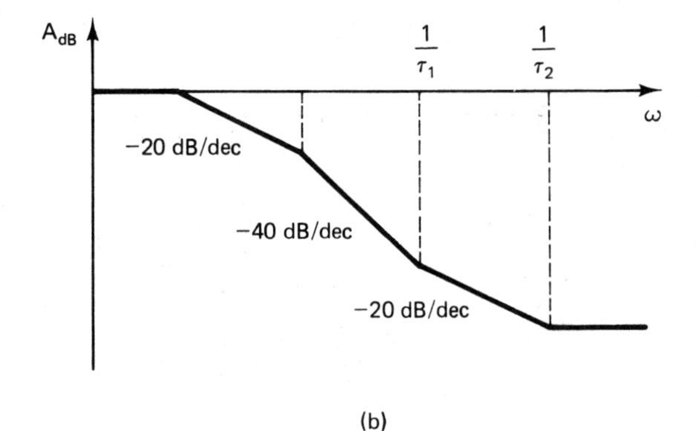


(b)

**Figure 9.67.** (a) 40-dB Lag Network
(b) Its Bode Plot

PROBLEM 9B-5

Determine the parameters of a phase-lead network that has a maximum phase-lead angle of 60° at $\omega_m = 2$ rad/sec.

PROBLEM 9B-6

The frequency response of a phase-lead network at $\omega = 3$ is given by

$$G(j3) = 0.1044 \underline{/14.98^\circ}$$

Find the parameters $\alpha$ and $\tau$.

PROBLEM 9B-7

The Bode diagram of a phase-lead network shows two corner frequencies, at $\omega_{c_1} = 5$ rad/sec and $\omega_{c_2} = 40$ rad/sec. Find the transfer function of the network, the maximum phase-lead angle, and the frequency at which this occurs.

PROBLEM 9B-8

Consider the open-loop transfer function

$$G_p(s) = \frac{K}{s(1+s)(1+0.2s)}$$

Determine the value of $K$ so that the dominant closed-loop poles have a damping ratio of 0.5. Assume that unity feedback control is used. Find the undamped natural frequency $\omega_n$.

PROBLEM 9B-9

Design a phase-lead network for cascade compensation of a unity feedback control system with open-loop transfer function

$$G_p(s) = \frac{K}{s^2}$$

The percentage overshoot of the system's transient response is to be less than 15%. To satisfy steady-state error requirements, take $K = 15$.

PROBLEM 9B-10

The open-loop transfer function of a type 1 system is given by

$$G_p(s) = \frac{K}{s(1+s/15)}$$

It is required to design a cascade lead compensator such that:

1. The steady-state error for sinusoidal inputs with frequency up to 12 rad/sec is less than 2%.
2. The phase margin is 45°.

PROBLEM 9B-11

Consider a closed-loop control system with an open-loop transfer function given by

$$G_p(s) = \frac{10(s+8)}{s(1+10s)}$$

Design a lead compensator so that the system has a 50° phase margin.

PROBLEM 9B-12

Use root-locus techniques to design a phase-lead compensator for the system of Problem 9B-9 such that the dominant closed-loop poles provide a damping ratio of 0.5 and have an undamped natural frequency $\omega_n = 50$ rad/sec.

PROBLEM 9B-13

Consider a closed-loop control system with open-loop transfer function given by

$$G_p(s) = \frac{10(s+10)}{s(1+10s)}$$

Design a lead compensator so that the system is critically damped ($\zeta = 1$).

PROBLEM 9B-14

Consider the third-order system with open-loop transfer function

$$G_p(s) = \frac{K}{s(s+4)(s+5)}$$

Design a phase-lag compensator to yield a damping ratio of 0.7 and velocity error constant of 30.

PROBLEM 9B-15

Consider the third-order system with open-loop transfer function

$$G_p(s) = \frac{80}{s(1+0.02s)(1+0.05s)}$$

(a) Find the gain and phase margins and the gain crossover frequency. Is the system stable?

(b) Design a phase-lag cascade compensator to achieve the following performance specifications:

1. Phase margin of approximately $30°$.
2. Gain margin of about 7 dB.

(c) Design a phase-lead compensator to achieve the specifications of part (b). (This obviously can be done easily using Bode procedures.)

PROBLEM 9B-16

Consider the open-loop transfer function given by

$$G_p(s) = \frac{15}{s(1 + s/15)(1 + s/25)}$$

It is required to design a lag compensator so that:

1. The velocity error constant $K_v = 75$.
2. The phase margin is at least $50°$. Use the first method of Example 9.7.

PROBLEM 9B-17

Repeat Problem 9B-16 using method 2 of Example 9.7.

PROBLEM 9B-18

Repeat Problem 9B-16 using the method of Problem 9A-5.

PROBLEM 9B-19

Repeat the design of Problem 9B-16, but instead of a phase margin requirement we require that $\zeta = 0.5$. (Emulate Example 9.8.)

PROBLEM 9B-20

Consider the open-loop transfer function given by

$$G_p(s) = \frac{K}{s(1 + s/10)}$$

Determine the gain $K$ to obtain a velocity error constant $K_v = 100$. Design a lag–lead compensator to achieve the following performance specifications:

1. Phase margin $= 60°$.
2. An error of less than 1% is obtained for sinusoidal inputs of frequencies of less than 1 rad/sec.

PROBLEM 9B-21

Design a lag compensator for the system with open-loop transfer function given by

$$G_p(s) = \frac{K}{s(1 + s/10)^2}$$

The phase margin is required to be 40°, and the velocity error constant is required to be $K_v = 50$.

PROBLEM 9B-22

Consider the open-loop transfer function given by

$$G_p(s) = \frac{50}{s(s + 10)}$$

Design a lead compensator to achieve the following specifications:

1. The velocity error constant $K_v = 100$.
2. Phase margin = 45°.
3. An error of less than 2% is achieved for sinusoidal input functions of frequency up to 1 rad/sec.

PROBLEM 9B-23

To stabilize a feedback control system, tachometer feedback is often used. Find the tachometer gain $K_T$ to achieve a damping ratio of 0.5 for the closed-loop system for the system shown in Figure 9.68.

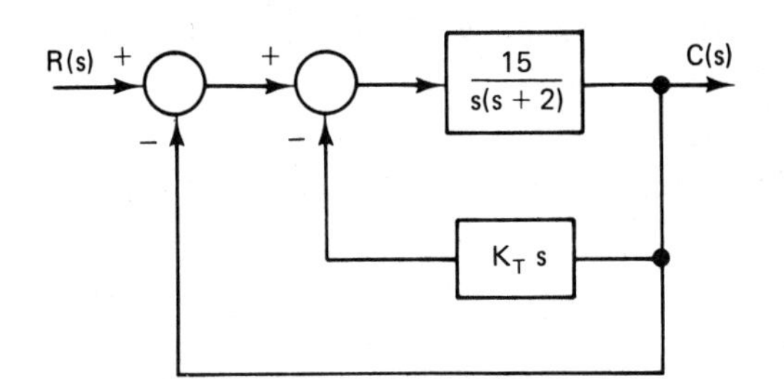


**Figure 9.68.** Block Diagram for Problem 9B-23

PROBLEM 9B-24

Design a lag–lead compensator for the system with open-loop transfer function

$$G_p(s) = \frac{K}{s(1 + 0.125s)(1 + 0.25s)}$$

The following performance specifications are available:

1. $K_v = 120$.
2. Closed-loop dominant damping ratio = 0.7.

PROBLEM 9B-25

Repeat the design of Problem 9B-24, replacing the damping ratio requirement by an equivalent requirement on the phase margin.

PROBLEM 9B-26

Consider the system of Example 9.11. The minor-loop feedback compensator transfer function is written as

$$G_H(s) = \frac{K_t s}{1 + \tau s}$$

For $K_t = 1$ and $\tau = 0.1$, we found that a phase margin of 35° is achieved. Redesign the compensator to achieve a phase margin of 45°.

PROBLEM 9B-27

Use the approximation method of Section 9.12 to solve Problem 9A-6.

PROBLEM 9B-28

Consider the uncompensated open-loop transfer function given by

$$G_{\text{MU}}(s) = \frac{K}{s(1 + s/10)}$$

Design a minor-loop feedback compensator of the form

$$G_H(s) = \frac{K_t s}{1 + \tau s}$$

to achieve the following specifications:

1. Velocity error constant $K_v = 10$.
2. Phase margin = 45°.

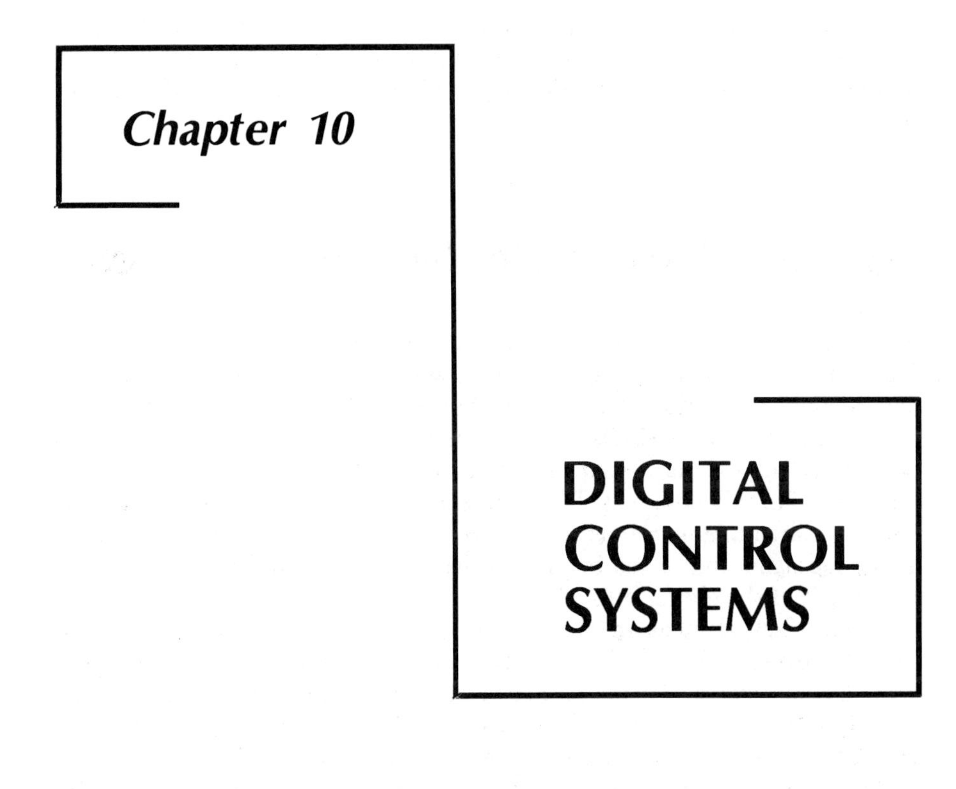


## 10.1 Introduction

A digital control system is a control system that incorporates a digital computer to perform compensation and/or control tasks within its structure. The introduction of the digital computer in the functioning of control systems has been enhanced by its accuracy, reliability, and most important, by its flexibility. Existing examples of digital control systems include autopilot control systems, rolling-mill regulating systems, and digital controllers for electric power generating units to name just a few. Advances in microprocessor technology continue to stimulate the application of digital control in new areas every day.

The purpose of this chapter is to provide an introduction to digital control for closed-loop systems where emphasis is on the dynamic response of the process. We start by a discussion of the basic elements of a digital control system.

# 10.2 Elements of Digital Control Systems

A typical digital control system configuration is shown in the schematic of Figure 10.1. The major block shown in dashed lines is the digital subsystem and contains elements of the system that are unique to the digital control function. In a manner similar to that employed in continuous control, the output $c(t)$ of the controlled process is sensed and a continuous (analog) error signal $e(t)$ is obtained. The error signal is the deviation of the output from a desired analog reference function $r(t)$. The control function $u(t)$ is an analog signal that is fed to the actuators of the controlled process and is the output of the digital subsystem.

The first element in the digital subsystem is an analog-to-digital (A/D) converter. The A/D converter acts on the continuous error signal $e(t)$ and converts it into a sequence of numbers (digital signal) that are fed into the digital computer. A clock sends a pulse every $T$ seconds to prompt the A/D to send a new output value every $T$ seconds. The output of the A/D is denoted by $m(kT)$ on the diagram and is referred to as a discrete-time signal, as it is available only at discrete times. The output of the A/D converter must be stored as a finite number of digits and as a result the A/D output is in practice quantized. A signal that is both discrete and quantized is called a digital signal. The digital computer executes a program with the digital signal as the input, to produce a digital output signal

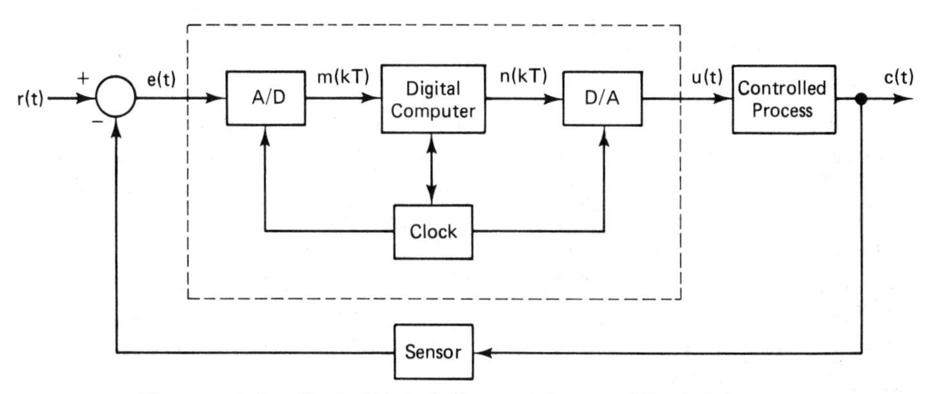


**Figure 10.1.** Basic Digital Control System Block Diagram

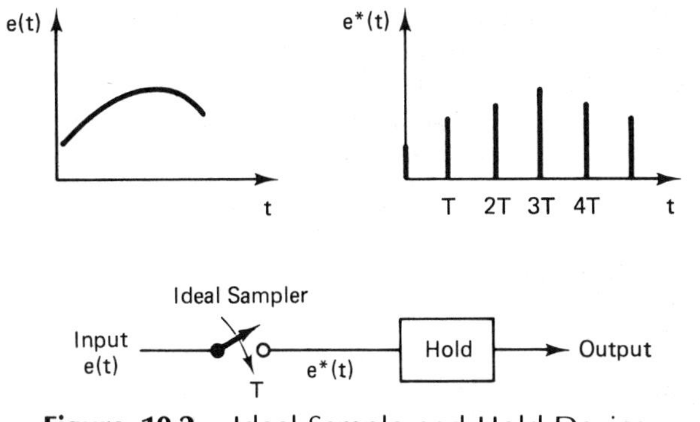


**Figure 10.2.** Ideal Sample-and-Hold Device

$n(kT)$. A digital-to-analog (D/A) converter, receives the signal $n(kT)$ and converts it into a continuous version $u(t)$, which is fed to the actuators of the controlled process.

## Sample-and-Hold Devices

A sampler is a device that converts an analog signal $e(t)$ into a train of pulses $e^*(t)$. A hold device maintains the value of the pulse for a prescribed time duration. A sample-and-hold (S/H) device performs both functions of sampling and holding. Figure 10.2 shows in block diagram form the function of an ideal sample-and-hold device.

## Analog-To-Digital Conversion

An analog (continuous) signal is transformed into a digital signal using an A/D device. A sequence of three operations is carried out as shown in Figure 10.3: sample-and-hold, quantization, and encoding. The sampling operation is carried out at periodic time intervals to sample the analog input. The sampled signal is held until the conversion is completed to

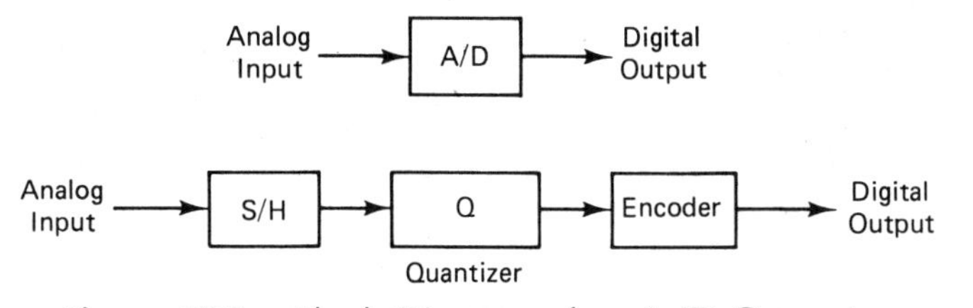


**Figure 10.3.** Block Diagram of an A/D Converter

reduce the effect of signal variation during conversion. Quantization is simply a rounding-off process performed on the analog number to convert it into the nearest digital level to conform with a finite digital word length. Encoding involves the conversion of the quantized discrete signals into digital form that is compatible with the computer used.

## Digital-To-Analog Conversion

A D/A device converts a digital signal into an analog signal of corresponding magnitude. From a functional point of view, the D/A converter consists of a decoder and a sample-and-hold unit as shown in Figure 10.4. The

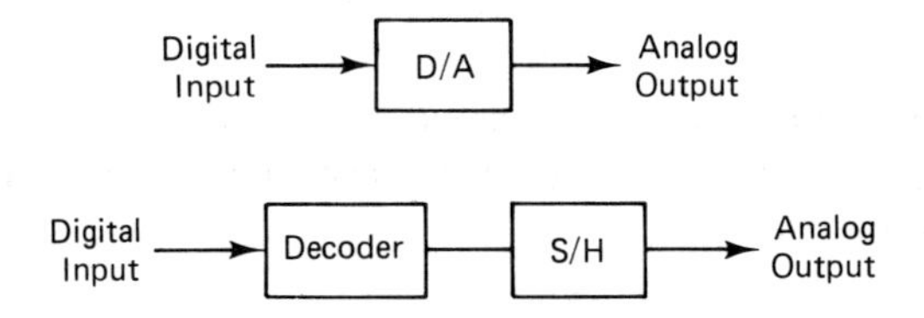


**Figure 10.4.** Block Diagram of a D/A Converter

decoder converts the digital word into a number or a pulse. The sampler in the S/H unit is not needed but is included in the function in recognition of the fact that commercially available units are of the S/H type. The hold unit reconstructs the discrete-time signal into a continuous version.

In studying digital control systems analytically, simplification of the A/D and D/A conversion models is desirable. The decoder and encoder functions are represented by ideal constant gains. The quantization effect is assumed negligible. As a result of this, we can reduce the functional block

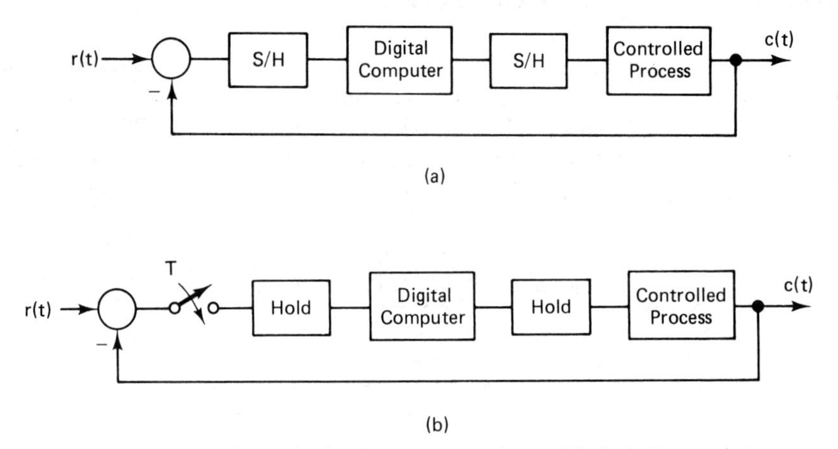


**Figure 10.5.** Analytical Block Diagrams for a Digital Control System

diagram of Figure 10.1 to that of Figure 10.5a, where the A/D and D/A conversion blocks are replaced by S/H devices. In Figure 10.5b, we replace the S/H devices by their idealizations.

# 10.3 Sample-and-Hold Functions

The discussion of the preceding section points out the importance of the sampling function performed by a sampler and the holding function performed by a hold device. In this section, analytical tools for describing these functions are developed.

## Sampling Process

Consider a continuous (or analog) signal $e(t)$ as shown in Figure 10.6a, which is discretized into the signal $e^*(t)$ shown in Figure 10.6b through the action of the ideal sampler shown in block diagram form in Figure 10.6c. The ideal sampler is a switch that closes every $T$ seconds to admit the signal $e(t)$. It is clear that for $t = kT$, the value of the signal $e^*(t)$ is simply $e(kT)$.

$$e^*(t) = e(kT)$$

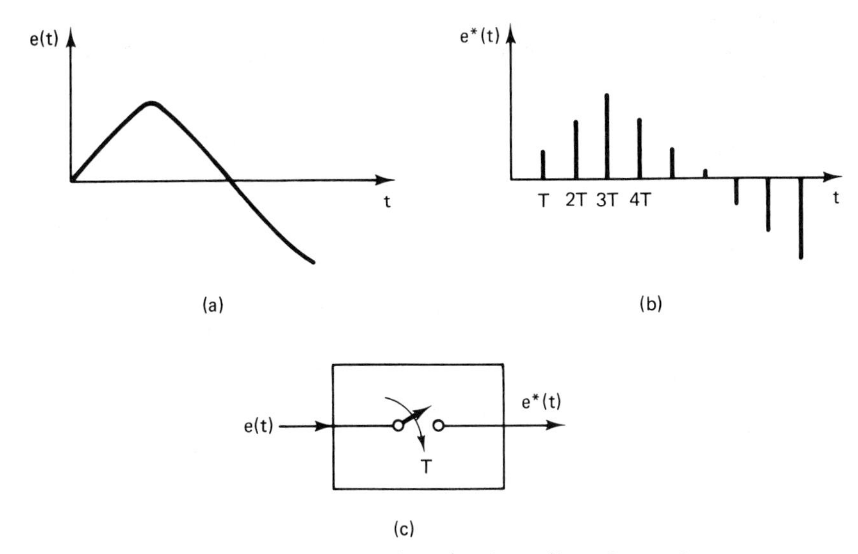


**Figure 10.6.** Representing the Sampling Operation

Recall that the unit impulse function $\delta(t - kT)$ is defined by

$$\begin{aligned} \delta(t - kT) &= 0 \qquad t \neq kT \\ &= 1 \qquad t = kT \end{aligned}$$

The signal $e^*(t)$ can thus be represented as the following sum:

$$e^*(t) = \sum_{k=0}^{\infty} e(kT)\delta(t - kT)$$

The Laplace transform of the sampler output $e^*(t)$ is denoted by $E^*(s)$ and is given by the Laplace transform of the equation above.

$$E^*(s) = \sum_{k=0}^{\infty} e(kT)e^{-kTs}$$

## Reconstruction of Discrete Signals

Most control system components are designed to accept analog (continuous data) signals rather than discrete signals. As a result, a filter (data reconstruction device) is often used as an interface between digital and analog components. The filter is commonly referred to as a hold device and is an integral part of A/D and D/A conversion devices, as discussed earlier.

To understand the underlying concept involved in the functioning of a hold device, reference is made to Figure 10.7. The discrete values of the function at the sampling instants are available up to and including that at $t = kT$. The role of the hold device is to provide the continuous-time signal $e_k(t)$ representing the signal in the time interval $kT \leq t \leq (k + 1)T$. Clearly, the requirement is for an extrapolating (prediction, or forecasting) expres-

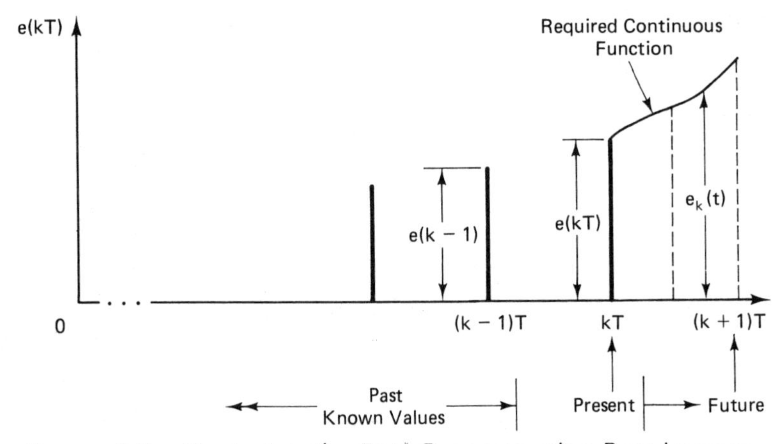


**Figure 10.7.** Illustrating the Data Reconstruction Requirement

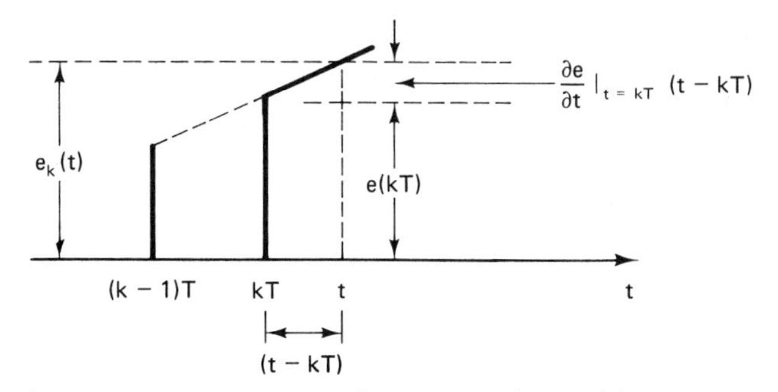


**Figure 10.8.** Concept of a First-Order Hold Device

sion. This is obtained using the following form of Taylor's expansion:

$$e_k(t) = e(kT) + \frac{\partial e}{\partial t}\bigg|_{t=kT}(t - kT) + \cdots$$

This formula is interpreted in Figure 10.8. The required predicted value $e_k(t)$ is the sum of an infinite number of components. The first component is the value $e(kT)$; the second component takes into account the slope (first derivative) of the function at the time instant $kT$. The third and higher components take into account the second- and higher-order derivatives. In practice, only the first two terms indicated are considered in a first-order device. The derivative required is obtained from the available record as

$$\frac{\partial e}{\partial t}\bigg|_{t=kT} = \frac{1}{T}\{e(kT) - e[(k-1)T]\}$$

The concept of a first-order hold device is illustrated in Figure 10.8.

A zero-order hold device simply takes the first term in the expansion. Thus we have for a zero-order hold as shown in Figure 10.9,

$$e_k(t) = e(kT) \qquad kT \le t \le (k+1)T$$

This equation defines the response of the zero-order hold device to a unit impulse as a rectangular pulse whose width is $T$ as shown in Figure 10.10.

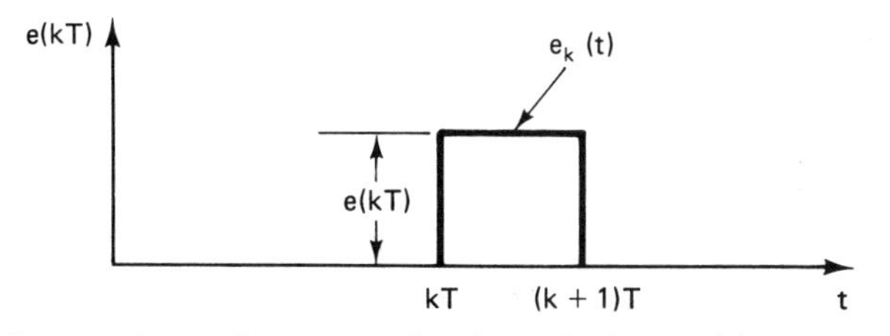


**Figure 10.9.** Concept of a Zero-Order Hold Device

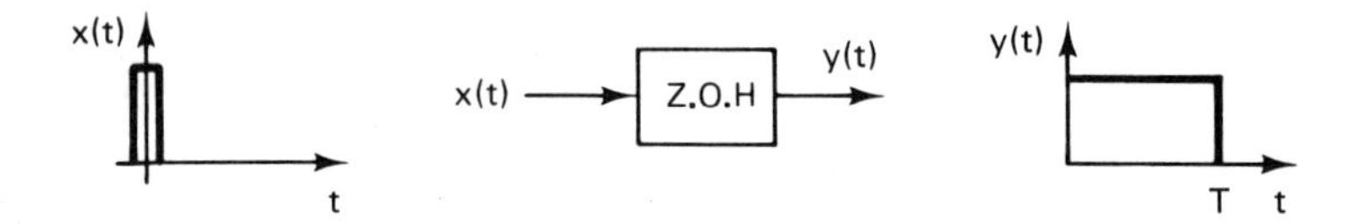


**Figure 10.10.** Illustrating the Impulse Response of a Zero-Order Hold Device

We can thus assert that the impulse response of a zero-order hold device is

$$g_{ho}(t) = u(t) - u(t - T)$$

where $u(t)$ is a unit step. As a result, the transfer function of a zero-order hold device is given by

$$G_{ho}(s) = \frac{1 - e^{-Ts}}{s}$$

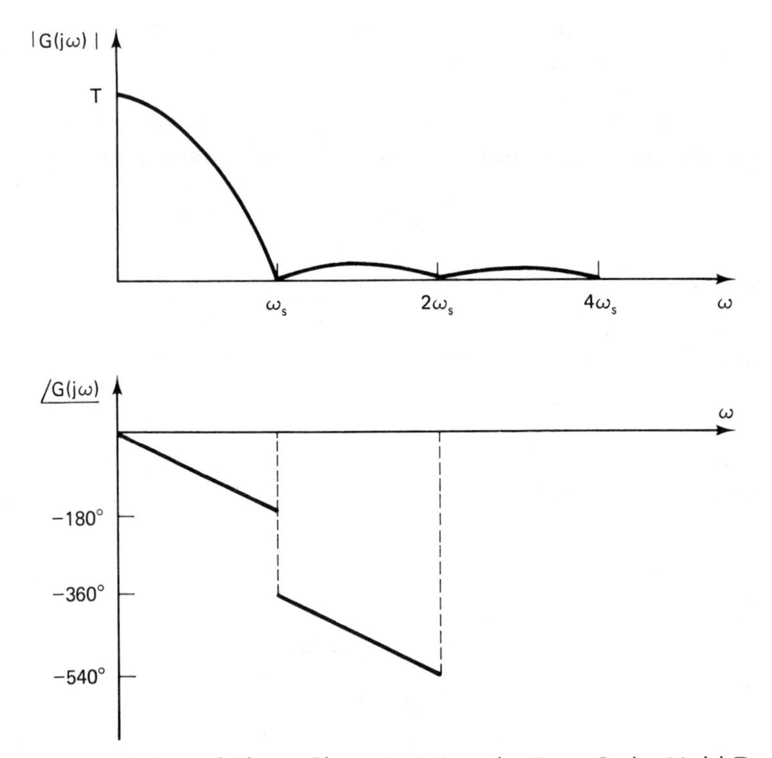


**Figure 10.11.** Gain and Phase Characteristics of a Zero-Order Hold Device

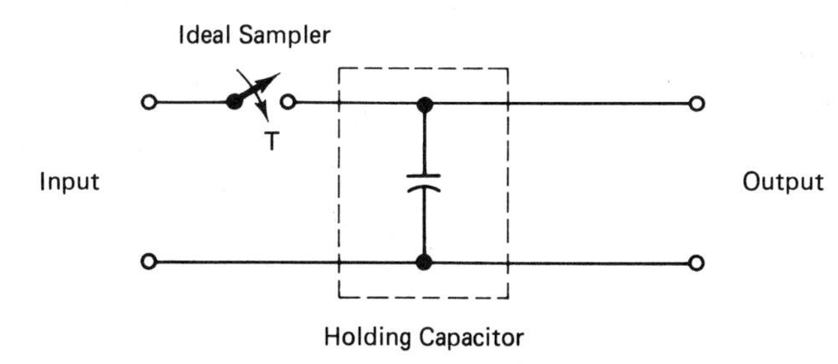


**Figure 10.12.** Simple Illustration of the Sample and Zero-Order Hold Function

We can also show that in the frequency domain, with $s = j\omega$, we have

$$G_{\text{ho}}(j\omega) = T\frac{\sin(\omega T/2)}{(\omega T/2)}e^{-j\omega T/2}$$

The gain and phase characteristics of a zero-order hold device are shown in Figure 10.11.

A simple circuit that provides the zero-order hold function is shown in Figure 10.12. The capacitor is assumed to be charged to $e(kT)$ when the sampling switch is on at the sampling instant. When the sampling switch is opened, the capacitor holds the charge until the next pulse comes along.

# *10.4 The z-Transform*

We have seen in the preceding section that the Laplace transform of the sampled time function $e^*(t)$ is given by

$$E^*(s) = \sum_{k=0}^{\infty} e(kT)e^{-kTs}$$

where $e(kT)$ denotes the values of $e(t)$ at the sampling instants. Since $E^*(s)$ contains exponential terms, difficulties arise in taking the inverse Laplace transform and partial fraction expansions. Desirable conveniences result if a change in variables is introduced to make the use of the exponential terms unnecessary.

The variable $z$ is introduced according to the definition

$$z = e^{Ts}$$

or inversely,

$$s = \frac{1}{T}\ln z$$

Note that for

$$s = \sigma + j\omega$$

we have

$$z = e^{(\sigma + j\omega)T}$$

Thus

$$\begin{aligned} z &= e^{\sigma T} e^{j\omega T} \\ &= e^{\sigma T}(\cos \omega T + j \sin \omega T) \end{aligned}$$

It is clear that $z$ is a complex variable.

Introducing $z$ in the expression of the Laplace transform of the sampled time function $E^*(s)$ yields

$$E^*\left(s = \frac{1}{T}\ln z\right) = \sum_{k=0}^{\infty} e(kT)z^{-k}$$

The left-hand side of the above relation is the $z$-transform of $e^*(t)$, denoted by $E(z)$. Thus

$$E(z) = \sum_{k=0}^{\infty} e(kT)z^{-k}$$

We will look now at a couple of examples.

EXAMPLE 10.1

The $z$-transform of a unit step $u(t)$ is obtained from the fundamental definition

$$\begin{aligned} U(z) &= \sum_{k=0}^{\infty} (1)z^{-k} \\ &= 1 + z^{-1} + z^{-2} + \cdots \\ &= \frac{z}{z - 1} \end{aligned}$$

Note that the above is valid for $|z^{-1}| < 1$ or for $|z| > 1$.

EXAMPLE 10.2

Consider the $z$-transform of the exponential function given in the time domain by

$$e(t) = e^{-at}$$

We have

$$z\{e^{-at}\} = \sum_{k=0}^{\infty} e^{-akT} z^{-k}$$
$$= \frac{z}{z - e^{-aT}}$$

This expression is valid for $|z^{-1}| < e^{aT}$.

Instead of proceeding with more examples, it is appropriate to list some important properties of the $z$-transform.

## Properties of the z-Transform

There are a number of results pertaining to the $z$-transform that can prove helpful in digital control applications. We state these results without proof.

1. The $z$-transform is a linear operation. Thus given two functions $e_1(t)$ and $e_2(t)$ and two scalars $a$ and $b$, we can say that

   $$z\{ae_1(t) + be_2(t)\} = aE_1(z) + bE_2(z)$$

   The $z$-transforms of $e_1(t)$ and $e_2(t)$ are denoted by

   $$z\{e_1(t)\} = E_1(z)$$
   $$z\{e_2(t)\} = E_2(z)$$

2. The real translation theorem states that if the $z$-transform of $e(t)$ is $E(z)$, then for an integer $n$, we have $z\{e(t - nT)\} = z^{-n}E(z)$ and

   $$z\{e(t + nT)\} = z^n\left[E(z) - \sum_{k=0}^{n-1} e(kT)z^{-k}\right]$$

3. The complex translation theorem states that $z\{e^{-at}e(t)\} = E(ze^{aT})$.

4. The initial-value theorem states that the behavior of the sampled function $e^*(t)$ as $t$ approaches zero is determined by the behavior of $E(z)$ as $z$ approaches infinity. We thus state that

   $$\lim_{t\to 0} e^*(t) = \lim_{z\to\infty} E(z)$$

   Provided that the limit in the right-hand side exists.

5. The final-value theorem states that the behavior of the sampled function $e^*(t)$ as $t$ approaches infinity is determined by the behavior of $E(z)$ as $z$ approaches unity, provided that

$E(z)$ does not have poles on or outside the unit circle in the $z$-plane.

$$\lim_{t \to 0} e^*(t) = \lim_{z \to \infty} e(nT)$$
$$= \lim_{z \to 1} (1 - z^{-1}) E(z)$$

Provided that the limit in the right-hand side exists.

6. The partial differentiation theorem deals with the function $e(t, a)$ with the $z$-transform denoted by $E(z, a)$, where $a$ is a parameter. The theorem is given by

$$z\left\{\frac{\partial}{\partial a}[e(t, a)]\right\} = \frac{\partial}{\partial a} E(z, a)$$

Let us now use property 1 to derive the $z$-transform of combinations of our just obtained $z$-transforms of Examples 10.1 and 10.2.

EXAMPLE 10.3

We have

$$\begin{aligned} z\{1 - e^{-at}\} &= z\{u(t)\} - z\{e^{-at}\} \\ &= \frac{z}{z-1} - \frac{z}{z - e^{-aT}} \\ &= \frac{z(1 - e^{-aT})}{(z-1)(z - e^{-aT})} \end{aligned}$$

EXAMPLE 10.4

The $z$-transform of the sinusoid can be obtained without much effort as follows:

$$\begin{aligned} z\{\sin \omega t\} &= z\frac{\{e^{j\omega t} - e^{-j\omega t}\}}{2j} \\ &= \frac{1}{2j}\{z(e^{j\omega t}) - z(e^{-j\omega t})\} \\ &= \frac{1}{2j}\left(\frac{z}{z - e^{j\omega T}} - \frac{z}{z - e^{-j\omega T}}\right) \\ &= \frac{z \sin \omega T}{z^2 - 2(\cos \omega T)z + 1} \end{aligned}$$

Table 10.1 lists some commonly used signals described in the time domain and the corresponding $z$-transforms. In the next section we look at extending the concept to transfer functions.

**TABLE 10.1**
**z-TRANSFORMS OF COMMONLY USED SIGNALS**

| TIME FUNCTION, $e(t), t > 0$ | z-TRANSFORM, $E(z)$ |
|---|---|
| $\delta(t)$ | $1$ |
| $\delta(t-kT)$ | $z^{-k}$ |
| $u(t)$ | $\dfrac{z}{z-1}$ |
| $t$ | $\dfrac{Tz}{(z-1)^2}$ |
| $t^2$ | $\dfrac{T^2z(z+1)}{(z-1)^3}$ |
| $t^{k-1}$ | $\lim_{a\to 0}(-1)^{k-1}\dfrac{\partial^{k-1}}{\partial a^{k-1}}\dfrac{z}{z-e^{-aT}}$ |
| $e^{-at}$ | $\dfrac{z}{z-e^{-aT}}$ |
| $te^{-at}$ | $\dfrac{Tze^{-aT}}{(z-e^{-aT})^2}$ |
| $t^ke^{-at}$ | $(-1)^k\dfrac{\partial^k}{\partial a^k}\dfrac{z}{z-e^{-aT}}$ |
| $1-e^{-at}$ | $\dfrac{z(1-e^{-aT})}{(z-1)(z-e^{-aT})}$ |
| $\dfrac{1}{b-a}(e^{-at}-e^{-bt})$ | $\dfrac{1}{b-a}\left(\dfrac{z}{z-e^{-aT}}-\dfrac{z}{z-e^{-bT}}\right)$ |
| $t-\dfrac{1}{a}(1-e^{-at})$ | $\dfrac{Tz}{(z-1)^2}-\dfrac{(1-e^{-aT})z}{a(z-1)(z-e^{-aT})}$ |
| $\dfrac{1}{2}\left(t^2-\dfrac{2}{a}t+\dfrac{2}{a^2}u(t)-\dfrac{2}{a^2}e^{-at}\right)$ | $\dfrac{T^2z}{(z-1)^3}+\dfrac{(aT-2)Tz}{2a(z-1)^2}+\dfrac{z}{a^2(z-1)}-\dfrac{z}{a^2(z-e^{-aT})}$ |
| $u(t)-(1+at)e^{-at}$ | $\dfrac{z}{z-1}-\dfrac{z}{z-e^{-aT}}-\dfrac{aTe^{-aT}z}{(z-e^{-aT})^2}$ |
| $t-\dfrac{2}{a}u(t)+\left(t+\dfrac{2}{a}\right)e^{-at}$ | $\dfrac{1}{a}\left[\dfrac{(aT+2)z-2z^2}{(z-1)^2}+\dfrac{2z}{z-e^{-aT}}+\dfrac{aTe^{-aT}z}{(z-e^{-aT})^2}\right]$ |
| $\sin\omega t$ | $\dfrac{z\sin\omega T}{z^2-2z\cos\omega T+1}$ |
| $\cos\omega t$ | $\dfrac{z(z-\cos\omega T)}{z^2-2z\cos\omega T+1}$ |
| $\sinh\omega t$ | $\dfrac{z\sinh\omega T}{z^2-2z\cosh\omega T+1}$ |
| $\cosh\omega t$ | $\dfrac{z(z-\cosh\omega T)}{z^2-2z\cosh\omega T+1}$ |
| $e^{-at}\sin\omega t$ | $\dfrac{ze^{-aT}\sin\omega T}{z^2-2ze^{-aT}\cos\omega T+e^{-2aT}}$ |
| $1-e^{-at}\sec\phi\cos(\omega t+\phi)$, $\phi=\tan^{-1}\dfrac{-a}{\omega}$ | $\dfrac{z}{z-1}-\dfrac{z^2-ze^{-aT}\sec\phi\cos(\omega T-\phi)}{z^2-2ze^{-aT}\cos\omega T+e^{-2aT}}$ |
| $e^{-at}\cos\omega t$ | $\dfrac{z^2-ze^{-aT}\cos\omega T}{z^2-2ze^{-aT}\cos\omega T+e^{-2aT}}$ |

# 10.5 The Pulse Transfer Function

Consider an open-loop system with transfer function $G(s)$. The response of the system to an input impulse of magnitude $r(0)$ applied at $t = 0$ is given by

$$c_0(t) = r(0)g(t)$$

The function $g(t)$ is the system's impulse response function. Assume that a fictitious sampler is placed at the output as shown in Figure 10.13. The discrete signal at the sampler's output can be written as

$$c_0^*(t) = r(0)g^*(t) = r(0) \sum_{k=0}^{\infty} g(kT)\delta(t - kT)$$

Taking the Laplace transform of the equation above, we get

$$C_0^*(s) = r(0) \sum_{k=0}^{\infty} g(kT)e^{-kTs}$$

or

$$G^*(s) = \sum_{k=0}^{\infty} g(kT)e^{-kTs}$$

Substituting $z = e^{Ts}$, we get

$$G(z) = \sum_{k=0}^{\infty} g(kT)z^{-k}$$

where

$$G(z) = G^*(s)\,|_{s=(1/T)\ln z}$$

$G(z)$ is the $z$-transfer function of the system.

Suppose now that the input is a sequence of discrete values $r(0)$, $r(T)$, as shown in Figure 10.14. The system response is clearly given by

$$c(t) = r(0)g(t) + r(T)g(t - T)\cdots$$

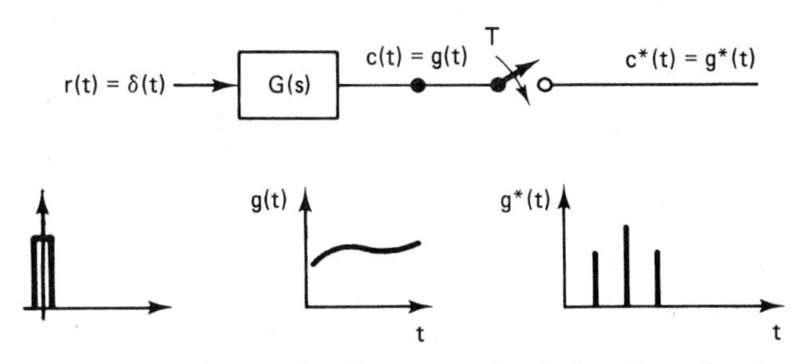


**Figure 10.13.** Deriving the Concept of a Pulse Transfer Function

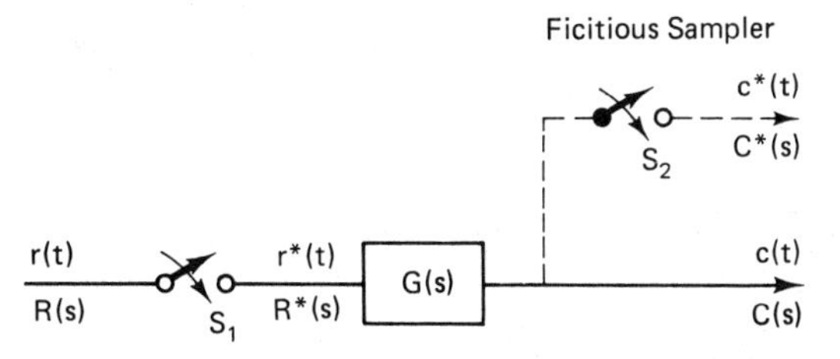


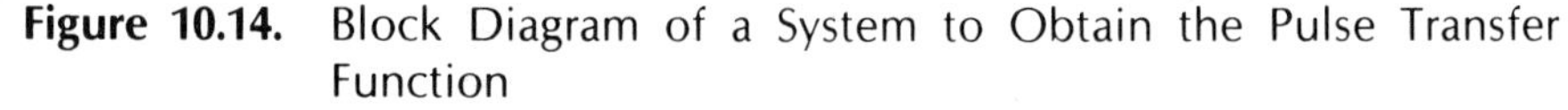

**Figure 10.14.** Block Diagram of a System to Obtain the Pulse Transfer Function

At $t = kT$, we have

$$c(kT) = r(0)g(kT) + r(T)g[(k-1)T] + \cdots + r(kT)g(0)$$
$$= \sum_{n=0}^{k} r(nT)g(kT - nT)$$

Taking the $z$-transform of both sides, we get

$$C(z) = R(z)G(z)$$

Note that in deriving the above, we need the real-convolution result given by

$$F_1(z)F_2(z) = z\left\{\sum_{n=0}^{k} f_1(nT)f_2(kT - nT)\right\}$$

As a conclusion to the discussion above, it is clear that the $z$-transfer function relates the $z$-transform of the output to the $z$-transform of the input in the same manner as its sister $s$-domain transfer function $G(s)$. There is an important difference, however, which will be discussed presently.

Consider the system shown in Figure 10.15, with two cascaded elements $G_1$ and $G_2$ separated by a sampler $S_2$. The transfer relations are

$$D(z) = G_1(z)R(z)$$
$$C(z) = G_2(z)D(z)$$

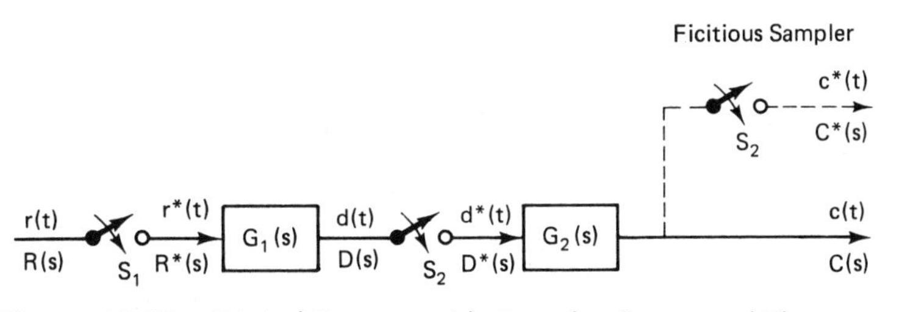


**Figure 10.15.** Digital System with Sampler-Separated Elements

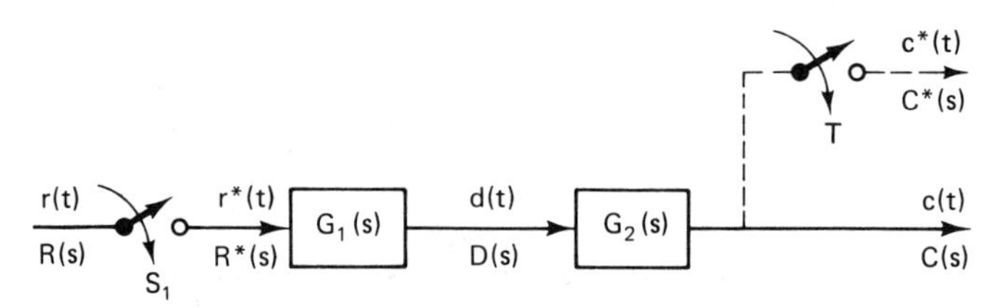


**Figure 10.16.** Digital System with Cascaded Elements

Thus

$$C(z) = G_1(z)G_2(z)R(z)$$

As a result, we can conclude that the $z$-transfer function of two linear elements separated by a sampler is equal to the product of the $z$-transfer functions of the two elements.

If the two cascade elements are not separated by a sampler, we should write

$$G_1G_2(z) = z\{G_1(s)G_2(s)\}$$

Notice that in general

$$G_1G_2(z) \neq G_1(z)G_2(z)$$

The input–output relationship for the system of Figure 10.16 is

$$C(z) = G_1G_2(z)R(z)$$

In order to obtain the $z$-transfer function for a given system, we often start with the $s$-domain transfer function. It is therefore helpful to use tables of the $z$-transform corresponding to the Laplace transfer functions. Table 10.2 gives a summary of some common pairs. A number of examples are in order now.

EXAMPLE 10.5

The transfer function of a zero-order hold is given by

$$G_{\text{ho}}(s) = \frac{1 - e^{-Ts}}{s}$$

The $z$-transform of $G_{\text{ho}}(s)$ is

$$G_{\text{ho}}(z) = z\left(\frac{1 - e^{-Ts}}{s}\right)$$

$$= (1 - z^{-1})z\left(\frac{1}{s}\right) = 1$$

Thus

$$G_{\text{ho}}(z) = 1$$

**TABLE 10.2**
**COMMON LAPLACE TRANSFORMS AND THE CORRESPONDING z-TRANSFORMS**

| *LAPLACE TRANSFORM* | *z-TRANSFORM* |
|---|---|
| $1$ | $1$ |
| $\frac{1}{s}$ | $\frac{z}{z-1}$ |
| $\frac{1}{1-e^{-Ts}}$ | $\frac{z}{z-1}$ |
| $\frac{1}{s^2}$ | $\frac{Tz}{(z-1)^2}$ |
| $\frac{1}{s^3}$ | $\frac{T^2z(z+1)}{2(z-1)^3}$ |
| $\frac{1}{s^{n+1}}$ | $\lim_{a\to 0}\frac{(-1)^n\partial^n}{n!\partial a^n}\frac{z}{z-e^{-aT}}$ |
| $\frac{1}{s+a}$ | $\frac{z}{z-e^{-aT}}$ |
| $\frac{1}{(s+a)^2}$ | $\frac{Tze^{-aT}}{(z-e^{-aT})^2}$ |
| $\frac{a}{s(s+a)}$ | $\frac{z(1-e^{-aT})}{(z-1)(z-e^{-aT})}$ |
| $\frac{\omega}{s^2+\omega^2}$ | $\frac{z\sin\omega T}{z^2-2z\cos\omega T+1}$ |
| $\frac{\omega}{(s+a)^2+\omega^2}$ | $\frac{ze^{-aT}\sin\omega T}{z^2-2ze^{-aT}\cos\omega T+e^{-2aT}}$ |
| $\frac{s}{s^2+\omega^2}$ | $\frac{z(z-\cos\omega T)}{z^2-2z\cos\omega T+1}$ |
| $\frac{s+a}{(s+a)^2+\omega^2}$ | $\frac{z^2-ze^{-aT}\cos\omega T}{z^2-2ze^{-aT}\cos\omega T+e^{-2aT}}$ |

EXAMPLE 10.6

Consider the transfer function given by

$$G(s)=\frac{K}{s(s+a)}$$

Performing a partial fraction expansion, we get

$$G(s)=\frac{K}{a}\left(\frac{1}{s}-\frac{1}{s+a}\right)$$

Taking the $z$-transform of both sides, we get

$$G(z)=\frac{K}{a}\left[z\left\{\frac{1}{s}\right\}-z\left\{\frac{1}{s+a}\right\}\right]$$

From the table of $z$-transforms, we get

$$G(z) = \frac{K}{a}\left(\frac{z}{z-1} - \frac{z}{z-e^{-aT}}\right)$$

This reduces to

$$G(z) = \frac{Kz(1-e^{-aT})}{(z-1)(z-e^{-aT})}$$

This conforms with the expression given in Table 10.2.

EXAMPLE 10.7

Consider the cascade of a zero-order hold and a system described by the transfer function of Example 10.6. Here we have

$$G_{\text{ho}}(s)G(s) = \frac{1-e^{-sT}}{s}\,\frac{K}{s(s+a)}$$

The $z$-transfer function is obtained as

$$G_{\text{ho}}G(z) = z\left\{\frac{K[1-e^{-sT}]}{s^2(s+a)}\right\}$$

$$= K(1-z^{-1})z\left\{\frac{1}{s^2(s+a)}\right\}$$

$$= \frac{K(1-z^{-1})}{a^2}\left[\frac{aTz}{(z-1)^2} - \frac{z}{z-1} + \frac{z}{z-e^{-aT}}\right]$$

Now let us look at the product using the results of Example 10.6.

$$G_{\text{ho}}(z)G(z) = (1)\frac{Kz(1-e^{-aT})}{(z-1)(z-e^{-aT})}$$

Clearly,

$$G_{\text{ho}}G(z) \neq G_{\text{ho}}(z)G(z)$$

The $z$-transfer function of the system reduces to

$$G_{\text{ho}}G(z) = \frac{K_2(z+C)}{(z-1)(z-e^{-aT})}$$

where

$$K_2 = \frac{K(aT-1+e^{-aT})}{a^2}$$

$$C = \frac{1-(1+aT)e^{-aT}}{aT-1+e^{-aT}}$$

## Closed Loop Systems

The derivation of the $z$-transfer function of a closed-loop digital system depends on the location of the sampler but follows the same rules as those used for an open-loop system. Consider the single-loop system shown in Figure 10.17. We can write the Laplace transform of the output as

$$C(s) = G(s)E^*(s)$$

The Laplace transform of the error is given by

$$E(s) = R(s) - H(s)C(s)$$

Thus

$$E(s) = R(s) - H(s)G(s)E^*(s)$$

Taking the pulse transform of both sides, we have

$$E^*(s) = R^*(s) - [H(s)G(s)]^*E^*(s)$$

As a result, we get the pulse transform of the error as

$$E^*(s) = \frac{R^*(s)}{1 + [H(s)G(s)]^*}$$

The pulse transform of the output is

$$C^*(s) = G^*(s)E^*(s)$$

As a result,

$$C^*(s) = \frac{G^*(s)}{1 + [H(s)G(s)]^*}R^*(s)$$

The $z$-transform of the output is thus given by

$$C(z) = \frac{G(z)}{1 + HG(z)}R(z)$$

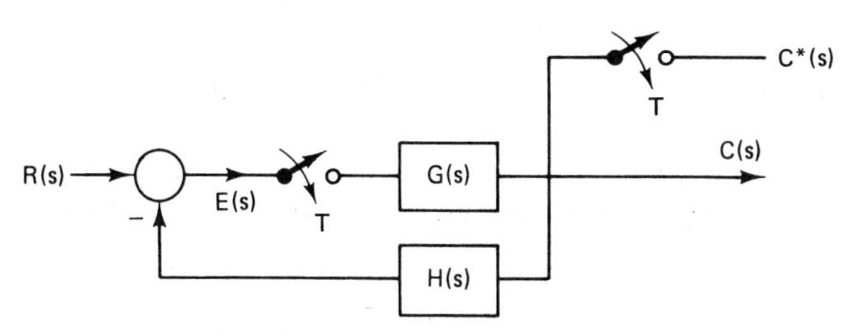


**Figure 10.17.** Closed-Loop Digital System

As a result, the $z$-transfer function of the closed-loop system is obtained as

$$\frac{C(z)}{R(z)} = \frac{G(z)}{1 + HG(z)}$$

Note again that $HG(z) \neq H(z)G(z)$. As an application of the above, we consider the following two examples.

EXAMPLE 10.8

Consider the plant of Example 10.6, where error sampling is employed as shown in Figure 10.18. The closed-loop $z$-transfer function is given by

$$\frac{C(z)}{R(z)} = \frac{G(z)}{1 + G(z)}$$

From Example 10.6 we conclude that

$$\frac{C(z)}{R(z)} = \frac{Kz(1 - e^{-aT})}{Kz(1 - e^{-aT}) + (z - 1)(z - e^{-aT})}$$

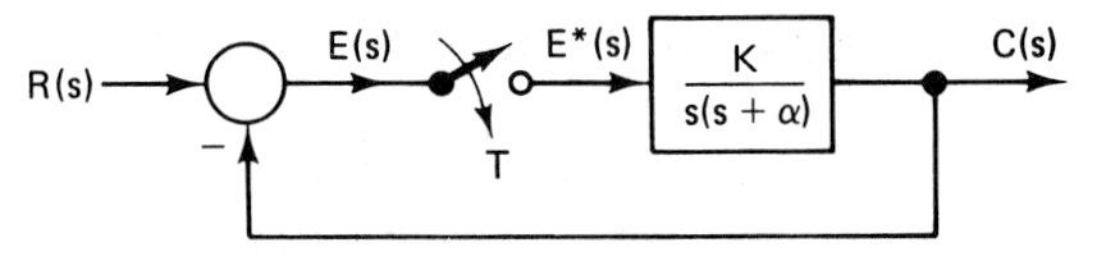


**Figure 10.18.** Closed-Loop System for Example 10.8

EXAMPLE 10.9

The plant of Example 10.8, including a zero-order hold, has a $z$-transfer function obtained as

$$G_T(z) = \frac{K_2(z + C)}{(z - 1)(z - e^{-aT})}$$

where $K_2$ and $C$ are defined in Example 10.7. The closed-loop $z$-transfer function is obtained as

$$\frac{C(z)}{R(z)} = \frac{K_2(z + C)}{K_2(z + C) + (z - 1)(z - e^{-aT})}$$

The system configuration is as shown in Figure 10.19.

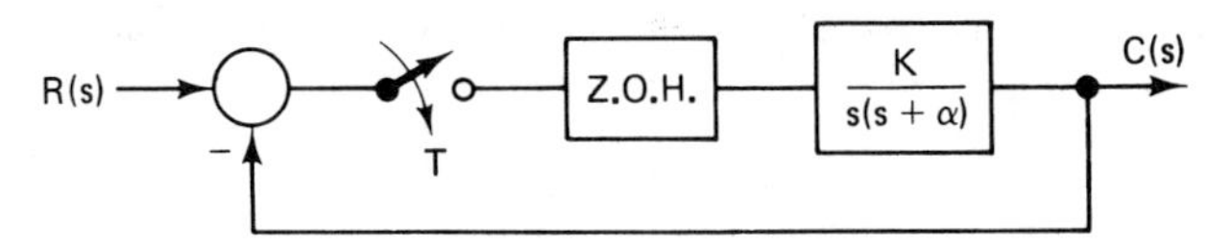


**Figure 10.19.** Closed-Loop System for Example 10.9

# 10.6 Time Response

The previous sections were concerned with some aspects of mathematical modeling of digital control systems. The present section deals with the analysis of transient response of such systems. Before dealing with this important topic, it is necessary to briefly discuss the inverse $z$-transform.

## The Inverse $z$-Transform

In a manner similar to that employed in continuous-time systems using the Laplace transform analysis, the time-domain response of digital systems must be obtained from the $z$-transform expressions. There are three methods for inverting the $z$-transform. Presently, we discuss only two such methods.

1. *Partial Fraction Method.* Given a $z$-transform $E(z)$, we first obtain

$$E_1(z) = \frac{E(z)}{z}$$

The function $E_1(z)$ is then expanded in partial fractions of the form

$$E_1(z) = \frac{A_1}{z + a_1} + \frac{A_2}{z + a_2} + \cdots$$

As a result, the expanded version of $E(z)$ is

$$E(z) = \frac{A_1 z}{z + a_1} + \frac{A_2 z}{z + a_2} + \cdots$$

The inverse $z$-transform of each of the elements of the right-hand side can be found in the tables.

An example is helpful here.

EXAMPLE 10.10

Consider the $z$-transform

$$E(z) = \frac{z(1 - e^{-aT})}{(z - 1)(z - e^{-aT})}$$

We have

$$E_1(z) = \frac{E(z)}{z}$$

Thus

$$E_1(z) = \frac{1 - e^{-aT}}{(z-1)(z - e^{-aT})}$$

A partial fraction expansion gives

$$E_1(z) = \frac{1}{z-1} - \frac{1}{z - e^{-aT}}$$

Thus

$$E(z) = \frac{z}{z-1} - \frac{z}{z - e^{-aT}}$$

From Table 10.1 we find that

$$e(kT) = 1 - e^{-aKT}$$

Thus the sampled time function is

$$e^*(t) = \sum_{k=0}^{\infty} (1 - e^{-aKT})\delta(t - kT)$$

The time function $e(t)$ cannot be determined from the inverse $z$-transform since its behavior between sampling instants is not available.

2. *Power-Series Method.* The idea here is to expand the function $E(z)$ into an infinite series in powers of $z^{-1}$ of the form

   $$E(z) = e(0) + e(T)z^{-1} + e(2T)z^{-2} + \cdots + e(kT)z^{-k} + \cdots$$

   The coefficients of the series can thus be seen to provide the values of $e(kT)$ for $k = 0, 1, \ldots$ (i.e., the sampling instants).

An example will illustrate the idea.

EXAMPLE 10.11

Consider the $z$-transform given by

$$E(z) = \frac{0.125z}{z^2 - 1.75z + 0.75}$$

Performing a long division, we get

$$E(z) = 0.125z^{-1} + 0.218z^{-2} + 0.288z^{-3} + 0.34z^{-4} + 0.379z^{-5} + \cdots$$

Thus

$$e(0) = 0$$
$$e(T) = 0.125$$
$$e(2T) = 0.218$$

and so on.

## An Application

We are now ready to deal with a system such as that of Example 10.9. We have seen that the $z$-transfer function of the system shown in Figure 10.19 is given by

$$\frac{C(z)}{R(z)} = \frac{K_2(z + C)}{z^2 - (1 + e^{-aT} - K_2)z + (e^{-aT} + K_2C)}$$

We note that the denominator is a second-order function of $z$. Since we are tempted to obtain a closed-form solution for the time response to an input $r(t)$, a quick look-up in the table of $z$-Laplace transform pairs leads us to define

$$b^2 = e^{-aT} + K_2C$$
$$2b\cos\omega T = 1 + e^{-aT} - K_2$$

Thus we can write the transfer function in a form close to that which is available in the tables.

$$\frac{C(z)}{R(z)} = \frac{K_2(z + C)}{z^2 - 2bz\cos\omega T + b^2}$$

When a step input is applied, the output transform is obtained as

$$C(z) = \frac{z}{z - 1}\,\frac{K_2(z + C)}{z^2 - 2bz\cos\omega T + b^2}$$

We will now perform a partial fraction expansion to obtain

$$C(z) = \frac{z}{z - 1} - \frac{z(z - e^{-aT})}{z^2 - 2bz\cos\omega T + b^2}$$

The table of pairs cannot be used directly for the second term. To focus our attention, we note that the following two pairs are available:

$$z\{e^{-aT}\cos\omega t\} = \frac{z(z - e^{-aT}\cos\omega T)}{z^2 - 2ze^{-aT}\cos\omega' T + e^{-2aT}}$$
$$z\{e^{-aT}\sin\omega t\} = \frac{ze^{-aT}\sin\omega T}{z^2 - 2ze^{-aT}\cos\omega T + e^{-2aT}}$$

If we let $b = e^{-aT}$, we can see that the denominator of the second term in the expression for $C(z)$ is in the required form. The numerator, of course, needs a little bit of treatment to fit. Consider the following identity:

$$z - e^{-aT} = z - b\cos\omega T + (b\cos\omega T - e^{-aT})$$

Using the definition of $b\cos\omega T$, we get

$$z - e^{-aT} = (z - b\cos\omega T) + \frac{1 - K_2 - e^{-aT}}{2}$$

We can further write

$$z - e^{-aT} = (z - b\cos\omega T) + \frac{1 - K_2 - e^{-aT}}{2b\sin\omega T} b\sin\omega T$$

Let us now define

$$M = \frac{1 - K_2 - e^{-aT}}{2b\sin\omega T}$$

As a result,

$$z - e^{-aT} = (z - b\cos\omega T) + Mb\sin\omega T$$

The transform of the output is now written as

$$C(z) = \frac{z}{z-1} - \frac{z(z - b\cos\omega T) + Mzb\sin\omega T}{z^2 - 2bz\cos\omega T + b^2}$$

The inverse $z$-transform is now applied to yield the time response of the system at the sampling instants as

$$c(kT) = 1 - e^{-\beta kT}(\cos\omega kT + M\sin\omega kT)$$

where

$$e^{-\beta kT} = b^k$$

The value of $\beta$ is obtained using

$$\beta = \frac{-1}{T}\ln(b)$$
$$= \frac{-1}{2T}\ln(e^{-aT} + K_2 C)$$

Note that $\beta$ is analogous to the damping ratio in the continuous case.

Having gone through the previous algebraic details, it is probably instructive to consider a specific numerical example which should help summarize the important features of the development.

EXAMPLE 10.12

Let us assume that

$$K = 1$$
$$a = 1$$
$$T = 1$$

From Example 10.7 we have

$$K_2 = \frac{K(aT - 1 + e^{-aT})}{a^2}$$

Thus with the given numerical values, we get

$$K_2 = e^{-1} = 0.368$$

We also have

$$C = \frac{1-(1+aT)e^{-aT}}{aT-1+e^{-aT}}$$

Thus numerically, we have

$$C = e - 2 = 0.718$$

We have subsequently defined

$$b^2 = e^{-aT} + K_2 C$$

Thus we have numerically

$$b^2 = 1 - e^{-1} = 0.632$$

from which

$$b = 0.795$$

We also have

$$2b\cos\omega T = 1 + e^{-aT} - K_2$$

Thus

$$\cos\omega T = \frac{1}{2b} = 0.629$$

From the above,

$$\omega T = 51.032°$$

The value of $\beta$ is given by

$$\beta = \frac{-1}{T}\ln b = 0.229$$

We also have

$$\begin{aligned} M &= \frac{1-K_2-e^{-aT}}{2b\sin\omega T} \\ &= \frac{1-2e^{-1}}{2(0.795)\sin 51.032^0} \\ &= 0.214 \end{aligned}$$

We can now substitute in the response expression given by

$$c(kT) = 1 - e^{-\beta kT}(\cos\omega kT + M\sin\omega kT)$$

Numerically, we have

$$c(kT) = 1 - e^{-0.229k}(\cos 51.032^0 k + 0.214\sin 51.032^0 k)$$

Clearly, substituting for $k = 0, 1, \ldots$ provides the required response at the sampling instants. The step response of this system is shown in Figure 10.20. It should be noted that the system exhibits underdamped oscillations, as might be expected.

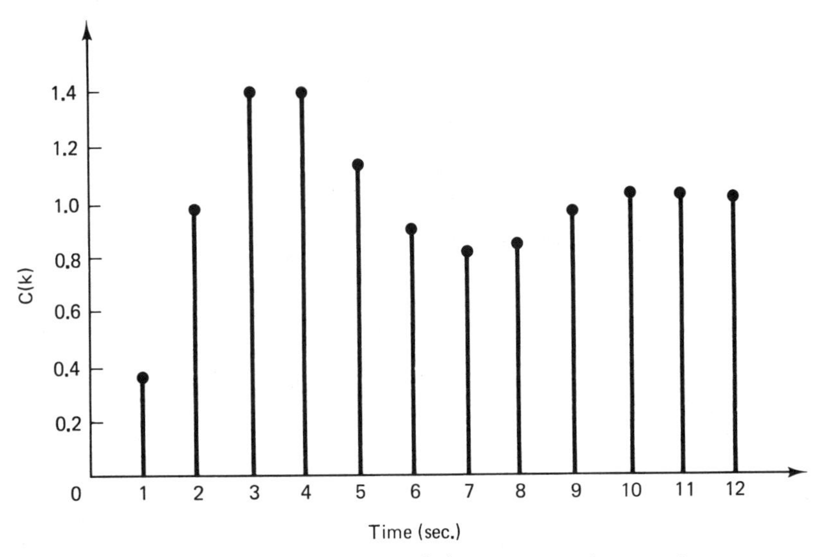


**Figure 10.20.** Step Response of the System of Example 10.12

# *10.7 Stability Analysis*

Earlier in this text we emphasized the importance of stability as the first concern in the analysis and design of continuous control systems. The situation with digital control systems is similar and we require for stability that the discrete response $c(kT)$ be bounded for a bounded input. The actual (continuous) output $c(t)$ must therefore be bounded for the system to be stable.

We have seen that for stability, the poles of $C(s)$ must all lie in the left-hand side of the $s$-plane. As we have developed analysis tools for digital control systems, based on the $z$-transform, it seems logical to seek an equivalance of the $s$-plane stability criteria that apply in the $z$-plane. Our first step is to investigate properties of the mapping from the $s$-plane into the $z$-plane.

## From the *s*-Plane to the *z*-Plane

The transformation from $s$ to $z$ is given by the fundamental relation

$$z = e^{Ts}$$

Let us assume that $s$ is expressed in rectangular form as

$$s = \sigma + j\omega$$

Thus

$$z = e^{T\sigma} e^{j\omega T}$$

or

$$z = e^{T\sigma} \underline{/\omega T}$$

Thus in the $z$-plane, the magnitude of $z$ varies with the real part of $s$, while the angle of $z$ varies with the angular frequency or imaginary part of $s$.

The sampling frequency $\omega_s$ associated with $T$ is given by

$$\omega_s = \frac{2\pi}{T}$$

Thus we write

$$z = e^{T\sigma} \underline{/2\pi(\omega/\omega_s)} = |z| \underline{/\psi}$$

We are now ready to state some important results on the basis of the following expressions:

$$|z| = e^{T\sigma}$$
$$\psi = 2\pi \frac{\omega}{\omega_s}$$

1. The imaginary axis of the $s$-plane maps into the unit circle in the $s$-plane. This follows since in this case, $\sigma = 0$, thus from a magnitude point of view,

   $$|z|_{\sigma=0} = 1$$

   Consider now the angle situation best illustrated using the following list:

   $$\omega = 0 \qquad \psi = 0$$
   $$\omega = \frac{\omega_s}{2} \qquad \psi = 180°$$
   $$\omega = \frac{3\omega_s}{2} \qquad \psi = 3(180°) \triangleq 180°$$
   $$\omega = \frac{-\omega_s}{2} \qquad \psi = -180° \triangleq 180°$$

   We can thus observe that the segment from $-\omega_s/2$ to $\omega_s/2$ on the imaginary axis of the $s$-plane maps into the unit circle as shown in Figure 10.21. The next segment, from $\omega_s/2$ to $3\omega_s/2$, maps into the same circle, as shown in Figure 10.22.

2. A constant damping locus in the $s$-plane is mapped into a circle in the $z$-plane. Consider the vertical line in the left half of the $s$-plane, for which

   $$\sigma = -\sigma_1$$

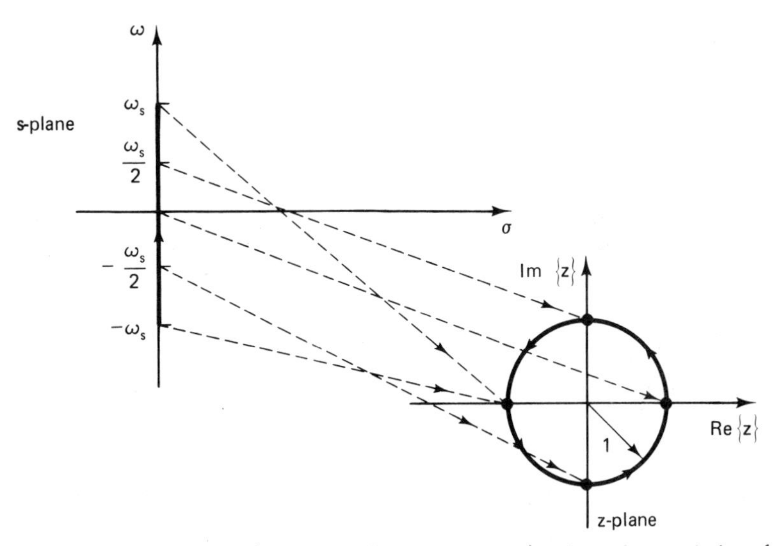

**Figure 10.21.** Mapping of Primary Segment on the Imaginary Axis of the *s*-Plane into the Unit Circle in the *z*-Plane

where $\sigma_1$ is given. It is clear that

$$|z| = e^{-\sigma_1 T} < 1$$

The corresponding locus is thus a circle whose radius is less than 1. For a vertical line in the right half of the *s*-plane, we have for a given $\sigma_2$,

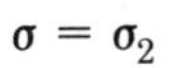

$$\sigma = \sigma_2$$

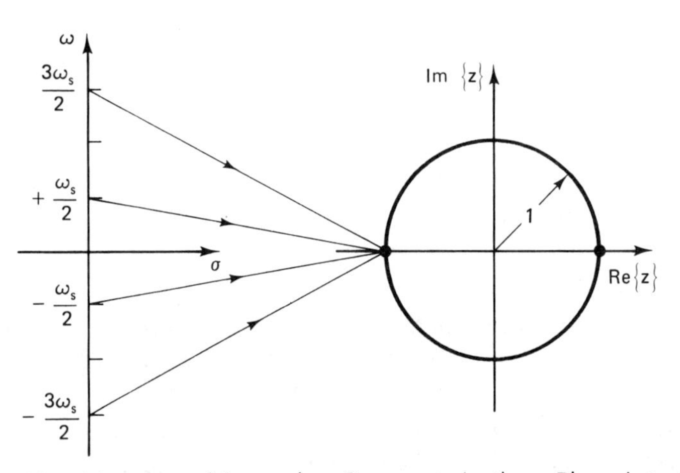


**Figure 10.22.** Mapping of Secondary Segments in the *s*-Plane into the Unit Circle in the *z*-Plane

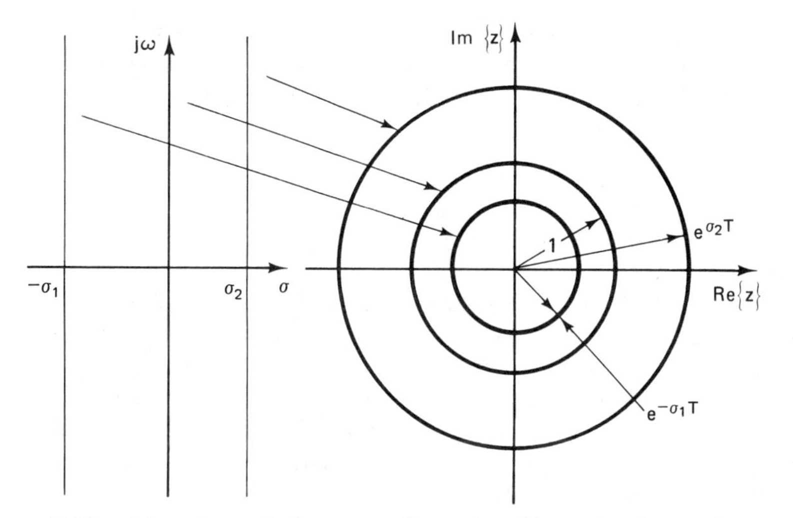


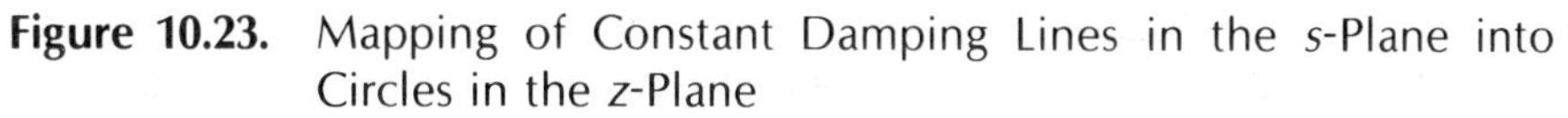

**Figure 10.23.** Mapping of Constant Damping Lines in the *s*-Plane into Circles in the *z*-Plane

Thus

$$|z| = e^{\sigma_2 T} > 1$$

Thus the locus in the $z$-plane is a circle whose radius is greater than 1, as illustrated in Figure 10.23.

3. The entire left side of the $s$-plane is mapped into the interior of the unit circle in the $z$-plane, as shown in Figure 10.24. The

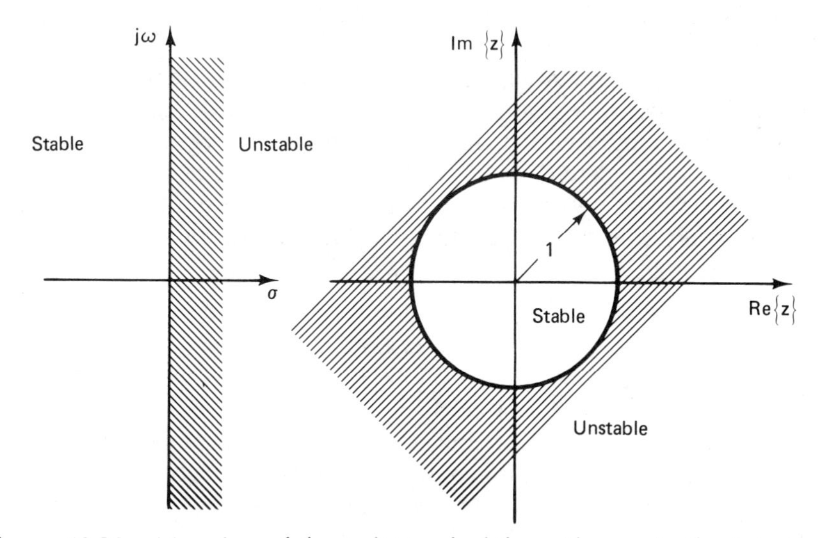


**Figure 10.24.** Mapping of the Left Hand of the *s*-Plane into the Interior of the Unit Circle in the *z*-Plane

validity of this statement is clear from the fact that as $\sigma_1$ varies from 0 to infinity in the $s$-plane, the radius of the circle in the $z$-plane varies from unity to zero as a consequence of result 2 above.

## Testing For Stability

It is clear from the previous discussion that the interior of the unit circle in the $z$-plane is the map of the left-hand side of the $s$-plane. As a result, a digital control system is stable if all the roots of the characteristic equation in $z$ lie inside the unit circle about the origin of the $z$-plane. In a manner similar to that corresponding to continuous-time systems, we should expect that graphical and analytical stability testing procedures exist. Indeed, an equivalent Nyquist stability criterion is available as a graphical aid for testing system stability. Analytical procedures similar to the Routh–Hurwitz criterion have been developed. It should be mentioned, however, that these criteria are a little bit more involved and will not be discussed in great detail at this level.

To put things in perspective, the stability of a linear closed-loop digital control system such as that shown in Figure 10.18 depends on the location of the roots of the characteristic equation

$$F(z) = 1 + GH(z) = 0$$

If all roots of $F(z)$ lie inside the unit circle in the $z$-plane, the system is stable.

The Schur–Cohn stability criterion uses the coefficients of the polynomial $F(z)$ to set up a sequence of determinants whose signs provide the required stability information. As mentioned earlier, we will refrain from detail. It is interesting to note, however, that a simple test for systems with a second-order characteristic equation can be derived from the criterion.

The result states that for stability, the following conditions are required:

$$|F(0)| < 1$$
$$F(1) > 0$$
$$F(-1) > 0$$

For higher-order systems, the Schur–Cohn criterion is cumbersome.

The ease of implementation of the Routh–Hurwitz criterion motivates our study of the concept of bilinear transforms. The quest is for an alternative plane ($w$-plane). The interior of the unit circle in the $z$-plane should map into the left half of our promised plane. Once there, we can apply the Routh–Hurwitz criterion to determine the system stability. Just in case you are wondering why we do not go back to the $s$-plane, remember that back there we encounter exponential terms in $s$ for digital control systems with their associated difficulties.

## The Extended $w$-Transform

Consider the transformation of the complex variable $z$ into a new complex variable $w$ defined by the relation

$$w = A\frac{z - a}{z + a}$$

where $A$ and $a$ are real constants. Assume that $z$ and $w$ are expressed in rectangular coordinates as

$$z = x + jy$$
$$w = u + jv$$

By direct substitution and some manipulations, we can show that the imaginary part of $w$ is given by

$$v = \frac{2aAy}{(x + a)^2 + y^2}$$

The real part of $w$ is given by

$$u = \frac{A(|z|^2 - a^2)}{(x + a)^2 + y^2}$$

Note that

$$|z|^2 = x^2 + y^2$$

Let us restrict our attention to cases where $A$ is positive and $a^2 = 1$. In the interior of the unit circle in the $z$-plane, we have $|z| < 1$, and we can see immediately that the real part of $w$ is negative in this case. Stated differently, the entire region of the $z$-plane that lies outside the unit circle is transformed into the right half of the $w$-plane. Similarly, the entire region that lies inside the unit circle of the $z$-plane is transformed into the left half of the $w$-plane.

Let us note here that the inverse relation for obtaining $z$ from $w$ is given by

$$z = \frac{a(A + w)}{A - w}$$

The transformation discussed above is an extension of the original $w$-transform concept and encompasses three forms described in digital control literature. The original $w$-transform is obtained by setting $A = 1$ and $a = 1$.

Thus

$$w_0 = \frac{z-1}{z+1}$$

and

$$z = \frac{1+w_0}{1-w_0}$$

The $r$-transform is obtained by setting $A = 1$ and $a = -1$. Thus we have

$$r = \frac{z+1}{z-1}$$

$$z = \frac{r+1}{r-1}$$

The modified $w$-transform is obtained by setting

$$A = \frac{2}{T}$$

$$a = 1$$

Thus we have

$$\tilde{w} = \frac{2}{T}\frac{z-1}{z+1}$$

$$z = \frac{2+T\tilde{w}}{2-T\tilde{w}}$$

It is to be noted that the modified $w$-transform involves a scale factor of $(2/T)$ to result in a $w$-plane transfer function that approaches that in the $s$-plane as $T$ approaches zero.

Let us consider now the transformation of the point $s = j\omega$, on the imaginary axis of the $s$-plane through $z$ defined by

$$z = e^{sT} = e^{j\omega T}$$

This point is transformed into the extended $w$-plane as

$$w = A\frac{e^{j\omega T} - a}{e^{j\omega T} + a}$$

With a little bit of algebra, we obtain

$$w = jAa\frac{\sin \omega T}{1 + a\cos \omega T}$$

This shows us that the imaginary axis of the $s$-plane which is transformed into the unit circle of $z$-plane is transformed again to the imaginary axis of the extended $w$-plane. To be specific, for the original and modified $w$-transforms, we have $a = 1$, and we get

$$w = jA\tan\frac{\omega T}{2}$$

For the original $w$-transform $A = 1$,

$$w_0 = j\tan\frac{\omega T}{2}$$

For the modified $w$-transform $A = 2/T$, and we get

$$\tilde{w} = j\frac{2}{T}\tan\frac{\omega T}{2}$$

For the $r$-transform, we get with $A = 1$ and $a = -1$,

$$r = j\cot\frac{\omega T}{2}$$

We will now look at two examples to illustrate the use of the extended $w$-transform to assert system stability.

EXAMPLE 10.13

We considered earlier the transfer function given by

$$G(s) = \frac{K}{s(s+a)}$$

The $z$-transform of $G(s)$ was shown in Example 10.6 to be given by

$$G(z) = \frac{K}{a}\frac{z(1-e^{-aT})}{(z-1)(z-e^{-aT})}$$

For the closed-loop system shown in Figure 10.25, the characteristic equation is obtained using

$$1 + G(z) = 0$$

After some manipulations, we get the characteristic equation in the form

$$z^2 + z\left[\frac{K}{a}(1-e^{-aT}) - (1+e^{-aT})\right] + e^{-aT} = 0$$

Let us use the original $w$-transform to map from the $z$-plane to the $w$-plane using

$$z = \frac{1+w}{1-w}$$

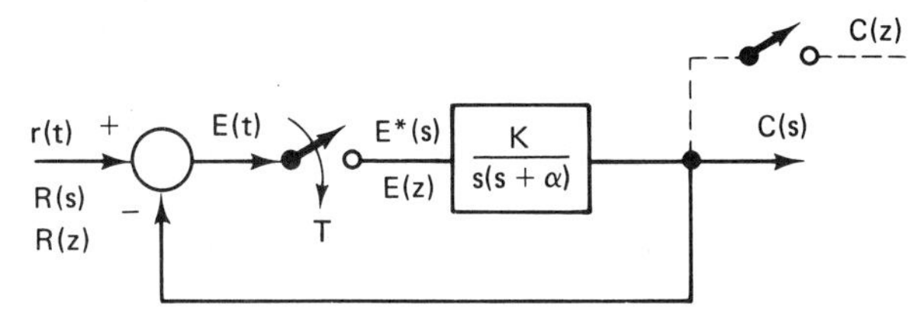


**Figure 10.25.** System For Example 10.13

As a result, we obtain the $w$-plane form of the characteristic equation given by

$$\left[\frac{2(1+e^{-aT})}{1-e^{-aT}}-\frac{K}{a}\right]w^2+2w+\frac{K}{a}=0$$

We can now use the Routh–Hurwitz criterion to comment on the stability of the system. The reader may want to construct the Routh array. A moment's reflection shows this to be unnecessary. This follows since a primary requirement is to have positive coefficients for the polynomial. Thus, we require in addition to $K/a > 0$ that

$$\frac{2(1+e^{-aT})}{1-e^{-aT}}-\frac{K}{a}>0$$

or

$$K<\frac{2a(1+e^{-aT})}{1-e^{-aT}}$$

It should be noted that if we use the $r$-transform instead, we get

$$\frac{K}{a}r^2+2r+\left[\frac{2(1+e^{-aT})}{1-e^{-aT}}-\frac{K}{a}\right]=0$$

Clearly, we arrive at the same conclusion requiring that for stability

$$K<\frac{2a(1+e^{-aT})}{1-e^{-aT}}$$

Our discussion of this system has concentrated so far on the mechanics of applying the Routh–Hurwitz criterion for digital control systems. We can use the results of this example to arrive at some conclusions relating sampling rate represented by $T$ to the stability of the system.

Let us note that the transfer function $G(s)$ can be written as

$$G(s)=\frac{K_1}{s(1+\tau s)}=\frac{K}{s(s+a)}$$

In this case, we have

$$K_1=\frac{K}{a}$$
$$a=\frac{1}{\tau}$$

As a result, the stability condition for this system is

$$K_1\leq\frac{2(1+e^{-T/\tau})}{1-e^{-T/\tau}}$$

Figure 10.26 shows the variation of the maximum gain for stability with the

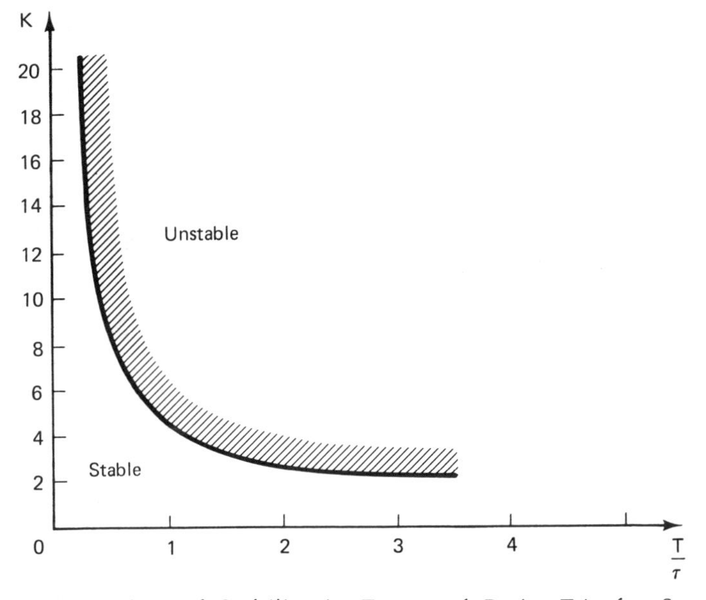


**Figure 10.26.** Region of Stability in Terms of Ratio $T/\tau$ for System of Example 10.13

ratio of sampling time $T$ to the time constant $\tau$. Note that the higher the ratio (i.e., the slower the sampling), the lower is the maximum gain for stability. Thus for a specific gain, say 20, the system becomes unstable for sampling rate $T > 0.2\tau$. Stated in a different way, for a sampling rate of, say, $T = \tau$, the system becomes unstable for gain values higher than $K = 4.33$.

In the next example, we introduce a zero-order-hold device between the sampler and the plant.

EXAMPLE 10.14

Consider the system of Example 10.13 with a zero-order-hold device inserted between the sampler and the plant, as shown in Figure 10.27. The

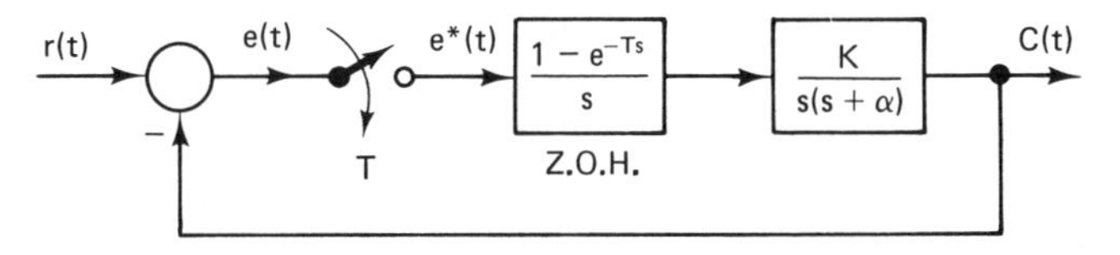


**Figure 10.27.** Block Diagram for the System of Example 10.14

forward loop transfer function is thus given by

$$G_{\text{ho}}G(s) = \frac{1 - e^{-sT}}{s} \frac{K}{s(s+a)}$$

Applying the $z$-transform, we obtain

$$\begin{aligned} G_{\text{ho}}G(z) &= \frac{K(1 - z^{-1})}{a^2} z\left\{\frac{a}{s^2} - \frac{1}{s} + \frac{1}{s+a}\right\} \\ &= \frac{K(1 - z^{-1})}{a^2}\left[\frac{aTz}{(z-1)^2} - \frac{z}{z-1} + \frac{z}{z - e^{-aT}}\right] \end{aligned}$$

The above reduces to

$$G_{\text{ho}}G(z) = \frac{K_2(z + C)}{(z-1)(z - e^{-aT})}$$

where

$$K_2 = \frac{K(aT - 1 + e^{-aT})}{a^2}$$

$$C = \frac{1 - (1 + aT)e^{-aT}}{aT - 1 + e^{-aT}}$$

These results were obtained in Example 10.7.

The characteristic equation of the closed-loop system is thus obtained as

$$z^2 + \left[K_2 - (1 + e^{-aT})\right]z + \left(K_2C + e^{-aT}\right) = 0$$

We apply the original $w$-transform to obtain after simplification.

$$a_2w^2 + a_1w + a_0 = 0$$

where

$$\begin{aligned} a_2 &= 2(1 + e^{-aT}) + K_2(C - 1) \\ a_1 &= 2\left(1 - e^{-aT} - K_2C\right) \\ a_0 &= 1 + e^{-aT} + K_2C \end{aligned}$$

Clearly, for stability, we require that each of the coefficients $a_2$, $a_1$, and $a_0$ be positive.

The first condition requiring that $a_2 > 0$ results in

$$K < K_{1_2}$$

where the limiting $K$ in this case is given by

$$K_{l_2} = \frac{2a^2(1 + e^{-aT})}{aT(1 + e^{-aT}) + 2(e^{-aT} - 1)}$$

The second condition requiring that $a_1 > 0$ results in

$$K < K_{l_1}$$

where the limiting $K$ in this case is given by

$$K_{l_1} = \frac{a^2(1 - e^{-aT})}{1 - (1 + aT)e^{-aT}}$$

The third condition is related to $a_0$. We can show that

$$a_0 = \left(1 + \frac{K}{K_{l_1}}\right) + \left(1 - \frac{K}{K_{l_1}}\right)e^{-aT}$$

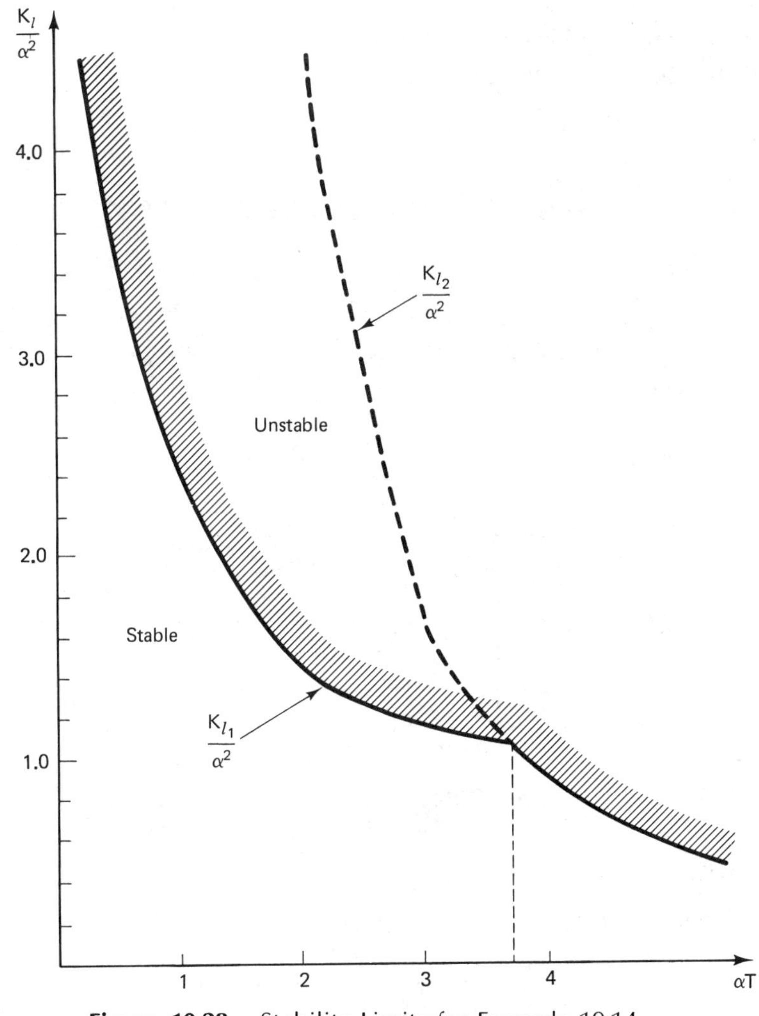

**Figure 10.28.** Stability Limits for Example 10.14

Clearly, if $K < K_{l_1}$, we automatically satisfy the requirement that $a_0 > 0$, provided that $K$, $T$, and $a$ are positive.

Figure 10.28 shows plots of $K_{l_1}/a^2$ and $K_{l_2}/a^2$ as $aT$ is varied. Note that $K_{l_1}/a^2$ provides a lower limit in the range $0 \leq aT \leq 3.72$. Above the value $aT = 3.72$, we see that $K_{l_2}/a^2$ provides a lower limit. Note that at $aT = 3.72$, we have $K_l/a^2 = 1.02$.

We can show that the stability limits for this system are less than the case without the zero-order hold. To do this, we consider the case for which $aT = 1$, with $a = 1$. The condition $a_2 > 0$ results in requiring that $K \leq 26.397$. The condition $a_1 > 0$ results in

$$K \leq 2.394$$

From Example 10.13 we found that the corresponding gain limit is $K = 4.33$ without a zero-order hold, which proves our point.

# Some Solved Problems

PROBLEM 10A-1

Find the $z$-transform of the unit alternating signal $u_A(k)$ shown in Figure 10.29.

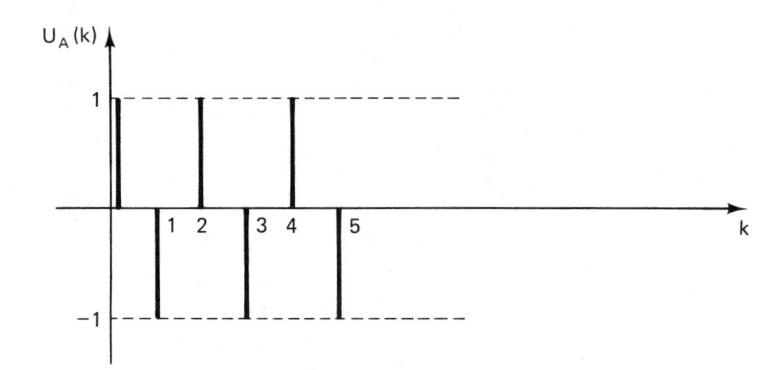


**Figure 10.29.** Unit Alternating Signal

SOLUTION

The signal can be defined as

$$\begin{aligned} u_A(k) &= (-1)^k \qquad \text{for } k \geq 0 \\ &= 0 \qquad \text{for } k < 0 \end{aligned}$$

From the basic definition

$$U_A(z) = \sum_{k=0}^{\infty} (-1)^k z^k$$

This turns out to be

$$U_A(z) = \frac{z}{z+1}$$

The series converges for $|z| > 1$.

PROBLEM 10A-2

The $z$-transform of the output response $C(z)$ of a digital control system due to a step input is given by

$$C(z) = \frac{z(z+0.5)}{(z+0.2)(z+0.4)(z-1)}$$

Find the response in the time domain at the sampling instants. Assume zero initial conditions.

SOLUTION

This is essentially a problem in finding the inverse $z$-transform. We thus have to deal with finding a partial fraction expansion of

$$C_1(z) = \frac{C(z)}{z} = \frac{z+0.5}{(z+0.2)(z+0.4)(z-1)}$$

The result of the expansion is

$$C(z) = 0.8929\frac{z}{z-1} + 0.3571\frac{z}{z+0.4} - 1.25\frac{z}{z+0.2}$$

Invoking the inverse $z$-transform, we obtain

$$c(kT) = 0.8929 + 0.3571(-0.4)^k - 1.25(-0.2)^k$$

Note that the steady-state output is obtained by letting $k$ approach infinity. The result is $C_{ss} = 0.8929$.

PROBLEM 10A-3

Use block diagram reduction techniques to obtain the $z$-transform of the output $C(z)$ for the two systems shown in Figure 10.30.

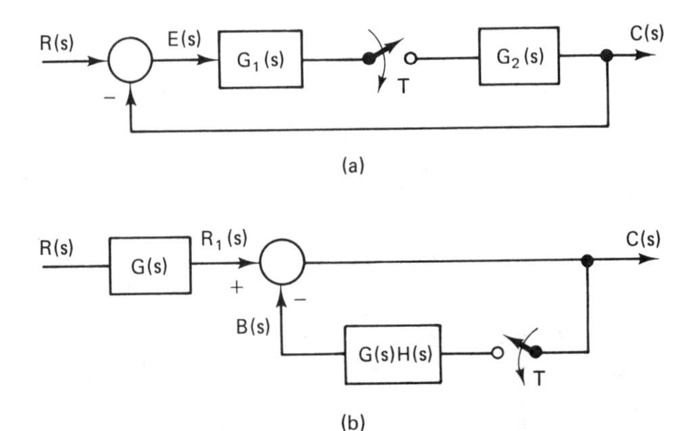


**Figure 10.30.** Block Diagrams for Problem 10A-3

SOLUTION

For the system of Figure 10.30a, we can obtain easily an alternative diagram as shown in Figure 10.31 by moving the summing point ahead. The output is then obtained as

$$C(z) = \frac{G_2(z)}{1 + G_1G_2(z)} R_1(z)$$

Note that

$$R_1(z) = RG_1(z)$$

Thus

$$C(z) = \frac{G_2(z)}{1 + G_1G_2(z)} RG_1(z)$$

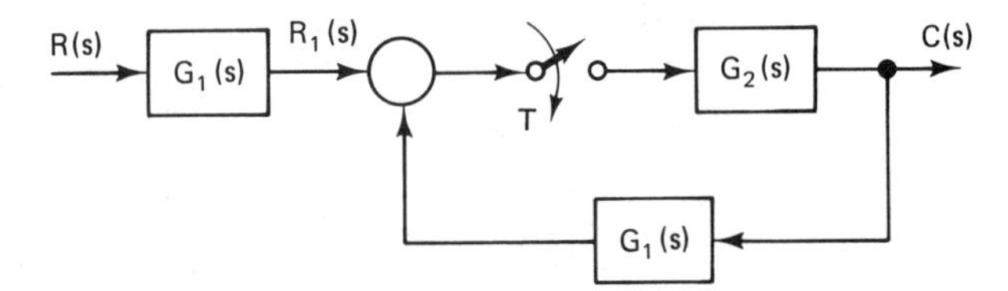


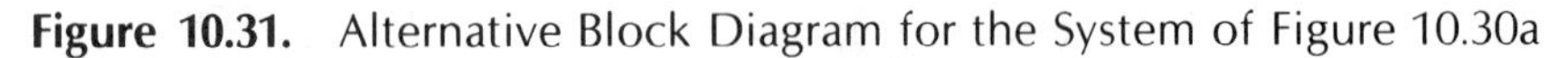

**Figure 10.31.** Alternative Block Diagram for the System of Figure 10.30a

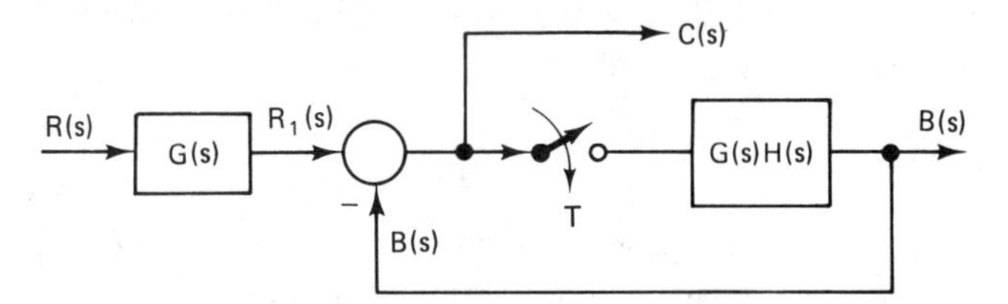


**Figure 10.32.** Alternative Block Diagram for the System of Figure 10.30b

For the system of Figure 10.30b, we have the alternative shown in Figure 10.32. We have

$$B(z) = \frac{GH(z)}{1 + GH(z)} R_1(z)$$

The output is

$$C(z) = R_1(z) - B(z)$$

The above reduces to

$$C(z) = \frac{GR(z)}{1 + GH(z)}$$

PROBLEM 10A-4

Obtain the $z$-transform of the system output for the block diagram of Figure 10.33.

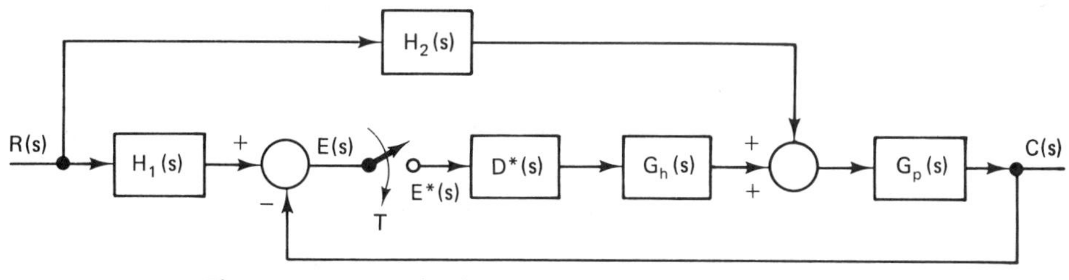


**Figure 10.33.** Block Diagram for Problem 10A-4

SOLUTION

We will use block diagram reduction to obtain an equivalent representation. In Figure 10.34a, the right-hand-side summing point and takeoff point of Figure 10.33 are interchanged. This necessitates multiplication of the feed-forward gain $H_2$ by $G_p(s)$. Note also that we should account for the portion of $C(s)$ corresponding to the feedforward path that is fed back to the summing point of Figure 10.33. This results in the parallel combination of $H_1$ and $G_pH_2(s)$ shown on the left-hand side of Figure 10.34a.

Figure 10.34b shows a reduced form of the block diagram of Figure 10.34a. From the diagram we can conclude that

$$C(z) = G_pH_2R(z) + \frac{D(z)G_hG_p(z)}{1 + D(z)G_hG_p(z)} \left[ H_1R(z) - G_pH_2R(z) \right]$$

This is the required answer.

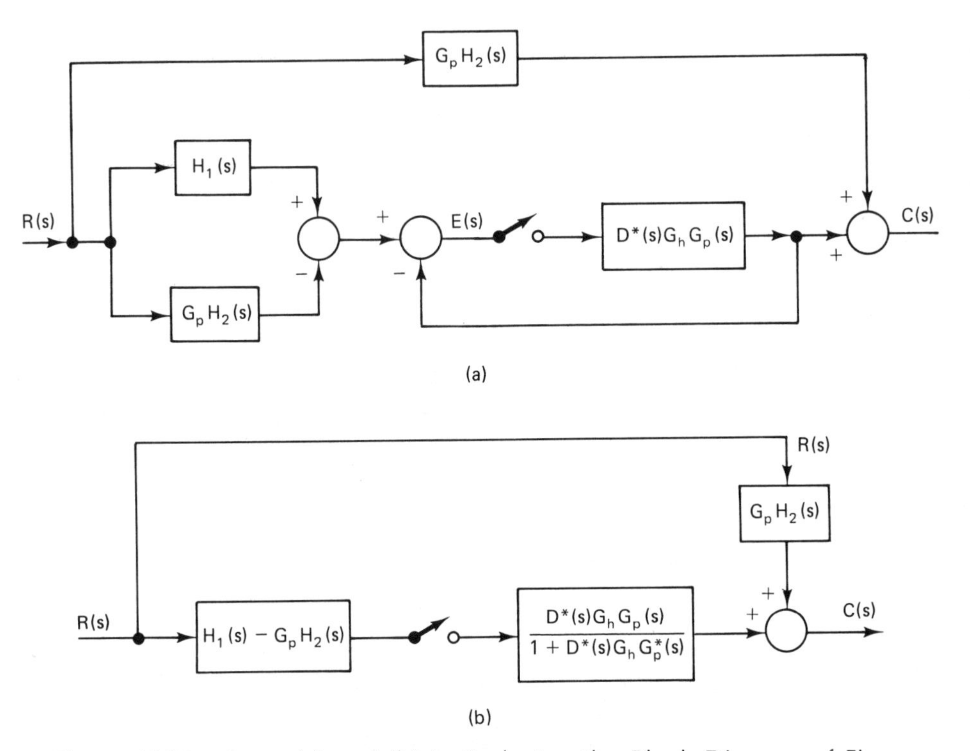


**Figure 10.34.** Steps (a) and (b) in Reducing the Block Diagram of Figure 10.33

PROBLEM 10A-5

Determine the maximum value of the gain $K$ for stability for the digital control system shown in Figure 10.35. Assume that the sampling period is $T = 0.1$ sec.

SOLUTION

The open-loop transfer function for the system is given by

$$G_{ho}G_p(s) = \frac{K(1 - e^{-Ts})}{s^2(1 + 0.1s)(1 + 0.05s)}$$

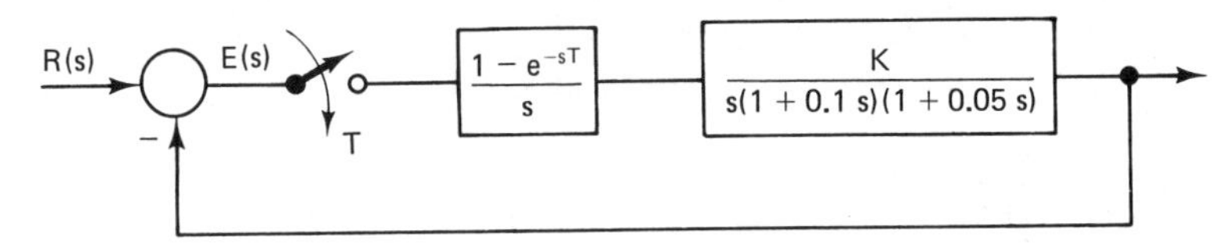

**Figure 10.35.** Block Diagram of the System of Problem 10A-5

The corresponding $z$-transfer function is obtained as

$$G_{ho}G_p(z) = K(1 - z^{-1})z\left\{\frac{1}{s^2(1 + 0.1s)(1 + 0.05s)}\right\}$$

We now use a partial fraction expansion to obtain

$$G_{ho}G_p(z) = K(1 - z^{-1})z\left\{\frac{1}{s^2} - \frac{0.15}{s} + \frac{0.2}{s + 10} - \frac{0.05}{s + 20}\right\}$$

From Table 10.2 we obtain,

$$G_{ho}G_p(z) = K(1 - z^{-1})\left[\frac{Tz}{(z - 1)^2} - \frac{0.15z}{z - 1} + \frac{0.2z}{z - e^{-10T}} - \frac{0.05z}{z - e^{-20T}}\right]$$

The above reduces to

$$G_{ho}G_p(z) = \frac{K(Az + Bz^2 + C)}{(z - 1)(z - e^{-10T})(z - e^{-20T})}$$

where

$$A = T - 0.15 + 0.2e^{-10T} - 0.05e^{-20T}$$
$$B = 0.15 - (T + 0.25)e^{-10T} + (0.25 - T)e^{-20T} - 0.15e^{-30T}$$
$$C = 0.05e^{-10T} - 0.2e^{-20T} + (T + 0.15)e^{-30T}$$

The characteristic equation of the system in terms of $z$ is given by

$$1 + \frac{K(Az^2 + Bz + C)}{(z - 1)(z - e^{-10T})(z - e^{-20T})} = 0$$

which leads to

$$K(Az^2 + Bz + C) + (z - 1)(z - e^{-10T})(z - e^{-20T}) = 0$$

This reduces to

$$z^3 + Dz^2 + Ez + F = 0$$

where

$$D = KA - (1 + e^{-10T} + e^{-20T})$$
$$E = KB + e^{-10T} + e^{-20T} + e^{-30T}$$
$$F = KC - e^{-30T}$$

We now employ the original $w$-transform given by

$$z = \frac{1 + w_0}{1 - w_0}$$

As a result, we obtain the characteristic equation in the $w$-domain as

$$a_3w_0^3 + a_2w_0^2 + a_1w_0 + a_0 = 0$$

where

$$a_3 = 1 - D + E - F$$
$$a_2 = 3 - D - E + 3F$$
$$a_1 = 3 + D - E - 3F$$
$$a_0 = 1 + D + E + F$$

The coefficients $a_3$, $a_2$, $a_1$, and $a_0$ can be expressed as functions of $K$ and $T$ by using the defining relations as

$$a_3 = Kb_3 + C_3$$
$$a_2 = Kb_2 + C_2$$
$$a_1 = Kb_1 + C_1$$
$$a_0 = Kb_0$$

The expressions for the $b$ and $C$ parameters are

$$b_3 = (0.3 - T) - (T + 0.5)e^{-10T} + (0.5 - T)e^{-20T} - (T + 0.3)e^{-30T}$$
$$C_3 = 2(1 + e^{-10T} + e^{-20T} + e^{-30T})$$
$$b_2 = -T + (T + 0.2)e^{-10T} + (T - 0.8)e^{-20T} + 3(T + 0.2)e^{-30T}$$
$$C_2 = 4(1 - e^{-30T})$$
$$b_1 = (T - 0.3) + (T + 0.3)(e^{-10T} + e^{-20T}) - 3(T + 0.1)e^{-30T}$$
$$C_1 = 2(1 - e^{-10T} - e^{-20T} + e^{-30T})$$
$$b_0 = T(1 - e^{-10T} - e^{-20T} + e^{-30T})$$

For $T = 0.1$ sec, the numerical values of the coefficients $a_3$, $a_2$, $a_1$, $a_0$ in terms of $K$ are obtained as

$$a_3 = 0.01349K + 3.106$$
$$a_2 = -0.0396K + 3.8009$$
$$a_1 = -0.0286K + 1.09314$$
$$a_0 = 0.05466K$$

Remembering that we are dealing with the application of Routh–Hurwitz criterion to the polynomial

$$a_3 w_0^3 + a_2 w_0^2 + a_1 w_0 + a_0 = 0$$

we thus require first that all the coefficients $a_i$ be present and positive. Thus a first condition due to $a_2$ is that

$$0 < K \leq 95.98$$

A second condition given by

$$K < 96.05$$

is due to $a_1$ and is implied by the first condition.

The Routh array is given by

$$\begin{array}{c|cc} w_0^3 & a_3 & a_1 \\ w_0^2 & a_2 & a_0 \\ w_0 & d & 0 \\ w_0^0 & a_0 & \end{array}$$

An additional requirement for stability is that $d > 0$. Here $d$ is given by

$$d = \frac{a_2 a_1 - a_0 a_3}{a_2}$$

Thus we require that

$$a_2 a_1 - a_0 a_3 > 0$$

In terms of $K$, we get after some algebra

$$K^2(b_1 b_2 - b_3 b_0) + K(b_1 C_2 + b_2 C_1 - b_0 C_3) + C_1 C_2 > 0$$

Numerically for $T = 0.1$, we get

$$3.952 \times 10^{-4} K^2 - 0.32177K + 4.155 > 0$$

This reduces to

$$K^2 - 814.197K + 1.051 \times 10^4 > 0$$

Finding the roots of the second-order expression results in requiring that

$$(K - 801.07)(K - 13.1243) > 0$$

Since $K < 95.98$, the value of the terms in the first set of parentheses in the inequality above is less than zero. We thus require that

$$K < 13.1243$$

As a result, the maximum value of the gain for stability is

$$K_{max} = 13.1243$$

# Problems

PROBLEM 10B-1

Find the $z$-transfer functions $G(z)$ associated with the $s$-domain transfer functions $G(s)$ given by

(a) $G(s) = \dfrac{K(b - a)}{(s + a)(s + b)}$

(b) $G(s) = \dfrac{K}{s(s + a)^2}$

(c) $G(s) = \dfrac{4}{s(1 + 0.1s + 0.01s^2)}$

PROBLEM 10B-2

Find the $z$-transform associated with the following time-domain functions.

(a) $e(t) = A \sin(\omega t + \theta)$

(b) $e(t) = A \sinh(\omega t + \theta)$

(c) $e(t) = A t \sin \omega t$

PROBLEM 10B-3

Find the inverse $z$-transforms for the following.

(a) $$C(z) = \frac{0.792z^2}{(z-1)(z^2 - 0.416z + 0.208)}$$

(b) $$C(z) = \frac{z^2(z+1)(1-\cos 20T)}{(z-1)^2(z^2 - 2z\cos 20T + 1)}$$

PROBLEM 10B-4

Find the $z$-transfer function for the system shown in the block diagram of Figure 10.36, assuming that $T = 0.1$ sec.

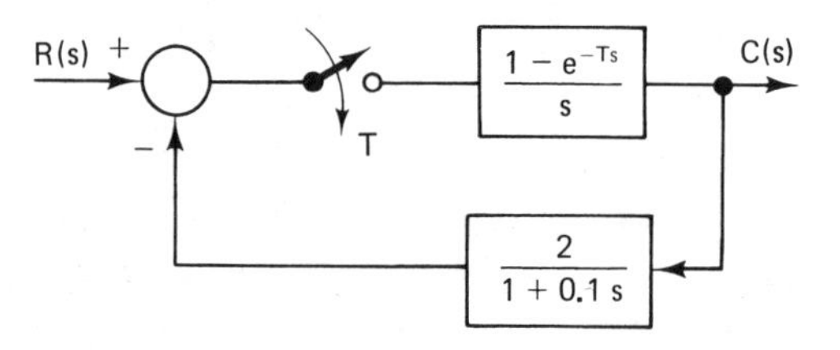


**Figure 10.36.** System for Problem 10B-4

PROBLEM 10B-5

Obtain the $z$-transfer function $C(z)/R(z)$ for the system shown in Figure 10.37. Determine the output $c(nT)$ for a unit step input. Assume that $T = 1$ sec.

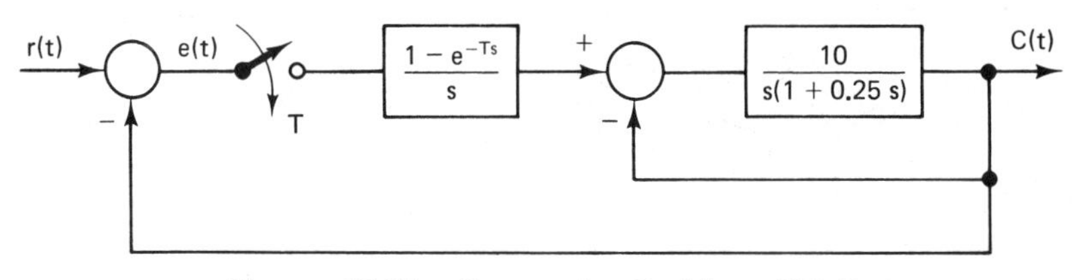


**Figure 10.37.** System for Problem 10B-5

PROBLEM 10B-6

The $z$-transform of the output of a digital control system due to a unit step function input is given by

$$C(z) = \frac{0.484z^{-1} + 0.516z^{-2}}{1 - z^{-1}}$$

Find the values of $c(nT)$ for $n = 0, 1, 2, \ldots$.

PROBLEM 10B-7

The characteristic equation of a digital control system in terms of the $z$-operator is given by

$$F(z) = 6z^2 + z - 1 = 0$$

Use the result of the Schur–Cohn criterion to check the stability of the system.

PROBLEM 10B-8

Repeat Problem 10B-7 for

$$F(z) = 2z^2 + 3z - 2 = 0$$

PROBLEM 10B-9

Assume that the sampling period is $T = 0.2$ for the system of Problem 10A-5. Verify the stability of the system for $K = 2$.

PROBLEM 10B-10

Find the maximum value of the gain $K$ for stability for the digital control system of Problem 10A-5 with $T = 0.2$ sec.

PROBLEM 10B-11

Assume that the sampling period is $T = 0.05$ sec for the system of Problem 10A-5. Will the system be stable for $K = 2$?

PROBLEM 10B-12

Find the maximum value of the gain $K$ for stability for the digital control system of Problem 10A-5 with $T = 0.05$ sec.

# INDEX